Biotechnology: From Science to Applications

Biotechnology: From Science to Applications

Editor: Suzy Hill

STATES
ACADEMIC PRESS
www.statesacademicpress.com

States Academic Press,
109 South 5th Street,
Brooklyn, NY 11249, USA

Visit us on the World Wide Web at:
www.statesacademicpress.com

ISBN: 978-1-63989-080-4 (Hardback)

Trademark Notice: Registered trademark of products or corporate names are used only for explanation and identification without intent to infringe.

Cataloging-in-Publication Data

Biotechnology : from science to applications / edited by Suzy Hill.
  p. cm.
Includes bibliographical references and index.
ISBN 978-1-63989-080-4
1. Biotechnology. 2. Genetic engineering. I. Hill, Suzy.
TP248.2 .B56 2022
660.6--dc23

Table of Contents

Preface

Over the recent decade, advancements and applications have progressed exponentially. This has led to the increased interest in this field and projects are being conducted to enhance knowledge. The main objective of this book is to present some of the critical challenges and provide insights into possible solutions. This book will answer the varied questions that arise in the field and also provide an increased scope for furthering studies.

Biotechnology is a broad area of biology. It involves harnessing biomolecular and cellular processes for the creation and development of new products. A few of its major subfields are genomics, recombinant gene techniques and applied immunology. It is also used for the development of pharmaceutical therapies and diagnostic tests. One of the most significant advancements of biotechnology is the production of therapeutic proteins and other drugs through genetic engineering. This book is a valuable compilation of topics, ranging from the basic to the most complex advancements in the field of biotechnology. There has been rapid progress in this field and its applications are finding their way across multiple industries. This book is appropriate for students seeking detailed information in this area as well as for experts.

I hope that this book, with its visionary approach, will be a valuable addition and will promote interest among readers. Each of the authors has provided their extraordinary competence in their specific fields by providing different perspectives as they come from diverse nations and regions. I thank them for their contributions.

Editor

Ex vivo evolution of human antibodies by CRISPR-X: from a naive B cell repertoire to affinity matured antibodies

Marie-Claire Devilder[1,2,3], Melinda Moyon[1,2], Laetitia Gautreau-Rolland[1,2], Benjamin Navet[1,2], Jeanne Perroteau[1,2], Florent Delbos[4], Marie-Claude Gesnel[1,2,3], Richard Breathnach[1,2*] and Xavier Saulquin[1,2*]

Keywords: Human antibodies, SHM, CRISPR, CRISPR-X, AID, Tetramers, HLA, Cytofluorimetry

Background Somatic hypermutation promotes affinity maturation of antibodies by targeting the cytidine deaminase AID to antibody genes, followed by antigen-based selection of matured antibodies. Given the importance of antibodies in medicine and research, developing approaches to reproduce this natural phenomenon in cell culture is of some interest.

Results We use here the CRISPR-Cas 9 based *CRISPR-X* approach to target AID to antibody genes carried by expression vectors in HEK 293 cells. This directed mutagenesis approach, combined with a highly sensitive antigen-associated magnetic enrichment process, allowed rapid progressive evolution of a human antibody against the Human Leucocyte Antigen A*02:01 allele. Starting from a low affinity monoclonal antibody expressed on Ag-specific naïve blood circulating B cells, we obtained in approximately 6 weeks antibodies with a two log increase in affinity and which retained their specificity.

Conclusion Our strategy for in vitro affinity maturation of antibodies is applicable to virtually any antigen. It not only allows us to tap into the vast naive B cell repertoire but could also be useful when dealing with antigens that only elicit low affinity antibodies after immunization.

Background

The human B cell repertoire constitutes a source of antibodies capable of recognizing virtually any antigen (Ag).

This is the result of a complex B lymphocyte maturation process. Newly produced B cells express B cell receptors (BCRs) generated by random somatic recombination of V (Variable), D (Diversity) and J (Junction) gene segments and which generally have a low affinity for their cognate Ag [1]. After exposure to an Ag, naïve B cells with Ag-specific BCRs undergo somatic hypermutation (SHM) catalyzed by the enzyme Activation induced cytidine deaminase (AID) [2–4]. This enzyme is targeted to the Ig-loci in B cells and deaminates cytosines, thus provoking point mutations, insertions and deletions in the variable domains of both the heavy and light chains. This process ultimately leads to antibody diversification and is followed by the selection of a matured B cell repertoire with higher affinity and specificity for the Ag. This allows the overall diversity of the BCR / antibody molecules to reach theoretically about 10^{13} different receptors in humans [5]. The repertoire thus constitutes an almost unlimited resource of antibodies.

For several decades, monoclonal antibodies (mAb) have been crucial tools in the treatment of diseases such as autoimmune diseases and cancer, or for the control of graft rejection. It is important to generate fully human mAbs because they have a lower risk of immune response induction in humans than the mouse, chimeric or humanized mAbs generally used hitherto. Various methods have been developed for isolating antibodies directly from a natural repertoire of human B lymphocytes. In general, they derive from two main approaches. The first of these is the high-throughput screening of mAb produced by B cell cultures or plasma cells [6, 7]. This is a very effective method for obtaining mAb against Ag to which an individual is exposed naturally or by vaccination. However, many Ag of therapeutic interest are

* Correspondence: Richard.Breathnach@univ-nantes.fr; xavier.saulquin@univ-nantes.fr
[1]CRCINA, INSERM, CNRS, Université d'Angers, Université de Nantes, Nantes, France

not encountered sufficiently frequently naturally, or exploitable in vaccine strategies in humans, to profit from this type of methodology. The second technique consists in isolating single Ag-specific B cells using fluorescent-tagged Ag, followed by cloning of their immunoglobulin genes and expression of recombinant antibodies in a cell line. This technique allows interrogation of both the immune/matured B cell repertoire and the naïve/germline repertoire of an individual with respect to any Ag available in purified form [8–10]. There is a limitation to the interrogation of a naive B cell repertoire however: the generally limited affinity of the corresponding recombinant antibodies, requiring identification of mutations that enhance affinity while maintaining specificity.

Antibody optimization currently relies heavily on the use of libraries generated by mutagenesis of antibody chains using error-prone PCR or degenerate primers. Libraries are screened using techniques such as ribosome, phage, yeast or mammalian display [11]. Co-expression of AID and antibody or non-antibody genes in various mammalian cell lines has also been used to initiate a mutagenic process mimicking SHM [12–20]. This approach circumvents the need to construct mutant libraries, but does not allow targeting of the AID enzyme to sequences encoding the antibody. In B cells, AID is targeted to the immunoglobulin locus by complex mechanisms not yet fully elucidated [21].

We wanted to develop a simple strategy for AID-targeting to antibody sequences in non-B cells to obtain mutated antibodies with increased affinity. Various CRISPR Cas9-based approaches using guide RNAs to target base editors such as APOBEC or AID fused to dead Cas9 (dCas9) to specific DNA sequences have been described recently [22, 23]. These approaches generally lead to mutations limited to a small part of the sequences corresponding to the guide RNA binding site. A variant approach (CRISPR-X) uses a complex containing dCas9 and a guide RNA containing bacteriophage MS2 coat protein binding sites to recruit a coat-AID fusion to DNA [24]. This leads to more extensive mutagenesis covering a window of approximately 100 bp around the guide RNA binding site.

In this work, we present a CRISPR-X based strategy for targeted *in cellulo* affinity maturation of low affinity human mAbs. We apply it to a low affinity mAb named A2Ab against HLA-A*02:01 which shows some cross-reactivity against other HLA-A alleles. A2Ab was isolated from circulating B cells of a naïve individual using a procedure recently developed by our group [8, 10]. We used CRISPR-X with multiplexed guide RNAs to target AID to the VDJ segment encoding the A2Ab heavy chain variable domain in HEK 293 cells co-expressing the light chain. This directed-mutagenesis approach, combined with mammalian surface expression display and a very

sensitive Ag-associated magnetic enrichment process, allowed us to identify mAbs with increased affinity and a sharpening of their specificity for HLA-A*02:01. Overall we describe a novel procedure for generation of high-affinity/optimized human mAbs that is applicable to both naïve and mature circulating human B cells, raising the possibility of generation of private antibodies from a particular individual.

Methods
Donors
Human peripheral blood samples were obtained from anonymous adult donors after informed consent in accordance with the local ethics committee (Etablissement Français du Sang, EFS, Nantes, procedure PLER NTS-2016-08).

Cell lines and culture conditions
Human embryonic kidney 293A cells were obtained from Thermo Fisher Scientific, San Jose, CA, USA (R70507). Cells were grown as adherent monolayers in DMEM (4.5 g/l glucose) supplemented with 10% FBS, 1% Glutamax (Gibco) and 1% penicillin (10,000 U/ml)/streptomycin (10,000 U/ml) (a mixture from Gibco). The BLCL cell lines HEN (HLA-A*02:01/ HLA-A*0101), B721.221 and stably transfected HLA-A2 B721.221 (B721.221 A2) were grown in suspension in RPMI medium supplemented with 10% FBS, 1% Glutamax (Gibco) and 1% penicillin (10,000 U/ml)/streptomycin (10,000 U/ml) (a mixture from Gibco).

Plasmid constructions
Plasmids for mutagenesis were obtained from Addgene: pGH335_MS2-AID*Δ-Hygro (catalogue n° 85.406), pX330S-2 to 7 from the Multiplex CRISPR/Cas9 Assembly System kit (n° 1.000.000.055) and pX330A_dCas9-1 × 7 from the multiplex CRISPR dCas9/Fok-dCas9 Accessory pack (n° 1.000.000.062). The sgRNA scaffolds in the seven latter plasmids were replaced by the sgRNA_2MS2 scaffold from pGH224_sgRNA_2xMS2_Puro (Addgene n° 85.413) and guide sequences then introduced into their BbsI sites before Golden Gate assembly. SgRNA design was performed online using Sequence Scan for CRISPR software (http://crispr.dfci.harvard.edu/SSC/). Final plasmids for mutagenesis thus obtained contain expression cassettes for dCas9 and seven sgRNAs. For production of antibodies, VH and VL regions from human antibodies were subcloned respectively in an IgG-Abvec expression vector (FJ475055) and an Iglambda −AbVec expression vector (FJ51647) as previously described [8]. For mammalian display of antibodies as IgG1, VH and VL regions were subcloned into home-made expression vectors derived from the OriP/EBNA1 based episomal vector pCEP4. The VH and VL expression vectors contain a

hygromycin B or Zeocin resistance marker respectively, and a transmembrane region encoding sequence exists in the C gamma constant region sequence.

IgG1 mammalian cell display

Heavy and light chain expression vectors were co-transfected into the 293A cell line at a 1:1 ratio using JetPEI (PolyplusTransfection, Cat. 101–10 N) and cultured for 48 h. Selection of doubly transfected cells was performed using Hygromycin B and Zeocin. Antibody surface expression on the selected cells was confirmed by flow cytometry analysis after staining with a PE-labeled goat-anti-human IgG Fc (Jackson ImmunoResearch).

Peptide MHC tetramer

The HLA-A*02:01–restricted peptides $Pp65_{495}$ (human CMV [HCMV], NLVPMVATV) and MelA27 (melanoma Ag, ELAGIGILTV) and the HLA-B*0702-restricted UV-sensitive peptide (AARGJTLAM; where J is 3-amino-3-(2-nitro)phenyl-propionic acid) were purchased from GL Biochem (Shanghaï, China). Soluble peptide MHC monomers used in this study carried a mutation in the α3 domain (A245V), that reduces CD8 binding to MHC class I. Biotinylated HLA-A*02:01/MelA$_{27}$ (HLA-A2/MelA), HLA-A*02:01/Pp65$_{495}$ (HLA-A2/Pp65), HLA-B*0702/UV sensitive peptide (HLA-B7/pUV) monomers were tetramerized with allophycocyanin (APC)-labeled premium grade streptavidins (Molecular Probes, Thermo Fischer Scientific, ref. S32362) at a molar ratio of 4:1. When applicable, the avidity of the tetramer for its specific antibody was decreased by mixing specific (ie peptide HLA-A2) and unspecific (ie peptide UV-sensitive HLA-B7) biotinylated monomers before tetramerization with APC-labeled streptavidins at different molar ratios.

Ag-specific B cell sorting from PBMC

B cell isolation was performed as previously described [8, 10]. Briefly, PBMCs were obtained by Ficoll density gradient centrifugation and incubated with PE-, APC and BV421-conjugated tetramers (10 µg/mL in PBS plus 2% FBS, for 30 min at room temperature). The tetramer-stained cells were enriched using anti-PE and-APC Ab-coated paramagnetic beads and then stained with anti-CD19-PerCpCy5.5 (BD Biosciences) mAbs. Stained samples were collected on an ARIA Cell Sorter Cytometer (BD Biosciences) and single $CD19^+$ $CD3^-$ PE^+ APC^+ $BV421^-$ tetramer cells were collected in individual PCR tubes.

Flow cytometry analysis

The specificity and avidity of IgG expressing HEK 293 cells was analysed by flow cytometry. Cells were first stained in PBS containing 0.5% BSA with Ag tetramers for 30 min at room temperature. Anti-PE human IgG

was then added at a 1/500 dilution for 15 min on ice without prior washing. The binding of mutant antibodies was evaluted on 150,000 BLCL cells. Cells were incubated with various concentrations of large-scale purified mAbs diluted in 25 ml of PBS containing 0.5% BSA for 30 min at room temperature. Anti-PE goat anti-human IgG was then added at a 1/500 dilution for 15 min on ice without prior washing.

Mutagenesis

4×10^6 anti HLA-A2 IgG-expressing cells were seeded the day before transfection in a 175 cm flask. For each round of mutation, cells were transiently transfected using JET-PRIME (PolyplusTransfection, Cat. 101–10 N) with pGH335_MS2-AID*Δ-Hygro together with two other plasmids allowing expression of a total of 9 different sgRNAs along with dCas9 at a ration 1: 1: 1.

Affinity-based cell selection and immunomagnetic enrichment

After a round of mutagenesis, transfected cells were expanded until confluency over a week. For selection, $10–20 \times 10^6$ cells were washed, resuspended in 0.2 mL of PBS containing 2% BSA and the antigen (i.e. APC HLA-A2 tetramers or mixed APC HLA-A2/HLA-B7 tetramers) and incubated for 30 min at room temperature. The tetramer-stained cells were then positively enriched using anti APC Ab-coated immunomagnetic beads and columns as previously described [8]. The resulting enriched fraction was stained with an anti human IgG-PE. IgG PE+ and tetramer APC+ cells were collected on an ARIA cell sorter. The adopted strategy for evolution of mAb A2Ab was as follows: 1) three rounds of mutagenesis; 2) magnetic enrichment with 3A2/1B7 tetramer; 3) FACS sorting of positive cells. Positively selected and sorted mutated HEK 293 underwent two new rounds of mutation using the same sgRNAs before selection with the 1A2/3B7 tetramer.

Antibody production

Antibody production was performed as previously described [8]. Briefly, 293A cell lines were transiently transfected with VH and VL expression vectors and cultured for 5 days in serum free medium in 175 cm2 flasks. Recombinant antibodies produced were purified from cell supernatant by Fast Protein Liquid Chromatography (FPLC) using a protein A column, and their concentration determined by absorbance measurement at 280 nm.

Elisa

96-well ELISA plates (Maxisorp, Nunc) were coated with HLA-A2 monomers (overnight at 4 °C, final concentration 2 µg/mL in a coating buffer 1X (Affymetrix)), saturated with a 10% FBS DMEM blocking buffer (Thermo

Fischer Scientific) for 2 h at 37 °C and (iii) incubated with serial dilutions of purified mAbs for 2 h at room temperature. Binding of mAbs was detected with an anti-human IgG-HRP Ab (BD Bioscience, 1 µg/mL, 1 h) and addition of a chromogenic substrate for 20 min at room temperature (Maxisorp, Nunc).

Anti–HLA antibody testing (Luminex)
A Single Antigen Flow Bead assay (LabScreen single-antigen LS1A04, One Lambda, Inc., Canoga Park, CA), was used to detect anti-HLA antibodies in donors and test the specificity of antibodies against 97 MHC-class I alleles. Analysis was performed with a Luminex 100 analyser (Luminex, Austin, TX) after removal of the background as previously described [10].

Surface Plasmon resonance
Surface Plasmon Resonance (SPR) experiments were performed on a Biacore 3000 apparatus (GE Healthcare Life Sciences, Uppsala, Sweden) on CM5 chips (GE Healthcare) as previously described [10]. Briefly, mAbs were immobilized at 10 µg/mL The sensor chip surface was then deactivated and various dilutions of HLA-A*02:01 peptide monomers were injected for 180 s at 40 µL/min.

Bioinformatics analysis
Amplicon preparation: total RNA was purified from 5×10^6 HEK 293 cells and 1 µg of total RNA was reverse transcribed using Superscript reverse transcriptase III (ThermoFisher). cDNA was subsequently amplified using Q5 DNA polymerase and primers targeting VH sequences. Sense and antisense primers include target sequences suitable for Nextera indexage. Barcodes were further introduced by PCR with indexed nextera and the amplicons were sequenced at the IRIC's Genomics Core Facility at Montreal. Paired-end MiSeq technology (Miseq Reagent Nano kit v2 (500 cycles) from Illumina, Inc. San Diego, CA, USA) was used, with a 2 × 250 bp setup.

Pretreatment and sequence clustering
For each chip generated, approximately one million reads were obtained for all the samples. The quality and length distribution of the reads were checked using the FASTQ tool (v0.11.7). After that, for each sample, the paired-end sequences were assembled using the PEAR software (v0.9.6) while keeping only the sequences whose Phred score was greater than 33 and whose overlap was at least 10 nucleotides. Then 30,000 sequences were randomly selected to normalize samples. Next, for each sample, full length VH sequences were grouped according to their identity and counted and clusters were formed as described in the text. Mutations observed in the mock control (gRNA only) experiment were then eliminated in order to distinguish site-directed mutations from RT-PCR or sequence errors. Only clusters representing more than 0.1% of the total number of sequences were retained.

Alignment and mutation analysis
For each sample, the generated clusters were annotated by aligning each sequence cluster against the reference sequence using Biostring library (v2.48.0) in a custom R script, to generate a counting table. The generated data were filtered by subtracting the mutations detected in the mock sample. A position matrix was then generated to create a Weblogo using the ggseqlogo library (v0.1). The data processing was performed using a custom R script.

Results
Isolation of a low affinity human antibody against HLA-A*02:01
A human HLA-A*02:01 molecule (hereafter referred to as HLA-A2) was selected as a target for antibody discovery and maturation as it is easy to obtain blood samples from donors not previously immunized against this MHC allele. In addition, various recombinant HLA molecules were readily available in our laboratory. PBMCs from three HLA-A2-negative donors with negative serology for HLA-A2 circulating antibodies (Additional file 1: Table S1) were tested for the presence of blood circulating B cells specific for HLA-A2. This was done by flow cytometry sorting of B cells that bound HLA-A2 tetramers labeled with two different fluorochromes but did not bind HLA-B7 tetramers, using a technique described previously [8, 10]. B lymphocytes stained specifically by HLA-A2 tetramers could be identified in PBMC from all three donors (see Fig. 1a for an example) and were isolated as single cells. We attempted RT-PCR amplification of sequences coding for the variable regions of the heavy and light chains of four B lymphocytes isolated from one donor (NO) using a recently published protocol [8, 10]. A pair of heavy and light chain V region coding sequences was obtained for one of the four cells. After cloning these gene segments into eukaryotic expression vectors in phase with human heavy and light chain constant domains, the corresponding antibody (A2Ab) was successfully produced in the supernatant of transfected HEK cells and tested for its specificity. A2Ab recognizes HLA-A2 but not HLA-B7 in ELISA tests and this recognition does not depend on the peptide loaded into the HLA pocket (Fig. 1b). A single HLA antigen flow bead assay analysis confirmed that A2Ab can recognize HLA-A*02:01, but also showed that A2Ab recognizes closely related alleles belonging to the HLA-A*02 supertype (HLA-A*02:03, A*02:06 and A*69:01) and weakly cross-reacts with other MHC A

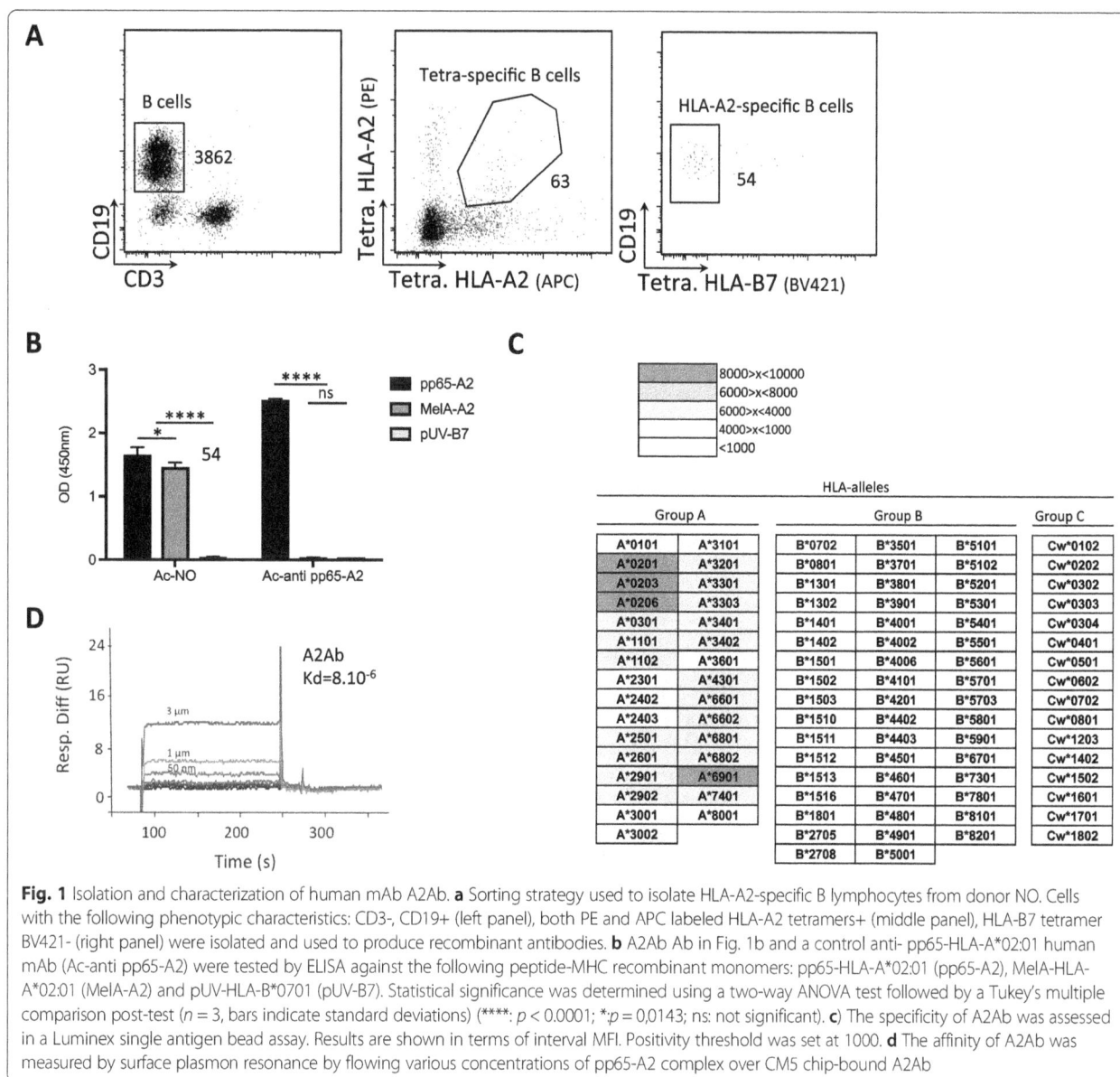

Fig. 1 Isolation and characterization of human mAb A2Ab. **a** Sorting strategy used to isolate HLA-A2-specific B lymphocytes from donor NO. Cells with the following phenotypic characteristics: CD3-, CD19+ (left panel), both PE and APC labeled HLA-A2 tetramers+ (middle panel), HLA-B7 tetramer BV421- (right panel) were isolated and used to produce recombinant antibodies. **b** A2Ab Ab in Fig. 1b and a control anti- pp65-HLA-A*02:01 human mAb (Ac-anti pp65-A2) were tested by ELISA against the following peptide-MHC recombinant monomers: pp65-HLA-A*02:01 (pp65-A2), MelA-HLA-A*02:01 (MelA-A2) and pUV-HLA-B*0701 (pUV-B7). Statistical significance was determined using a two-way ANOVA test followed by a Tukey's multiple comparison post-test ($n = 3$, bars indicate standard deviations) (****: $p < 0.0001$; *$p = 0,0143$; ns: not significant). **c)** The specificity of A2Ab was assessed in a Luminex single antigen bead assay. Results are shown in terms of interval MFI. Positivity threshold was set at 1000. **d** The affinity of A2Ab was measured by surface plasmon resonance by flowing various concentrations of pp65-A2 complex over CM5 chip-bound A2Ab

alleles. However, B or C alleles are not recognized (data not shown, results summarized in Fig. 1c). Finally, the affinity of A2Ab for the pp65/HLA-A2 complex was determined by surface plasmon resonance (SPR) to be in the low micromolar range (Kd = 8.10^{-6}, Fig. 1d). This is consistent with the HLA-A2-specific B cells being isolated from a naive/non-immune blood circulating B cell repertoire. The full nucleotide sequences of the heavy and light chains are provided in Additional file 1: Table S2.

CRISPR-X targeted mutagenesis of A2Ab and screening for higher avidity antibodies

We used the CRISPR-X approach [24] (Fig. 2a) to mutate the A2Ab sequence. Our overall procedure using iterative mutation and selection is summarized in Fig. 3a.

HEK 293 cells were engineered to express cell surface A2Ab by stable transfection of episomal vectors expressing its heavy and light chains (HC and LC, respectively). For induction of mutations, these cells were then transiently transfected with a plasmid coding for AID*Δ fused to MS2 coat protein, and plasmids coding for dCas9 and nine different sgRNAs (Additional file 1: Table S3) spanning the sequence coding for the A2Ab HC variable domain (Fig. 2b). AID*Δ is an AID mutant with increased SHM activity whose Nuclear Export Signal (NES) has been removed [24]. It has significantly increased mutation activity compared to wild-type AID without a NES [24]. Three successive transient transfections were performed before cells were screened for expression of mutant antibodies with increased avidity for HLA-A2.

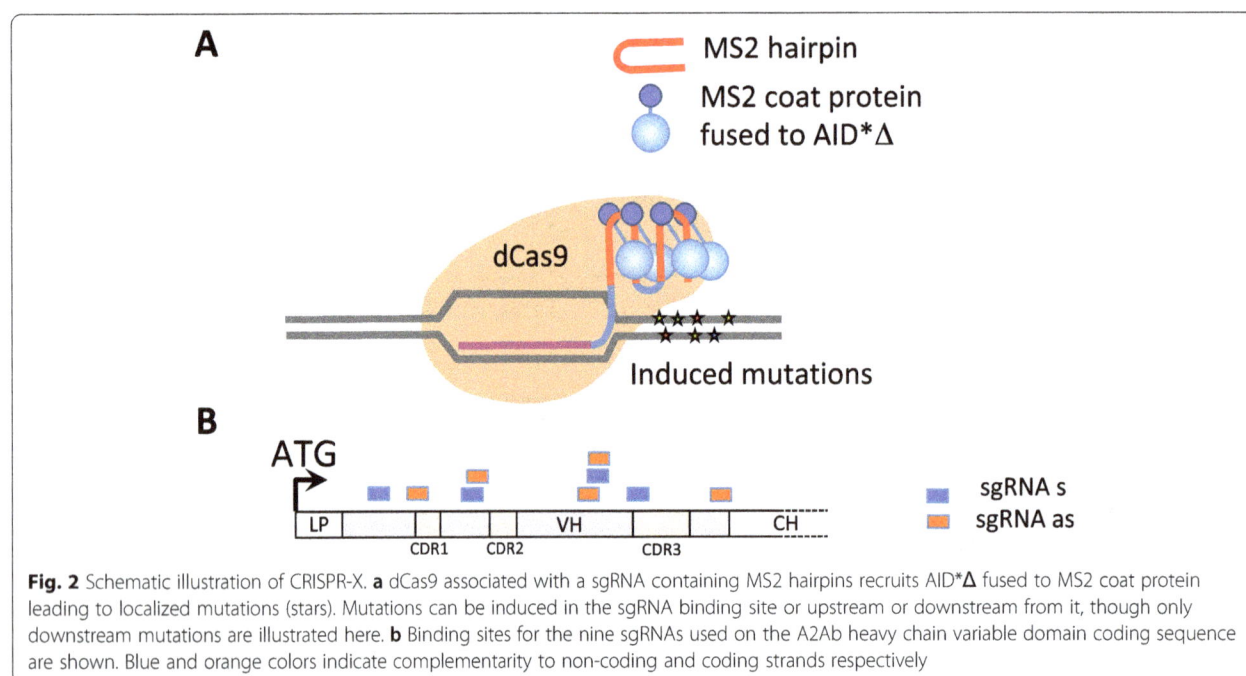

Fig. 2 Schematic illustration of CRISPR-X. **a** dCas9 associated with a sgRNA containing MS2 hairpins recruits AID*Δ fused to MS2 coat protein leading to localized mutations (stars). Mutations can be induced in the sgRNA binding site or upstream or downstream from it, though only downstream mutations are illustrated here. **b** Binding sites for the nine sgRNAs used on the A2Ab heavy chain variable domain coding sequence are shown. Blue and orange colors indicate complementarity to non-coding and coding strands respectively

Cells we started from stably expressed cell surface A2Ab and thus were able to bind tetramers comprising four HLA-A2 molecules. These cells were subjected to three successive transfections. We expected cells expressing higher avidity antibodies post-mutagenesis to be able to bind tetramers containing fewer HLA-A2 molecules. We thus sought to identify cells in the mutated polyclonal population using labeling with a tetramer made up of 3 HLA-A2 molecules and one B7 molecule (3A2/1B7). As shown in Fig. 3b, we were unable to detect any 3A2/1B7-labeled cells in the mutated polyclonal population by flow cytometry, while all cells expressing IgG were labeled with the initial tetramer (4A2) as expected.

We suspected that 3A2/1B7-labeled cells might be too rare to be detectable in the fraction of the mutated polyclonal population we tested, so we tried to enrich them before analysis. The mutated polyclonal population was first incubated with the 3A2/1B7 tetramer coupled to APC, then subjected to positive selection using paramagnetic beads coupled to anti-APC antibodies. After magnetic enrichment, we observed a small proportion of cells clearly labeled by the 3A2/1B7 tetramer (Fig. 3c, left dot-plot). Notably, no such cells were detected when our protocol was carried out using A2Ab-expressing HEK 293 cells transfected with a hyperactive non-guided AID (Fig. 3c, middle dot-plot), or with guide RNAs alone ("mock", Fig. 3c, right dot-plot). This first "positive" population (R1) was purified by cell sorting and expanded in vitro to yield population R1+ (> 95% pure). In marked contrast to the starting population, the R1+ population bound tetramers with just 3 HLA-A2 molecules (3A2/1B7, Fig. 3d, upper left dot-plot).

To complete a further round of mutagenesis/selection, we exposed the R1+ population to two successive transfections for mutagenesis using the same batch of sgRNAs as above, before selection was performed. This time we used a more stringent enrichment process with tetramers containing only one HLA-A2 molecule (1A2/3B7). A new population of tetramer positive cells was obtained (R2+), with a 2.2 fold increase in the 3A2/1B7 tetramer mean fluorescence intensity compared to R1+ (Fig. 3d, bottom left dot-plot). The R2+ population was also stained by tetramer 1A2/3B7, in marked contrast to R1+ cells (Fig. 3d, compare upper and lower right dot-plots). Each round of mutation and selection thus increases the avidity of the antibodies.

Antibody sequence evolution during mutagenesis and selection rounds

As described above, we were unable to detect cells capable of binding to the 3A2/1B7 tetramer after one round of mutagenesis until we used magnetic enrichment. This enrichment generated the R1 population. FACS sorting of this population yielded the R1+ population capable of binding 3A2/1B7 tetramers and the R1- population incapable of binding this tetramer. We used next generation sequencing (NGS) to search for heavy chain sequences enriched in the R1+ population relative to the R1- population and which could contain mutations responsible for the increased affinity of the R1+ population antibodies. 30,000 randomly selected reads from each population were analyzed. Reads represented more than 50 times were placed into a read-specific cluster, while

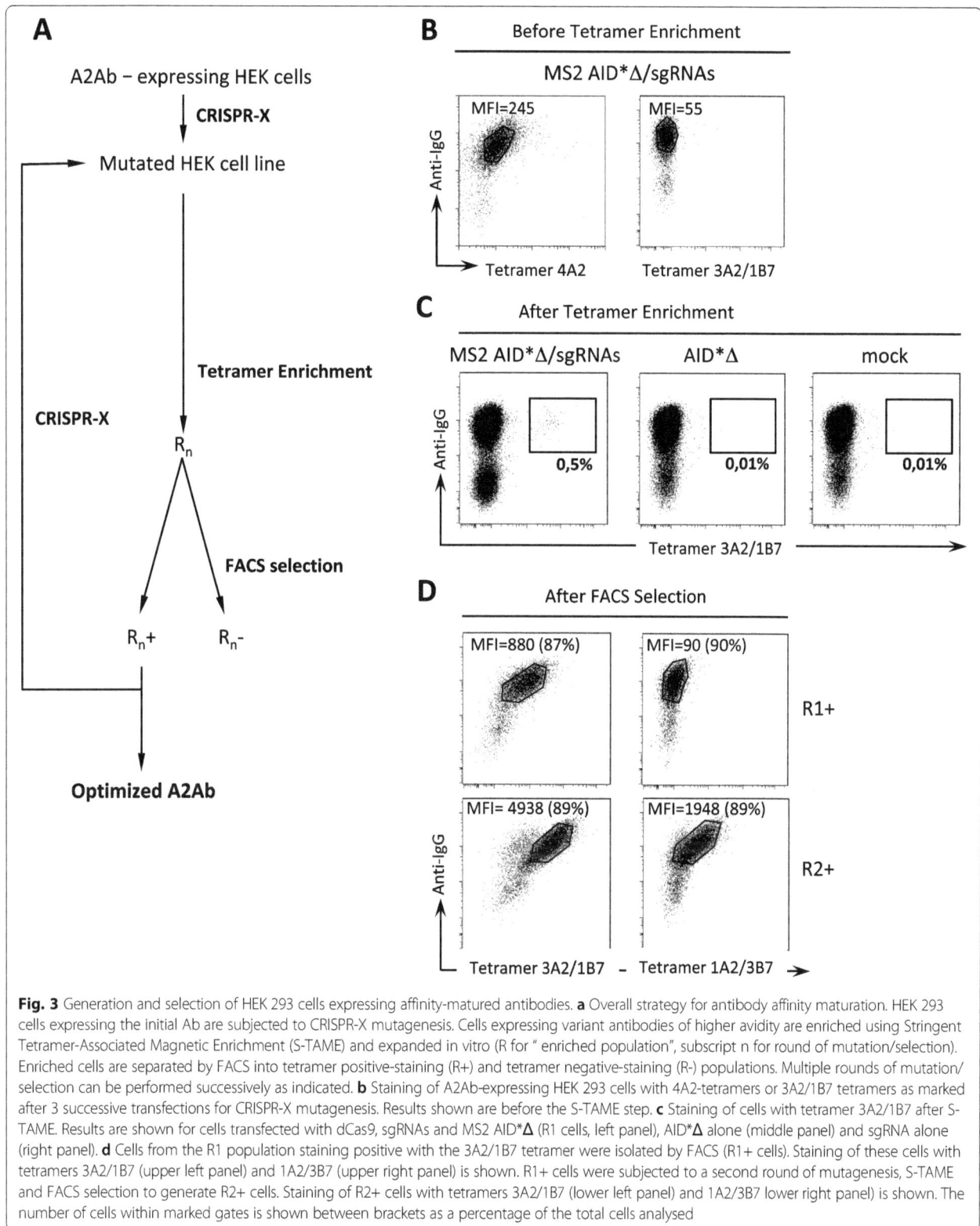

Fig. 3 Generation and selection of HEK 293 cells expressing affinity-matured antibodies. **a** Overall strategy for antibody affinity maturation. HEK 293 cells expressing the initial Ab are subjected to CRISPR-X mutagenesis. Cells expressing variant antibodies of higher avidity are enriched using Stringent Tetramer-Associated Magnetic Enrichment (S-TAME) and expanded in vitro (R for " enriched population", subscript n for round of mutation/selection). Enriched cells are separated by FACS into tetramer positive-staining (R+) and tetramer negative-staining (R-) populations. Multiple rounds of mutation/selection can be performed successively as indicated. **b** Staining of A2Ab-expressing HEK 293 cells with 4A2-tetramers or 3A2/1B7 tetramers as marked after 3 successive transfections for CRISPR-X mutagenesis. Results shown are before the S-TAME step. **c** Staining of cells with tetramer 3A2/1B7 after S-TAME. Results are shown for cells transfected with dCas9, sgRNAs and MS2 AID*Δ (R1 cells, left panel), AID*Δ alone (middle panel) and sgRNA alone (right panel). **d** Cells from the R1 population staining positive with the 3A2/1B7 tetramer were isolated by FACS (R1+ cells). Staining of these cells with tetramers 3A2/1B7 (upper left panel) and 1A2/3B7 (upper right panel) is shown. R1+ cells were subjected to a second round of mutagenesis, S-TAME and FACS selection to generate R2+ cells. Staining of R2+ cells with tetramers 3A2/1B7 (lower left panel) and 1A2/3B7 lower right panel) is shown. The number of cells within marked gates is shown between brackets as a percentage of the total cells analysed

reads represented less than 50 times were grouped together in a category we termed "small clusters". For the R1+ population, two large clusters representing together 42.5% of reads were detected, in addition to a third large cluster representing WT sequences (Table 1). Six other clusters representing together 5.2% of reads were also detected, together with numerous reads in the small cluster category. Seven of these eight non-WT clusters were clearly under-represented in the R1- population, where the WT cluster and small clusters predominated. Mutations observed in the seven clusters were located in the FRW3 and CDR3 regions (Fig. 4). They were often shared between different clusters, suggesting that they contribute to the increased affinity of R1+ population antibodies.

That WT and small cluster sequences represent 52.6% of R1+ reads might seem surprising. However, in the HEK 293 cells subjected to mutagenesis, antibody genes are present on episomal vectors, with several vector copies per cell [25]. Cells selected with the 3A2/1B7 tetramer may contain only one gene copy with a mutation leading to an antibody of increased affinity. All the other copies could contain either no mutation or neutral or even deleterious mutations, yet they will be co-enriched with the copy carrying the affinity-increasing mutation.

The second round of mutation/selection led to a drastic decline in WT reads (from 13.6% for R1+, to 0% for R2+), while in the R2+ population a cluster representing nearly half of the NGS reads emerged, corresponding to HCs accumulating six mutated amino acids: D74H/S80 T/W102 L/M112I/G121D/R124P (Table 2, Fig. 4). Interestingly, the CDR2 D74H mutation was not detected in the R1+ population. Nine of the thirteen R2+ clusters (a cluster contains more than 50 reads of the cluster-specific sequence) differ only very slightly from this main sequence, underlining a strong convergence of most of the R2+ clusters. The W102, M112I, G121D and R124P mutations were already well represented in the R1+ population (Table 1). The second round of mutation/selection led to emergence of two new R2 + –specific mutations: D74H in the CDR2 and S80 T in the FRW3 region.

Characterization of evolved antibodies against HLA-A2

The R2+ antibodies C4.4 and C4.18 (Tables 1 and 2) were produced as recombinant proteins for comparison of their affinity and specificity to those of the initial A2Ab. As shown in Fig. 5a, C4.4 and C4.18 mAbs show clearly increased reactivity against HLA-A*02:01 compared to A2Ab in an ELISA. We next determined C4.18's affinity for HLA-A*02:01 by SPR: Kd = 10^{-7} (Fig. 5b). This is an almost two log increase over that of the initial A2Ab (Kd = 8×10^{-6}). We were unable to make enough C4.4 for SPR studies.

These results demonstrate that our matured antibodies bind with higher affinity to antigen than A2Ab in fully

Table 1 CRISPR-X-mediated evolution of A2Ab: NGS analysis, round 1

R1			
Cluster name	mAb name	%R1+ (counts)	%R1- (counts)
G121E	C3.2	31.8 (9542)	0.3 (94)
WT	A2Ab	13.6 (4103)	52.6 (15788)
W102 L//M112I//G121D//R124P	C3.9	10.7 (3197)	0
G121E//V140 L		1.1 (340)	0
G121D	C3.3	1.1 (316)	0.3 (95)
S103 N//G121D	C3.5	0.9 (261)	0
W102 L//D109A//M112I//G121D//R124P		0.8 (239)	0
M112I//G121D//R124P		0.7 (209)	0
V140 L		0,6 (168)	1.5 (448)
R117S		0	0.5 (148)
Y114S		0	0.4 (124)
D109A		0	0.4 (119)
S103R		0	0.2 (67)
S108A		0	0.2 (66)
G137R		0	0.2 (61)
R119S		0	0.2 (60)
P60A		0	0.2 (54)
V123G		0	0.2 (54)
small clusters R1+ (number)	C3.4	38.7 (11625)	
small clusters R1- (number)			42.7 (12817)
total		100 (30000)	100 (30000)

in vitro tests. But can they bind to antigen expressed on the surface of cells, a prerequisite for biological activity? The initial A2Ab was not of sufficient affinity to bind to two HLA-A2 expressing cell lines tested, 721.221 B cells made HLA-A2 positive by transfection (721.221(A2)), and naturally HLA-A2 expressing BLCL HEN. However, the increased affinity of C4.4 and C4.18 led to ready detection of such binding (Fig. 5c). Binding to 721.221(A2) B cells was HLA-A2 dependent, as no binding was observed to the parental HLA-A2 negative 721.221 B cells. A single HLA antigen flow bead assay analysis confirmed that C4.4 and C4.18 had higher affinity than A2Ab for HLA-A*02:01 and also showed a gain in specificity, as they had significantly less crossreactivity against other HLA-A alleles (compare Fig. 5d to Fig. 1c).

Discussion

We show that starting from a low affinity antibody, CRISPR-X targeting of AID to antibody genes can be used to obtain affinity-matured human antibodies in cellulo in about 6 weeks. Thus we increased the affinity of a

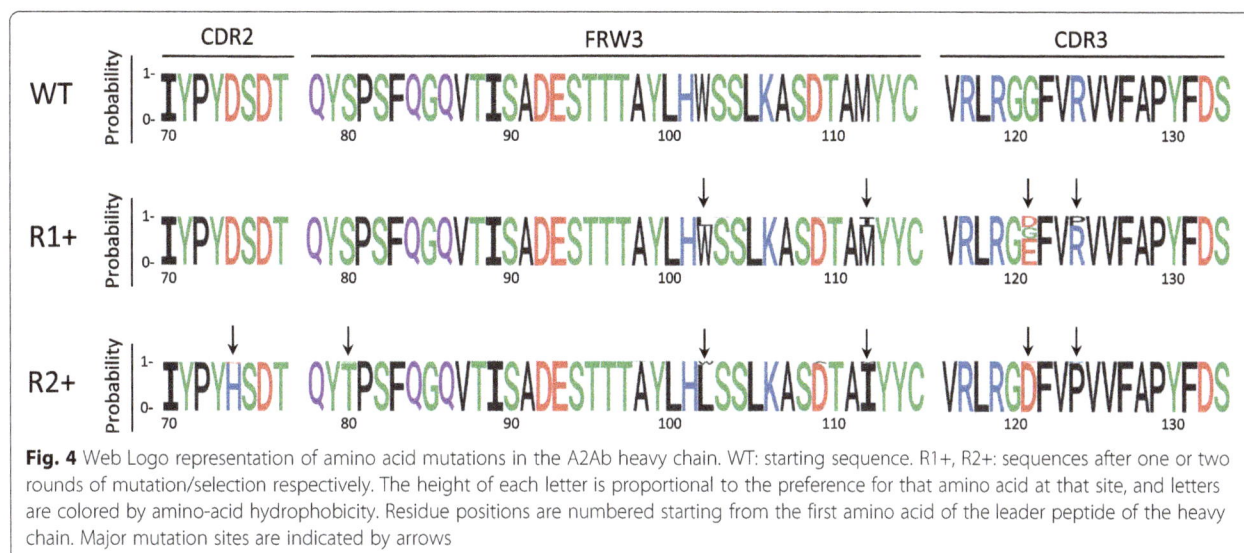

Fig. 4 Web Logo representation of amino acid mutations in the A2Ab heavy chain. WT: starting sequence. R1+, R2+: sequences after one or two rounds of mutation/selection respectively. The height of each letter is proportional to the preference for that amino acid at that site, and letters are colored by amino-acid hydrophobicity. Residue positions are numbered starting from the first amino acid of the leader peptide of the heavy chain. Major mutation sites are indicated by arrows

fully human anti-HLA-A*02:01 mAb to sufficient levels for biological activity and without loss of specificity in just 2 cycles of mutation/selection (each cycle consisting of several successive mutagenesis transfections prior to the selection steps). The low affinity antibody we started from was expressed by naive B cells. Our procedure thus mimics in vitro antibody maturation in secondary lymphoid organs, where naive B lymphocytes stimulated by Ag recognition via specific BCRs of limited affinity go on to generate receptors optimized for Ag recognition.

Using SHM for in vitro affinity maturation of antibodies is an attractive strategy and has been used previously in a variety of cell lines [2, 26–29]. Some recently described technologies to affinity-mature antibodies in vitro rely on the integration of a library of CDR3 domains using CRISPR Cas9 technology [30] or mutagenesis of only the most permissive CDR positions [31]. Prior to these approaches, the Bowers group pioneered the coupling of AID-induced somatic hypermutation with mammalian cell surface display in the easily transfectable HEK 293 cells for in vitro maturation of mAbs [15]. We have extended this latter approach to include specific targeting of AID to the immunoglobulin genes to be mutated using a combination of dCas9-AID fusions and specific guide RNAs. We have also introduced a magnetic enrichment step prior to FACS sorting of mutated cells to facilitate isolation of cells expressing higher affinity antibodies. These modifications proved necessary to obtain our affinity matured anti-HLA antibodies after only 2 rounds of mutation/selection. Indeed, we were unable to detect any cells carrying higher affinity antibodies when AID activity was not targeted to the Ig sequences, and we could only detect and isolate them after the first mutation round if magnetic enrichment preceded FACS sorting.

While this manuscript was in preparation, Liu et al. described a variety of diversifying base editors and showed that they retained their intrinsic nucleotide preferences when recruited to DNA as MS2 coat fusions [32]. They also demonstrated that it was possible to use diversifying base editors to affinity mature a previously studied murine anti-4-hydroxy-3-nitrophenylacetyl (NP) antibody called B1–8 [32]. The matured antibodies they obtained contained various mutations that had already been observed after subjecting B1–8 to SHM in a mouse in vivo immunization model. The effect of these point mutations was tested separately, and it was not clear whether any of their antibodies contained multiple mutations. In our study, we define previously unknown combination of mutations that are required to increase the affinity of a human antibody against HLA-A2, without loss of specificity. As might be expected, "beneficial" mutations could be found in the CDR2 and CDR3. Interestingly, CDR3 mutations appeared after the first round of mutation/selection, while CDR2 mutations only appeared after the second round. In addition to the CDR2 and CDR3 mutations, some mutations also appeared in the FRW3. In particular, the C4.18 mAb obtained after the second round of mutagenesis differs from the first round C3.9 mAb by only two additional mutated amino acids located in FRW3. This is interesting as antibody in vitro evolution studies have suggested that mutations leading to higher affinity often correspond to residues distant from the antigen binding site and that affinity maturation of antibodies occurs most effectively by changes in second sphere residues rather than contact residues [33, 34]. It is also interesting to note that increasing the affinity of our antibodies for HLA-A*02:01 also led to an increase in their

Table 2: CRISPR-X-mediated evolution of A2Ab: NGS analysis, round 2

R2

Cluster name	mAb name	%R2+ (counts)	%R2- (counts)
D74H//S80 T//W102 L//M112I//G121D//R124P	C4.4	49.2 (14755)	9.2 (2756)
D74H//S80 T//W102 L//D109A//M112I//G121D//R124P		2.4 (733)	0.2 (73)
D74H//S80 T//M112I//G121D//R124P		2.2 (650)	0
G121E		1.7 (496)	38.9 (11670)
D74H//S80 T//F83S//W102 L//M112I//G121D//R124P		0.7 (223)	0.4 (112)
D74H//S80 T//A98P//W102 L//M112I//G121D//R124P		0.6 (182)	0
D74H//S80 T//W102 L//M112I//G121D//R124P//V140 L		0.6 (173)	0
D74H//W102 L//M112I//G121D//R124P	C4.18	0.5 (163)	0
G121D//R124P		0.2 (74)	0.9 (274)
D74H//S80 T//W102 L//S104 T//M112I//G121D//R124P		0.2 (72)	0
D74H//S80 T//W102 L//G121E		0.2 (68)	0
W102 L//M112I//G121D//R124P		0.2 (63)	4.2 (1247)
D74H//S80 T//W102 L//L105R//M112I//G121D//R124P		0.2 (59)	0
W52C//G121E		0	1.1 (333)
G121E//V140 L		0	0.7 (209)
R47S//R57H//G121E		0	0.7 (204)
W102 L//G121E		0	0.7 (203)
WT		0	0.6 (193)
M112I//G121D//R124P	A2Ab	0	0.5 (137)
W102 L//M112I//G121E		0	0.4 (128)
H101Q//G121E		0	0.4 (122)
I39M//H101Q//G121E		0	0.4 (119)
P60S//G121E		0	0,4 (81)
C41Y//G121E		0	0.2 (66)
I39M//G121E		0	0.2 (56)
W102C//G121E		0	0.2 (56)
small clusters R2+ (number)		41 (12289)	
small clusters R2- (number)			39.9 (11961)
total		100 (30000)	100 (30000)

specificity: they progressively lost their crossreactivity against non-HLA-A*02 alleles.

The progressive evolution of A2Ab we observed, with a gradual accumulation of combinations of mutations, is probably necessary for the maturation of the affinity of most antibodies. The combination of CDR and FRW mutations could result from CRISPR-X allowing simultaneous targeting of multiple sites all along the Ig variable sequence and potentially represents an important advantage over other recently described technologies limiting mutagenesis to the CDR3 [30] or to the most permissive CDR positions [31].

Our CRISPR-X based approach can readily be developed further to increase the potential for antibody diversification. We used the same 9 gRNAs for both rounds

of mutagenesis. Further rounds of mutagenesis could be carried out using different gRNAs. The CRISPR-X approach using *S. pyogenes* dCas9 requires the presence of an NGG PAM immediately downstream from the gRNA binding site. Cas9 variants with relaxed PAM requirements could also be used in this approach, including the recently described variant using a PAM reduced to NG. This would lift almost all constraints on gRNA choice. We focused on mutating the Ig heavy chain gene alone, but both heavy and light chain genes were present in cells subjected to mutagenesis. We did not detect any light chain mutations after transfection of the heavy chain gRNAs (data not shown), demonstrating the specificity of the targeting approach. However, AID could be targeted simultaneously to both heavy and light chain

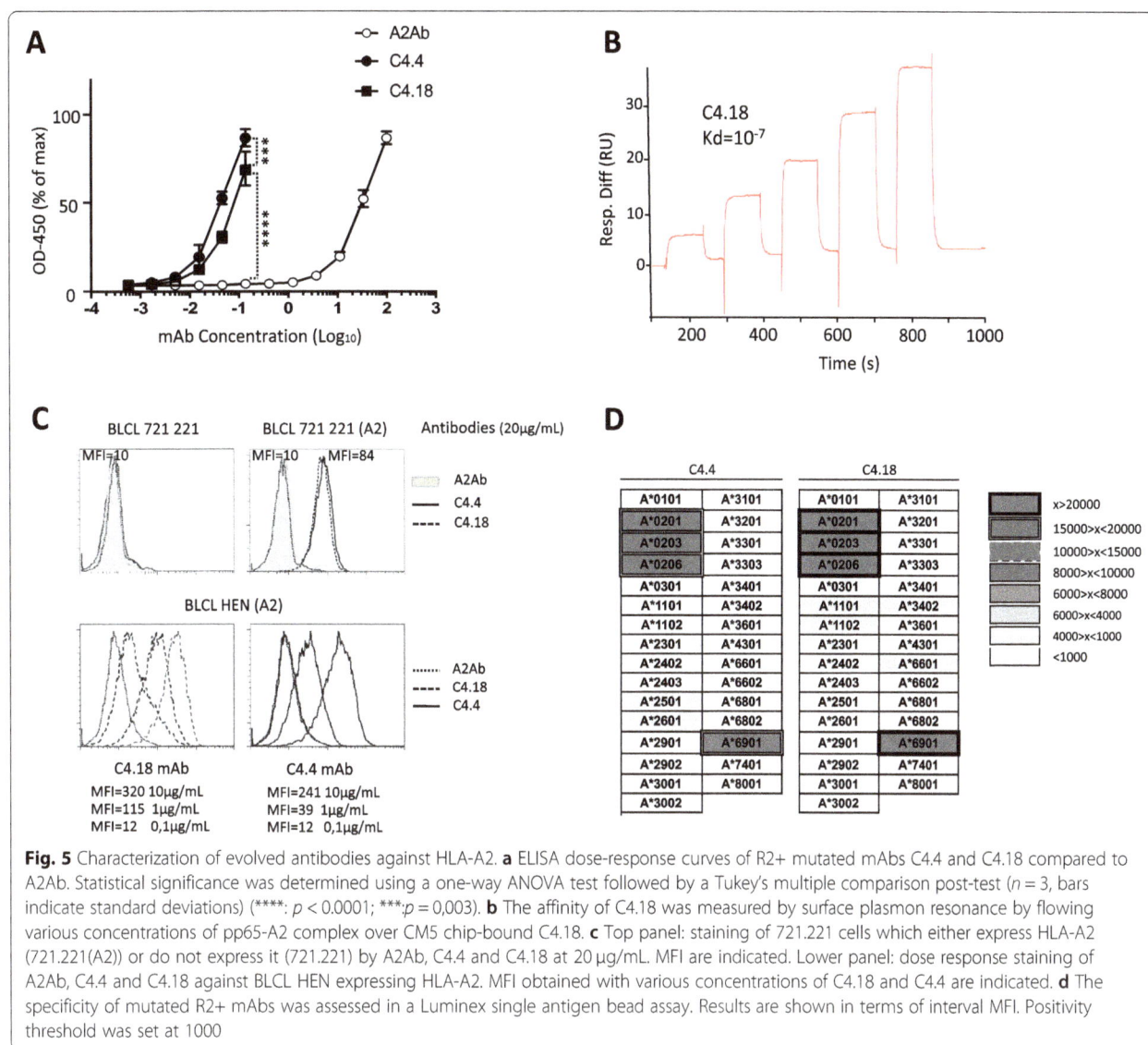

Fig. 5 Characterization of evolved antibodies against HLA-A2. **a** ELISA dose-response curves of R2+ mutated mAbs C4.4 and C4.18 compared to A2Ab. Statistical significance was determined using a one-way ANOVA test followed by a Tukey's multiple comparison post-test ($n = 3$, bars indicate standard deviations) (****: $p < 0.0001$; ***:$p = 0.003$). **b** The affinity of C4.18 was measured by surface plasmon resonance by flowing various concentrations of pp65-A2 complex over CM5 chip-bound C4.18. **c** Top panel: staining of 721.221 cells which either express HLA-A2 (721.221(A2)) or do not express it (721.221) by A2Ab, C4.4 and C4.18 at 20 μg/mL. MFI are indicated. Lower panel: dose response staining of A2Ab, C4.4 and C4.18 against BLCL HEN expressing HLA-A2. MFI obtained with various concentrations of C4.18 and C4.4 are indicated. **d** The specificity of mutated R2+ mAbs was assessed in a Luminex single antigen bead assay. Results are shown in terms of interval MFI. Positivity threshold was set at 1000

genes by cotransfecting cells with a mixture of heavy and light chain gRNAs, increasing the diversification possibilities by association of mutated heavy and light chains in different combinations.

The A2Ab mAb used here served as an initial proof of concept for antibody maturation in vitro using CRISPR-X. However, the fully human mAbs specific for the HLA-A*02:01 allele we generated could have direct clinical applications, notably in the context of mismatch HLA-A2 organ transplantation. Two recent studies described the efficacy of anti-HLA-A2-specific CARs of murine origin in the control of graft rejection in animal models [35, 36]. Using fully human antibodies could be an important step forward for implementation of such strategies to humans. Furthermore, the availability of a series of mAbs of increasing affinity (derived from different rounds of mutation/selection) could be useful to study the impact of CAR affinity on biological activity and could also help to improve predictive algorythms for antibody maturation.

Conclusions

We describe here a new approach for progressive and controlled antibody evolution. This procedure should allow us to obtain antibodies of high affinity and specificity against virtually any Ag, if available in a recombinant form, starting directly from circulating naïve B cells, which represent a vast pool of Ag-specific antibodies to tap into. Our approach may prove particularly useful when fully human antibodies are required: when first isolated from non-immunized individuals, they are often of insufficient affinity for therapeutic or research purposes. Many Ag of interest for the treatment of pathologies such as cancer are in this category and thus represent potential targets for this approach. In addition, our approach can be adapted to optimize antibody specificity by addition of a simple negative selection step to eliminate antibodies with undesired interactions. This could be useful for improving the specificity of currently existing murine, chimeric or humanized antibodies.

Additional file

Additional file 1: Table S1. Isolation of human anti-HLA-A2 B lymphocytes from the PBMC of grafted patients.This table indicates the number of HLA-A2-specific B cells isolated from each donors.**Table S2.** Full nucleotide sequences.This table indicates the nucleotide sequences of the variable segments of the heavy and light chains of A2Ab and of the heavy chain of the various R1+ or R2+ mutants.**Table S3.** gRNA sequences binding to the Ig gene sense (s) or antisense (as) strands.This table indicates the nucleotide sequence of the gRNAs which are numbered according to their position from the ATG (A corresponding to nucleotide number 1) of the A2Ab variable heavy chain sequence (see Additional file 1: **Table S2**). (DOCX 51 kb)

Abbreviations

Ag: Antigen; APC: Allophycocyanin; BCR: B-cell receptor; BV: Brillant violet; CDR: Complementarity determining region; FRW: Framework region; HC: Heavy chain; LC: Light chain; mAbs: Monoclonal antibodies; PBMC: Peripheral blood mononuclear cells; PE: Phycoerythrin; pMHC: Peptide-major histocompatibility complex; SPR: Surface plasmon resonance

Acknowledgments

We thank the Cytometry Facility "CytoCell" (SFR Santé, Biogenouest, Nantes) for expert technical assistance. We thank also all the staff of recombinant protein production (P2R) and of IMPACT platforms (UMR-S892, SSFR Santé, Biogenouest, Nantes) for their technical support.
We thank Dr. Anne Cesbron, head of the HLA laboratory (EFS, Nantes) where luminex were performed.
We thank Dr. Magali Giral for access to the DIVAT cohort.

Funding

This work was financially supported by the IHU-Cesti project funded by the « Investissements d'Avenir » French Government program, managed by the French National Research Agency (ANR) (ANR-10-IBHU-005). The IHU-Cesti project is also supported by Nantes Métropole and Région Pays de la Loire. This work was realized in the context of the LabEX IGO program supported by the National Research Agency via the investment of the future program ANR-11-LABX-0016-01. These funds were used for the design of the study, the generation and analysis of mutant antibodies, the interpretation and the writting/publication of the manuscript.

Authors' contributions

MCD, MM, LG, MCG, JP performed the experiments. BN performed the bioinformatic studies. FD performed the luminex analysis. MCD, RB and XS wrote the manuscript, designed the experiments, analyzed and interpreted the data and supervised the working program. All authors read and approved the final manuscript.

Ethics approval and consent to participate

Blood samples were collected from donors with written informed consents. The procedure and the cohort of donors (named « DIVAT ») was approved by the local ethic committee « CPP Grand Ouest IV » reference number: MESR DC-2017-2987.

Consent for publication
Not applicable.

Competing interests
The authors declare that they have no competing interests.

Author details
[1]CRCINA, INSERM, CNRS, Université d'Angers, Université de Nantes, Nantes, France. [2]LabEx IGO "Immunotherapy, Graft, Oncology", Nantes, France. [3]Centre Hospitalier Universitaire Hôtel-Dieu, Nantes, France. [4]HLA Laboratory, EFS Centre Pays de la Loire, Nantes, France.

References

1. Schatz DG, Ji Y. Recombination centres and the orchestration of V(D)J recombination. Nat Rev Immunol. 2011;11(4):251–63.
2. Martin A, Scharff MD. Somatic hypermutation of the AID transgene in B and non-B cells. Proc Natl Acad Sci U S A. 2002;99(19):12304–8.
3. Muramatsu M, Kinoshita K, Fagarasan S, Yamada S, Shinkai Y, Honjo T. Class switch recombination and hypermutation require activation-induced cytidine deaminase (AID), a potential RNA editing enzyme. Cell. 2000;102(5): 553–63.
4. Williams SC, Frippiat JP, Tomlinson IM, Ignatovich O, Lefranc MP, Winter G. Sequence and evolution of the human germline V lambda repertoire. J Mol Biol. 1996;264(2):220–32.
5. Calis JJ, Rosenberg BR. Characterizing immune repertoires by high throughput sequencing: strategies and applications. Trends Immunol. 2014; 35(12):581–90.
6. Corti D, Langedijk JP, Hinz A, Seaman MS, Vanzetta F, Fernandez-Rodriguez BM, Silacci C, Pinna D, Jarrossay D, Balla-Jhagjhoorsingh S, et al. Analysis of memory B cell responses and isolation of novel monoclonal antibodies with neutralizing breadth from HIV-1-infected individuals. PLoS One. 2010;5(1):e8805.
7. Corti D, Voss J, Gamblin SJ, Codoni G, Macagno A, Jarrossay D, Vachieri SG, Pinna D, Minola A, Vanzetta F, et al. A neutralizing antibody selected from plasma cells that binds to group 1 and group 2 influenza a hemagglutinins. Science. 2011;333(6044):850–6.
8. Devilder MC, Moyon M, Saulquin X, Gautreau-Rolland L. Generation of discriminative human monoclonal antibodies from rare antigen-specific B cells circulating in blood. J Vis Exp. 2018;(132).
9. Franz B, May KF Jr, Dranoff G, Wucherpfennig K. Ex vivo characterization and isolation of rare memory B cells with antigen tetramers. Blood. 2011;118(2): 348–57.
10. Ouisse LH, Gautreau-Rolland L, Devilder MC, Osborn M, Moyon M, Visentin J, Halary F, Bruggemann M, Buelow R, Anegon I, et al. Antigen-specific single B cell sorting and expression-cloning from immunoglobulin humanized rats: a rapid and versatile method for the generation of high affinity and discriminative human monoclonal antibodies. BMC Biotechnol. 2017;17(1):3.
11. Hoogenboom HR. Selecting and screening recombinant antibody libraries. Nat Biotechnol. 2005;23(9):1105–16.
12. Akamatsu Y, Pakabunto K, Xu Z, Zhang Y, Tsurushita N. Whole IgG surface display on mammalian cells: application to isolation of neutralizing chicken monoclonal anti-IL-12 antibodies. J Immunol Methods. 2007;327(1–2):40–52.
13. Al-Qaisi TS, Su YC, Roffler SR. Transient AID expression for in situ mutagenesis with improved cellular fitness. Sci Rep. 2018;8(1):9413.
14. An L, Chen C, Luo R, Zhao Y, Hang H. Activation-induced cytidine deaminase aided in vitro antibody evolution. Methods Mol Biol. 2018;1707:1–14.
15. Bowers PM, Horlick RA, Neben TY, Toobian RM, Tomlinson GL, Dalton JL, Jones HA, Chen A, Altobell L 3rd, Zhang X, et al. Coupling mammalian cell surface display with somatic hypermutation for the discovery and maturation of human antibodies. Proc Natl Acad Sci U S A. 2011;108(51): 20455–60.
16. Ho M, Nagata S, Pastan I. Isolation of anti-CD22 Fv with high affinity by Fv display on human cells. Proc Natl Acad Sci U S A. 2006;103(25):9637–42.
17. Ho M, Pastan I. Display and selection of scFv antibodies on HEK-293T cells. Methods Mol Biol. 2009;562:99–113.
18. McConnell AD, Do M, Neben TY, Spasojevic V, MacLaren J, Chen AP, Altobell L 3rd, Macomber JL, Berkebile AD, Horlick RA, et al. High affinity humanized antibodies without making hybridomas; immunization paired

with mammalian cell display and in vitro somatic hypermutation. PLoS One. 2012;7(11):e49458.

19. Su YC, Al-Qaisi TS, Tung HY, Cheng TL, Chuang KH, Chen BM, Roffler SR. Mimicking the germinal center reaction in hybridoma cells to isolate temperature-selective anti-PEG antibodies. mAbs. 2014;6(4):1069–83.

20. Wang L, Jackson WC, Steinbach PA, Tsien RY. Evolution of new nonantibody proteins via iterative somatic hypermutation. Proc Natl Acad Sci U S A. 2004;101(48):16745–9.

21. Hwang JK, Alt FW, Yeap LS: Related Mechanisms of Antibody Somatic Hypermutation and Class Switch Recombination. Microbiol Spectr 2015, 3(1):MDNA3–0037-2014.

22. Hess GT, Tycko J, Yao D, Bassik MC. Methods and applications of CRISPR-Mediated Base editing in eukaryotic genomes. Mol Cell. 2017;68(1):26–43.

23. Rees HA, Liu DR. Base editing: precision chemistry on the genome and transcriptome of living cells. Nat Rev Genet. 2018;19(12):770–88.

24. Hess GT, Fresard L, Han K, Lee CH, Li A, Cimprich KA, Montgomery SB, Bassik MC. Directed evolution using dCas9-targeted somatic hypermutation in mammalian cells. Nat Methods. 2016;13(12):1036–42.

25. Yates JL, Warren N, Sugden B. Stable replication of plasmids derived from Epstein-Barr virus in various mammalian cells. Nature. 1985;313(6005):812–5.

26. Cumbers SJ, Williams GT, Davies SL, Grenfell RL, Takeda S, Batista FD, Sale JE, Neuberger MS. Generation and iterative affinity maturation of antibodies in vitro using hypermutating B-cell lines. Nat Biotechnol. 2002;20(11):1129–34.

27. Delker RK, Fugmann SD, Papavasiliou FN. A coming-of-age story: activation-induced cytidine deaminase turns 10. Nat Immunol. 2009;10(11):1147–53.

28. Maul RW, Gearhart PJ. AID and somatic hypermutation. Adv Immunol. 2010;105:159–91.

29. Seo H, Hashimoto S, Tsuchiya K, Lin W, Shibata T, Ohta K. An ex vivo method for rapid generation of monoclonal antibodies (ADLib system). Nat Protoc. 2006;1(3):1502–6.

30. Mason DM, Weber CR, Parola C, Meng SM, Greiff V, Kelton WJ, Reddy ST. High-throughput antibody engineering in mammalian cells by CRISPR/Cas9-mediated homology-directed mutagenesis. Nucleic Acids Res. 2018;46(14):7436–49.

31. Tiller KE, Chowdhury R, Li T, Ludwig SD, Sen S, Maranas CD, Tessier PM. Facile affinity maturation of antibody variable domains using natural diversity mutagenesis. Front Immunol. 2017;8:986.

32. Liu LD, Huang M, Dai P, Liu T, Fan S, Cheng X, Zhao Y, Yeap LS, Meng FL: Intrinsic nucleotide preference of Diversifying Base editors guides antibody ex vivo affinity maturation. Cell Rep 2018, 25(4):884–892 e883.

33. Boder ET, Midelfort KS, Wittrup KD. Directed evolution of antibody fragments with monovalent femtomolar antigen-binding affinity. Proc Natl Acad Sci U S A. 2000;97(20):10701–5.

34. Persson H, Kirik U, Thornqvist L, Greiff L, Levander F, Ohlin M. In vitro evolution of antibodies inspired by in vivo evolution. Front Immunol. 2018;9:1391.

35. MacDonald KG, Hoeppli RE, Huang Q, Gillies J, Luciani DS, Orban PC, Broady R, Levings MK. Alloantigen-specific regulatory T cells generated with a chimeric antigen receptor. J Clin Invest. 2016;126(4):1413–24.

36. Noyan F, Zimmermann K, Hardtke-Wolenski M, Knoefel A, Schulde E, Geffers R, Hust M, Huehn J, Galla M, Morgan M, et al. Prevention of allograft rejection by use of regulatory T cells with an MHC-specific chimeric antigen receptor. Am J Transplant. 2017;17(4):917–30.

2

A simple method for in vitro preparation of natural killer cells from cord blood

Yong Xu Mu[1†], Yu Xia Zhao[2†], Bing Yao Li[3], Hong Jing Bao[2], Hui Jiang[2], Xiao Lei Qi[2], Li Yun Bai[2], Yun Hong Wang[4,5], Zhi Jie Ma[6*] and Xiao Yun Wu[4,5*]

Abstract

Background: Cord Blood (CB) has been considered a promising source of natural killer (NK) cells for cellular immunotherapy. However, it is difficult to expand the large numbers of highly pure NK cells from CB without cell sorting and feeder cells/multiple cytokines. In this study, we try to develop a simple, safe and economical method for ex vivo expansion and purification of NK cells from CB without cell sorting and feeder cells/multiple cytokines.

Results: The large numbers (mean: 1.59×10^{10}) of highly pure (≥90%) NK cells from CB could be obtained through interleukin-2, group A streptococcus and zoledronate stimulation of mononuclear cells using the 21-day culture approach. When compared to resting NK cells, expanded NK cells were a higher expression of activating receptors CD16, NKG2D, NKp30, NKp44, NKp46 and activating markers CD62L and CD69, while the inhibitory receptors, CD158a and CD158b remained largely unchanged. In addition, these cells showed a higher concentration of IFN-γ, TNF-α and GM-CSF secretion and cytotoxicity to K562 cells and acute myeloid leukemia targets than resting NK cells.

Conclusion: We develop a simple, safe and economical method to obtain high yield, purity, and functionality NK cells from CB without cell sorting and feeder cells/multiple cytokines.

Keywords: Cord blood, Natural killer cells, Expansion, Cytotoxicity, Immunotherapy

Background

Allogeneic natural killer (NK) cell infusion is promising for cancer immunotherapy because of the "missing self" hypothesis [1]. Cord blood (CB), serves as an immediate "off-the-shelf" source of NK cells, has been considered an attractive source of allogeneic NK cells for therapeutic infusion [2, 3]. However, a major challenge of cell therapy with NK cells is to attain sufficient amount of highly pure cells (> 70% pure, > 1×10^9) because of the low frequency and number (<20% pure, <1×10^8) of NK cells in the CB [3, 4]. To provide allogeneic NK cells with high yield, purity and functionality, some methods have been developed to purify and expand NK cells from CB ex vivo [5–10].

To date, most methods for in vitro preparation of NK cells from CB require to selecte NK cells with immune-selection techniques because of low frequency [11]. In order to avoid the limitations in low number and imma-ture state of NK cells in CB, ex vivo expansion and acti-vation is necessary [12]. NK cells are generally isolated from CB through immunomagnetic beads selection pro-tocols to enrich CD56-positive cells and/or deplete CD3-positive cells, and then cultured for functional ex-pansion and activation using feeder cells, such as Epstein-Barr virus-transformed lymphoblastoid cell lines, mesenchymal stromal cells, gene-modified K562 cells ex-pressing 4-1BB ligand and IL-15, and other irradiated tumor cell lines [5, 13]. In addition, NK cells are origin-ally generated from CD34$^+$ hematopoietic stem cells (HSCs), some studies have described an alternative method to generate NK cells with high yield, purity and functionality from CB-derived CD34$^+$ HSCs under feeder cells-based conditions [10, 14–16]. Recently, a feeder cells-free method has been successfully performed for the generation of NK cells from CB-derived CD34$^+$ HSCs [7, 17]. However, it needs delicate culture

* Correspondence: 13811647091@163.com; stemcells@foxmail.com
†Yong Xu Mu and Yu Xia Zhao contributed equally to this work.
⁶Department of Pharmacy, Beijing Friendship Hospital, Capital Medical University, Beijing, China
⁴Department of Technology, Stem Cell Medicine Engineering & Technology Research Center of Inner Mongolia, Huhhot, Inner Mongolia, China
Full list of author information is available at the end of the article

regimens and multiple cytokine cocktails, which may lead to high cost-effectiveness. Generally, these methods require a complicated technology of cell sorting in an initial step, and it may increase the risk of cell trauma and contamination. Furthermore, the use of feeder cells or multiple cytokines during longer-term cultures would lead to NK cell apoptosis in vivo when optimum culturing conditions are eliminated after adoptive transfer [18]. In addition, these methods are also more costly because of complex operations and supplements.

Although several methods have been proposed to generate clinically relevant NK cell products (mean: 2×10^9 cells) with high purity (> 90%) from CB [13, 19], it is still difficult to obtain the sufficient numbers of highly pure NK cells from CB without cell sorting and feeder cells/ multiple cytokines [13]. Previously, we had found that zoledronate could increase enrichment, expansion and activation of NK cells from CB-derived mononuclear cells (MNCs) [20]. Some studies have reported that interleukin (IL)-2 expansion could recruit and activate key regulators involved in lytic immunological synapse formation of CB-derived NK cells, enabling effective cytotoxicity against killing of acute myeloid leukemia (AML) cells in vitro and in vivo [21, 22]. Group A streptococcus preparation, which is widely used as an immunopotentiator with considerable success in patients with malignant diseases, strongly augmented human NK cell activity in vivo as well as in vitro [23]. Therefore, we try to use develop a simple method with the capability of generating NK cells with high yield, purity and functionality from CB through using zoledronate, group A streptococcus and IL-2 stimulation of MNCs without cell sorting and feeder cells/multiple cytokines.

Results

Preparation of NK cells from CB

After the isolation process by Ficoll, an average of 5.02% CD56$^+$CD3$^-$ NK cells (range, 1.92 to 9.66%) was obtained in MNCs, whereas CD56$^-$CD3$^+$ T cells constituted 84.53% (range, 72.98 to 96.34%). Expansion of CD56$^+$CD3$^-$ NK cells was much higher compared with other types of cells, so NK cells dominated at the end of the culture, reaching on average 92.37% of the total cell populations (range: 88.91 to 96.37%; Additional file 1) by day 21. The frequency of CD56$^+$CD3$^+$ NKT cells remained largely unchanged before and after culture (day 0: 1.05%, day 21: 1.35%), whereas the frequency of CD56$^-$CD3$^+$ T cells declined, decreasing to an average of 4.56% (range: 1.30 to 9.63%; Fig. 1a and b). By day 21, the total cell count had expanded on average 101-fold (range: 65–137-fold) and, among these, CD56$^+$CD3$^-$ NK cell population had expanded on average 1561-fold (range: 695–2387-fold), reaching on average 1.59×10^{10} (range: $0.84–2.23 \times 10^9$; Fig. 1c). At this time point the

expansion potential reached a plateau, thus the cells reached a quiescence phase when measured later on day 28. These data demonstrate that NK cells with high yield and purity could be expanded efficiently from CB-derived MNCs ex vivo using the method described here.

.

Surface expression of NK receptors and markers

The expression of receptors and markers in expanded NK cells was assessed and compared to unexpanded NK cells. Among activating receptors, CD16 (day 0: 31.40%, day 21: 66.91%; $P < 0.001$), NKp30 (day 0: 42.99%, day 21: 82.61%; $P < 0.001$), NKp44 (day 0: 11.72%, day 21: 82.19%; $P < 0.001$), NKp46 (day 0: 6.68%, day 21: 44.20%; $P < 0.001$) and NKG2D (day 0: 5.14%, day 21: 72.20%; $P < 0.001$) significantly increased during the expansion while expression of the inhibitory receptors, CD158a (day 0: 6.33%, day 21: 6.68%; $P > 0.05$) and CD158b (day 0: 9.40%, day 21: 10.91%; $P > 0.05$) remained largely unchanged during the culture period. Also, activation markers of the NK cells including CD62L (day 0: 21.66%, day 21: 64.91%; $P < 0.001$) and CD69 (day 0: 7.47%, day 21: 85.43%; $P < 0.001$) were increased on the surface of expanded NK cells (Fig. 2). These data demonstrate that NK cells are markedly activated after expansion.

Cytokines production of expanded NK cells

The secretion cytokines including granulocyte-macrophage colony-stimulating factor (GM-CSF), interferon-gamma (IFN-γ), and tumor necrosis factor-alpha (TNF-α) were also assessed before and after culture. The very low level of IFN-γ (mean: 5.21 pg/mL, range: 1.37–8.45 pg/mL), TNF-α (mean: 2.99 pg/mL, range: 1.16–5.04 pg/mL) and GM-CSF (mean: 6.95 pg/mL, range: 4.38–9.81 pg/mL) secretion was observed in unexpanded NK cells. Expanded NK cells exhibited significantly increased IFN-γ (mean: 124.67 pg/mL, range: 57.95–183.12 pg/mL), TNF-α (mean: 853.94 pg/mL, range: 479.67–1201.26 pg/mL) and GM-CSF (mean: 81.67 pg/mL, range: 55.43–108.42 pg/mL) secretion compared to unexpanded NK cells (all $P < 0.001$, Fig. 3). These results demonstrate that the patterns of cytokine secretion are markedly increased after culture.

Cytotoxicity of expanded NK cells

The cytotoxicity levels of expanded vs. unexpanded NK cells against leukemia cell line (K562) and primary patient AML blasts were evaluated. The unexpanded NK cells were poor cytotoxicity against K562 cells (range: 1.91–5.86%) and primary patient AML blasts (range: 1.24–2.89%) at a variety of E:T ratios, but the expanded NK cells showed high levels of cytotoxicity against K562 (range: 52.63–82.18%) and AML targets (range: 15.28–24.07%). NK cells expanded in culture consistently exhibited a high level of cytotoxicity against K562 and AML targets

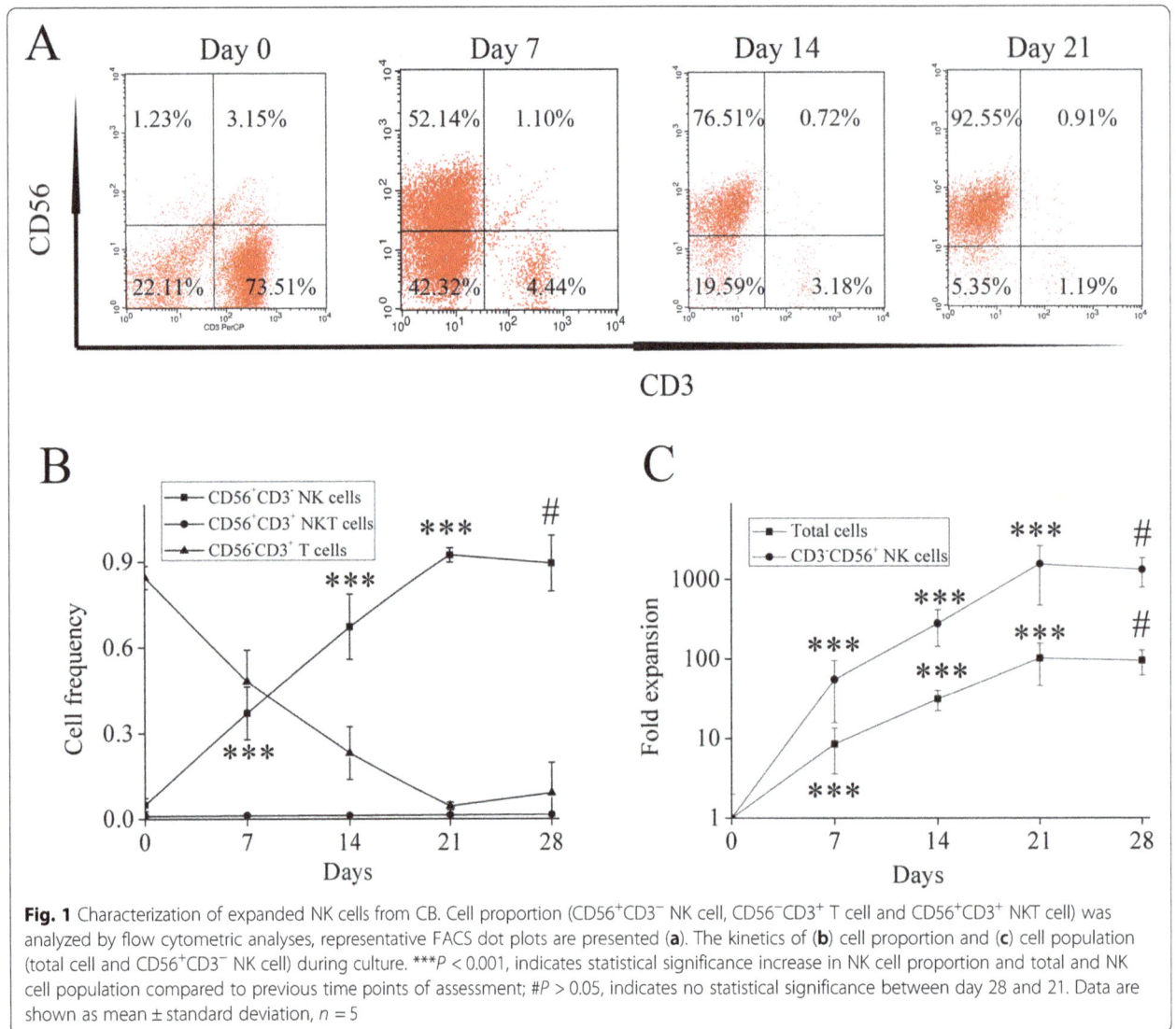

Fig. 1 Characterization of expanded NK cells from CB. Cell proportion (CD56+CD3− NK cell, CD56−CD3+ T cell and CD56+CD3+ NKT cell) was analyzed by flow cytometric analyses, representative FACS dot plots are presented (**a**). The kinetics of (**b**) cell proportion and (**c**) cell population (total cell and CD56+CD3− NK cell) during culture. ***$P < 0.001$, indicates statistical significance increase in NK cell proportion and total and NK cell population compared to previous time points of assessment; #$P > 0.05$, indicates no statistical significance between day 28 and 21. Data are shown as mean ± standard deviation, $n = 5$

compared to unexpanded NK cells at a wide range of E:T ratios (all $P < 0.001$, Fig. 4). These results show that cytotoxicity of NK cells are markedly elevated after expansion.

Discussion

In this study, we develop a simple and economical method for in vitro preparation of NK cells through IL-2, group A streptococcus and zoledronate stimulation of CB-derived MNCs. it is a practically advantageous method that does not need cell sorting and feeder cells/multiple cytokines, which could lead to reduction in cost. Furthermore, IL-2, streptococcus A group and zoledronate have also been approved for human use, which could increase the safety. In addition, this simple method is more easily standardized and exportability. To our knowledge, this is the first report of a simple method for in vitro preparation of NK cells from CB without cell sorting and feeder cells/multiple cytokines.

Using this method, a mean number of 1.59×10^{10} NK cells with an average purity of 92.37% are generated after 21 days of culture. Previously, we found that zoledronate had a favorable effect on expansion of NK cells from CB-derived MNCs [20]. Furthermore, the addition of group A streptococcus induced a higher increase in the expansion fold (1561 vs. 1286) of NK cells, showing that both have the synergism. A study has reported a multiple cytokines-based delicate culture method for up to 6 weeks to generate therapeutic NK cell products (mean: 2×10^9 cells) with high purity (> 90%) from CB-derived CD34+ HSCs [9]. Compared with this method, our method yields much higher number (8-times) of NK cells with less time. We hypothesize that this is related to very low proportion of CD34+ HSCs in CB and cumbersome procedures including HSCs expansion and NK cell differentiation and expansion. Other group has reported that a feeder cell-based NK cultivation method is

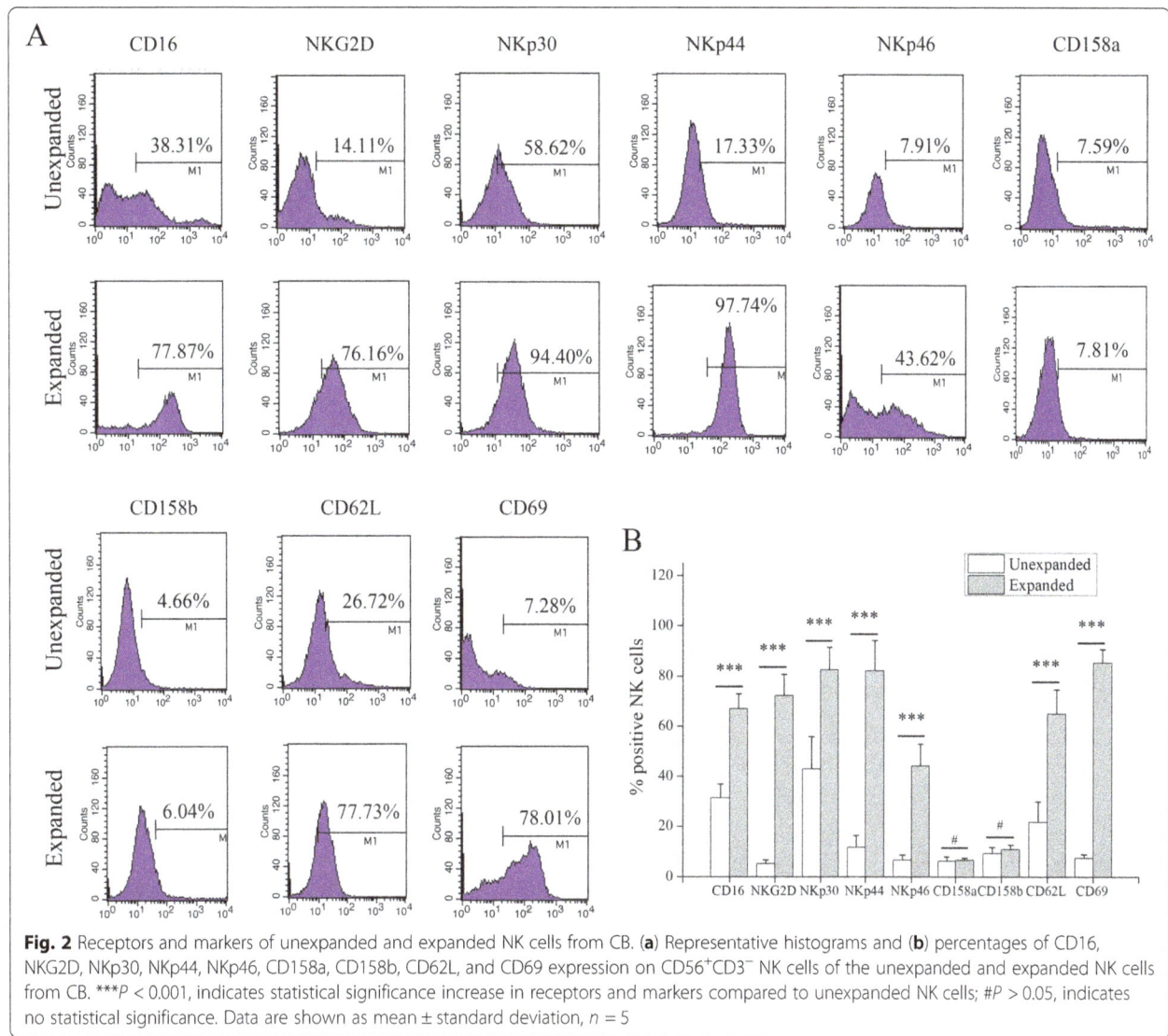

Fig. 2 Receptors and markers of unexpanded and expanded NK cells from CB. (**a**) Representative histograms and (**b**) percentages of CD16, NKG2D, NKp30, NKp44, NKp46, CD158a, CD158b, CD62L, and CD69 expression on CD56+CD3− NK cells of the unexpanded and expanded NK cells from CB. ***$P < 0.001$, indicates statistical significance increase in receptors and markers compared to unexpanded NK cells; #$P > 0.05$, indicates no statistical significance. Data are shown as mean ± standard deviation, $n = 5$

established with the capability of generating a clinically relevant dose (mean: 1.2×10^9 cells) with a purity of > 80% of CB-derived NK cells [6]. Compared with this method, our method also yields much higher number (15-times) of NK cells with the same time. A possible explanation might be that accessory cells such as monocytes in CB-derived MNCs can support NK cell expansion [24]. Therefore, A large amount of NK cells generated in vitro by our method can be, allowed for multiple infusions of NK cells.

Purity, as one of crucial release criteria for NK cell-based therapies, has not yet been standardized, but > 90% CD56+CD3− for allogenic NK cells have been suggested in recently published study [9, 25]. Previously, we found that zoledronate had a favorable effect on enrichment of NK cells from CB-derived MNCs [20]. Furthermore, the addition of group A streptococcus induced a greater increase in the frequency (92.37% vs. 80.46%) of NK cells,

showing that both have the synergism. The purity of NK cells produced by our method is comparable to cytokine-based culture method (> 90%) [9], but higher compared to feeder cell-based method (> 80%) [6].

Another crucial obstacle is the immaturity state of NK cells in CB [3]. It has been known that NK cell cytotoxicity is regulated by the complex balance between activating and inhibitory receptors. NK cells expended by our method exhibit high expression level of activating receptors CD16, NKG2D, NKp30, NKp44 and NKp46, but unchanged expression level of inhibitory receptors CD158a and CD158b compared with unexpanded NK cells. This leads to high-level expression of activating markers CD62L and CD69, showing that NK cells expanded in this study have activated a cellular mechanism controlling effector function. A recent study also reported no changes in expression of CD158a and CD158b by CB-derived NK cells after incubation with

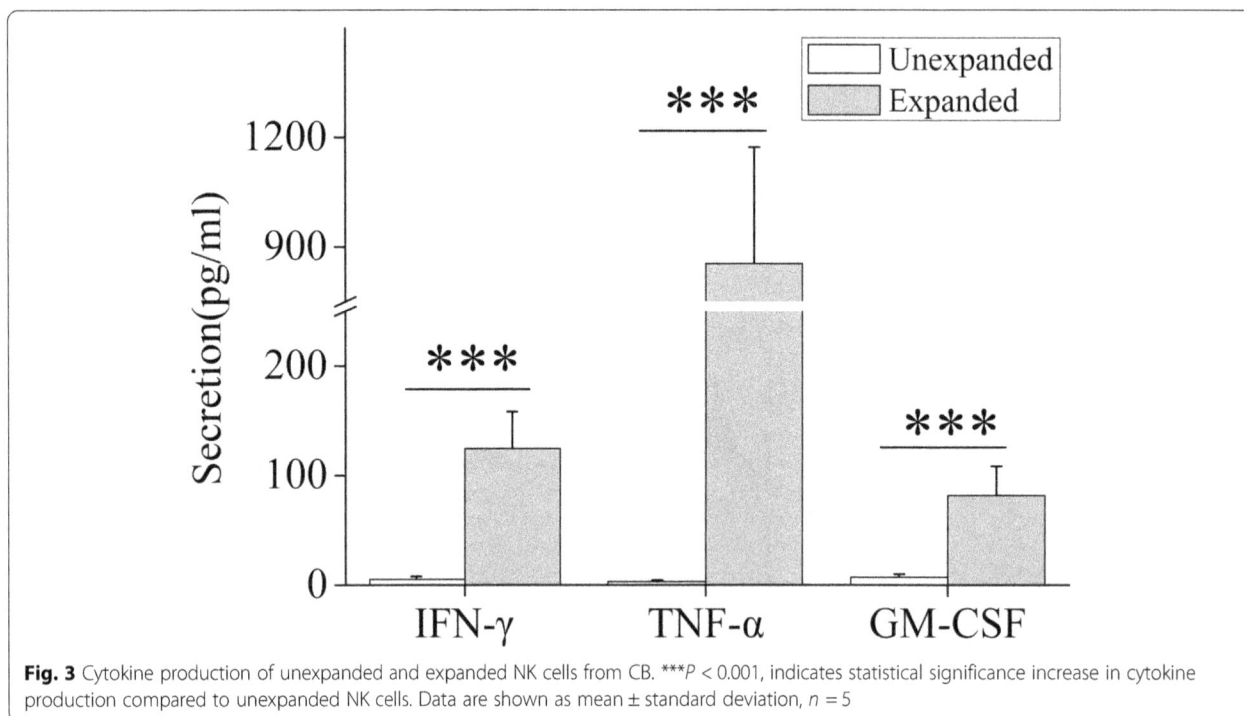

Fig. 3 Cytokine production of unexpanded and expanded NK cells from CB. ***$P < 0.001$, indicates statistical significance increase in cytokine production compared to unexpanded NK cells. Data are shown as mean ± standard deviation, $n = 5$

IL-2, IL-12 or IL-18 [26]. This confirms that killer cell Ig-like receptor acquisition does not require group A streptococcus and zoledronate activation.

Previous studies have shown that IL-2-activated NK cells from CB exhibited significantly increased frequency of IFN-γ and TNF-α secretion compared with unexpended NK cells [21, 26]. In previous study, we showed that zoledronate increased IFN-γ, TNF-α and GM-CSF secretion by IL-2-activated NK cells from CB [20]. Early study showed that group A streptococcus stimulated human lymphocytes to produce IFN (both a- and γ-types) and IL-2 and that both factors were primarily responsible for the NK augmentation by group A streptococcus [23]. It has been previously shown that cytokines including IFN-γ, TNF-α and GM-CSF play key roles in NK cell functions [27]. To evaluate the impact of IL-2, zoledronate and group A streptococcus on the secretion of cytokines by CB NK cells, cytokines secretion was measured in the supernatants of unexpended and expanded NK cell cultures. The result shows that NK cells expanded by our method can produce robust inflammatory cytokine IFN-γ, TNF-α and GM-CSF. We further

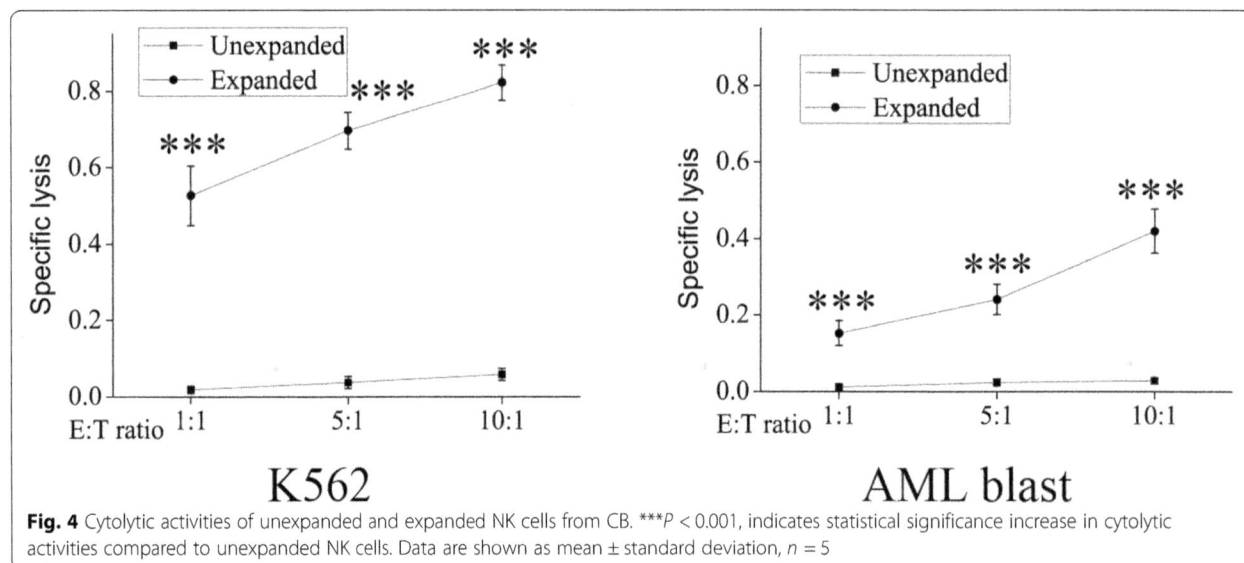

Fig. 4 Cytolytic activities of unexpanded and expanded NK cells from CB. ***$P < 0.001$, indicates statistical significance increase in cytolytic activities compared to unexpanded NK cells. Data are shown as mean ± standard deviation, $n = 5$

investigate that NK cells expanded by our method can induce lytic function against K562 and primary AML targets compared with unexpended NK cells, showing that ex vivo generated NK cells from CB could contribute greatly to the elimination of tumor during adoptive NK cell immunotherapy. These results are comparable to data from zoledronate only, showing no synergy between them. This might be related to MNCs rather than purified NK cells. However, in vivo functional studies are needed to explore in the next study.

Conclusion

We have developed a simple, safe and economical method to obtain high yield, purity, and functionality NK cells from CB through IL-2, streptococcus A group and zoledronate stimulation of MNCs without cell sorting and feeder cells/ multiple cytokines. Adoptive transfer of NK cells generated from CB by this method may therefore provide a promising new paradigm for the treatment of patients with AML and other hematological malignancies. This simple method satisfies the standards for in vitro preparation of NK cells as a potential therapeutic product, is more accessible for clinical practice, and may hold valuable potential for adoptive cellular immunotherapy for hematological malignancies like AML.

Methods

NK cells preparation

All frozen CB samples were drawn from the Shandong Cord Blood Bank. MNCs were isolated from CB using Ficoll-Hypaque density gradient centrifugation. For activation, MNCs were cultured for 3 days in T175 flasks at 2×10^6 cells/mL in AIM-V serum free media (Life Technologies) supplemented with 2000 IU IL-2/mL (Four Rings Biopharma, China), 0.01 KE/mL group A streptococcus (Lu Ya Pharma, China) and 5 μM zoledronate (Novartis Pharma). For expansion, the fresh medium containing 2000 IU IL-2/mL was added every 2 to 3 days for 21 days. An outline of the preparation protocol is summarized in Fig. 5.

Expansion kinetics assessment

Total cell counts were determined by trypan blue staining on days 0, 7, 14, 21 and 28. Absolute cell numbers were calculated by multiplying the total cell count by the percentage of each cell type determined by flow cytometry. Cell expansion efficiency was expressed as "fold expansion" which was determined by dividing the number of total cells or absolute NK cells on days 7, 14, 21 and 28 by the number on day 0.

Flow Cytometr

The cells were labeled with fluorochrome-conjugated mouse mAbs against human cluster of differentiation CD3, CD56, CD16, CD158a, CD158b and NKG2D (BD Biosciences); NKp30, NKp44 and NKp46 (Beckman Coulter); CD69, CD62L (Miltenyi Biotec). Isotype-matched antibodies were used as controls. Flow cytometry analysis was performed using a MACSQuant Analyzer (Miltenyi Biotec). Data were analyzed with MACSQuantify Software.

Cytokines release assay

The cell culture supernatants were collected, the level of cytokines including IFN-γ, TNF-α and GM-CSF in supernatants was analyzed using enzyme-linked immunosorbent assay kits (eBioscience, Inc) according to the manufacturer's instructions.

Cytotoxicity assay

The cytotoxicity of NK cells was analyzed using the standard 4-h ^{51}Cr release assay. The NK-sensitive leukemia cell line (K562) and primary AML blasts (target) were labeled with ^{51}Cr sodium chromate and incubated with NK cells (effector) at different target-to-effector (E:T) ratios (1:1, 5:1, and 10:1) in U-bottomed 96-well plates. ^{51}Cr-release was determined in the supernatant after co-culture for 4 h. The maximum release was measured by treating target cells with 2% Triton X-100 and spontaneous release was determined with medium alone. The specific lysis was calculated according to the formula: (experimental release - spontaneous release) / (maximum release - spontaneous release).

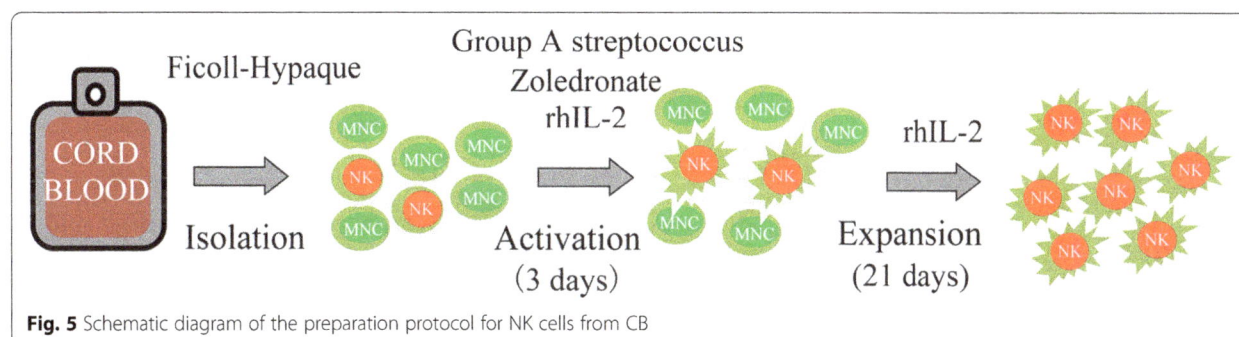

Fig. 5 Schematic diagram of the preparation protocol for NK cells from CB

Statistical analysis

Data are presented as the mean ± standard deviation. Statistical difference was determined using a student's t test when comparing two groups, or a one-way ANOVA analysis when more than two groups (Graphpad Prism 5.0). $P < 0.05$ was considered significant.

Abbreviations

AML: Acute myeloid leukemia; CB: Cord blood; E:T: Effector-to-target; GM-CSF: Granulocyte-macrophage colony-stimulating factor; HSCs: Hematopoietic stem cells; IFN-γ: Interferon-gamma; IL: Interleukin; MNCs: Mononuclear cells; NK: Natural killer; TNF-α: Tumor necrosis factor-alpha

Acknowledgements

We acknowledged the kind support from Department of Technology, Beijing.
JingMeng Stem Cell Technology. Co. Ltd., Beijing, China.

Authors' contributions

All authors participated in the design, interpretation of the results, and review of the manuscript; YXM and XYW were involved in the experimentation, and wrote the manuscript; YXZ, HJB and BYL were involved in the analysis of the data. HJ, XLQ and LYB generated the figures and contributed to editing the manuscript, the study was conceived by YHW and ZGL who also edited the manuscript. All authors read and approved the final manuscript.

Funding

This work was supported by National Natural Science Foundation of China (No. 81860157) and Natural Science Foundation of Inner Mongolia (No. 2017MS0314).

Ethics approval and consent to participate

All frozen CB samples were obtained with prior consent and ethical committee approval from the Shandong Cord Blood Bank. The study had full ethical approval from the institutional ethics committee of Beijing Friendship Hospital.

Consent for publication

Not applicable.

Competing interests

A patent application for the composition for expanding NK cells and the use of them has been filed with Xiao Yun Wu as a first inventor. Other authors declare that they have no conflict of interest.

Author details

[1]Interventional Department, the First Affiliated Hospital of Baotou Medical College, Inner Mongolia University of Science and Technology, Baotou, Inner Mongolia, China. [2]Department of Blood, the People's Hospital of Xing'an League, Xing'an League, Inner Mongolia, China. [3]Department of Medicine, Chifeng Cancer Hospital, Chifeng, Inner Mongolia, China. [4]Department of Technology, Stem Cell Medicine Engineering & Technology Research Center of Inner Mongolia, Huhhot, Inner Mongolia, China. [5]Department of Research and Development, Beijing Jingmeng Stem Cell Technology CO., LTD, Beijing, China. [6]Department of Pharmacy, Beijing Friendship Hospital, Capital Medical University, Beijing, China.

References

1. Lim O, Jung MY, Hwang YK, Shin EC. Present and future of allogeneic natural killer cell therapy. Front Immunol. 2015;6:286.
2. Shaim H, Yvon E. Cord blood: a promising source of allogeneic natural killer cells for immunotherapy. Cytotherapy. 2015;17(1):1–2.
3. Mehta RS, Shpall EJ, Rezvani K. Cord blood as a source of natural killer cells. Front Med. 2015;2:93.
4. Klingemann H. Challenges of cancer therapy with natural killer cells. Cytotherapy. 2015;17(3):245–9.
5. Hosseini E, Ghasemzadeh M, Kamalizad M, Schwarer AP. Ex vivo expansion of CD3depleted cord blood-MNCs in the presence of bone marrow stromal cells; an appropriate strategy to provide functional NK cells applicable for cellular therapy. Stem Cell Res. 2017;19:148–55.
6. Kang L, Voskinarian-Berse V, Law E, Reddin T, Bhatia M, Hariri A, Ning Y, Dong D, Maguire T, Yarmush M, et al. Characterization and ex vivo expansion of human placenta-derived natural killer cells for Cancer immunotherapy. Front Immunol. 2013;4:101.
7. Cany J, van der Waart AB, Tordoir M, Franssen GM, Hangalapura BN, de Vries J, Boerman O, Schaap N, van der Voort R, Spanholtz J, et al. Natural killer cells generated from cord blood hematopoietic progenitor cells efficiently target bone marrow-residing human leukemia cells in NOD/SCID/IL2Rg(null) mice. PLoS One. 2013;8(6):e64384.
8. Dezell SA, Ahn YO, Spanholtz J, Wang H, Weeres M, Jackson S, Cooley S, Dolstra H, Miller JS, Verneris MR. Natural killer cell differentiation from hematopoietic stem cells: a comparative analysis of heparin- and stromal cell-supported methods. Biol Blood Marrow Transplant. 2012;18(4):536–45.
9. Spanholtz J, Preijers F, Tordoir M, Trilsbeek C, Paardekooper J, de Witte T, Schaap N, Dolstra H. Clinical-grade generation of active NK cells from cord blood hematopoietic progenitor cells for immunotherapy using a closed-system culture process. PLoS One. 2011;6(6):e20740.
10. Frias AM, Porada CD, Crapnell KB, Cabral JM, Zanjani ED, Almeida-Porada G. Generation of functional natural killer and dendritic cells in a human stromal-based serum-free culture system designed for cord blood expansion. Exp Hematol. 2008;36(1):61–8.
11. Becker PS, Suck G, Nowakowska P, Ullrich E, Seifried E, Bader P, Tonn T, Seidl C. Selection and expansion of natural killer cells for NK cell-based immunotherapy. Cancer Immunol Immunother. 2016;65(4):477–84.
12. Lapteva N, Szmania SM, van Rhee F, Rooney CM. Clinical grade purification and expansion of natural killer cells. Crit Rev Oncog. 2014;19(1–2):121–32.
13. Chabannon C, Mfarrej B, Guia S, Ugolini S, Devillier R, Blaise D, Vivier E, Calmels B. Manufacturing natural killer cells as medicinal products. Front Immunol. 2016;7:504.
14. Pinho MJ, Punzel M, Sousa M, Barros A. Ex vivo differentiation of natural killer cells from human umbilical cord blood CD34+ progenitor cells. Cell Commun Adhes. 2011;18(3):45–55.
15. Boissel L, Tuncer HH, Betancur M, Wolfberg A, Klingemann H. Umbilical cord mesenchymal stem cells increase expansion of cord blood natural killer cells. Biol Blood Marrow Transplant. 2008;14(9):1031–8.
16. Kao IT, Yao CL, Kong ZL, Wu ML, Chuang TL, Hwang SM. Generation of natural killer cells from serum-free, expanded human umbilical cord blood CD34+ cells. Stem Cells Dev. 2007;16(6):1043–51.
17. Domogala A, Blundell M, Thrasher A, Lowdell MW, Madrigal JA, Saudemont A. Natural killer cells differentiated in vitro from cord blood CD34+ cells are more advantageous for use as an immunotherapy than peripheral blood and cord blood natural killer cells. Cytotherapy. 2017;19(6):710–20.
18. Childs RW, Berg M. Bringing natural killer cells to the clinic: ex vivo manipulation. Hematology Am Soc Hematol Educ Program. 2013;2013:234–46.
19. Koehl U, Kalberer C, Spanholtz J, Lee DA, Miller JS, Cooley S, Lowdell M, Uharek L, Klingemann H, Curti A, et al. Advances in clinical NK cell studies: donor selection, manufacturing and quality control. Oncoimmunology. 2016;5(4):e1115178.
20. Ma Z, Wang Y, Kang H, Wu X. Zoledronate increases enrichment, activation and expansion of natural killer cells from umbilical cord blood. Hum Cell. 2018;31:310–2.
21. Xing D, Ramsay AG, Gribben JG, Decker WK, Burks JK, Munsell M, Li S, Robinson SN, Yang H, Steiner D, et al. Cord blood natural killer cells exhibit impaired lytic immunological synapse formation that is reversed with IL-2 exvivo expansion. J Immunother. 2010;33(7):684–96.
22. Khaziri N, Mohammadi M, Aliyari Z, Soleimani Rad J, Tayefi Nasrabadi H, Nozad Charoudeh H. Cord blood mononuclear cells have a potential to produce NK cells using IL2Rg cytokines. Adv Pharm Bull. 2016;6(1):5–8.

23. Wakasugi H, Kasahara T, Minato N, Hamuro J, Miyata M, Morioka Y. In vitro potentiation of human natural killer cell activity by a streptococcal preparation, OK-432: interferon and interleukin-2 participation in the stimulation with OK-432. J Natl Cancer Inst. 1982;69(4):807–12.

24. Koehl U, Brehm C, Huenecke S, Zimmermann SY, Kloess S, Bremm M, Ullrich E, Soerensen J, Quaiser A, Erben S, et al. Clinical grade purification and expansion of NK cell products for an optimized manufacturing protocol. Front Oncol. 2013;3:118.

25. Berg M, Childs R. Ex-vivo expansion of NK cells: what is the priority--high yield or high purity? Cytotherapy. 2010;12(8):969–70.

26. Alnabhan R, Madrigal A, Saudemont A. Differential activation of cord blood and peripheral blood natural killer cells by cytokines. Cytotherapy. 2015; 17(1):73–85.

27. Yoon SR, Kim TD, Choi I. Understanding of molecular mechanisms in natural killer cell therapy. Exp Mol Med. 2015;47:e141.

3

Establishment of an erythroid progenitor cell line capable of enucleation achieved with an inducible c-Myc vector

Steven Mayers[1,2], Pablo Diego Moço[2], Talha Maqbool[2], Pamuditha N. Silva[2], Dawn M. Kilkenny[2] and Julie Audet[1,2*]

Abstract

Background: A robust scalable method for producing enucleated red blood cells (RBCs) is not only a process to produce packed RBC units for transfusion but a potential platform to produce modified RBCs with applications in advanced cellular therapy. Current strategies for producing RBCs have shortcomings in the limited self-renewal capacity of progenitor cells, or difficulties in effectively enucleating erythroid cell lines. We explored a new method to produce RBCs by inducibly expressing c-Myc in primary erythroid progenitor cells and evaluated the proliferative and maturation potential of these modified cells.

Results: Primary erythroid progenitor cells were genetically modified with an inducible gene transfer vector expressing a single transcription factor, c-Myc, and all the gene elements required to achieve dox-inducible expression. Genetically modified cells had enhanced proliferative potential compared to control cells, resulting in exponential growth for at least 6 weeks. Inducibly proliferating erythroid (IPE) cells were isolated with surface receptors similar to colony forming unit-erythroid (CFU-Es), and after removal of ectopic c-Myc expression cells hemoglobinized, decreased in cell size to that of native RBCs, and enucleated achieving cultures with 17% enucleated cells. Experiments with IPE cells at various levels of ectopic c-Myc expression provided insight into differentiation dynamics of the modified cells, and an optimized two-stage differentiation strategy was shown to promote greater expansion and maturation.

Conclusions: Genetic engineering of adult erythroid progenitor cells with an inducible c-Myc vector established an erythroid progenitor cell line that could produce RBCs, demonstrating the potential of this approach to produce large quantities of RBCs and modified RBC products.

Keywords: Erythroid progenitor cell, Induced proliferation, Red blood cell, c-Myc, Enucleation, Cell manufacturing

Background

Red blood cells (RBCs) are an ideal platform for novel cellular therapies as they are physically stable, universally biocompatible (type O, Rhesus factor negative), contain no nucleus, and can potentially be packed and coated with biologically active molecules. It has already been shown that RBCs can be engineered to carry therapeutic molecules in multiple ways [1–3], and pre-clinical studies have shown proof of concept for RBC based immunotherapy [3–5]. Production of human RBCs from hematopoietic stem cells has been achieved with 60,000 fold expansion from cord blood stem cells including safe transfusion into human patients, signifying the potential for the clinical translation of in vitro cultured RBCs [6].

RBCs are produced by erythropoiesis, which is an erythropoietin (EPO)-dependant cell development process in which a nucleus-containing erythroid precursor is differentiated to become a hemoglobin-containing enucleated RBC [7]. The earliest committed erythroid progenitor is the burst forming unit-erythroid (BFU-E) [8]. BFU-Es differentiate to form CFU-Es (c-kit$^+$CD71^{high}Ter119$^-$) [9–11], which usually divide three to five times over two to 3 days as they hemoglobinize, and undergo a decrease in cell size [8]. As these cells develop, they exit the cell cycle, repress transcription,

* Correspondence: julie.audet@utoronto.ca
[1]Department of Chemical Engineering and Applied Chemistry, University of Toronto, Toronto, Canada
[2]Institute of Biomaterials and Biomedical Engineering (IBBME), University of Toronto, Toronto, Canada

condense their chromatin, and finally extrude their nucleus to form a reticulocyte [12]. After about 2 days, the reticulocyte shows loss of reticulin as it terminally differentiates into a RBC [13].

Induced proliferation of erythroid progenitor cells is a promising strategy for producing RBCs that has been explored with some success. It was shown that the introduction of HPV16-E6/E7 genes into erythroid progenitor cells expressed by an inducible promoter could achieve enhanced proliferation while retaining the ability for cells to terminally differentiate [14, 15]. This approach was applied to adult bone marrow-derived cells and although a robust RBC production platform was reported, significant cell losses were observed during differentiation (approx. 80%), and of those viable cells, enucleation was observed in about 25% of the remaining cells [14]. Another group has reported the creation of inducibly proliferating erythroid (IPE) progenitor cells from human embryonic stem (ES) cells by over-expression of both c-Myc and bcl-xl [16]. In this approach, with c-Myc removal, cells hemoglobinized, but with only 0.36% enucleation efficiency. Bcl-xl expression suppresses p53, which is an regulator of apoptosis and genomic integrity [17], potentially limiting the genomic stability and developmental potential of these cells.

Induced proliferation of many cell types has been shown by ectopic expression of the transcription factor Myc alone [18–20]. Expression level is important, where supraphysiological levels have been shown to be required for driving induced erythroid progenitor cell proliferation [18]. The transient up-regulation of v-Myc has been shown to induce reversible proliferation of primary neural progenitors with no observed tumorigenic impact [19]. In a β-cell in vivo model, the selective expression of c-Myc induced a proliferative state accompanied by broad reversible gene expression changes, indicating the reversible nature of c-Myc induced proliferation [20].

The transcription factor c-Myc activates many genes involved in cell proliferation, promoting growth and inhibiting cell cycle arrest [21–24]. C-Myc is a G_0/G_1 transition regulator which promotes cell cycle entry [25]. Over-expression of c-Myc can induce apoptosis through a p53 tumour suppression pathway, which eliminates cells that are inappropriately bypassing the G_1-S checkpoint [26]. The effect of ectopic c-Myc expression on apoptosis depends on its level of expression [24], as well as the state of the cell and its physiological status [27]. In vitro, the effect of c-Myc on bcl-2 family proteins and cytochrome C release may be blocked by the survival factor insulin like growth factor 1 (IGF-1) [28]. Also, apoptosis induced by c-Myc over-expression can also be avoided by complementary signal transduction pathways that result from the presence of mitogens [29].

C-Myc-induced sensitization to apoptosis presents a challenge when inducing proliferation, where the ideal expression would be just enough to induce proliferation accompanied by sufficient mitogenic survival signals to prevent triggering apoptosis.

C-Myc has been shown to positively regulate histone acetyl transferases (HATs) which expose DNA through chromatin remodelling [30]. In erythroid cell development, histone deacetylation, which reverses HAT activity, is critical for chromatin condensation and enucleation [18]. In erythroid cells in which c-Myc has been ectopically expressed, HAT up-regulation results in an inhibition of nuclear condensation [18]. These observations outline the importance of complete removal of c-Myc expression to allow for histone deacetylation, chromatin condensation, and enucleation of erythroid progenitors.

In attempts to develop a new method to produce large quantities of RBCs, inducible over-expression of c-Myc in primary erythroid progenitors was investigated. The proliferative capacity of modified cells expressing ectopic c-Myc was evaluated, as well as their ability to terminally differentiate upon ectopic expression removal. Our goal was to establish an erythroid progenitor cell line capable of extensive self renewal and terminal differentiation into enucleated RBCs.

Results

Tightly controlled ectopic expression of functional c-Myc

An all-in-one lentiviral gene transfer vector (Fig. 1 and Additional file 1: Figure S1) was developed to achieve dox-inducible expression of the transcription factor c-Myc in primary cells. The vector contained the c-Myc (mouse) gene with an N-terminus FLAG-tag under the third-generation doxycycline/tetracycline responsive element (TRE3G) transcriptional promoter which has minimal background expression [31]. To achieve full gene expression control with a single vector, the reverse tetracycline transactivator (rtTA) gene was included in the vector under constitutive expression. The gene vector also contains a puromycin resistance gene under constitutive expression to allow for selection of genetically modified cells by culturing them in media containing puromycin.

This all-in-one gene transfer vector was validated in the c-Myc$^{-/-}$ knock-out fibroblast cell line HO15.19 which grows slower without c-Myc [32]. A purified population of HO15.19 cells, carrying the TRE3G-cMyc transfer vector, had a dox-dependant expression of c-Myc. This was shown by western blot (Fig. 2), and a cell proliferation assay (Fig. 3). Together, these assays showed a clear dynamic range of dox-induced c-Myc expression that increased the growth rate of vector containing cells proportionally to dox concentration. Negligible c-Myc background expression with no dox

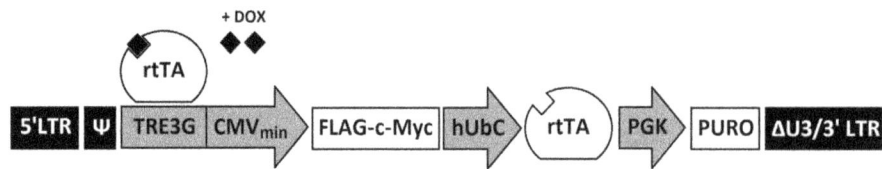

Fig. 1 Simplified DNA plasmid map of lentivirus transfer vector TRE3G-cMyc. The vector is a third-generation lentivirus transfer vector containing the third-generation tetracycline responsive element (TRE3G) controlling the expression of FLAG tagged c-Myc (mouse) transcription factor, constitutive expression of the rtTA, and a puromycin (puro) resistance gene. Depicted is the rtTA transcription factor activated by soluble dox binding to the TRE3G promoter

was confirmed by western blot, and the growth rate of modified cells with no dox matching the knock out cell line (Additional file 1: Figure S2). The western blot confirmed that the size of the c-Myc protein matched the endogenously produced transcription factor (50 kDa). At high dox concentrations a larger protein band is visible which is consistent with a phosphorylated form of c-Myc [33].

Induced proliferation of c-Myc transduced hematopoietic progenitor cells

Three separate independent experiments were done on different preparations of primary mouse bone marrow cells purified to have the surface receptor profile lineage negative $(Lin^-)Ter119^-Mac-1^-Gr-1^-c-Kit^+CD71^{(low/-)}$ (Additional file 1: Figure S3), which is known to be enriched with BFU-Es [9]. The presence of BFU-Es was confirmed by colony forming cell (CFC) assay (Additional file 1: Figure S4). Cells were either genetically modified with the TRE3G-cMyc vector, or treated with a phosphate buffered saline (PBS) control, then cultured in the presence of growth factors, with or without dox, and analysed over a period of 6 weeks (Fig. 4). For the PBS control condition, the total number of cells increased rapidly to 100-fold over a period of two weeks and then stopped proliferating. For both conditions where cells were modified with the TRE3G-cMyc vector, substantial cell death occurred as a result of the

puromycin selection step. Modified cells with no dox present showed only a modest (2-fold) net increase, which was still substantial growth considering the losses from puromycin selection. In contrast, in cultures with dox, and therefore ectopic c-Myc expression, cell number increased extensively after the puromycin selection step; cells proliferated continuously for at least 6 weeks reaching a 22,000-fold net expansion.

To further investigate c-Myc-induced proliferation, fractions of the cell cultures were analysed using CFC assays to assess for clonogenic erythroid progenitor cells. Colonies derived from modified cells expressing c-Myc ectopically were atypical for mouse bone marrow hematopoietic cells, containing what appeared to be a large number of undifferentiated (blast) cells even after two weeks of incubation in methylcellulose (with dox and growth factors). The appearance of these colonies resembled human BFU-Es [34] (Additional file 1: Figure S5), and were described as 'BFU-E like' colonies. As shown in Fig. 5, BFU-E-enriched cells transduced with the TRE3G-cMyc vector cultured in the presence of dox showed an increase in 'BFU-E like' colonies that reached at least a 100-fold net expansion over a period of 3 weeks. This sustained expansion was not observed in modified cultures without dox, or non-transduced cells clearly showing the positive effect of ectopic c-Myc expression on self-renewal of CFC. Greater proliferative potential and CFC expansion was achieved in all three experiments where c-Myc was ectopically expressed compared with all control conditions.

Isolation of IPE cells

After 3.5 weeks of culture, a purified population of IPE cells was isolated by flow cytometry with the surface marker expression profile $Mac-1^-Gr-1^-c-Kit^{(high/+)}CD71^+$ (Additional file 1: Figure S6) which in enriched in CFU-Es [9, 35]. For the other cell isolation experiments, the cultures contained a majority of $Mac-1^+$ and $Gr-1^+$ positive cells after 3.5 weeks (Additional file 1: Figure S7) and erythroid cells were too rare to successfully isolate and culture. The isolated 'CFU-E like' cell line was cultured further in the presence of growth factors (IPE Media) and 2 µg/ml dox for at least an

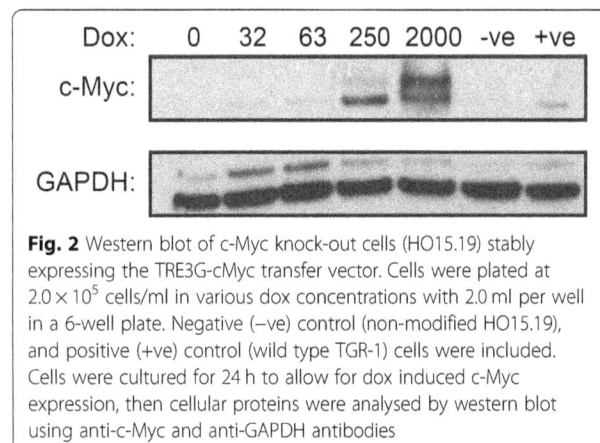

Fig. 2 Western blot of c-Myc knock-out cells (HO15.19) stably expressing the TRE3G-cMyc transfer vector. Cells were plated at 2.0×10^5 cells/ml in various dox concentrations with 2.0 ml per well in a 6-well plate. Negative (−ve) control (non-modified HO15.19), and positive (+ve) control (wild type TGR-1) cells were included. Cells were cultured for 24 h to allow for dox induced c-Myc expression, then cellular proteins were analysed by western blot using anti-c-Myc and anti-GAPDH antibodies

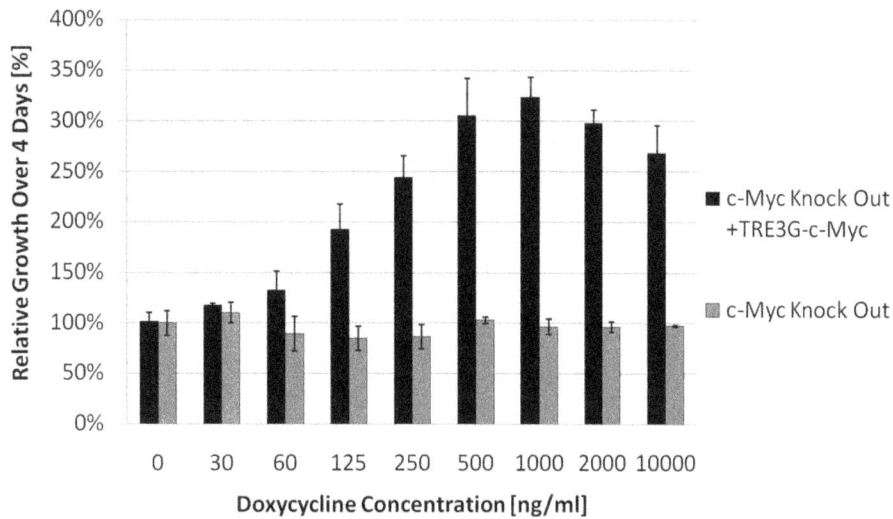

Fig. 3 Functional validation of TRE3G-cMyc transfer vector in c-Myc knock-out cells HO15.19. Cells modified and purified with puromycin were cultured for 4 days with various dox concentrations. Growth for all conditions was measured relative to the no vector control with no dox. Relative growth was measured as relative fluorescence using the viability assay reagent AlamarBlue at 4 days which assumes AlamarBlue conversion is linearly proportional to cell number

additional 23 days after isolation. The cells maintained expression of CD71 and c-Kit without giving rise to Gr-1$^+$ or Mac-1$^+$ cells (Additional file 1: Figure S6), indicating that c-Myc was capable of inducing self-renewal of erythroid progenitors far beyond the expected self-renewal capacity of only a few divisions [36].

Cytokine dependence of IPE cells typical of erythroid cells
IPE cells responded to both erythropoeitin (EPO) and stem cell factor (SCF), each cytokine having a positive

effect on cell proliferation (Fig. 6). Culture with both EPO and SCF achieved the greatest proliferation of cells, reaching over a 60-fold expansion after 6 days of culture. In contrast, significant cell death was observed for cells cultured in the absence of both EPO and SCF.

Colony-forming ability of IPE cells
CFC assays were performed on IPE cells in the presence of varying concentrations of dox. Both the frequency and the cell output (colony size) per clonogenic cell was

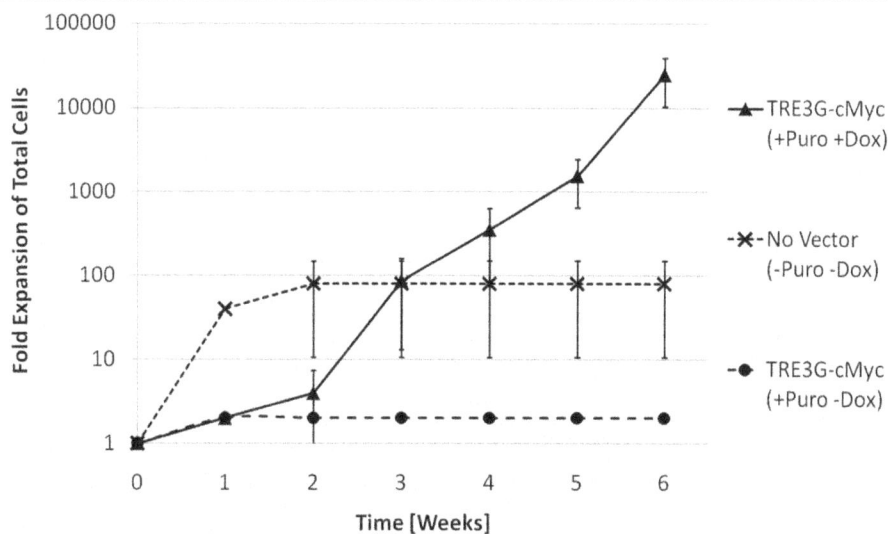

Fig. 4 Induced proliferation of lin$^-$Ter119$^-$Mac-1$^-$Gr-1$^-$c-Kit$^+$CD71$^{(low/-)}$ mouse bone marrow cells by over-expression of the transcription factor c-Myc. Shown are average measurements from 3 separate in vitro experiments where cells were modified by stable gene transfer of the TRE3G-cMyc vector, cultured in puromycin to eliminate non-modified cells (except wild type condition), then cultured with 2000 ng/ml dox for expression of the vector gene of interest or no dox as a control. Passage number and split ratio were used to determine fold expansion of total cells from the starting well and shown for 6 weeks of culture

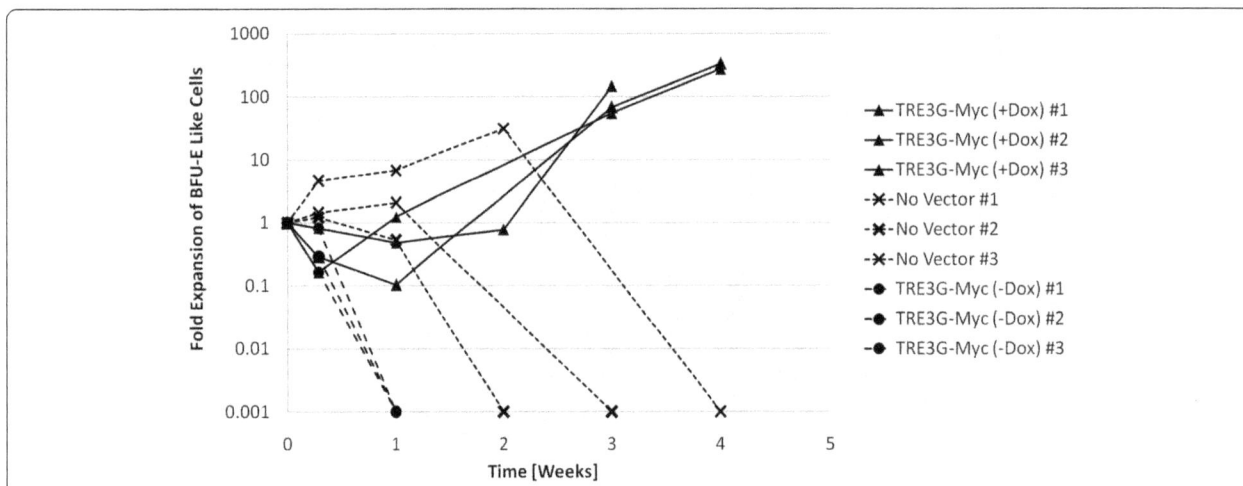

Fig. 5 Induced proliferation of 'BFU-E like' cells by over-expression of the transcription factor c-Myc. Shown are measurements from 3 separate in vitro experiments where cells were modified by stable gene transfer of the TRE3G-cMyc vector, cultured in puromycin to eliminate non-modified cells (except wild type condition), then cultured with 2000 ng/ml dox for expression of the vector gene of interest or no dox as a control

dependent on the concentration of dox (Fig. 7a), where higher dox concentrations resulted in a higher frequency of colony formation and larger colonies. The maximum CFC frequency obtained was low (2.5% at 2000 ng/ml dox) which suggests that the cell population was heterogeneous. This could be because erythroid progenitors at different stages of differentiation are self-renewing in culture, or a small population of BFU-E/CFU-E progenitors are self-renewing to a certain extent but also differentiating to some degree. The colonies formed at high concentrations of dox appeared largely composed of

clustered undifferentiated (blast) cells. The phenotype of the colonies formed with dox concentrations of 63 ng/ml and 125 ng/ml appeared very similar to CFU-Es, comprising small clusters of small cells. Interestingly, at 0 ng/ml or 32 ng/ml dox, the few colonies that formed appeared unhealthy and underdeveloped with too few cells to be considered a CFU-E [37].

Hemoglobinization with reduced c-Myc expression

Culture of IPE with reduced ectopic c-Myc expression resulted in red pigmentation observed in cell pellets

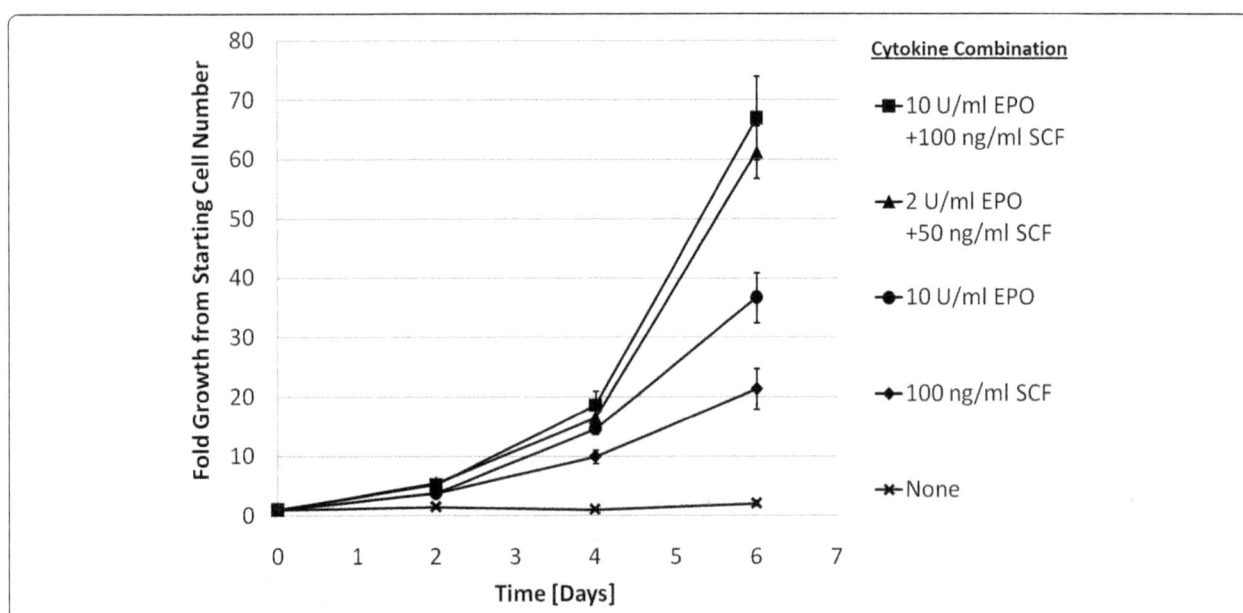

Fig. 6 Cytokine dependence of IPE cell expansion. IPE cells were plated at 2.5 × 10⁵ cells/ml in 200 µl IPE base media (without EPO and SCF) and cultured for 6 days with various combinations of EPO and SCF cytokines. Cells were passaged 1:4 every 2 days, and each data point is the average of three technical replicates

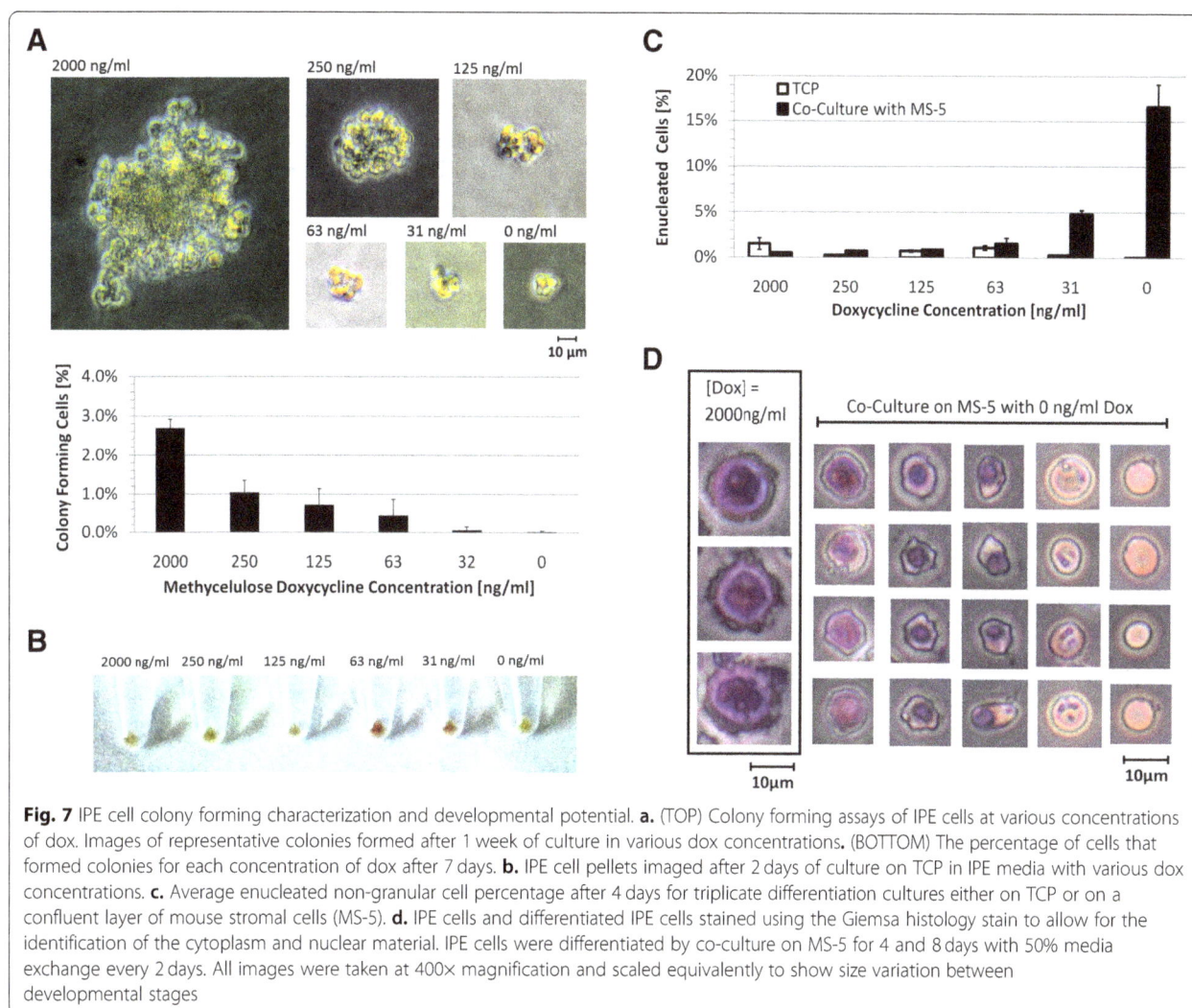

Fig. 7 IPE cell colony forming characterization and developmental potential. **a.** (TOP) Colony forming assays of IPE cells at various concentrations of dox. Images of representative colonies formed after 1 week of culture in various dox concentrations. (BOTTOM) The percentage of cells that formed colonies for each concentration of dox after 7 days. **b.** IPE cell pellets imaged after 2 days of culture on TCP in IPE media with various dox concentrations. **c.** Average enucleated non-granular cell percentage after 4 days for triplicate differentiation cultures either on TCP or on a confluent layer of mouse stromal cells (MS-5). **d.** IPE cells and differentiated IPE cells stained using the Giemsa histology stain to allow for the identification of the cytoplasm and nuclear material. IPE cells were differentiated by co-culture on MS-5 for 4 and 8 days with 50% media exchange every 2 days. All images were taken at 400× magnification and scaled equivalently to show size variation between developmental stages

after centrifugation, indicating the presence of hemoglobin [15]. This development of red pigmentation (Fig. 7b) was more pronounced for cells cultured at an intermediate level of dox (63 ng/ml dox) and c-Myc expression, suggesting that a reduction (and not complete removal of ectopic c-Myc expression) was optimal for promoting differentiation into hemoglobin-producing cells. The presence of hemoglobin was confirmed by tetramethylbenzidine staining [38–40] (Additional file 1: Supplementary Methods and Figure S8). These results indicate that IPE cells can produce hemoglobin to some degree.

Enucleation potential

To promote IPE cell differentiation and enucleation, dox was removed from the culture media to stop ectopic c-Myc expression in attempts to restore normal erythroid development. The enucleation methods included co-culture on a confluent layer of mouse stromal cells (MS-5) which has been shown to promote enucleation

of erythroid cells in vitro [41]. After 4 days of co-culture on MS-5 cells at 0 ng/ml dox, a clear population of nucleus-free cells was observed which was quantified to be $17 \pm 2\%$ of the total cells when measured by flow cytometry (Fig. 7c). The same high levels of enucleated cells was not observed in cultures without MS-5, where after 4 days of culture there was no significant increase in proportion of enucleated cells.

The enucleation potential of IPE cells was further assessed using the Giemsa histology stain combined with cytospin (Additional file 1: Supplementary Methods) in which erythroid cells at different stages of normal erythropoeisis are distinguishable (Fig. 7d) [42]. After dox removal, nuclear condensation and polarisation was observed, indicated by the presence and position of the cell nucleus [12]. Enucleated reticulocytes were visible based on their lack of nucleus and presence of residual RNA [43]. Enucleated cells without visible residual RNA were also present which correspond to terminally differentiated RBCs.

Decrease in cell size

The size of enucleated cells produced from IPE cells before and after culture on MS-5 with 0 ng/ml dox were measured with phase contrast microscopy and compared with fresh mouse RBCs. The diameter of IPE cells decreased from $14.5 \pm 0.4\,\mu m$ to $6.7 \pm 0.9\,\mu m$ after enucleation, which is similar to wild type RBCs which have a diameter of $5.4 \pm 0.5\,\mu m$ (Additional file 1: Figure S9).

IPE cells with intermediate levels of ectopic c-Myc expression

By culturing IPE cells on tissue culture plastic (TCP) at various concentrations of dox, it was found that cells transition to a unique cell phenotype when exposed to intermediate concentrations of dox (between 63 ng/ml and 125 ng/ml) as shown in Fig. 8a Reducing the dox concentration from 2000 ng/ml to 125 ng/ml promotes transition to a different cellular phenotype compared with both high and no dox. This cell phenotype is characterized by a lower proliferative rate, and better cell survival than when dox is completely removed (Fig. 8a). In this assay, cells were plated for 4 days at a high seeding density and, as a result, cultures with high levels of dox (250 ng/ml and 2000 ng/ml) reached excessive cell

densities (5.0×10^6 cells/ml) and significant cell death was observed which was atypical of routine passage.

IPE cell development by gradual reduction of ectopic c-Myc expression

A two-stage differentiation method was investigated to understand if a transitional stage with intermediate levels of ectopic c-Myc expression, before complete dox removal, could improve the development of IPE cells into enucleated RBCs. Cells cultured for 2 days at a reduced level of ectopic c-Myc expression (125 ng/ml dox) and subsequently with complete removal (0 ng/ml dox) in co-culture with MS-5 cells for 4 days appeared to undergo maturation more readily than those with no intermediate stage of ectopic expression (Fig. 8b). The transferrin receptor CD71 was completely lost in the two-stage approach in contrast to only moderate loss in the immediate transition approach. A major difference was the presence of a large homogeneous population of very small, non-granular nucleus containing (DRAQ5-- positive) cells with the two-stage approach. As well, this two-stage process resulted in a more homogenous population of small, low granularity, and nucleus- free cells which seems to more closely recapitulate normal erythropoiesis.

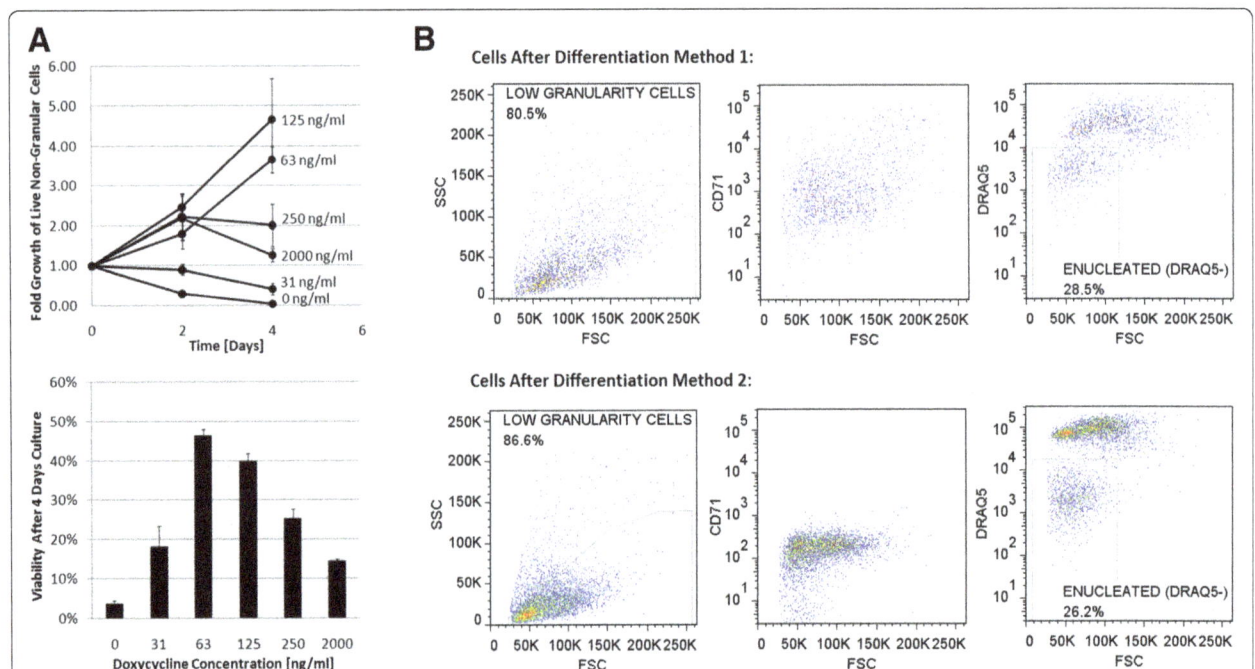

Fig. 8 IPE cell dox dependence assay and optimized differentiation results **a**. (TOP) Fold growth, and (BOTTOM) cell viability of IPE cells cultured with various dox concentrations on TCP. For each time point triplicate 200 μl cultures of IPE cells were plated at 5×10^5 cells/ml in a single well of a 96-well plate. Cells were cultured under static conditions at 37 °C and 5% CO_2 without passage, media change, or agitation. For each measurement, cells were stained with 7AAD and mixed with cell counting beads for analysis with flow cytometry. **b**: IPE cells after two different differentiation methods. Method 1 (TOP): 4 days of co-culture on MS-5 with 0 ng/ml dox. Method 2 (BOTTOM): 2 days of culture on TCP with 125 ng/ml dox, and then 4 days of co-culture on MS-5 with 0 ng/ml dox. For each condition the final cells were analyzed using flow cytometry measuring forward (FSC) and side (SSC) scatter, 7AAD, CD71 (transferrin receptor), and DRAQ5 (nuclear stain)

Discussion

Increased progenitor cell proliferation with c-Myc

The ectopic expression of high levels of c-Myc consistently induced extensive proliferation of primary hematopoietic progenitor cells, far exceeding the proliferative capacity of wild type cells. Ectopic expression of c-Myc promoted extensive cell proliferation (Fig. 4), and self-renewal of CFC (Fig. 5) with a unique morphology which is reminiscent of human BFU-E colonies (Additional file 1: Figure S5). The isolation of IPE cells modified with c-Myc alone is consistent with previous reports of c-Myc inducing proliferation of primary erythroid mouse cells [18], but differs from published work in human erythroid progenitor cells, where c-Myc-induced proliferation was observed to stop after a few weeks [16]. This discrepancy may be explained by the difference in species, but may also be the result of differences in the level of ectopic c-Myc expression achieved, the cell culture strategy, the medium composition, or a combination of these factors. In this work, intercellular signaling was optimized by maintaining the cellular aggregates that formed during routine cell passage. Also, in this work media was supplemented with IGF-1, which is known to suppress p53 induced apoptosis [28, 44], and this may have promoted cell survival. The genomic stability of IPE cells was not determined as karyotype analysis was not performed. However, mutagenesis is not considered to be the mechanism of c-Myc-induced proliferation. The reproducibility of increasing proliferative potential with c-Myc expression suggests that increased proliferative capacity is the result of increased transcription factor-associated gene expression and not a reproducible c-Myc-induced genomic mutation (mutagenesis is expected to be more random and less reproducible). The susceptibility of IPE cells to mutagenesis in long term culture should however be assessed if the use of IPE cells requires a stable cell line.

Isolating IPE cells

Although ectopic c-Myc expression enhanced cell expansion in all three experiments, isolation of induced proliferating erythroid cells was only achieved in one experiment. In the experiments where erythroid cell lines could not be isolated, the majority of the cells had become Gr-1 or Mac-1 positive, indicating the presence of predominantly macrophage and granulocyte progenitors. Although the starting population was enriched for BFU-Es, mixed hematopoietic colonies were observed, and the culture conditions included SCF which supports the survival and proliferation of progenitor cells in other lineages. Challenges in isolating erythroid cell lines may also be the result of inhibitory factors in the culture media. IPE media contained FBS which may have contained transforming growth factor beta (TGF-β) [45], a known inhibitor of

erythropoiesis [46]. The potential contribution of inhibitory factors is supported by the observation that the expansion of wild type BFU-Es was lower than expected [9]. Regardless of the challenges encountered when culturing purified erythroid cells, IPE cells were successfully generated, isolated, and further characterized.

IPE cells respond to cytokines

The individual and combined effects of EPO and SCF cytokines on IPE cell proliferation shown in Fig. 6 are similar to what has been reported for non-transduced mouse CFU-E [35, 47] but with much higher yields. In the absence of both EPO and SCF, there was significant cell death which is also observed in wild type cell cultures [35]. The observed cytokine-mediated cell survival is an indication of conserved apoptosis regulation in IPE cells. Therefore, it seems likely that evasion of c-Myc-induced apoptosis during IPE proliferation is the result of complementary signal transduction pathways affecting cell survival [29], and not a mutation affecting the cells ability to undergo apoptosis. This is supported by the known role of both SCF and EPO in activating bcl-xl [48–50] which binds to bim and blocks its apoptotic activity [51]. Inducing proliferation with overexpression of c-Myc without the apoptosis regulation disrupting genetic modifications used in other approaches [16, 52] is likely to result in a more stable cell line, with its genome still monitored by the p53 [17].

IPE cell differentiation into RBCs

IPE cells were reproducibly shown to differentiate with a reduction of ectopic c-Myc expression, signifying the reversibility of c-Myc-induced proliferation. As shown in Fig. 7, upon reducing ectopic c-Myc expression, IPE cells hemoglobinized, decreased in cell size and enucleated. Enucleation was best achieved with co-culture with MS-5, which has been shown to promote erythroid enucleation [41]. Measured enucleation of mouse erythroid cells modified with c-Myc alone allowed for 17% of cells to become enucleated, which far exceeds reported enucleation results reported from human cells modified with both c-Myc and bcl-xl where the enucleated fraction of differentiated cells was reported to be just 0.36% [16]. An improved differentiation strategy based on staged c-Myc expression was developed where cells hemoglobinized and formed colonies in methylcellulose in an optimal level when subjected to an intermediate level of ectopic c-Myc expression. It was found that some level of ectopic c-Myc expression was required for survival which is in agreement with previous reports suggesting that cells with c-Myc over-expression can become 'addicted' to c-Myc and where turning off conditional c-Myc can cause apoptosis [53]. By staging the removal of ectopic c-Myc expression, better development through erythropoiesis was observed (Fig. 8b). The enucleated RBC fraction contained predominantly small non-granular cells, and was more homogenous than RBCs produced by a one-step removal. The remaining nucleus-containing cells were also more homogenous, with a unique cluster of small, non-granular, CD71$^-$ viable cells, indicating the potential for further enucleation in cultures with longer incubation times. Further work is required to optimize the staged differentiation approach, but these preliminary results indicate it promoted erythroid maturation, establishing a new strategy to produce RBCs from induced proliferating erythroid progenitor cells.

Conclusions

A functionally validated all-in-one lentiviral gene transfer vector was developed and made it possible to achieve a tightly controlled dox-inducible expression of a 'FLAG' tagged c-Myc transcription factor. Genetic engineering of primary bone marrow-derived erythroid progenitor cells with this vector enabled the creation of IPE cells. These cells proliferated beyond the limited potential of wild type cells and their number increased over a period of at least 6 weeks. These cells maintained a surface receptor profile and cytokine dependence similar to wild type CFU-Es.

IPE cells were shown to maintain their developmental potential with ectopic c-Myc removal; they produced hemoglobin, decreased in size and enucleated. It was discovered that transition of IPE cells to low levels of ectopic expression, instead of complete removal, allowed for CFU-E like colony formation in methylcellulose, greater hemoglobinization and greater cell survival. Considering that complete removal of c-Myc was important for chromatin condensation and enucleation, an optimized two-stage differentiation strategy was developed, and was shown to promote IPE cell differentiation. Using this two-stage approach, after 6 days of culture, a final cell population was obtained which contained 16% ± 3% of enucleated cells. Further work is required to optimize this differentiation protocol to achieve higher enucleation efficiency.

This work demonstrates a new method to create inducibly proliferating erythroid progenitor cells which can maintain their ability to subsequently produce enucleated RBCs. This method is potentially useful for the production of large quantities of clinically-compatible enucleated cells. Furthermore, this strategy for RBC production presents a valuable platform to produce engineered enucleated cells expressing and containing biological molecules for novel cellular therapies.

Methods

TRE3G-c-Myc transfer vector molecular cloning and lentivirus production

The reverse tetracyclin responsive element (rtTA) with the human Ubiquitin C (hUbC) constitutive promoter from the FUW-M2rtTA plasmid (a gift from Rudolf Jaenisch, Addgene ref.: 20342) [54] was amplified with primer set 1 (Additional file 1: Supplementary Methods) to create the DNA fragment NotI-HUbC-rtTA-EcoRV. This was cloned into the pENTR1 Gateway Entry plasmid (a gift from Eric Campeau, Addgene ref.: 19364) [55] resulting in the pENTR1-GFP-HUbC-rtTA plasmid. Template DNA for mouse c-Myc (Isoform 2) was amplified from TetO-FUW-cMyc (a gift from Rudolf Jaenisch, Addgene ref.: 20324) [56] using primer set 2 (Additional file 1: Supplementary Methods) introducing a FLAG epitope (Peptide: DYKDDDDK) on the N–Terminus of c-Myc gene, creating the DNA fragment SacI-FLAG-c-Myc-NotI. This was cloned into the pENTR1-GFP-HUbC-rtTA creating the pENTR1-FLAG-cMyc-HUbC-rtTA plasmid. This was then cloned into the pLenti CMVTRE3G Puro Destination plasmid (a gift from Eric Campeau, Addgene ref.: 27565) [55] creating the final transfer vector pLenti-TRE3G-FLAG-cMyc-HUbC-rtTA-PGK-Puro plasmid (TRE3G-cMyc). Transfer vectors were packaged in HEK293T cells (ATCC, ref.: CRL-3216) with plasmids psPAX2 and VSV-G (gifts from Didier Trono, Addgene refs: 12260 and 12259), then concentrated by ultracentrifugation, and stored at − 80 °C. Titres were quantified using puromycin in colony-forming assays [57]. Additional information about lentivirus production and concentration is available in Additional file 1: Supplementary Methods.

Vector validation assay with c-Myc$^{-/-}$ cell line

Rat fibroblast cells TGR-1 [32], the c-Myc$^{-/-}$ knock out variant HO15.19 [32], and HO15.19 modified with the TRE3G-cMyc vector were passaged under normal cell culture conditions (Additional file 1: Supplementary methods) and plated at low densities of 1000 cells per well in a 96-well plate (Sarstedt, ref.: 83.1835) at various dox concentrations. After 4 days, to detect relative cell density cells were incubated in 5% v/v AlamarBlue (AbDSerotec, ref.: BUF012B) for 1 h under normal culture conditions. The quantity of reaction product was detected using a SpectraMAX GeminiXS (Molecular Devices) at an excitation frequency of 530 nm and detection above 590 nm [58]. Relative growth was a direct comparison of reaction products and doubling times were calculated using an exponential growth model (Additional file 1: Supplementary Methods).

Western blot

Cells were collected in 1X phosphate buffered saline (PBS), centrifuged and re-suspended in lysis buffer (1% Triton-X-100, 100 mM sodium chloride, 50 mM HEPES, 5% glycerol, protease inhibitor mixture and Phospho-STOP phosphatase inhibitor (Roche Applied Science) as previously described [59]. Concentrations of whole cell lysate protein were measured by Bradford assay. Twenty micrograms of total protein samples were prepared in Laemmli buffer, boiled (100 °C; 5 min) and loaded into a 10% SDS-PAGE gel. Proteins were transferred to a nitrocellulose membrane, blocked with with 5% bovine serum albumin (BSA)/Tris buffered saline with Tween 20 (TBST) (1 h at RT) and incubated overnight with anti-c-Myc antibody (Cell Signalling Technology; 1:1000; 4 °C). Blots were incubated with anti-rabbit horseradish peroxidase-conjugated secondary antibody (1:2000, Cell Signalling Technology; 45 min; RT). Blots were stripped, blocked 5% BSA/TBST (1 h; RT) and re-probed with anti-GAPDH antibody (Santa Cruz Biotechnology; 1:3000; 40 min; RT) and anti-mouse horseradish peroxidase-conjugated secondary antibody (1:2000, Cell Signalling Technology; 45 min; RT). Protein was detected with enhanced chemiluminescent substrate.

Generation and isolation of IPE cells

Female Crl:CD1(ICR) mice (strain referred to as CD-1) were purchased from Charles River Laboratories. Adult mice between 6 and 16 weeks of age were euthanized by exposure to CO_2 prior to tissue collection. Cervical dislocation was performed after CO_2 asphyxiation to ensure euthanasia of mice. In each experiment primary bone marrow cells were isolated from femurs and tibias of 6–8 sacrificed mice by flushing with Hank's balanced salt solution (HBSS) (Gibco, ref.: 14025092) supplemented with 2% fetal bovine serum (FBS) (Gibco, ref.: 12483–020), using a 23-gauge needle (BD Biosciences, ref.: 305145). The cells were pooled, centrifuged and re-suspended in 2% FBS HBSS. Existing RBCs were lysed with ammonium chloride solution (Stem Cell Technologies, ref.: 07800) and vortexing for 5 s, then incubating on ice for 5 min. The cells were then lineage-depleted (lin⁻) using a hematopoietic isolation kit (Stem Cell Technologies, ref.: 19756A) at a cell density of 2×10^8 cells/ml. Purified cells were then sorted using fluorescence activated cell sorting (FACS) with conjugated antibodies and a FACSAria cell sorter (BD Biosciences) to obtain a BFU-E enriched population with surface marker profile Mac-1⁻Gr-1⁻Ter119⁻c-Kit⁺CD71(low/−) [9].

BFU-E enriched cells were re-suspended in IPE culture media at 2.5×10^5 cells/ml and plated in 200 µl volumes at 37 °C and 5% CO_2 in a 96-well tissue culture plastic (TCP) plate (Sarstedt, ref.: 83.1835) to recover. One day after plating cells were transduced with the TRE3G-c-Myc lentivirus vector. One day after transduction, cells were diluted with fresh media in a 1:2 split passage, and dox (2000 ng/ml) and puromycin (1000 ng/ml) were added. Cell cultures were then passaged at a split ratio of 1:4 every 2 days, being careful not to disrupt cellular aggregates. Twenty-five days after harvest, cells

were sorted by FACS using the surface marker expression profile Mac-1⁻Gr-1⁻c-Kit(high/+)CD71⁺. Purified cells were continuously cultured in IPE media following the same split ratio of 1:4 every 2 days in 200 µl volumes of 96 well plates making fresh IPE media with dox and puromycin for each passage. A flowchart of the IPE cell establishment and differentiation is available in the supplementary materials (Additional file 1: Figure S11).

IPE cell differentiation assay

IPE cells were washed three times in 2% FBS in Iscove's Modified Dulbecco's Medium (IMDM) by centrifugation, then either plated in a 96 well TCP plate at 5×10^5 cells/ml in 200 µl of IPE media or in a 6 well plate on a confluent layer of mouse stromal cells (MS-5) [41] at 1.25×10^5 cells/ml in 800 µl of IPE media. The cells were left under static conditions for 2 or 4 days with no media changes before analysis. Under the two-stage differentiation protocol, the cells were transferred after 2 days of culture on TCP with low levels of dox (125 ng/ml) to a confluent layer of mouse stromal cells with 800 µl of IPE media and no dox after triple washing in 2% FBS IMDM.

IPE media

Hematopoietic cells and IPE cells were cultured in IPE media which consists of IMDM (Gibco, ref.: 12440053) supplemented with 20% FBS (Gibco, ref.: 12483–020), 10 U/ml mouse EPO (R&D Systems, ref.: 959-ME-010), 100 ng/ml mouse SCF (R&D Systems, ref.: 455-MC-010), 500 µg/ml human holo-transferrin, 40 ng/ml human IGF-1 (R&D Systems, ref.: 291-G1–200), 10 µg/ml human insulin (Sigma, ref.: I9278), and 2x Pen/Strep (Gibco, ref.: 15070063). If necessary media was supplemented with 2 µg/ml dox (Biobasic, ref.: DB00889) to induce c-Myc expression, and/or puromycin (1000 ng/ml) (Sigma, ref.: P8833) for selection of cells containing the transfer vector.

Lentiviral transduction of primary hematopoietic cells

Concentrated lentivirus, media supplements, and polybrene (Sigma, ref.: 107689-10G) were added to existing cell cultures at a ratio of 1:1 to achieve a final multiplicity of infection of 5, polybrene concentration of 8 µg/ml, and full IPE media supplement concentrations. The plates were centrifuged in a 32 °C pre-heated centrifuge at 1200×g for 2 h, then incubated at 37 °C and 5% CO_2.

Hematopoietic colony-forming cell (CFC) assays

Colony-forming cell assays [37] were performed to enumerate hematopoietic progenitor cells using Methocult M3434 (Stem Cell Technologies, ref.: 03434). For transduced cells, dox (2 µg/ml) was added and mixed when plating. In this assay, cells were mixed with M3434 solution by pipetting slowly using a blunt end needle and a syringe. The cells were then plated in a 6-well nontissue culture treated plate (Costar, ref.: 3736). Colonies were scored according to company protocol according to images provided and colony descriptions.

Cytokine dependence assay

IPE cells were plated at 2.5×10^5 cells/ml in 200 µl IPE base media (without EPO and SCF) and supplemented with various combinations of EPO and SCF in the presence of 2 µg/ml dox. Experiments were done in triplicate and cells were passaged 1:4 every 2 days. Cells were counted with a hemocytometer and trypan blue vital stain (Gibco, ref.: 15250061).

Flow cytometry

Conjugated anti-mouse antibodies used in this study were PE-CD71 (BD, ref.: 553267), APC-c-Kit (BD, ref.: 553356), PE-Cy/7-Ter119 (BD, ref.: 557853), FITC-CD71 (BD, ref.: 553266), FITC-CD11b (Mac-1) (BioLegend, ref.: 101205), and APC-Cy/7-Ly-6G/Ly-6c (Gr-1) (Biolegend, ref.: 108423). The following nuclear dyes were used: 7-aminoactinomycin D (7-AAD) (Invitrogen, ref.: A1310), and DRAQ5 (Cell Signalling Technology, ref.: 4084S). For analytical staining, approximately 100,000 cells or less were washed with 2% FBS HBSS and incubated in a 20 µl staining volume at 4 °C between 15 min to 20 min containing relevant antibodies, each at a dilution of 1:100 in 2% FBS in HBSS. To determine viability, 7-AAD was included at a dilution of 1:500, and nuclei were stained with DRAQ5 at a dilution of 1:5000 (1 µM). After staining, cells were washed, re-suspended in 200 µl of 2% FBS HBSS, and run on a FACSCanto flow cytometer (BD Biosciences). When counting beads were used, an equal volume containing approximately 1500 blank beads (Spherotech, ref.: ACBP-20-10) was added to each sample. Fluorescence spectral overlap compensation was done using single stain compensation controls and algorithms within FlowJo (Tree Star, San Carlos, CA). A general gating strategy (Additional file 1: Figure S10) included selection of cells and beads based on their forward scatter (FSC) and side scatter (SSC), excluding highly granular cells and small debris, followed by a live/dead exclusion applied using 7-AAD.

Additional files

Additional file 1: Figure S1. Detailed DNA plasmid map of lentivirus transfer vector TRE3G-cMyc. Figure S2. Doubling time of c-Myc knock-out HO15.19 cells modified with the TRE3G-cMyc transfer vector compared with wild-typ type cells. Figure S3. Fluorescence activating cell sorting gates used to isolate erythroid enriched populations from fresh lineage depleted bone marrow. Figure S4. Colony forming cell (CFC) results for isolated Lin⁻c-Kit⁺CD71$^{(low/-)}$ mouse bone marrow cells. Figure S5. Comparing normal BFU-E and 'BFU-E like' colonies formed from normal and genetically modified hematopoietic cells. Figure S6. Cell surface protein profile of IPE cells at isolation and after culture. Figure S7. Cell surface protein profile of Lin⁻c-Kit⁺CD71$^{(low/-)}$ cells modified with the TRE3G-cMyc and cultured for three weeks. Supplementary Methods. Figure S8. Tetramethylbenzidine (TMB) staining of IPE cells after 48 h of culture with 0 ng/ml dox. Figure S9. Relative size of cells comparing IPE cells, RBCs derived from IPE cells (IPE -> RBC), and fresh RBCs. Figure S10. Example of gating strategy based on FSC and SSC and then on 7-AAD. Figure S11. Flowchart for IPE cell establishment, isolation, and optimised differentiation protocol. (DOCX 3550 kb)

Abbreviations

7-AAD: 7-aminoactinomycin D; BFU-E: Burst-forming unit-erythroid; BSA: Bovine serum albumin; CFC: Colony-forming cell; CFU-E: Colony forming unit-erythroid; DNA: Deoxyribonucleic acid; dox: doxycycline; EPO: Erythropoietin; ES: Embryonic stem; FACS: Fluorescence activated cell sorting; FBS: Fetal bovine serum; FSC: Forward scatter; GFP: Green fluorescent protein; HAT: Histone acetyl transferase; HBSS: Hanks balanced salt solution; HPV: Human papillomavirus; hUbC: Human ubiquinase C; IGF-1: Insulin like growth factor 1; IMDM: Iscove's modified Dulbecco's medium; IPE: Inducibly proliferating erythroid; Lin-: Lineage negative; Pen: Penicillin; Puro: Puromycin; RBC: Red blood cell; RNA: Ribonucleic acid; RT: Room temperature; rtTA: reverse tetracycline transactivator; SCF: Stem cell factor; SSC: Side scatter; strep: streptomycin; TBST: Tris-buffered saline with Tween 20; TCP: Tissue culture plate; TGF-β : Transforming growth factor beta; TRE: Tetracycline responsive element; TRE3G: Third generation TRE; U: Unit

Acknowledgements

We would like to thank Weijia Wang for training in animal protocols, Linda Penn for generously gifting the c-Myc knock-out cell line and wild type control, Peter Zandstra for generously gifting the MS-5 cell line, and Jason Moffat for generously gifting HEK293T cells. We would also like to thank Dieder Trono, Rudolf Jaenisch, and Eric Campeau for generously gifting plasmid DNA, and Addgene for organising material transfers.

Funding

Funding for this work was provided from the Ontario government as part of an Ontario Graduate Scholarship to SM as well as from a Natural Sciences and Engineering Council of Canada (NSERC) Discovery Grant to JA. These funding organizations were not involved in the design of the study and collection, analysis, and interpretation of data, or in writing the manuscript.

Authors' contributions

SM and JA designed experiments and interpreted findings together. SM planned all experiments and carried out most experiments. TM performed some molecular cloning experiments. PDM performed some cell line experiments. PNS performed the western blot with support from DMK. SM wrote the manuscript. All authors read, revised, and approved the manuscript.

Ethics approval

The animal use and experimental protocols were approved by the University of Toronto Animal Care Committee in accordance with the Guidelines of the Canadian Council on Animal Care.

Consent for publication

Not Applicable.

Competing interests

The authors declare that they have no competing interests.

References

1. Shi J, Kundrat L, Pishesha N, Bilate A, Theile C, Maruyama T, et al. Engineered red blood cells as carriers for systemic delivery of a wide array of functional probes. Proc Natl Acad Sci. 2014;111:10131–6.

2. Banz A, Cremel M, Rembert A, Godfrin Y. In situ targeting of dendritic cells by antigen-loaded red blood cells: a novel approach to cancer immunotherapy. Vaccine. 2010;28:2965–72.

3. Kontos S, Kourtis IC, Dane KY, Hubbell JA. Engineering antigens for in situ erythrocyte binding induces T-cell deletion. Proc Natl Acad Sci U S A. 2013; 110:E60–8.

4. Banz A, Cremel M, Mouvant A, Guerin N, Horand F, Godfrin Y. Tumor growth control using red blood cells as the antigen delivery system and poly(I:C). J Immunother. 2012;35:409–17.

5. Pishesha N, Bilate AM, Wibowo MC, Huang N-J, Li Z, Dhesycka R, et al. Engineered erythrocytes covalently linked to antigenic peptides can protect against autoimmune disease. Proc Natl Acad Sci. 2017;114:3157–62.

6. Giarratana MC, Rouard H, Dumont A, Kiger L, Safeukui I, Le Pennec PY, et al. Proof of principle for transfusion of in vitro-generated red blood cells. Blood. 2011;118:5071–9.

7. Wu H, Liu X, Jaenisch R, Lodish HF. Generation of committed erythroid BFU-E and CFU-E progenitors does not require erythropoietin or the erythropoietin receptor. Cell. 1995;83:59–67.

8. Hattangadi SM, Wong P, Zhang L, Flygare J, Lodish HF. From stem cell to red cell: regulation of erythropoiesis at multiple levels by multiple proteins, RNAs, and chromatin modifications. Blood. 2011;118:6258–68.

9. Flygare J, Estrada VR, Shin C, Gupta S, Lodish HF. HIF1α synergizes with glucocorticoids to promote BFU-E progenitor self-renewal. Blood. 2011;117: 3435–44.

10. Stumpf M, Waskow C, Krötschel M, van Essen D, Rodriguez P, Zhang X, et al. The mediator complex functions as a coactivator for GATA-1 in erythropoiesis via subunit Med1/TRAP220. Proc Natl Acad Sci U S A. 2006; 103:18504–9.

11. Zhang J, Socolovsky M, Gross AW, Lodish HF. Role of Ras signaling in erythroid differentiation of mouse fetal liver cells: functional analysis by a flow cytometry-based novel culture system. Blood. 2003;102:3938–46.

12. Keerthivasan G, Wickrema A, Crispino JD. Erythroblast enucleation. Stem Cells Int. 2011;2011:139851.

13. Elliott S, Pham E, Macdougall IC. Erythropoietins: a common mechanism of action. Exp Hematol. 2008;36:1573–84.

14. Trakarnsanga K, Griffiths RE, Wilson MC, Blair A, Satchwell TJ, Meinders M, et al. An immortalized adult human erythroid line facilitates sustainable and scalable generation of functional red cells. Nat Commun. 2017;8:1–7.

15. Kurita R, Suda N, Sudo K, Miharada K, Hiroyama T, Miyoshi H, et al. Establishment of immortalized human erythroid progenitor cell lines able to produce enucleated red blood cells. PLoS One. 2013;8:e59890.

16. Hirose SI, Takayama N, Nakamura S, Nagasawa K, Ochi K, Hirata S, et al. Immortalization of erythroblasts by c-MYC and BCL-XL enables large-scale erythrocyte production from human pluripotent stem cells. Stem Cell Rep. 2013;1:499–508.

17. Albrechtsen N, Dornreiter I, Grosse F, Kim E, Wiesmüller L, Deppert W. Maintenance of genomic integrity by p53: complementary roles for activated and non-activated p53. Oncogene. 1999;18:7706–17.

18. Jayapal SR, Lee KL, Ji P, Kaldis P, Lim B, Lodish HF. Down-regulation of Myc is essential for terminal erythroid maturation. J Biol Chem. 2010;285:40252–65.

19. Kim KS, Lee HJ, Jeong HS, Li J, Teng YD, Sidman RL, et al. Self-renewal induced efficiently, safely, and effective therapeutically with one regulatable gene in a human somatic progenitor cell. Proc Natl Acad Sci U S A. 2011; 108:4876–81.

20. Lawlor ER, Soucek L, Brown-Swigart L, Shchors K, Bialucha CU, Evan GI. Reversible kinetic analysis of Myc targets in vivo provides novel insights into Myc-mediated tumorigenesis. Cancer Res. 2006;66:4591–601.

21. Adhikary S, Eilers M. Transcriptional regulation and transformation by Myc proteins. Nat Rev Mol Cell Biol. 2005;6:635–45.

22. Dang CV. MYC on the path to cancer. Cell. 2012;149:22–35.

23. Eilers M, Eisenman RN. Myc's broad reach. Genes Dev. 2008;22:2755–66.

24. Murphy DJ, Junttila MR, Pouyet L, Karnezis A, Shchors K, D a B, et al. Distinct thresholds govern Myc's biological output in vivo. Cancer Cell. 2008;14:447–57.

25. Sahar S, Sassone-Corsi P. Metabolism and cancer: the circadian clock connection. Nat Rev Cancer 2009 9;886–96.

26. Hermeking H, Eick D. Mediation of c-Myc-induced apoptosis by p53. Science. 1994;265:2091–3.

27. Hoffman B, D a L. Apoptotic signaling by c-MYC. Oncogene. 2008;27: 6462–72.

28. Juin P, Hueber A, Littlewood T, Evan G. C-Myc-induced sensitization to apoptosis is mediated through cytochrome c release. Genes Dev. 1999;13: 1367–81.

29. Evan GI, Wyllie AH, Gilbert S, Littlewood TD, Land H, Brooks M, et al. Induction of apoptosis by c-myc protein in fibroblasts. Cell. 1992;69:119–28.

30. Morrish F, Neretti N, Sedivy JM, Hockenbery DM. The oncogene c-Myc coordinates regulation of metabolic networks to enable rapid cell cycle entry. Cell Cycle. 2008;7:1054–66.

31. Loew R, Heinz N, Hampf M, Bujard H, Gossen M. Improved Tet-responsive promoters with minimized background expression. BMC Biotechnol. 2010; 10:1–13.

32. Mateyak MK, Obaya AJ, Adachi S, Sedivy JM. Phenotypes of c-Myc-deficient rat fibroblasts isolated by targeted homologous recombination. Cell Growth Differ. 1997;8:1039–48.

33. Cao Z, Fan-Minogue H, Bellovin DI, Yevtodiyenko A, Arzeno J, Yang Q, et al. MYC phosphorylation, activation, and tumorigenic potential in hepatocellular carcinoma are regulated by HMG-CoA reductase. Cancer Res. 2011;71:2286–97.

34. Eaves C, Lambie K. Atlas of human hematopoietic colonies. Vancouver: STEMCELL Technologies Inc; 1995.

35. Wang W, Akbarian V, Audet J. Biochemical measurements on single erythroid progenitor cells shed light on the combinatorial regulation of red blood cell production. Mol BioSyst. 2013;9:234–45.

36. Hattangadi SM, Wong P, Zhang L, Flygare J, Lodish HF. From stem cell to red cell: regulation of erythropoiesis at multiple levels by multiple proteins, RNAs, and chromatin modifications. Blood. 2011;118:6258–68.

37. Miller CL, Dykstra B, Eaves CJ. Characterization of mouse hematopoietic stem and progenitor cells. Curr Protoc Immunol. 2008;80:22B.2.1–22B.2.31.

38. Kapralov A, Vlasova II, Feng W, Maeda A, Walson K, Tyurin VA, et al. Peroxidase activity of hemoglobin·haptoglobin complexes. Covalent aggregation and oxidative stress in plasma and macrophages. J Biol Chem. 2009;284:30395–407.

39. Liem HH, Cardenas F, Tavassoli M, Poh-Fitzpatrick MB, Muller-Eberhard U. Quantitative determination of hemoglobin and cytochemical staining for peroxidase using 3,3,5,5-tetramethylbenzidine dihydrochloride, a safe substitute for benzidine. Anal Biochem. 1979;98:388–93.

40. Reynolds M, Lawlor E, McCann SR, Temperley I. Use of 3,3',5,5'-tetramethylbenzidine (TMB) in the identification of erythroid colonies. J Clin Pathol. 1981;34:448–9.

41. Giarratana M-C, Kobari L, Lapillonne H, Chalmers D, Kiger L, Cynober T, et al. Ex vivo generation of fully mature human red blood cells from hematopoietic stem cells. Nat Biotechnol. 2005;23:69–74.

42. Chen K, Liu J, Heck S, J a C, An X, Mohandas N. Resolving the distinct stages in erythroid differentiation based on dynamic changes in membrane protein expression during erythropoiesis. Proc Natl Acad Sci U S A. 2009; 106:17413–8.

43. Lee E, Choi HS, Hwang JH, Hoh JK, Cho Y-H, Baek EJ. The RNA in reticulocytes is not just debris: it is necessary for the final stages of erythrocyte formation. Blood Cells Mol Dis. 2014;53:1–10.

44. Peruzzi F, Prisco M, Dews M, Salomoni P, Grassilli E, Romano G, et al. Multiple signaling pathways of the insulin-like growth factor 1 receptor in protection from apoptosis. Mol Cell Biol. 1999;19:7203–15.

45. Oida T, Weiner HL. Depletion of TGF-β from fetal bovine serum. J Immunol Methods. 2010;362:195–8.

46. Zermati Y, Fichelson S, Valensi F, Freyssinier JM, Rouyer-Fessard P, Cramer E, et al. Transforming growth factor inhibits erythropoiesis by blocking proliferation and accelerating differentiation of erythroid progenitors. Exp Hematol. 2000;28:885–94.

47. Wang W, Horner DN, Chen WLK, Zandstra PW, Audet J. Synergy between erythropoietin and stem cell factor during erythropoiesis can be quantitatively described without co-signaling effects. Biotechnol Bioeng. 2008;99:1261–72.

48. Dolznig H, Habermann B, Stangl K, Deiner EM, Moriggl R, Beug H, et al. Apoptosis protection by the Epo target Bcl-X(L) allows factor-independent differentiation of primary erythroblasts. Curr Biol. 2002;12:1076–85.

49. Kapur R, Zhang L. A novel mechanism of cooperation between c-kit and erythropoietin receptor: stem cell factor induces the expression of Stat5 and erythropoietin receptor, resulting in efficient proliferation and survival by erythropoietin. J Biol Chem. 2001;276:1099–106.

50. Socolovsky M, Nam H, Fleming MD, Haase VH, Brugnara C, Lodish HF. Ineffective erythropoiesis in Stat5a−/−5b−/− mice due to decreased survival of early erythroblasts. Blood. 2001;98:3261–73.

51. O'Connor L, Strasser A, O'Reilly LA, Hausmann G, Adams JM, Cory S, et al. Bim: a novel member of the Bcl-2 family that promotes apoptosis. EMBO J. 1998;17:384–95.

52. Huang X, Shah S, Wang J, Ye Z, Dowey SN, Tsang KM, et al. Extensive ex vivo expansion of functional human erythroid precursors established from umbilical cord blood cells by defined factors. Mol Ther. 2014;22:451–63.

53. Arvanitis C, Felsher DW. Conditional transgenic models define how MYC initiates and maintains tumorigenesis. Semin Cancer Biol. 2006;16:313–7.

54. Hockemeyer D, Soldner F, Cook EG, Gao Q, Mitalipova M, Jaenisch R. A drug-inducible system for direct reprogramming of human somatic cells to pluripotency. Cell Stem Cell. 2008;3:346–53.

55. Campeau E, Ruhl VE, Rodier F, Smith CL, Rahmberg BL, Fuss JO, et al. A versatile viral system for expression and depletion of proteins in mammalian cells. PLoS One. 2009;4:1–18.

56. Brambrink T, Foreman R, Welstead GG, Lengner CJ, Wernig M, Suh H, et al. Sequential expression of pluripotency markers during direct reprogramming of mouse somatic cells. Cell Stem Cell. 2008;2:151–9.

57. Burns JC, Friedmann T, Driever W, Burrascano M, Yee JK. Vesicular stomatitis virus G glycoprotein pseudotyped retroviral vectors: concentration to very high titer and efficient gene transfer into mammalian and nonmammalian cells. Proc Natl Acad Sci U S A. 1993;90:8033–7.

58. Li S, Lin W, Tchantchou F, Lai R, Wen J, Zhang Y. Protein kinase C mediates peroxynitrite toxicity to oligodendrocytes. Mol Cell Neurosci. 2011;48:62–71.

59. Silva PN, Altamentova SM, Kilkenny DM, Rocheleau JV. Fibroblast growth factor receptor Like-1 (FGFRL1) interacts with SHP-1 phosphatase at insulin secretory granules and induces beta-cell ERK1/2 protein activation. J Biol Chem. 2013;288:17859–70.

4

Integrating vectors for genetic studies in the rare Actinomycete *Amycolatopsis marina*

Hong Gao[1,3]* , Buvani Murugesan[1], Janina Hoßbach[1], Stephanie K. Evans[1], W. Marshall Stark[2] and Margaret C. M. Smith[1]

Abstract

Background: Few natural product pathways from rare Actinomycetes have been studied due to the difficulty in applying molecular approaches in these genetically intractable organisms. In this study, we sought to identify more integrating vectors, using phage *int/attP* loci, that would efficiently integrate site-specifically in the rare Actinomycete, *Amycolatopsis marina* DSM45569.

Results: Analysis of the genome of *A. marina* DSM45569 indicated the presence of *attB*-like sequences for TG1 and R4 integrases. The TG1 and R4 *attB*s were active in in vitro recombination assays with their cognate purified integrases and *attP* loci. Integrating vectors containing either the TG1 or R4 *int/attP* loci yielded exconjugants in conjugation assays from *Escherichia coli* to *A. marina* DSM45569. Site-specific recombination of the plasmids into the host TG1 or R4 *attB* sites was confirmed by sequencing.

Conclusions: The homologous TG1 and R4 *attB* sites within the genus *Amycolatopsis* have been identified. The results indicate that vectors based on TG1 and R4 integrases could be widely applicable in this genus.

Keywords: Rare Actinomycetes, *Amycolatopsis*, Integrating vectors, TG1 integrase, R4 integrase

Background

Streptomyces bacteria are widely exploited for their abundant bioactive natural products [1]. However, after decades of exploitation, the rate of discovery of new *Streptomyces*-derived bioactive products has declined, and interest has grown in other potential non-Streptomycete sources, such as the rare Actinomycetes [2, 3].

Amongst rare Actinomycetes, the *Amycolatopsis* genus is of particular interest for its production of critically important antibiotics such as vancomycin [4] and rifamycin [5], as well as a diverse range of active natural products [6–8]. The publicly available NCBI database contains nearly 90 genomes of *Amycolatopsis* strains, covering more than 40 species from this genus. Similar to *Streptomyces*, the genome of each *Amycolatopsis* contains averagely over 20 secondary metabolic gene clusters [9]. The mining of these metabolic clusters offers excellent potential for novel antibiotic discovery.

Phage-encoded serine and tyrosine integrases catalyse site-specific integration of a circularised phage genome into the host chromosome as part of the process to establish a lysogen. DNA integration mediated by serine integrases occurs between short (approximately 50 bp) DNA substrates that are located on the phage genome (the phage attachment site *attP*), and the host genome (the bacterial attachment site *attB*). The product of *attP* x *attB* recombination is an integrated phage genome flanked by two new sites, *attL* and *attR*, each of which contains half-sites from *attP* and *attB*. During phage induction, integrase in the presence of a recombination directionality factor (RDF) again mediates site-specific recombination, but this time between *attL* and *attR*, to excise the phage genome, which can then be replicated during a lytic cycle. The mechanism of recombination

* Correspondence: gaohong221@gmail.com; h.gao@tees.ac.uk
[1]Department of Biology, University of York, York, North Yorkshire YO10 5DD, UK
[3]Present address: School of Science, Engineering & Design, Teesside University, Middlesbrough TS1 3BX, UK
Full list of author information is available at the end of the article

and the factors that control integration versus excision have been elucidated in recent years [10–12].

Integrating vectors based on the *Streptomyces* phage ϕC31 integrase and *attP* locus are best known and most widely used in Actinomycete genome engineering [13–16], and in addition to the phage recombination machinery (*int/attP*), integrating vectors contain a replicon for maintenance in *Escherichia coli*, an *oriT* for conjugal transfer and a marker or markers for selection in *E. coli* and the recipient. They are powerful genome engineering tools that act in an efficient, highly controllable and predictable way [17].

Using serine integrase-mediated recombination, these integrating vectors require no additional phage or host functions for integration, which is an especially important feature when they are used in other organisms that cannot be infected by the phages. This property makes serine integrase-based vectors promising tools for use in various systems [10, 18]. However, the use of these integration vectors has not been fully explored in rare Actinomycetes, e.g. *Amycolatopsis*. There is one reported example of a conjugation system based on ϕC31 integrase in *Amycolatopsis japonicum* MG417-CF17 [19], and it has been reported that other *Amycolatopsis* species lack ϕC31 *attB* sites in their chromosomes [20]. The ϕBT1 *attB* sites have been more commonly identified in *Amycolatopsis*. A vector based on ϕBT1 *int/attP* has been successfully transferred into *Amycolatopsis mediterranei* [21]. Furthermore, electroporation remains the most widely applied method for transfer of integrative plasmids into this genus, rather than conjugation [20, 21].

In this paper, we chose to study *A. marina* DSM45569, a species isolated from an ocean-sediment sample collected in the South China Sea [22]. Since the marine environment has been assumed to offer an as yet mostly untapped treasure of chemical biodiversity [23], we are quite interested in natural product discovery from *A. marina*. We explored the application of bacterial genetic engineering using serine integrases and developed conjugative and integrating vectors for use in this species. We present evidence suggesting that these vectors could be applied to other species in this genus, thus opening up the prospect for versatile genetic manipulation of *Amycolatopsis*.

Results

Identification of *attB*-like sequences from the genome of *A. marina* DSM45569

The primers used in this study were listed in Table 1. The sequences of *attB* sites recognised by a variety of integrases (ϕC31 [24], ϕJoe [25], Bxb1 [26], R4 [27], SPBc [28], SV1 [29], TG1 [30] and TP901 [31]) were used in BLAST searches of the genome sequence of *A. marina* DSM45569 (NCBI Genome Database

NZ_FOKG00000000). The most significant hits for R4 and TG1 *attB* sites had the highest identities and lowest *E*-value (Table 2). The recognised R4 *attB*-like site is located within a gene predicted to encode a fatty-acyl-CoA synthase (SFB62308.1), and the TG1 *attB* site is located within a gene predicted to code for a putative succinyldiaminopimelate transaminase (WP_091671332.1). The BLAST analysis was extended to other species of *Amycolatopsis* to assess the conservation of these *attB* sites in the genus (Fig. 1). Both R4 and TG1 *attB* sites were highly conserved relative to the *attB* sites originally identified from *Streptomyces parvulus* [32] (84% for R4 integrase) and *Streptomyces avermitilis* [30] (62% for TG1 integrase).

A. marina attB-like sequences for TG1 and R4 are both active in in vitro recombination

In each recombination reaction, substrates containing *attP* and the putative *attB* site were mixed in cognate pairs with different concentrations of purified R4 or TG1 integrase in the corresponding buffer and incubated overnight at 30 °C, as described in Methods. The expected recombination events and the nature of the products are shown in Fig. 2a. TG1 catalysed recombination between the substrates more efficiently than R4 (Fig. 2b). As expected because neither phage is an *Amycolatopsis* phage, the recombination efficiencies for each integrase were observably better when the *Streptomyces attB* sites were used (Fig. 2c) compared to the *A. marina attB* sites (Fig. 2b), particularly for TG1 integrase. Nevertheless, the presence of recombination activity indicated that both *A. marina attB* sites were functional and were likely to be active integration sites for integrative conjugation vectors.

In vivo integration

A. marina DSM45569 is unable to grow in the presence of apramycin, so integrating plasmids pHG4 and pJH1R4, containing the apramycin resistance determinant *aac(3)IV*, were constructed. Following the standard *Streptomyces* conjugation protocol (see Methods), a frequency of approximately 160 exconjugants/10^8 spores was obtained for the transfer of pHG4 (encoding TG1 integrase), while the conjugation efficiency of pJH1R4 (R4 integrase) was only 20 exconjugants/10^8 spores (Table 3). For each integration, six exconjugants were picked at random and streaked on SM (soya mannitol) agar containing apramycin. Genomic DNA was then prepared and used as the template in PCR reactions, in which the primer pairs of TG1-attL-Am-for/rev and R4-attL-Am-for/rev were used to test for the occurrence of recombination at the expected TG1 and R4 *attB* sites (Fig. 3). All PCR reactions using exconjugants as templates gave the expected band sizes. Sequencing (GATC, Germany) of the PCR products with the primers

Table 1 Oligonucleotides used in this study

Oligonucleotide	Sequence (5'-3')
pHG1A-for	CGAACGCATCGATTAATTAAGGAGGATCGTATGACGACCGTTCCCG
pHG1A-rev	CGTGGTGGGCGCTAGCCTCCTCTAGTCATCCGTCG
pHG1-for	ACTAGAGGAGGCTAGCTTCAATGGAGGAGATGATCGAGG
pHG1-rev	GCAGGTCGACTCTAGATCTCGCTACGCCGCTACG
pHG4-for	CGAACGCATCGATTAATTAAGCGGCCGCCATATGGAATTCGGTACCGCATGCAGATCTAGGAACTTCGAAGTTCCCGC
pHG4-rev	TGATTACGCCAAGCTTTCGACTCTAGAGTAAGCGTCACGG
pJH1R4-for	CTAGCGATTGCCATGACGTCGGAGCTGCTTACCAATGTC
pJH1R4-rev	AAGAGGCCCGCACCGATTCCAAGAGGCCGGCAACTAC
TG1-attB-Am-for	<u>TCGATCTCCAGTGCGGGCAAGACGTTCAACTGCACCGGCTGGAAGATCGGGACCACCGGACGAACGCA</u>
TG1-attB-Sa-for	<u>TCGATCAGCTCCGCGGGCAAGACCTTCTCCTTCACGGGGTGGAAGGTCGGCGGTGGAGCTCGGAGA</u>
R4-attB-Am-for	<u>GGTTGCCCATCACCATGCCGAAGCAGTGATAGAAGGGAACCGGGATGCAGGTGAGAAGGTGCTCGTGT</u>
R4-attB-Sp-for	<u>AGTTGCCCATGACCATGCCGAAGCAGTGGTAGAAGGGCACCGGCAGACAC</u>GGTGAGAAGGTGCTCGTGT
attB-rev	CTGCATCTCAACGCCTTCCGG
TG1-attP-for	AACCTTCACGCTCATGCC
TG1-attP-rev	GTCGAGATTCTCCGTCTCCTG
R4-attP-for	GATCGGTCTTGCCTTGCTC
R4-attP-rev	ACCCGCAGAGTGTACCCA
TG1-attL-Am-for	ACAACCCCACCGGCACCGTCTTCA
TG1-attL-Am-rev	AGTATAGGAACTTCGAAGCAGCTC
R4-attL-Am-for	CGGCCGGTGATGTTGACGT
R4-attL-Am-rev	TCGGCCGTCACGATGGTCA

The *attB* sequences are shown underlined

TG1-attL-Am-for and R4-attL-Am-for confirmed that the plasmids had integrated into the predicted *attB* sites for TG1 or R4 integrase within *A. marina* DSM45569 (Fig. 4).

Discussion

The lack of effective genetic engineering tools is considered one of the greatest hindrances in the search for new natural products from rare Actinomycetes [33–35]. Previous studies in rare Actinomycete species have focused mainly on the use of the well-characterised φC31-based integration vectors, and have mostly overlooked tools based on other phage integrases [36–38]. Additionally, the easy-handling conjugation methods used widely in *Streptomyces* gene transfer have shown little success in rare Actinomycetes, including species in the genus *Amycolatopsis*, so direct transformation with plasmids [39–41], or electroporation, has been the long-preferred method of gene transfer for species in this genus [5, 42–44]. However, the growing interest in the use of serine integrases for synthetic biology applications [10] has led to further research into expanding the pool of available enzymes and their potentials as genetic tools [45–47]. Therefore, within this study, we explored whether integrating vectors based on eight serine

integrases could be employed for the genetic engineering of *A. marina* DSM45569. Sequence analysis of the *A. marina* DSM45569 genome identified close matches to the *attB* sites used by TG1 and R4 integrases. Although conjugation frequencies were relatively low, integrating plasmids based on the TG1 and R4 recombination systems have been successfully integrated into the expected *attB* sites in *A. marina* DSM45569. Conservation between the *attB* sites for TG1 and R4 in a number of *Amycolatopsis* species is high, suggesting that plasmids with the integration systems from these phages should be widely useful in this genus, including the species which have garnered much interest as natural product producers, such as *Amycolatopsis balhimycina* [40], *Amycolatopsis orientalis* [20], and *A. mediterranei* [39].

As is common with serine integrase-mediated recombination, the *attB* sites in *A. marina* are located within open reading frames and potentially disrupt the gene. The TG1 *attB^{Am}* site is located within a gene predicted to encode a putative succinyldiaminopimelate transaminase (WP_091671332.1), and the R4 *attB^{Am}* site is located within a gene predicted to code for a fatty-acyl-CoA synthase (SFB62308.1). Compared to the wild-type (unintegrated) strain, the strains with integrated pHG4 or pJH1R4 did not show any difference in

Table 2 The original *attB* sites for integrases and results of BLAST search

Integrase	*attB* sites and the best hit from BLAST	Homology (%)	*E*-value
φ31	S.coelicolor / hit from A.marina	41	0.015
φJoe	S.venezuelae / hit from A.marina	30	0.60
Bxb1	Mycobacterium smegma / hit from A.marina	48	0.014
R4	S.parvulus / hit from A.marina	84	3e-11
SPBc	No hit		
SV1	S.venezuelae / hit from A.marina	32	0.17
TG1	S.avermitilis / hit from A.marina	62	0.001
TP901	Lactococcus lactis s / hit from A.marina	25	8.1

growth. However, further study is required to investigate the effects of TG1 or R4 plasmid recombination on both primary and secondary metabolism as, for example, the integration of φC31 integrase-based plasmids has been shown to have pleiotropic effects on bacterial physiology [48].

Currently, the following methods have been used to establish a gene transfer system in *Amycolatopsis* species: protoplast transformation, direct transformation of mycelia, electroporation, electroduction, and conjugation [41]. Among them, direct transformation and electroporation are most popular. While for the conjugation methods which have been widely used in *Streptomyces* species, there are few publications on conjugative transfer of vectors based on serine integrases in *Amycolatopsis*: pSET152 based on φC31 into *A. japonicum* MG417-CF17 (conjugation frequency = 2.4 × 10^4 exconjugants/10^8 spores) [19] and pDZL802 based on φBT1 into *A. mediterranei* U32 (4 × 10^3 exconjugants/10^8 spores) [21]. In this study, we successfully integrated plasmids into the *attB* sites for TG1 and R4 integrases by conjugation, which supplements the potential gene transfer methods that could be used in the genus *Amycolatopsis*, broadens the applicability of gene transfer systems except for the ones based on φC31 and φBT1 in

previous publications, and will definitely facilitate the genetic manipulation of *Amycolatopsis*. Although the recombination efficiencies were lower for TG1 and R4, the conjugation conditions could be further optimised to achieve better conjugation results, or the application of integration based vectors for direct transformation of mycelia could be explored since the integrative vectors, for example, pMEA100 [39] and pMEA300 [49], used in direct transformation are based on integrase and corresponding *attP* site as well.

Conclusions

In conclusion, we have identified highly conserved sequences of the *attB* sites for TG1 and R4 integrases within the genus *Amycolatopsis* and demonstrated their use in conjugative DNA transfer. The *A. marina* DSM45569 *attB* sites showed slightly lower recombination efficiencies in vitro than the previously identified *attB* sites from *Streptomyces* spp. However, this slight reduction is not enough by itself to explain the order of magnitude reductions in conjugation frequencies observed with *A. marina* compared to *Streptomyces* spp. (Table 3). Optimising conjugation conditions could increase the conjugation frequencies further. Alternatively,

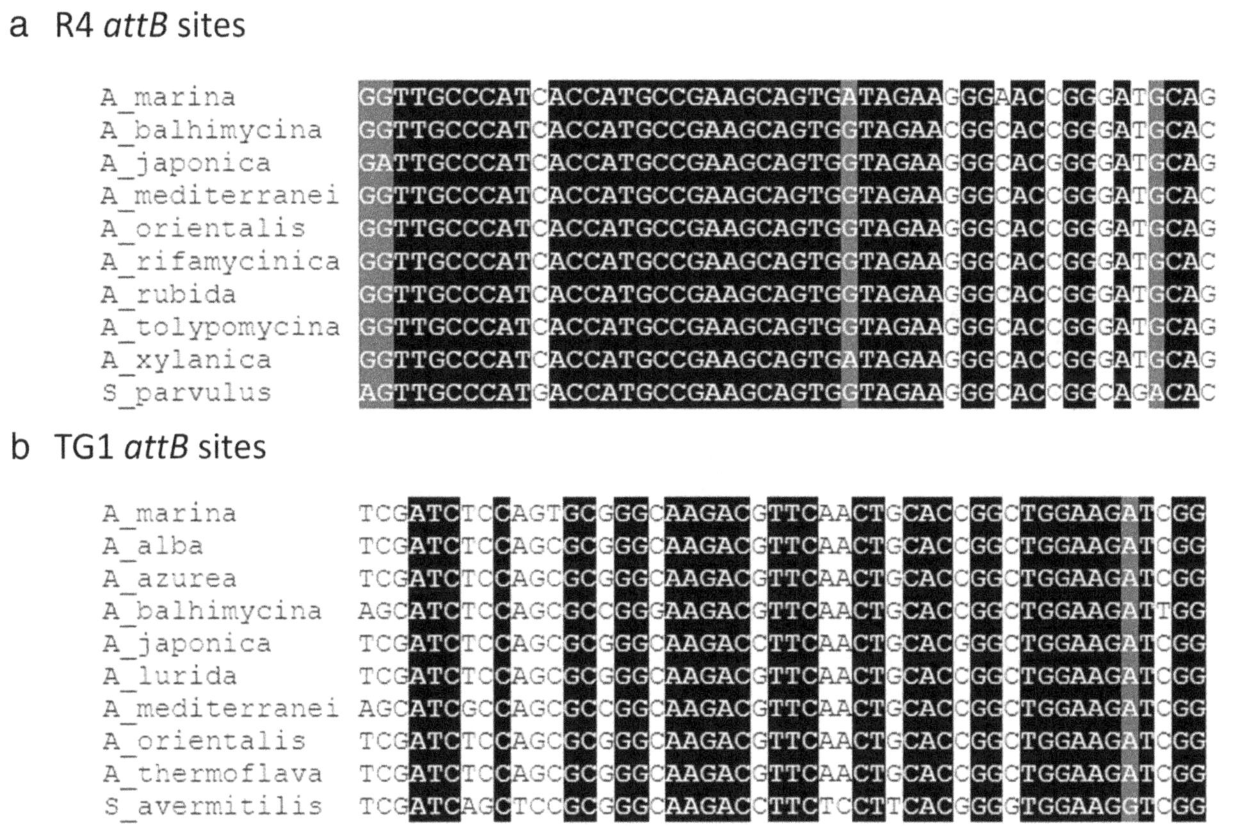

Fig. 1 Alignment of R4 and TG1 *attB* sites in *A. marina* DSM45569 and other *Amycolatopsis* species. **a**) GenBank accession nos. of DNA sequences: *Amycolatopsis balhimycina* (ARBH01000005.1), *Amycolatopsis japonica* (NZ_CP008953.1), *Amycolatopsis mediterranei* (NC_022116.1), *Amycolatopsis orientalis* (NZ_CP016174.1), *Amycolatopsis rifamycinica* (NZ_JMQI01000006.1), *Amycolatopsis rubida* (NZ_FOWC01000001.1), *Amycolatopsis tolypomycina* (NZ_FNSO01000004.1), *Amycolatopsis xylanica* (NZ_FNON01000002.1), and *S. parvulus* (CP015866.1); **b**) GenBank accession nos. of DNA sequences: *Amycolatopsis alba* (NZ_KB913032.1), *Amycolatopsis azurea* (MUXN01000005.1), *A. balhimycina* (ARBH01000007.1), *A. japonica* (NZ_CP008953.1), *Amycolatopsis lurida* (FNTA01000004.1), *A. mediterranei* (NC_022116.1), *A. orientalis* (NZ_CP01674.1), *Amycolatopsis thermoflava* (AXBH01000004.1), and *S. avermitilis* (NC_003155.5)

efficiently used *attB* sites for the widely used vectors, such as those based on φC31 *int/attP* could be incorporated into the *Amylcolatopsis* genome using TG1 or R4 integrating plasmids as described here. In short, this work shows that integrative vectors are viable and promising tools for the genetic engineering of rare Actinomycetes.

Methods

Bacterial strains and culture conditions

Plasmid propagation and subcloning was conducted using *E. coli* Top10 (F- *mcrA* Δ(*mrr-hsdRMS-mcrBC*) φ80*lacZ*ΔM15 Δ*lacX74 nupG recA1 araD139* Δ(*ara-leu*)7697 *galE15 galK16 rpsL*(Str^R) *endA1* λ$^-$). Plasmid conjugations from *E. coli* to *A. marina* DSM45569 were carried out using *E. coli* ET12567(pUZ8002) containing the plasmid to be transferred as the donor [50, 51], and conjugations from *E. coli* to *S. coelicolor* and *S. lividans* were used as control. *E. coli* strains were grown in Luria-Bertani broth (LB) or on LB agar at 37 °C.

A. marina DSM45569 was purchased from the German Collection of Microorganisms and Cell Cultures (DSMZ, Germany), and maintained on SM agar plates at 30 °C. Harvested spores were maintained long-term in 20% glycerol at -80 °C. Conjugations were plated on SM agar plates containing 10 mM MgCl$_2$, and ISP2 medium [52] was used for the preparation of genomic DNA [51].

DNA manipulation

E. coli transformation and gel electrophoresis were carried out as described previously [53]. Genomic DNA preparation from *Streptomyces* was performed following the salting out procedure in the *Streptomyces* manual [51]. Plasmids from *E. coli* were prepared using QIAprep® Spin Miniprep Kit (Qiagen, Germany) following the manufacturer's instructions. Polymerase Chain Reaction (PCR) was carried out using Phusion® High-Fidelity DNA Polymerase (NEB, USA) according to the manufacturer's instructions. The primers used in this study

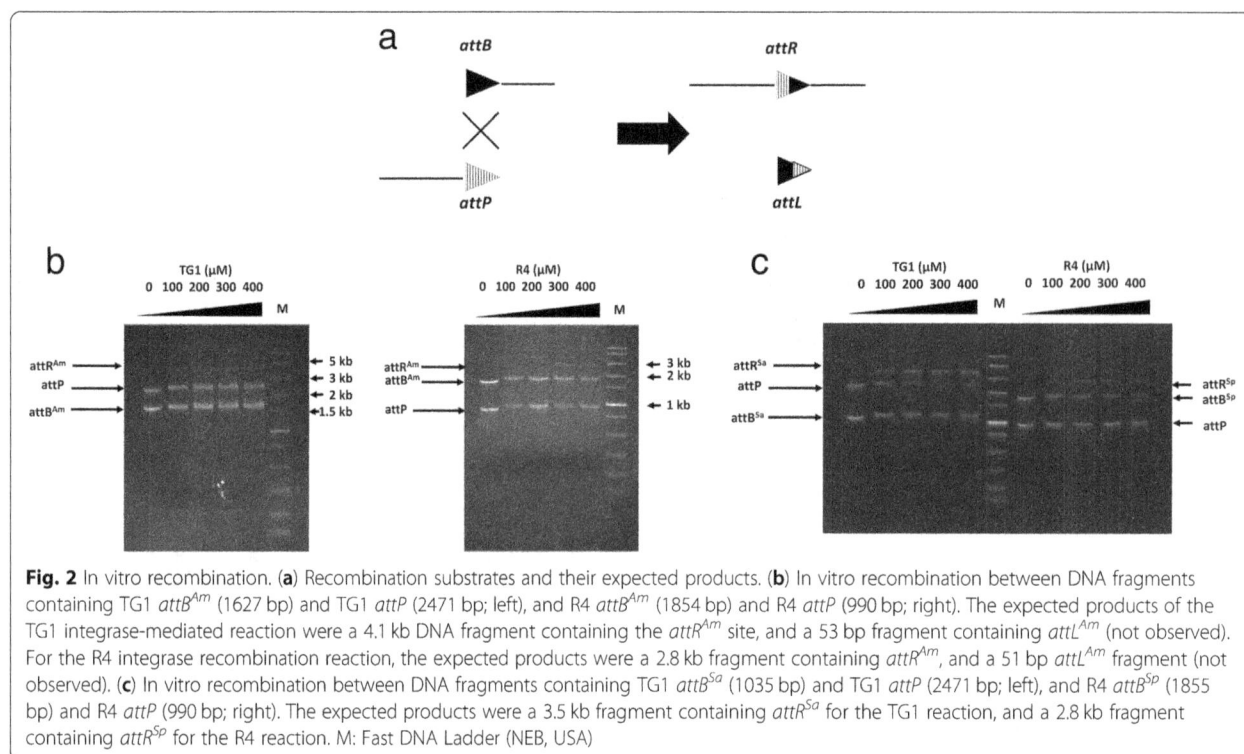

Fig. 2 In vitro recombination. (**a**) Recombination substrates and their expected products. (**b**) In vitro recombination between DNA fragments containing TG1 *attB^{Am}* (1627 bp) and TG1 *attP* (2471 bp; left), and R4 *attB^{Am}* (1854 bp) and R4 *attP* (990 bp; right). The expected products of the TG1 integrase-mediated reaction were a 4.1 kb DNA fragment containing the *attR^{Am}* site, and a 53 bp fragment containing *attL^{Am}* (not observed). For the R4 integrase recombination reaction, the expected products were a 2.8 kb fragment containing *attR^{Am}*, and a 51 bp *attL^{Am}* fragment (not observed). (**c**) In vitro recombination between DNA fragments containing TG1 *attB^{Sa}* (1035 bp) and TG1 *attP* (2471 bp; left), and R4 *attB^{Sp}* (1855 bp) and R4 *attP* (990 bp; right). The expected products were a 3.5 kb fragment containing *attR^{Sa}* for the TG1 reaction, and a 2.8 kb fragment containing *attR^{Sp}* for the R4 reaction. M: Fast DNA Ladder (NEB, USA)

are listed in Table 1. DNA samples were purified by the QIAquick Gel Extraction Kit (Qiagen, Germany).

Plasmid construction

The integrating plasmid pHG4 contains the TG1 *int/attP* locus and the apramycin-resistance gene (*aac(3)IV*) for selection (Fig. 5a). The fragment containing *oriT*, *aac(3)IV* and TG1 *int/attP* was amplified from plasmid pBF20 [54] using the primer pair pHG4-for/pHG4-rev. The fragment was joined via In-Fusion cloning to the 3344 bp HindIII-PacI fragment from pBF22 [54] (containing the *E. coli* plasmid replication origin, the *bla* gene encoding resistance to ampicillin and the *actII-orf4/act1p* expression cassette) to form the plasmid pHG4.

To construct the integrating plasmid pJH1R4 (Fig. 5a), pSET152 [55] was cut with AatII and PvuI to remove the φC31 *attP* site and integrase gene. R4 phage lysate was used as the template in a PCR with the primers pJH1R4-for and pJH1R4-rev to amplify the R4 *attP* site and integrase coding region. The PCR product was

Table 3 Conjugation efficiency of pHG4 and pJH1R4 in different species

Exconjugants/10^8 spores	pHG4	pJH1R4
A. marina	160	20
Streptomyces coelicolor	1.47×10^3	3.28×10^4
Streptomyces . lividans	1.56×10^3	3.33×10^4

joined to the AatII-PvuI fragment from pSET152 via In-Fusion cloning.

The plasmid pHG1 (Fig. 5c) was used as the template in PCR to amplify *attB*-containing sequences (Fig. 5d) for in vitro recombination assays. This plasmid was initially constructed for the expression of *EryF*. The *eryF* gene was amplified from *Saccharopolyspora erythraea* BIOT-0666 genomic DNA using the primer pair pHG1A-for/pHG1A-rev, and inserted by In-Fusion cloning into pBF20 [54] cut with NheI and PacI to form the plasmid pHG1A. The 3785 bp fragment containing the φC31 *int/attP* and hygromycin resistance gene was amplified from plasmid pBF27C [54], using the primer pair pHG1-for and pHG1-rev. Plasmid pHG1A was digested with XbaI and NheI, and the 5668 bp fragment was ligated with the 3785 bp PCR fragment from pBF27C by In-Fusion cloning to give the plasmid pHG1.

In vitro recombination assays

In vitro recombination assays were performed using PCR-amplified DNA fragments containing the *attB* and *attP* attachment sites located at the ends. Recombination between the *attP* and *attB* sites joined the two fragments to give a product whose length was almost the sum of the substrates (Fig. 2a). To generate the *attB*-containing substrates, the forward primer, TG1-attB-Am-for, contained the closest match in the *A. marina* genome to the characterised TG1 *attB* site from *S. avermitilis*, TG1 *attB^{Sa}* [30] (Fig. 1). TG1-attB-Am-for also had a

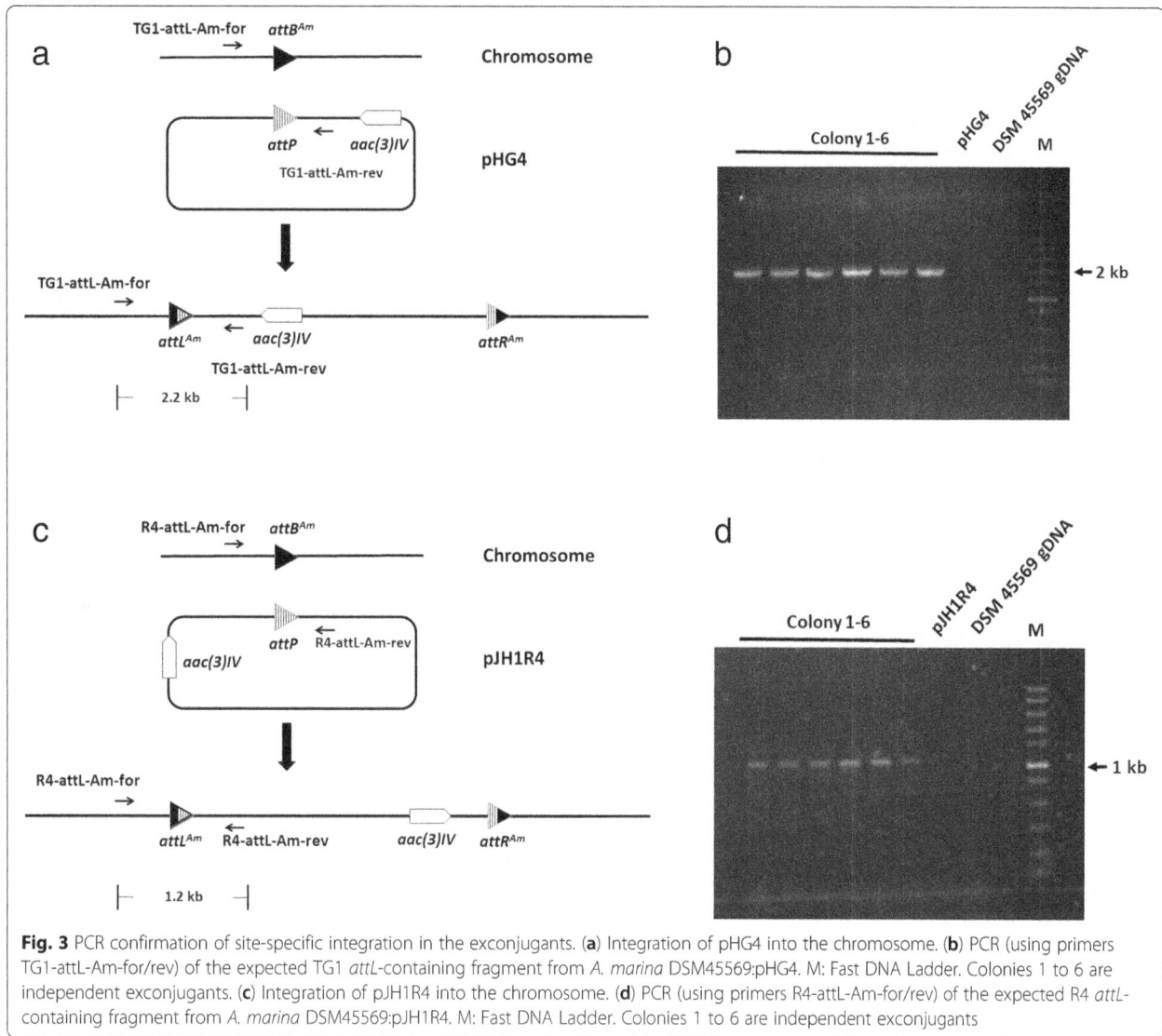

Fig. 3 PCR confirmation of site-specific integration in the exconjugants. (**a**) Integration of pHG4 into the chromosome. (**b**) PCR (using primers TG1-attL-Am-for/rev) of the expected TG1 *attL*-containing fragment from *A. marina* DSM45569:pHG4. M: Fast DNA Ladder. Colonies 1 to 6 are independent exconjugants. (**c**) Integration of pJH1R4 into the chromosome. (**d**) PCR (using primers R4-attL-Am-for/rev) of the expected R4 *attL*-containing fragment from *A. marina* DSM45569:pJH1R4. M: Fast DNA Ladder. Colonies 1 to 6 are independent exconjugants

sequence identical to the 3′ end of the act1p element from plasmid pHG1, which was used as a template for PCR (Fig. 5c). Similarly, the forward primer R4-attB-Am-for contained the closest match in the *A. marina* genome to the characterised R4 *attB* site from *S.*

parvulus, R4 *attB^Sp* [32] (Fig. 1). R4-attB-Am-for also had a sequence identical to the 3′ end of ActII-orf4 element from the template plasmid pHG1 (Fig. 5d). Forward primers TG1-attB-Sa-for and R4-attB-Sp-for were used to create positive control recombination substrates

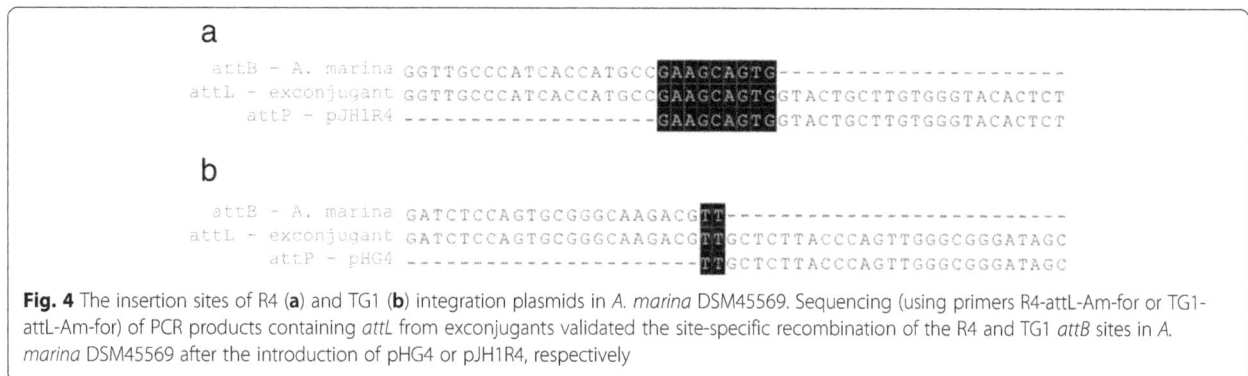

Fig. 4 The insertion sites of R4 (**a**) and TG1 (**b**) integration plasmids in *A. marina* DSM45569. Sequencing (using primers R4-attL-Am-for or TG1-attL-Am-for) of PCR products containing *attL* from exconjugants validated the site-specific recombination of the R4 and TG1 *attB* sites in *A. marina* DSM45569 after the introduction of pHG4 or pJH1R4, respectively

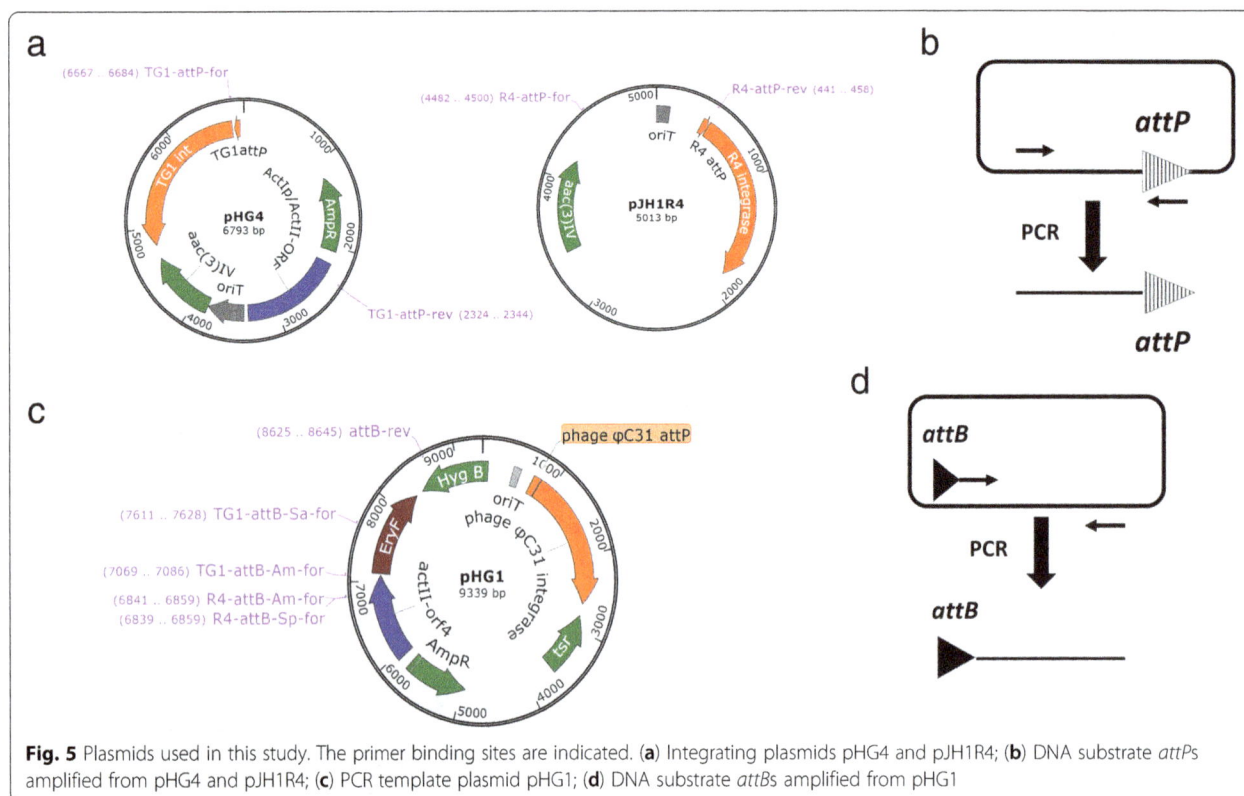

Fig. 5 Plasmids used in this study. The primer binding sites are indicated. (**a**) Integrating plasmids pHG4 and pJH1R4; (**b**) DNA substrate *attP*s amplified from pHG4 and pJH1R4; (**c**) PCR template plasmid pHG1; (**d**) DNA substrate *attB*s amplified from pHG1

containing the TG1 and R4 *attB* sites originally found in *S. avermitilis* [30] and *S. parvulus* [32] respectively. The reverse primer used to generate all the *attB*-containing substrates (attB-rev) was located within the *hyg* gene of pHG1; the amplified products were 1627 bp (TG1 *attB*^{Am}), 1035 bp (TG1 *attB*^{Sa}), 1854 bp (R4 *attB*^{Am}) and 1855 bp (R4 *attB*^{Sp}). The DNA fragments containing the *attP* sites were prepared as follows; the TG1-*attP* fragment (2471 bp) was amplified using the primer pair TG1-attP-for/TG1-attP-rev with pHG4 as the template, and the R4-*attP* fragment (990 bp) was amplified using the primer pair R4-attP-for/R4-attP-rev with pJH1R4 as the template (Fig. 5b). Note that other than the *attB* and *attP* sites, none of the substrates contained any DNA that should interact specifically with the integrases. Moreover, each fragment was designed to be easily identifiable by molecular weight.

The integrases were purified as described previously [27, 56]. All recombination reactions were in 20 μl final volume. Recombination reactions of TG1 substrates were carried out in TG1 RxE buffer (20 mM Tris [pH 7.5], 25 mM NaCl, 1 mM dithiothreitol [DTT], 10 mM spermidine, 10 mM EDTA, 0.1 mg/ml bovine serum albumin [BSA]) [57], and recombination reactions of R4 substrates were carried out in buffer containing 20 mM Tris-HCl (pH 7.5), 50 mM NaCl, 10 mM spermidine, 5 mM $CaCl_2$ and 50 mM DTT [27]. Integrase was added at the concentrations indicated. Recombination

substrates were used at 50 ng each per reaction. Reactions were incubated at 30 °C overnight and then heated (10 min, 75 °C) to denature integrase. The reaction mixtures were loaded on a 0.8% agarose gel in Tris/Borate/EDTA (TBE) buffer (90 mM Tris base, 90 mM boric acid and 2 mM EDTA) containing ethidium bromide for electrophoretic separation.

Abbreviations

BSA: Bovine serum albumin; PCR: Polymerase chain reaction; RDF: Recombination directionality factor; SM: Soya Mannitol; TBE: Tris/Borate/EDTA

Acknowledgements
Not applicable.

Funding
This work was supported by the Biotechnology and Biological Sciences Research Council project grant BB/K003356/1, and Buvani Murugesan acknowledges the receipt of a summer studentship from the Department of Biology, University of York. These funding organisations were not involved in the design of the study and collection, analysis, and interpretation of data, or in writing the manuscript.

Authors' contributions
HG designed the study, performed the experiments and wrote the manuscript. BM participated in TG1 experiments. JH constructed pJH1R4 and purified R4 integrase. SKE purified TG1 integrase. WMS and MCMS revised the manuscript. All authors read and approved the final manuscript.

Ethics approval and consent to participate
Not applicable.

Consent for publication

Not applicable.

Competing interests

The authors declare that they have no competing interests.

Author details

[1]Department of Biology, University of York, York, North Yorkshire YO10 5DD, UK. [2]Institute of Molecular, Cell and Systems Biology, University of Glasgow, Glasgow G12 8QQ, UK. [3]Present address: School of Science, Engineering & Design, Teesside University, Middlesbrough TS1 3BX, UK.

References

1. Jones SE, Ho L, Rees CA, Hill JE, Nodwell JR, Elliot MA. *Streptomyces* exploration is triggered by fungal interactions and volatile signals. Elife. 2017;6:e21738.
2. Zarins-Tutt JS, Barberi TT, Gao H, Mearns-Spragg A, Zhang L, Newman DJ, Goss RJM. Prospecting for new bacterial metabolites: a glossary of approaches for inducing, activating and upregulating the biosynthesis of bacterial cryptic or silent natural products. Nat Prod Rep. 2016;33(1):54–72.
3. Jose PA, Jebakumar SRD. Non-streptomycete actinomycetes nourish the current microbial antibiotic drug discovery. Front Microbiol. 2013;4:240.
4. Jung HM, Kim SY, Moon HJ, Oh DK, Lee JK. Optimization of culture conditions and scale-up to pilot and plant scales for vancomycin production by *Amycolatopsis orientalis*. Appl Microbiol Biotechnol. 2007; 77(4):789–95.
5. Li C, Liu X, Lei C, Yan H, Shao Z, Wang Y, Zhao G, Wang J, Ding X. RifZ (AMED_0655) is a pathway-specific regulator for rifamycin biosynthesis in *Amycolatopsis mediterranei*. J Appl Environ Microbiol. 2017;83(8):e03201–16.
6. Xu X, Han L, Zhao L, Chen X, Miao C, Hu L, Huang X, Chen Y, Li Y. Echinosporin antibiotics isolated from *Amycolatopsis* strain and their antifungal activity against root-rot pathogens of the *Panax notoginseng*. Folia Microbiol. 2018;64(2):171–5.
7. Hashizume H, Iijima K, Yamashita K, Kimura T, S-i W, Sawa R, Igarashi M. Valgamicin C, a novel cyclic depsipeptide containing the unusual amino acid cleonine, and related valgamicins a, T and V produced by *Amycolatopsis* sp. ML1-hF4. J Antibiot. 2018;71(1):129–34.
8. Li X, Wu X, Zhu J, Shen Y. Amexanthomycins A–J, pentangular polyphenols produced by *Amycolatopsis mediterranei* S699Δ rifA. Appl Microbiol Biotechnol. 2018;102(2):689–702.
9. Adamek M, Alanjary M, Sales-Ortells H, Goodfellow M, Bull AT, Winkler A, Wibberg D, Kalinowski J, Ziemert N. Comparative genomics reveals phylogenetic distribution patterns of secondary metabolites in *Amycolatopsis* species. BMC Genomics. 2018;19(1):426.
10. Fogg PC, Colloms S, Rosser S, Stark M, Smith MC. New applications for phage integrases. J Mol Biol. 2014;426(15):2703–16.
11. Colloms SD, Merrick CA, Olorunniji FJ, Stark WM, Smith MC, Osbourn A, Keasling JD, Rosser SJ. Rapid metabolic pathway assembly and modification using serine integrase site-specific recombination. Nucleic Acids Res. 2013; 42(4):e23.
12. Smith MC. Phage-encoded serine integrases and other large serine recombinases. In: Craig NL, Chandler M, Gellert M, Lambowitz AM, Rice PA, Sandmeyer SB, editors. Mobile DNA III. Washington, DC: ASM Press; 2015. p. 253–72.
13. Saha S, Zhang W, Zhang G, Zhu Y, Chen Y, Liu W, Yuan C, Zhang Q, Zhang H, Zhang L. Activation and characterization of a cryptic gene cluster reveals a cyclization cascade for polycyclic tetramate macrolactams. Chem Sci. 2017;8(2):1607–12.
14. Sosio M, Giusino F, Cappellano C, Bossi E, Puglia AM, Donadio S. Artificial chromosomes for antibiotic-producing actinomycetes. Nat Biotechnol. 2000; 18(3):343–5.
15. Hong Y, Hondalus MK. Site-specific integration of *Streptomyces* ΦC31 integrase-based vectors in the chromosome of *Rhodococcus equi*. FEMS Microbiol Lett. 2008;287(1):63–8.
16. Thorpe HM, Smith MC. *In vitro* site-specific integration of bacteriophage DNA catalyzed by a recombinase of the resolvase/invertase family. Proc Natl Acad Sci U S A. 1998;95(10):5505–10.
17. Rutherford K, Van Duyne GD. The ins and outs of serine integrase site-specific recombination. Curr Opin Struct Biol. 2014;24:125–31.
18. Baltz RH. *Streptomyces* temperate bacteriophage integration systems for stable genetic engineering of actinomycetes (and other organisms). J Ind Microbiol Biotechnol. 2012;39(5):661–72.
19. Stegmann E, Pelzer S, Wilken K, Wohlleben W. Development of three different gene cloning systems for genetic investigation of the new species *Amycolatopsis japonicum* MG417-CF17, the ethylenediaminedisuccinic acid producer. J Biotechnol. 2001;92(2):195–204.
20. Lei X, Zhang C, Jiang Z, Li X, Shi Y, Xie Y, Wang L, Hong B. Complete genome sequence of *Amycolatopsis orientalis* CPCC200066, the producer of norvancomycin. J Biotechnol. 2017;247:6–10.
21. Li C, Zhou L, Wang Y, Zhao G, Ding X. Conjugation of φBT1-derived integrative plasmid pDZL802 in *Amycolatopsis mediterranei* U32. Bioengineered. 2017;8(5):549–54.
22. Bian J, Li Y, Wang J, Song FH, Liu M, Dai HQ, Ren B, Gao H, Hu X, Liu ZH. *Amycolatopsis marina* sp. nov., an actinomycete isolated from an ocean sediment. Int J Syst Evol Microbiol. 2009;59(3):477–81.
23. Romano S, Jackson S, Patry S, Dobson A. Extending the "one strain many compounds"(OSMAC) principle to marine microorganisms. Mar Drugs. 2018; 16(7):244.
24. Combes P, Till R, Bee S, Smith MC. The *Streptomyces* genome contains multiple pseudo-attB sites for the φC31-encoded site-specific recombination system. J Bacteriol. 2002;184(20):5746–52.
25. Fogg PC, Haley JA, Stark WM, Smith MC. Genome integration and excision by a new *Streptomyces* bacteriophage, φJoe. Appl Environ Microbiol. 2016; 83(5):e02767–16.
26. Ghosh P, Pannunzio NR, Hatfull GF. Synapsis in phage Bxb1 integration: selection mechanism for the correct pair of recombination sites. J Mol Biol. 2005;349(2):331–48.
27. Takamasa M, Yayoi H, Yang YZ, Tomoyasu N, Munehiko A, Hideo T, Makoto S. *In vivo* and *in vitro* characterization of site-specific recombination of actinophage R4 integrase. J Gen Appl Microbiol. 2011; 57(1):45–57.
28. Lazarevic V, Düsterhöft A, Soldo B, Hilbert H, Mauel C, Karamata D. Nucleotide sequence of the *Bacillus subtilis* temperate bacteriophage SPβc2. Microbiology. 1999;145(5):1055–67.
29. Fayed B, Younger E, Taylor G, Smith MC. A novel Streptomyces spp. integration vector derived from the *S. venezuelae* phage, SV1. BMC Biotechnol. 2014;14(1):51.
30. Morita K, Yamamoto T, Fusada N, Komatsu M, Ikeda H, Hirano N, Takahashi H. *In vitro* characterization of the site-specific recombination system based on actinophage TG1 integrase. Mol Gen Genomics. 2009;282:607–16.
31. Christiansen B, Johnsen M, Stenby E, Vogensen F, Hammer K. Characterization of the lactococcal temperate phage TP901-1 and its site-specific integration. J Bacteriol. 1994;176(4):1069–76.
32. Shirai M, Nara H, Sato A, Aida T, Takahashi H. Site-specific integration of the actinophage R4 genome into the chromosome of *Streptomyces parvulus* upon lysogenization. J Bacteriol. 1991;173(13):4237–9.
33. Li X, Zhou X, Deng Z. Vector systems allowing efficient autonomous or integrative gene cloning in *Micromonospora* sp. strain 40027. Appl Environ Microbiol. 2003;69(6):3144–51.
34. Marcone GL, Carrano L, Marinelli F, Beltrametti F. Protoplast preparation and reversion to the normal filamentous growth in antibiotic-producing uncommon actinomycetes. J Antibiot. 2010;63:83–8.
35. Wolf T, Gren T, Thieme E, Wibberg D, Zemke T, Pühler A, Kalinowski J. Targeted genome editing in the rare actinomycete *Actinoplanes* sp. SE50/110 by using the CRISPR/Cas9 system. J Biotechnol. 2016;231:122–8.
36. Wagner N, Oßwald C, Biener R, Schwartz D. Comparative analysis of transcriptional activities of heterologous promoters in the rare actinomycete *Actinoplanes friuliensis*. J Biotechnol. 2009;142:200–4.
37. Anzai Y, Iizaka Y, Li W, Idemoto N, Tsukada SI, Koike K, Kinoshita K, Kato F. Production of rosamicin derivatives in *Micromonospora rosaria* by introduction of d-mycinose biosynthetic gene with ΦC31-derived integration vector pSET152. J Ind Microbiol Biotechnol. 2009;36:1013–21.
38. Kim DY, Huang YI, Choi SU. Cloning of metK from *Actinoplanes teichomyceticus* ATCC31121 and effect of its high expression on antibiotic production. J Microbiol Biotechnol. 2011;21(12):1294–8.
39. Madoń J, Hütter R. Transformation system for *Amycolatopsis (Nocardia) mediterranei*: direct transformation of mycelium with plasmid DNA. J Bacteriol. 1991;173(20):6325–31.

40. Kilian R, Frasch H-J, Kulik A, Wohlleben W, Stegmann E. The VanRS homologous two-component system VnlRSAb of the glycopeptide producer *Amycolatopsis balhimycina* activates transcription of the vanHAXSc genes in *Streptomyces coelicolor*, but not in *A. balhimycina*. Microb Drug Resist. 2016;22(6):499–509.

41. Malhotra S, Lal R. The genus *Amycolatopsis*: indigenous plasmids, cloning vectors and gene transfer systems. Indian J Microbiol. 2007;47(1):3–14.

42. Kumari R, Singh P, Lal R. Genetics and genomics of the genus *Amycolatopsis*. Indian J Microbiol. 2016;56(3):233–46.

43. Shen Y, Huang H, Zhu L, Luo M, Chen D. Type II thioesterase gene (ECO-orf27) from *Amycolatopsis orientalis* influences production of the polyketide antibiotic, ECO-0501 (LW01). Biotechnol Lett. 2012;34(11):2087–91.

44. Lee K, Lee B, Ryu J, Kim D, Kim Y, Lim S. Increased vancomycin production by overexpression of MbtH-like protein in *Amycolatopsis orientalis* KFCC 10990P. Lett Appl Microbiol. 2016;63(3):222–8.

45. Mandali S, Dhar G, Avliyakulov NK, Haykinson MJ, Johnson RC. The site-specific integration reaction of *Listeria* phage A118 integrase, a serine recombinase. Mob DNA. 2013;4(2).

46. Yoon B, Kim I, Nam JA, Chang HI, Ha CH. *In vivo* and *in vitro* characterization of site-specific recombination of a novel serine integrase from the temperate phage EFC-1. Biochem Biophys Res Commun. 2016;473:336–41.

47. Yang L, Nielsen AAK, Fernandez-Rodriguez J, McClune CJ, Laub MT, Lu TK, Voigt CA. Permanent genetic memory with> 1-byte capacity. Nat Methods. 2014;11:1261–6.

48. Talà A, Damiano F, Gallo G, Pinatel E, Calcagnile M, Testini M, Fico D, Rizzo D, Sutera A, Renzone G. Pirin: a novel redox-sensitive modulator of primary and secondary metabolism in *Streptomyces*. Metab Eng. 2018;48:254–68.

49. Vrijbloed J, Madoń J, Dijkhuizen L. Transformation of the methylotrophic actinomycete *Amycolatopis methanolica* with plasmid DNA: stimulatory effect of a pMEA300-encoded gene. Plasmid. 1995;34(2):96–104.

50. MacNeil D. Characterization of a unique methyl-specific restriction system in *Streptomyces avermitilis*. J Bacteriol. 1988;170(12):5607–12.

51. Kieser T, Bibb M, Buttner M, Chater K, Hopwood D. Practical *Streptomyces* genetics. Norwich: the John Innes Foundation; 2000.

52. Shirling E, Gottlieb D. Methods for characterization of *Streptomyces* species. Int J Syst Evol Microbiol. 1966;16(3):313–40.

53. Sambrook J, Russell DW. Molecular cloning: a laboratory manual. New York: Cold Spring Harbor Laboratory Press; 2001.

54. Fayed B, Ashford DA, Hashem AM, Amin MA, El Gazayerly ON, Gregory MA, Smith MC. Multiplexed integrating plasmids for engineering the erythromycin gene cluster for expression in *Streptomyces* and combinatorial biosynthesis. Appl Environ Microbiol. 2015;81(24):8402–13.

55. Wilkinson CJ, Hughes-Thomas ZA, Martin CJ, Bohm I, Mironenko T, Deacon M, Wheatcroft M, Wirtz G, Staunton J, Leadlay PF. Increasing the efficiency of heterologous promoters in actinomycetes. J Mol Microbiol Biotechnol. 2002;4(4):417–26.

56. Rowley P, Smith M, Younger E, Smith MC. A motif in the C-terminal domain of φC31 integrase controls the directionality of recombination. Nucleic Acids Res. 2008;36(12):3879–91.

57. Morita K, Morimura K, Fusada N, Komatsu M, Ikeda H, Hirano N, Takahashi H. Site-specific genome integration in alphaproteobacteria mediated by TG1 integrase. Appl Microbiol Biotechnol. 2012;93(1):295–304.

5

Genome-wide sequencing and metabolic annotation of *Pythium irregulare* CBS 494.86: understanding Eicosapentaenoic acid production

Bruna S. Fernandes[1,2*], Oscar Dias[2], Gisela Costa[2], Antonio A. Kaupert Neto[3], Tiago F. C. Resende[2], Juliana V. C. Oliveira[3], Diego M. Riaño-Pachón[4], Marcelo Zaiat[5*], José G. C. Pradella[6] and Isabel Rocha[2*]

Abstract

Background: *Pythium irregulare* is an oleaginous Oomycete able to accumulate large amounts of lipids, including Eicosapentaenoic acid (EPA). EPA is an important and expensive dietary supplement with a promising and very competitive market, which is dependent on fish-oil extraction. This has prompted several research groups to study biotechnological routes to obtain specific fatty acids rather than a mixture of various lipids. Moreover, microorganisms can use low cost carbon sources for lipid production, thus reducing production costs. Previous studies have highlighted the production of EPA by *P. irregulare*, exploiting diverse low cost carbon sources that are produced in large amounts, such as vinasse, glycerol, and food wastewater. However, there is still a lack of knowledge about its biosynthetic pathways, because no functional annotation of any *Pythium* sp. exists yet. The goal of this work was to identify key genes and pathways related to EPA biosynthesis, in *P. irregulare* CBS 494.86, by sequencing and performing an unprecedented annotation of its genome, considering the possibility of using wastewater as a carbon source.

Results: Genome sequencing provided 17,727 candidate genes, with 3809 of them associated with enzyme code and 945 with membrane transporter proteins. The functional annotation was compared with curated information of oleaginous organisms, understanding amino acids and fatty acids production, and consumption of carbon and nitrogen sources, present in the wastewater. The main features include the presence of genes related to the consumption of several sugars and candidate genes of unsaturated fatty acids production.

Conclusions: The whole metabolic genome presented, which is an unprecedented reconstruction of *P. irregulare* CBS 494.86, shows its potential to produce value-added products, in special EPA, for food and pharmaceutical industries, moreover it infers metabolic capabilities of the microorganism by incorporating information obtained from literature and genomic data, supplying information of great importance to future work.

Keywords: Eicosapentaenoic acid, Metabolic annotation, *Pythium irregulare*, unsaturated fatty acids, whole-genome sequence

* Correspondence: brunasofer@hotmail.com; zaiat@sc.ups.br; irocha@deb.uminho.pt
[1]Department of Civil and Environmental Engineering, Federal University of Pernambuco, Recife, PE, Brazil
[5]Biological Processes Laboratory, Center for Research, Development and Innovation in Environmental Engineering, São Carlos School of Engineering (EESC), University of São Paulo, São Carlos, SP, Brazil
[2]Centre of Biological Engineering, Universidade do Minho, Braga, Portugal

Background

Pythium irregulare is an oleaginous diploid Oomycete, a microscopic Stramenopiles [1, 2] and pathogen of various crops [1], including *Arabidopsis* plants [3]. *P. irregulare* has the potential to be industrially used to produce lipids because it is able to accumulate a large amount of these compounds, including Eicosapentaenoic acid (EPA) [4]. EPA $(C_{20}H_{30}O_2)$ is a 20-carbon polyunsaturated fatty acid with five *cis* double bonds, with the first double bond located at the third carbon from the omega end, which justifies its classification as an omega-3 fatty acid. The Food and Agriculture Organization of the United Nations recommends ingestion up to 500 mg per day of EPA and DHA (Docosahexaenoic acid) in the early years of life and for prevention of cardiovascular diseases [5], as it is not naturally synthesized in humans. Omega-3 fatty acids are important dietary supplements, with high selling prices (US$ 600 – US$ 4000 per kg of omega-3) [6], and a promising and very competitive market [7]. The expected omega-3 revenue is estimated at US$ 2.7 billion by 2020, with a Compound Annual Growth Rate (CAGR) of 17.5% (2014–2020), just in the pharmaceutical market [8]. This scenario has prompted several groups to search for alternative ways to produce omega-3, particularly EPA. Microorganisms are very attractive sources of EPA, because they can be driven to produce specific fatty acids rather than a mixture of various lipids, using low cost carbon sources without presence of heavy metals in the cultivated medium. This can reduce the cost of lipid extraction and purification and help to reduce the dependence on fish-oil. Some microorganisms have been studied with this goal, such as *Mortierella alpine, Mortierella elongate, Monochrysis luteri, Pseudopedinella sp., Coccolithus huxleyi, Cricosphaera carterae, Monodus sub-terraneous, Nannochlorus sp., Porphyrium cruentum, Cryptomonas muculata, Cryptomonas sp., Rhodomonas leans,* and *Pythium irregulare* [9–12].

Some studies have indicated the possibility of producing EPA using *P. irregulare*, exploiting diverse abundant low-cost carbon sources including wastewaters such as vinasse from corn-meal ethanol production, glycerol, wastewater from the food industry, and several sugars [4, 13, 14]. However, there is still a lack of knowledge about the biosynthetic pathways for EPA in this microorganism, and Stramenopiles, in general. This taxonomical order covers very diverse ecological niches and lifestyles ranging from photosynthetic diatoms and brown algae to filamentous saprophytic and pathogenic oomycetes [15].

Hereto, *Pythium irregulare* DAOM BR486 is the only *P. irregulare* strain sequenced and annotated at the National Center for Biotechnology Information - NCBI database (Bioproject number: PRJNA169053). Its annotation was performed automatically using MAKER v.203 tool [16] and was based on *Pythium ultimum* Genome database [17] [18]. However, it aimed at evaluating the pathogenicity of oomycetes, disregarding the annotation of metabolic functions. Moreover the automatic annotation can produce false positive and erroneous data [19].

The goal of this work was to identify the key genes and pathways related to EPA biosynthesis, as well as other metabolites of biotechnological importance, in *P. irregulare* strain CBS 494.86, including amino acids and fatty acids production, and consumption of carbon and nitrogen sources, present in the wastewater. As this strain was unexplored and unpublished, its genome was thus sequenced and annotated, with a special emphasis in metabolic functions that were thoroughly manually curated. Its possible application was examined through a biotechnological perspective, using as carbon sources vinasse from bioethanol production process a low cost wastewater produced globally in high amount, such as vinasse, from bioethanol process production, glycerol, from biodiesel wastewater (obtained in the biodiesel production), and several food and beverage wastewaters (Additional file 1: Figure S1).

The genome-wide functional annotation, presented in this manuscript, was corroborated with evidences from literature, thus allowing its use as the basis for the reconstruction of a genome scale metabolic model.

Results

Whole-genome sequencing

The species classification of the isolate selected for genome sequencing was confirmed by Sanger sequencing and analysis of the cytochrome oxidase I gene (COI) and internal transcribed spacer regions (ITS1 and ITS2), which were aligned, using the nucleotide Basic Local Alignment Search Tool (BLAST) [20], with the NCBI genomic database. The ITS sequence was 98% identical to *P. irregulare* CBS 250.28 (sequence ID: AY598702.2) with coverage of 86%, wheras the COI sequence was 99% identical to *P. irregulare* CBS 493.86 / CBS 250.28 (sequence ID: GU071821.1) with coverage of 99%.

The genomic DNA from the cultivated *P. irregulare* strain CBS 494.86 was extracted and sequenced on a HiSeq2500 using a single paired-end library (2x100bp). The HiSeq2500 produced 58,990,406 sequenced fragments 2x100bp which were used for assembly; 43,436, 209 of these remained after quality control. Genome assembly resulted in 9658 scaffolds larger than 500 bp, with an N50 of 13.460 bp (6.653 scaffolds longer than 1 Kbp) and a total genome size of 47.121.789 bp (Table 1). The coverage evaluation of the gene space by our assembly was performed using BUSCÒ v3 [21] with two sets of conserved genes, one for all eukaryotes with 303 conserved genes and one for protists with 215 conserved

Table 1 *Pythium irregulare* CBS 494.86 genome statistics

Assembly statistics for genome	
Estimated genome size	47.1 Mb
Number of scaffolds	9658
Number of scaffolds (≥ 1000 bp)	6653
Number of scaffolds (≥ 5000 bp)	2204
Number of scaffolds (> = 10,000 bp)	1192
Number of scaffolds (≥ 25,000 bp)	334
Number of scaffolds (≥ 50,000 bp)	64
Total length	45,784,433
Total length (≥1000 bp)	43,603,753
Total length (≥ 5000 bp)	33,709,764
Total length (≥ 10,000 bp)	26,550,523
Total length (≥ 25,000 bp)	13,047,962
Total length (≥ 50,000 bp)	4,077,371
Largest scaffolds	191,247
GC (%)	53.43
Scaffolds N50	13,460
Scaffolds N75	4686
Scaffolds L50	878
Scaffolds L75	2334
# N's per 100 kbp	262.38
Number of genes	17,758
Number of mRNAs	17,727
Number of tRNAs	29
Number of rRNAs	2
Total CDS length	23,740,968
Total Gene length	23,750,652
Average gene length	1337.54
Longest gene	26,310
Shortest gene	42
% of genome covered by genes	51.87
% of genome covered by CDS	51.85
Average number of exons	2.58
Max number of exons	42

genes. For both datasets our assembly showed over 90% of coverage of complete BUSCÒs (Additional file 2). The gene space coverage observed in this project assembly is similar to that of the published genome sequence of *P. irregulare* strain DAOM BR486, and the size of their haploid genomes is also similar [21] (Additional file 2). Gene prediction was carried out with Augustus [22], which was trained by exploiting available data from the Buell lab of another strains of *P. irregulare and P. ultimum* [23] [17], and resulted in the prediction of 17,008 protein-coding genes and 29 tRNA genes. The different copies of the ribosomal operon were collapsed into a

single copy. In strain CBS 493.86, 95.4% of the predicted genes can be mapped to the genome of strain CBS 805.95. The sequenced genome was deposited in NCBI (Bioproject number: PRJNA371716).

Metabolic annotation

The *merlin* software (metabolic models reconstruction using genome-scale information) [24] was used for the functional annotation of proteins with metabolic functions encoded in the genome of *P. irregulare*.

For the analysis and interpretation of the results from the semi-automatic annotation performed by *merlin*, each candidate metabolic gene was inspected and accepted, or rejected, according to a developed annotation pipeline, reported in the Methods section.

The manual curation of *merlin* results began by inspecting the information in different databases for each candidate and identified homologues, prioritizing UniProt's reviewed information [25]. As the functional annotation of *P. irregulare* is yet to be described and most annotations in the Stramenopiles lineage are not reviewed in UniProt, the annotation pipeline took into account phylogeny [26], to retrieve the closest organisms with reviewed information at Swiss-Prot.

Arabidopsis thaliana was defined as the organism of reference in the annotation process, since Stramenopiles are the closest relatives of Viridiplantae [27]. Moreover *Arabidopsis thaliana* has been widely used in genomic studies, for example by Arabidopsis Genome Initiative, AGI, since 1996 [28], affording high consistency in the *P. irregulare* annotation.

From the 17,727 candidate genes provided by the genome sequencing, 5213 were found to have homologies with metabolic genes. From those, 2622 candidates (50.3%) had very high confidence level, meaning that these genes have a very high probability of being correctly classified, because there was consistency in the Enzyme Commission (EC) numbers found in the similarity search conducted. On the other hand, there were 1404 gene candidates with very low confidence level, which means that these might have been erroneously assigned with metabolic functions, weakening the classification confidence, and consequently they were rejected from the set of metabolic genes. There were also 1187 candidates with high, medium and low confidence level, which were manually curated, according to the developed pipeline described in the Methods section (Fig. 8).

Figure 1 displays the distribution of organisms with at least one homologous gene found during the enzymatic annotation, and its domain or kingdom of origin. A total of 9338 organisms were mapped in the homologous gene analysis, with 470 of them being reported in the enzymatic annotation, 154 of which were reported at least twice in the annotation. Among the 470 organisms,

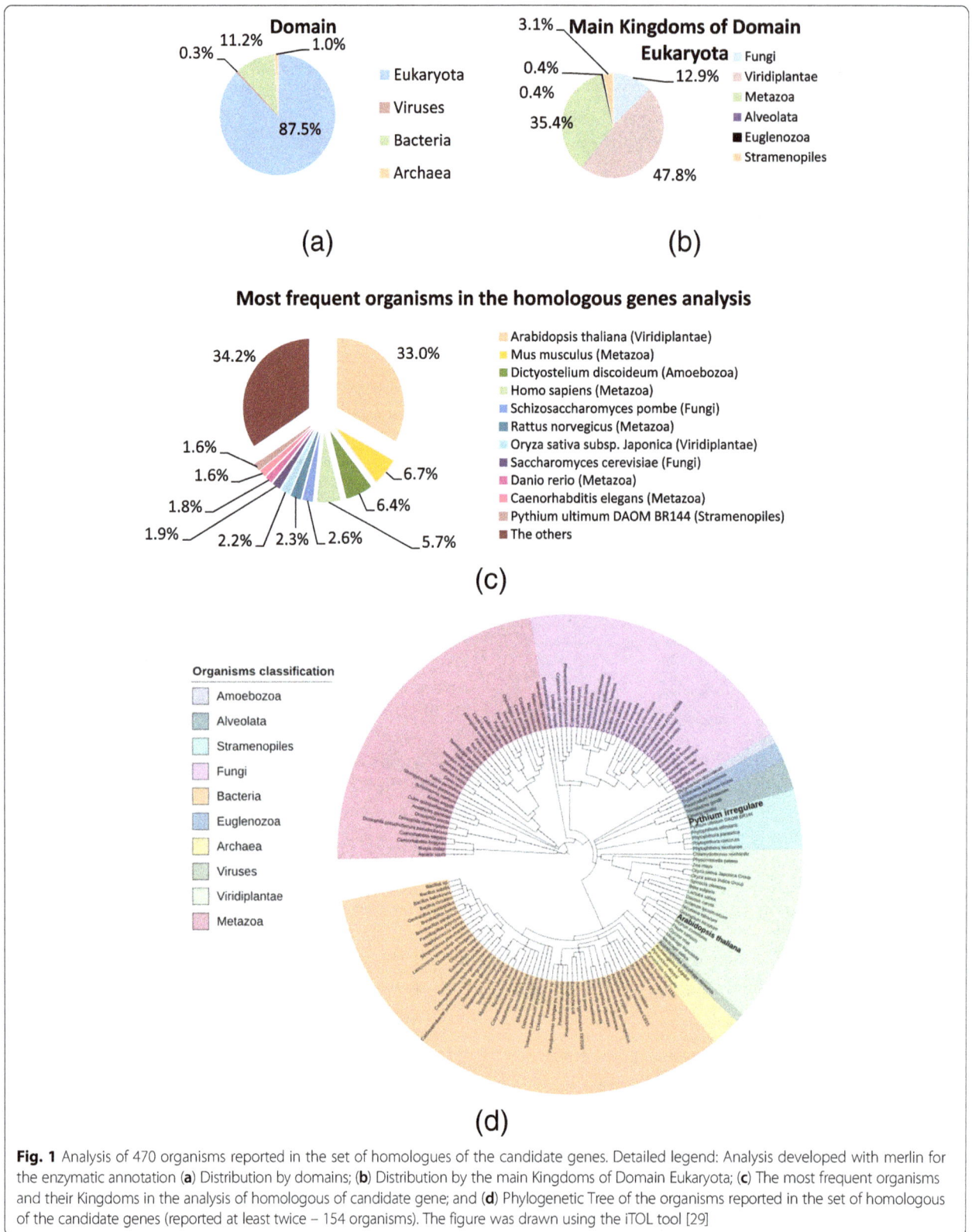

Fig. 1 Analysis of 470 organisms reported in the set of homologues of the candidate genes. Detailed legend: Analysis developed with merlin for the enzymatic annotation (**a**) Distribution by domains; (**b**) Distribution by the main Kingdoms of Domain Eukaryota; (**c**) The most frequent organisms and their Kingdoms in the analysis of homologous of candidate gene; and (**d**) Phylogenetic Tree of the organisms reported in the set of homologous of the candidate genes (reported at least twice – 154 organisms). The figure was drawn using the iTOL tool [29]

87.5% were from Eukaryota Domain (Fig. 1.a), predominating the Viridiplantae Kingdom (Fig. 1.b), with 34.3% of homologue genes coming from *Arabidopsis thaliana*. Other organisms listed do not exceed 7% of frequency; for example, *P. ultimum*, a Stramenopiles microorganism, was only listed in 1.6% of the cases (Fig. 1.c). These results corroborated the choice of *Arabidopsis thaliana* in the pipeline development. Moreover, the Phylogenetic tree represented in Fig. 1.d shows high taxonomic similarity between Viridiplantae and Stramenopiles, according to NCBI taxonomy identifiers.

According to the developed pipeline, 3852 EC numbers were manual and automatic assigned (Fig. 2). The enzyme class distribution is described in Fig. 2. Transferases and hydrolases are the biggest enzyme groups with 35 and 34% of annotated EC numbers, respectively. On the other hand, lyases, isomerases, and ligases, only encompass 4, 3, and 7% of the annotated EC numbers, respectively. Figure 2 shows that 24.4% of the annotated EC numbers are partial, with hydrolases as the group having the largest amount of incomplete EC numbers (11.2%).

Regarding the annotation of transporters, *merlin's* TRIAGE [30] independent module identified candidate transporter proteins encoding genes and, for the genes that fulfilled certain conditions, automatically created transport reactions. From this annotation, 945 candidate genes were identified to encode membrane transporter proteins associated with the transport of 860 metabolites. These data are available in the Availability of data and materials (Additional file 4). TRIAGE classified 39.9% genes as Electrochemical Potential-driven Transporters (transporter classification (TC) TC2), 24.4% as Channel/Pores (TC1), 20.2% as Primary Active Transporters (TC3), 9% as Incomplete Characterized Transport Systems (TC9), 4% as Accessory Factors Involved in Transport (TC8), 1% as Group Translocators (TC4), and 1% Transmembrane Electron Carriers (TC15) (Table 2).

Analysis of the functional annotation
Carbon source metabolism

The functional annotation of *P. irregulare* CBS 494.86 was compared with curated information for *Arabidopsis thaliana*, *Saccharomyces cerevisiae*, *Yarrowia lipolytica*, and *Mortierella alpina* [28, 31–37], because there is no curated metabolic annotation available for any *Pythium* strain. The reason for choosing *A. thaliana* was already mentioned above, while *S. cerevisiae*, *Y. lipolytica*, and *M. alpina* are relevant producers of lipids of commercial interest, are well characterized in the literature and are considered promising EPA producers [9]. Additionally, Oomycetes are generally not employed for lipids production, except *P. irregulare*. As this is the first curated functional annotation developed for *Pythium irregulare*, all metabolic pathways presented in the present article are mostly based on the findings obtained through the metabolic annotation and crossed with evidence from the literature.

In general, there were high similarity between glycolysis, pentose phosphate, and tricarboxylic acid (TCA) cycle pathways for *P. irregulare*, *A. thaliana*, *S. cerevisiae*, *Y. lipolytica*, and *M. alpina*. None of these organisms are able to perform the Entner-Doudoroff pathway, although this has been described for some Stramenopiles microorganism [38].

According to the illustration of the metabolic annotation shown in Fig. 3, the process of fatty acids biosynthesis starts with transport of some carbon source, such as sucrose, glucose, fructose, cellulose or glycerol. In the sucrose metabolism, for example, sucrose is degraded extracellularly by an irreversible reaction into D-fructose and D-glucose (by maltase-glucoamylase; EC:3.2.1.20 or beta-fructofuranosidase; EC:3.2.1.26) which will then be transported into the cell. Cellulose, as carbon source, is metabolized in cellobiose (by cellulose 1,4-beta-cellobiosidase; EC:3.2.1.91) and converted into D-glucose (by

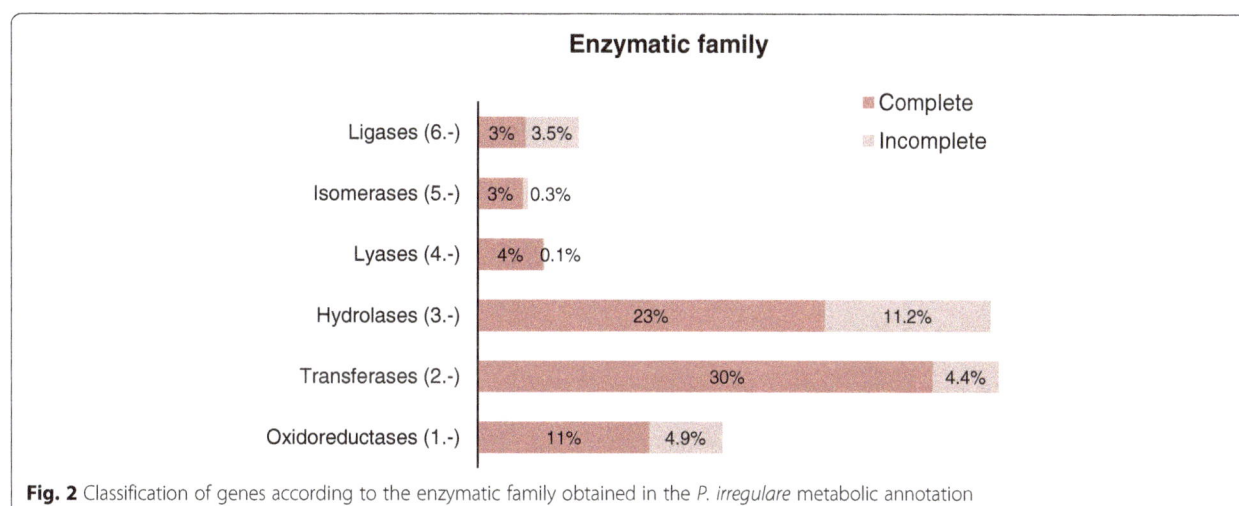

Fig. 2 Classification of genes according to the enzymatic family obtained in the *P. irregulare* metabolic annotation

Table 2 Transporters level and classification (TC) associated with candidate genes identified by *merlin*'s TRIAGE

TC level	Classification	Number of TCG[a]	Percentage (%)
TC1	Channels/Pores	231	24.4
TC2	Electrochemical Potential-driven Transporters	377	39.9
TC3	Primary Active Transporters	191	20.2
TC4	Group Translocators	8	0.9
TC5	Transmembrane Electron Carriers	13	1.4
TC8	Accessory Factors Involved in Transport	36	3.8
TC9	Incompletely characterized Transport Systems	89	9.4
Total		945	100

[a]*TCG* Transporter Candidate Genes

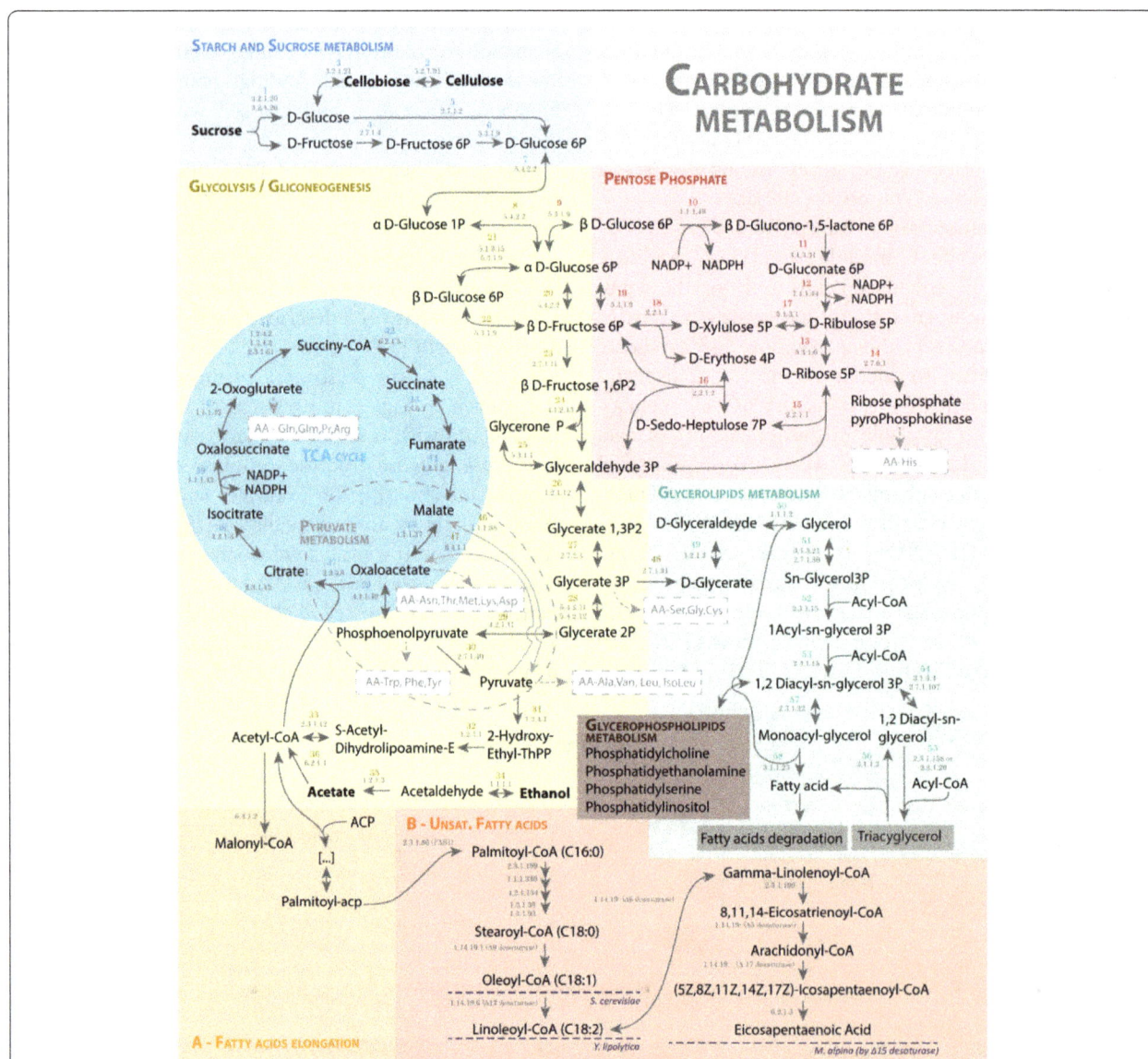

Fig. 3 Carbon source consumption and Eicosapentaenoic acid biosynthesis simplified pathway based on the metabolic annotation. Detailed legend: The figure represents metabolites, reactions and enzyme or transporter annotated for each reaction. The figure is divided by pathways, and the pathways marked with A (Fatty acids biosynthesis) and B (Unsaturated Fatty acids biosynthesis) are described in Figura 7. The dashed blue lines (pathway B) highlights the end of fatty acids production by *Saccharomyces cerevisiae*, *Yarrowia lipolytica*, and *Mortierella alpina*

beta-glucosidase; EC:3.2.1.21), which is then phosphory-lated to D-glucose-6P (by glucokinase; EC:2.7.1.2), entering into the Glycolysis pathway. Cellulose and cellobiose are not consumed by *Saccharomyces cerevisiae* or *Yarrowia lipolytica*. Glycerol is an example of a carbon source that comes from the glycolipid metabolism pathway (by glycerol kinase, EC: 2.7.1.30), which is converted by some reactions into 3-Phospho-D-glycerate to enter Glycolysis.

P. irregulare is a microorganism that could transport and use carbon sources available in different waste streams, namely, vinasse, from first or second generation ethanol production, composed by sucrose, D-glucose, D-fructose, glycerol, acetate, cellulose and others (Additional file 1: Figure S1) [4, 39–41]; glycerol, a by-product in biodiesel production process; and wastewaters from several food and beverage industries [42–44]. The results regarding the enzymes and transporters annotations are in agreement with the carbon sources reported in the literature [4, 13, 14, 29, 38] and presented in Table 3, with the exception of D-Galactose, L-rhamnose, and D-mannose. Though there is experimental evidence [14] regarding the use of these carbon sources by *P. irregulare*, and transporter proteins were identified for these sugars, no further consuming enzymes have been found in the metabolic annotation (Table 3).

Cellulose consumption is associated with the pathogenicity of *P. irregulare*, which is able to degrade plant cell walls of a wide range of plants (corn, soybeans, wheat, fruit trees, vegetables, cereals, and others). *P. irregulare* invades and forms haustoria within living plant cells, consuming its nutrients [23, 29, 45, 46]. The mapped candidate genes for cellulose consumption are important to understand the genetic basis of its pathogenicity.

In summary, most carbon sources (Fig. 3) are converted and guided to the Glycolysis pathway providing pyruvate, and subsequently acetyl-CoA, which, together with NADPH produced in the pentose phosphate pathway are critical precursors of fatty acid synthesis. Furthermore, other pathways can participate in acetyl-CoA's availability, mainly consumption of amino acids from the host. Those pathways, together with the biosynthetic pathways for amino acids are reported in the following subsections.

Amino acids

This subsection aims to present the metabolic annotation obtained for amino acid production and highlight the main pathways involved in the fatty acid production.

General view Figure 4 represents the possible pathways of amino acid production in *P. irregulare* according to its functional annotation developed in this research. In general, there are only a few differences in the amino acid production pathways reported for *A. thaliana* [28], some diatoms (*Thalassiosira pseudonana* and *Phaeodactylum tricornutum*), Stramenopiles microorganisms [47], and *S. cerevisiae* [35]. The divergences are in the pathways B – Glycine and Serine, C - Tyrosine, and E – Arginine, reported in this item (Fig. 4); and B – Cysteine and F –Lysine pathways, both described in item 2.3.2.1.1. Finally, a reflection regarding the pathways associated with fatty acids metabolism such as Leucine, Isoleucine, and Lysine degradation and Cysteine and Lysine biosynthesis is described in item 2.3.2.2 **Amino acids associated with metabolism of fatty acids (FAs)** .

B – Glycine and Serine pathways *A. thaliana* and Diatoms have glycine and serine synthesis as part of photorespiration and non-photorespiration [47]. Only the non-photorespiratory pathway of serine synthesis is observed in *P. irregulare*, based on the functional annotation, similarly to what is observed in *S. cerevisiae* [35] and *Y. lipolytica* [34], as well as other Stramenopiles [47] (Fig. 4).

C - Tyrosine pathway Phenylalanine, tyrosine, and tryptophan are aromatic amino acids and central molecules in *Arabidopsis thaliana* metabolism, serving as precursors for a variety of hormones, but they are not classified as essential in plants [48]. In this functional annotation, the shikimate pathway from erythrose4-phosphate and phosphoenolpyruvate to chorismic acid is a common pathway adopted by *S. cerevisiae* [49] and *A. thaliana* [48] and *Pythium irregulare* (Fig. 4). Tyrosine in *P. irregulare* comes from phenylalanine as found in Diatoms [47] and *M. alpine* [37], instead of 4-Hydroxyphenylpyruvate pathways found in *S. cerevisiae* [49] and *Y. lipolytica* [34] and from L-Arogenate pathway or Phenylalanine in *A. thaliana* [28]. According to Wang et al. [50], this degradation reaction by phenylalanine hydroxylase (EC:1.14.16.1) from phenylalanine to pyrosine is functionally relevant in lipid metabolism, including the sequential reactions to acetyl-CoA (Fig. 4).

E – Arginine pathway According to this functional annotation, glutamate, glutamine, proline, and arginine pathways are derived from 2-oxoglutarate, a product of the TCA cycle. In *P. irregulare*, these pathways can use ammonia, nitrate, or nitrite as a nitrogen source Fig. 4). This provides a great versatility for this microorganism, unlike *S. cerevisiae* and *Y. lipolitica* which cannot consume nitrate and nitrite as a nitrogen sources [34, 35]. Nitrate is present in great amounts in some wastewaters such as vinasse [51].

Besides this difference, the biosynthetic pathways for these amino acids are similar in all studied organisms,

Table 3 Comparison betwwen carbon source consumption based on the Metabolic Annotation and experimental data

Carbon Source	Encoded Genes (Enzymes)	Enzymes	Example of Encoded genes (Transporters)	Example of family of Transporters	Metabolic Annotation	Experimental data	References
Sucrose	PIR_09067.1	Maltase-glucoamylase, EC:3.2.1.20	PIR_09911.1, PIR_09803.1, PIR_14514.1, PIR_14733.1, PIR_09802.1, PIR_11721.1, PIR_09912.1, PIR_11722.1, PIR_14347.1, PIR_12515.1, PIR_09914.1	2.A.123#	√	√	(4,14)
	PIR_00611.1	Beta-fructofuranosidase; EC:3.2.1.26					
D-Glucose	PIR_07992.1	Glucokinase, EC:2.7.1.2	PIR_03670.1, PIR_12595.1, PIR_14733.1, PIR_12158.1, PIR_14021.1, PIR_08620.1	2.A.1#	√	√	(4,14)
D-Frutose	PIR_16062.1	Fructokinase, EC:2.7.1.4	PIR_09911.1, PIR_14514.1, PIR_14733.1, PIR_11721.1, PIR_09913.1	2.A.123#	√	√	(4,14)
Glycerol	PIR_00969.1,PIR_08514.1	Glycerol kinase, EC:2.7.1.30	PIR_09911.1, PIR_14514.1, PIR_14733.1, PIR_11721.1, PIR_09913.1, PIR_09912.1	2.A.123#	√	√	(4,14)
Lactose	PIR_02725.1	Beta-galactosidase, EC:3.2.1.23	PIR_09911.1, PIR_14514.1, PIR_14733.1, PIR_11721.1, PIR_09913.1	2.A.123#	√	√	(14)
D-Xylose	PIR_05464.1	Xylose isomerase, EC:5.3.1.5	PIR_09911.1, PIR_14514.1, PIR_14733.1, PIR_11721.1, PIR_09913.1	2.A.123.#	√	√	(14)
Cellulose	PIR_10473.1,PIR_10474.1, PIR_10834.1,PIR_11849.1, PIR_12055.1,PIR_13142.1, PIR_13232.1,PIR_13655.1, PIR_13859.1,PIR_14441.1, PIR_14503.1	Cellulose 1,4-beta-cellobiosidase, EC:3.2.1.91	PIR_09911.1, PIR_14514.1, PIR_14733.1, PIR_11721.1, PIR_09913.1, PIR_09912.1, PIR_11722.1	2.A.123#	√	√	(14)
Cellobiose	PIR_00491.1,PIR_00492.1, PIR_10473.1,PIR_10474.1, PIR_10834.1,PIR_11849.1, PIR_12055.1,PIR_13107.1, PIR_13488.1,PIR_14069.1, PIR_16235.1	Beta-glucosidase, EC:3.2.1.21	PIR_09911.1, PIR_14514.1, PIR_14733.1, PIR_11721.1, PIR_09913.1, PIR_09912.1, PIR_11722.1, PIR_14347.1, PIR_14694.1, PIR_09803.1, PIR_09802.1	2.A.123#	√	√	(14)
Starch	PIR_02673.1	Alpha-amylase, EC:3.2.1.1	PIR_09911.1, PIR_14514.1, PIR_14733.1, PIR_11721.1, PIR_09913.1, PIR_09912.1, PIR_11722.1, PIR_14347.1, PIR_14694.1	2.A.123#	√	√	(14)
	PIR_13416.1	Beta-amylase, EC:3.2.1.2					
L-arabinose	PIR_09630.1	Alpha-N-arabinofuranosidase, EC:3.2.1.55	PIR_09911.1, PIR_14514.1, PIR_14733.1, PIR_11721.1, PIR_09913.1, PIR_09912.1	2.A.123.#	√	√	(14)
D-Galactose	-	-	PIR_09803.1, PIR_14514.1, PIR_14733.1, PIR_09802.1, PIR_11721.1, PIR_09912.1	2.A.123.#	√	√	(14)
L-rhamnose	-	-	PIR_09911.1, PIR_14514.1, PIR_14733.1, PIR_11721.1, PIR_09913.1, PIR_09912.1, PIR_11722.1, PIR_14347.1, PIR_14694.1	2.A.123.#	√	√	(14)
D-mannose	-	-	PIR_09911.1, PIR_14514.1, PIR_14733.1, PIR_11721.1, PIR_09913.1, PIR_09912.1, PIR_11722.1, PIR_14022.1, PIR_14347.1	2.A.123#	√	√	(14)

except for reaction 16, in pathway E – L-Glutamate, L-Glutamine, Proline, and Arginine described in Fig. 4, which catalyzes the oxidation of Arginine into Citrulline in the presence of NADPH and O_2 by the family of enzymes named nitric-oxide synthases (NOSs, EC: 1.14.13.39) These are found in *P. irregulare*'s metabolic annotation and *A. thaliana* [28, 52], but not in *S. cerevisiae* [35] and *Y. lipolytica* Arginine is a major storage and transport form of organic nitrogen in plants. Additionally, it has a role in protein synthesis, as a precursor of nitric oxide, polyamines, besides its importance as a pathway in pathogen resistance mechanism [52]. *P.*

irregulare, using *A. thaliana as a* host, could develop invasion strategies, including enzyme production to interfere in the metabolic targets common to its host [3], in this case, probably reducing Arginine availability in plants by converting it into Citrulline (by nitric oxide synthase, EC:1.14.13.39) (Fig. 4).

Amino acids associated with metabolism of fatty acids (FAs) Amino acid biosynthesis and degradation play an important role in the biomass development and fatty acids (FAs) biosynthesis, by providing Acetyl-CoA [53], obtained in the degradation of branched-chain

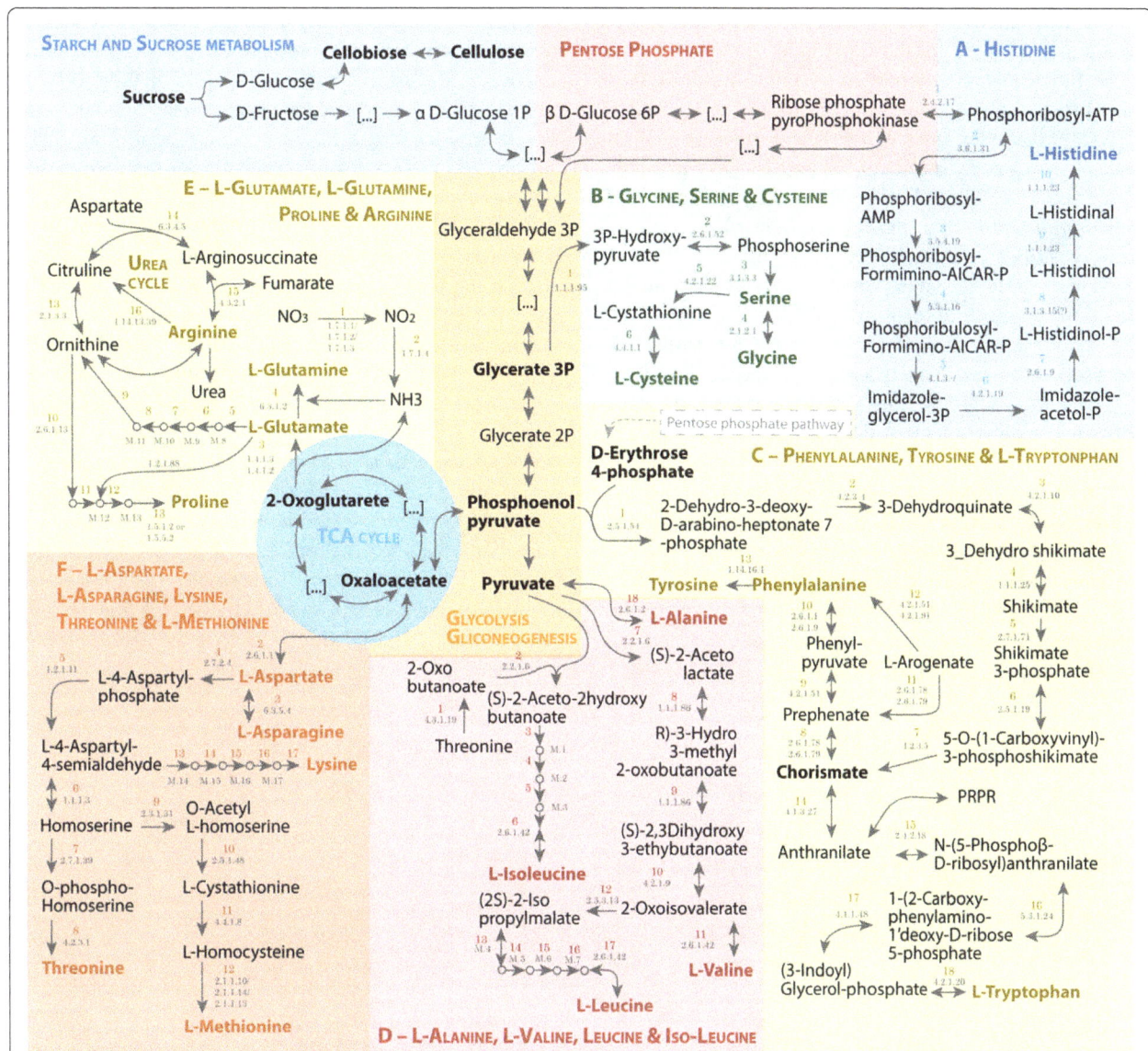

Fig. 4 Prediction of amino acid biosynthesis according to the metabolic annotation of *P. irregulare* CBS 494.86. The amino acid pathways are listed from A to F. The reactions are enumerated inside each specific pathway, followed by the encoded EC number. The EC numbers with a question mark (?) are cases of reactions not predicted by the metabolic annotation. Some metabolites and reactions are marked with M (metabolites) and/or numbers decoded as follows: in D- L-Alanine, L-Valine, Leucine & Iso-Leucine pathway, M.1 = (R)-3-Hydro 3-methyl 2-oxopentanoate, M.2 = (S)-2,3 Dihydroxy 3-methypentanoate, M.3 = (S)-3-Methyl 2-oxopentanoate, M.4 = 2-Isopropylmaleate, M.5 = (2R,3S)-3-Isopropylmalate, M.6 = (2S)-2-Isopropyl 3-oxopentanoate, M.7 = 4-Methyl 2-oxopentanoate; 3 = EC:1.1.1.86; 4 = EC: 1.1.1.86; 5 = EC: 4.2.1.9; 13 = EC: 2.3.3.13, 14 = EC: 4.2.1.33; 15 = EC: 1.1.1.85; 16 = Spontaneous; in E – L-Glutamate, L-Glutamine, Proline & Arginine pathway, M.8 = N-Acetyl-glutamate, M.9 = N-Acetyl-glutamyl-P, M.10 = N-Acetylglutamate semialdehyde, M.11 = N-Acetyl-ornithine, 5 = EC: 2.3.1.1; 6 = EC: 2.7.2.8; 7 = EC: 1.2.1.38; 8 = EC: 2.6.1.11; 9 = EC: 3.5.1.14; in E – L-Glutamine, L-Glutamate, Proline & Arginine pathway, M.12 = L-Glutamate 5-semialdehyde; M13 = (S)-1-Pyrroline-5-carboxylate;11 = non-enzymatic; 12 = EC:1.5.1.2; in F – L-Aspartate, L-Asparagine, Lysine, Threonine & L-Methionine pathway, M.14 = (2S,4S)-4-hydroxy 2,3,4,5 tetrahydro-dipicolinate, M.15 = L-2,3,4,5- Tetrahydro-dipicolinate, M.16 = L-L-2,6 Diamino-pimelate, M.17 = meso2,6 Diamino-pimelate, 13 = EC: 4.3.3.7;14 = EC: 1.17.1.8; 15 = EC: 2.6.1.83 (?); 16 = EC: 5.1.1.7 (?); 17 = EC: 4.1.1.20

amino acid such as leucine (EC:2.3.1.9, EC:4.1.3.4), isoleucine (EC:2.3.1.16), and lysine (EC:1.5.1.8, EC:2.3.1.9) (Fig. 4), as well as in the lysine [54] and L-cysteine biosynthesis [55].

The leucine, isoleucine, and lysine degradation pathways (Fig. 4) are identified in *Y. lipolytica*, an oleaginous microorganism [36], but not in *S. cerevisiae* [35]. Such pathways can be associated to the pathogen's resistance mechanism, maximizing the energy storage through the accumulation of lipids.

Lysine biosynthesis, in *P. irregulare*, comes from the diaminopimelate (DAP) pathway, observed in plants [56] and in other Stramenopiles [47], instead of the alphaaminoadipate (AAA) pathway found in *S. cerevisiae* and

Y. lipolytica. In DAP, the acetyl-CoA precursor of fatty acids biosynthesis is not consumed [54] (Fig. 5).

The functional annotation of cysteine biosynthesis is guided to the cystathionine (CT) pathway, also observed in *Phytophthora infestans* which is an oomycete [55], instead of the O-acetylserine (OAS) pathway like in *Arabidopsis thaliana* [28] and other Stramenopiles, such as *T. pseudonana* and *P. tricornutum* [47] (Fig. 6). In the OAS pathway, sulfide is an important metabolite produced in the sulfate assimilation process and its reduction, obtained in plants and diatoms prokaryotes, fungi, and photosynthetic organisms [57]. Acetyl-CoA is used in the production of o-acetylserine (L-Serine + Acetyl-CoA < => O-Acetyl-L-serine + CoA), reducing its avaiability for the biosynthesis of fatty acids (Fig. 6).

Fatty acids, unsaturated fatty acids, including EPA

Several Oomycetes, in which *P. irregular* is included, are plant pathogens. They are persistent to several pesticides, due to their ability to store energy in the form of lipids, as their degradation provides acetyl-CoA for further catabolism by the TCA cycle [58]. For this reason, lipids metabolism, including fatty acids biosynthesis, has been thoroughly

studied [59]. Moreover, fatty acids are essential compounds in the cell structure of Oomycetes and play an important function in the cell membrane due to the hydrophobic nature of acyl chains, which create subcellular compartments. Moreover, some fatty acids, like polyunsaturated fatty acids, can be used as precursors of eicosanoids that regulate inflammatory and immune responses [60].

FAs are the basic elements of complex lipids (phospholipids, triacylglycerols, sphingolipids, sterol esters) [61]. In the biosynthesis of fatty acids, up to C16 or C18 saturated FAs are produced. This pathway involves two enzymatic systems: type I fatty acid synthase (I FAS), in which enzymes are encoded by a distinct gene, as occurs in most bacteria as well as in the organelles of prokaryotic ancestry, mitochondria and chloroplasts [62]; and type II FAS [63], an enzymatic complex, composed of two subunits, Fas1 (Fasβ) and Fas2 (Fasα) [64], which are found in mammals and lower eukaryotes.

The fatty acid biosynthesis in *P. irregulare* reveals high similarity with *S. cerevisiae*, *Y. lipolytica*, and *M. alpina*, which also possess type I and II FAS enzyme systems, instead of *A. thaliana* that possess only I FAS enzymes [28, 31–37].

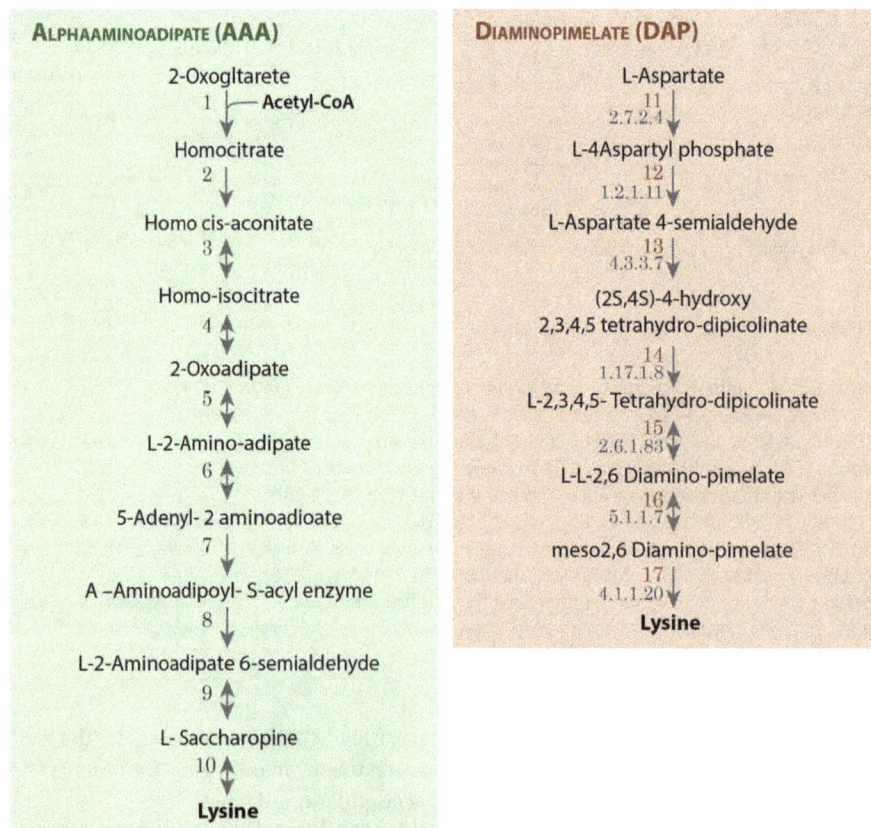

Fig. 5 Prediction of Lysine biosynthesis pathways according to the metabolic annotation of *P. irregulare* CBS 494.86. Detailed legend: Diaminopimelate (DAP) possible Lysine biosynthesis pathways with EC numbers provided in the metabolic annotation of *P. irregulare* and Alphaaminoadipate (AAA) pathway available for *S. cerevisiae* [50, 51]. In the DAP pathway enzymes 15 and 16 was not confirmed by the current annotation

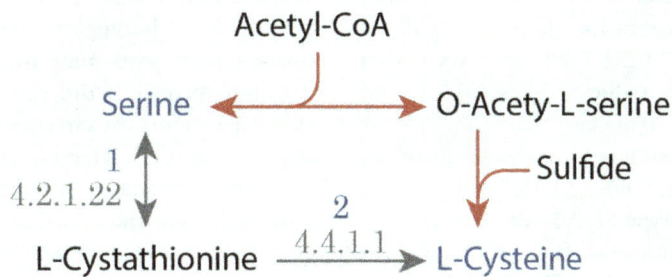

Fig. 6 Prediction of Cysteine biosynthesis pathways according to the metabolic annotation of *P. irregulare* CBS 494.86. Detailed legend: The black arrows represent cystathionine (CT) pathway, the most probable pathway applied by *P irregulare* CBS 494.86 according to its metabolic annotation; red arrows represent the O-acetylserine (OAS) pathway

In fatty acid biosynthesis, acetyl-CoA is carboxylated into malonyl-CoA (acetyl-CoA carboxylase; EC:6.4.1.2, Gene: PIR_06001.1), then malonyl is transferred to an acyl-carrier protein (ACP) (fatty acid synthase subunit beta, fungi type, EC:2.3.1.86 or ACP S-malonyltransferase, EC:2.3.1.39). The fatty acid biosynthesis involves 4 reactions, in cyclic steps, which allow producing saturated fatty acids with 16 and 18 carbons (Fig. 7), described next: **initiation/elongation** (3-oxoacyl-[acyl-carrier-protein] synthase I, EC:2.3.1.41, or fatty acid synthase subunit

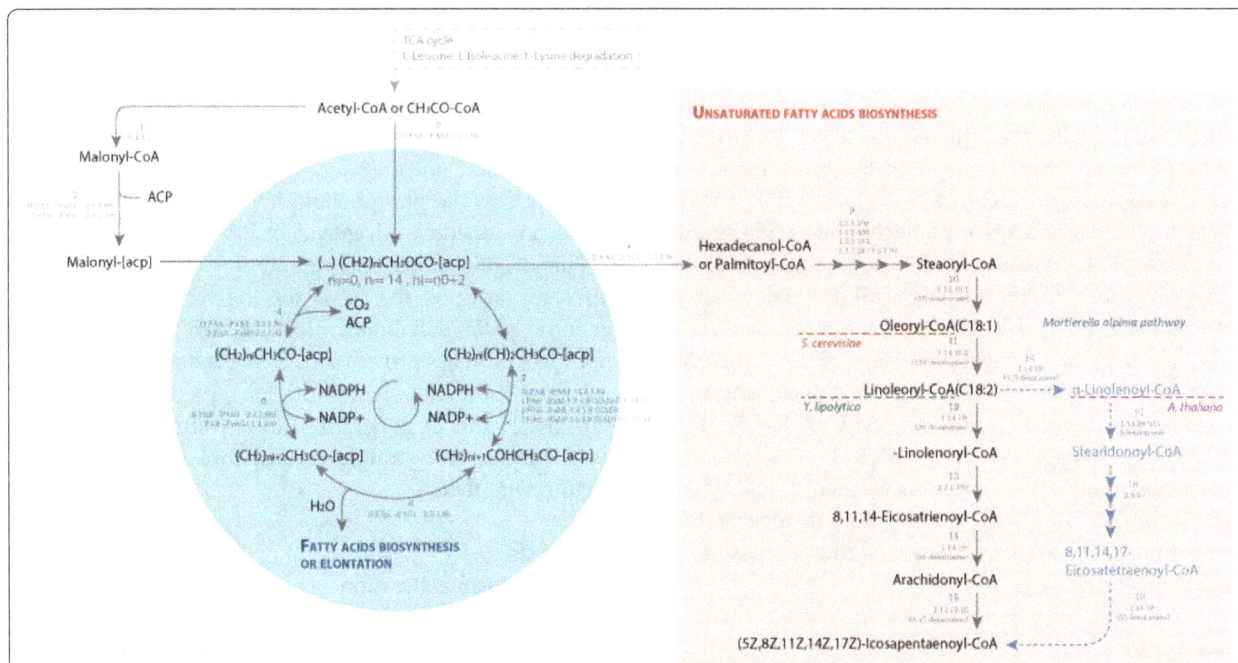

Fig. 7 Biosynthesis and unsaturation of fatty acids until EPA predicted from the metabolic annotation. Detailed legend: The biosynthesis predicted from the metabolic annotation in *P. irregulare* CBS 494.86 occurs in the cyclic process by elongation from 2C (Acetyl-acp) to 16C (Hexadecanol-acp), where two carbons from Malonyl-acp are incorporated in the fatty acids per cycle. The process involves fatty acid enzymes from types I (IFAS) and II (II FAS). The enzymatic complex II FAS is composed of two subunits Fas1 (or Fasβ) and Fas2 (or Fasα). The cyclic process involves 4 steps: initiation/elongation (where two carbons of Malonyl-acp are incorporated); reduction (where the first NADPH is reduced to NADP+); dehydration (where there is H2O liberation); and reduction (where the second NADPH is reduced to NADP+). The cyclic process is concluded when Hexadecanol-acp is produced, and converted into Hexadecanol-CoA by the FAS2 enzyme, fatty acid synthase subunit beta (2.3.1.86). The next steps involve unsaturation, by elongation and desaturation processes. The representative figure shows (5Z,8Z,11Z,14Z,17Z)-Icosapentaenoyl-CoA conversion into EPA by long-chain-fatty-acid-CoA ligase and its export by 2.A.126# transporter family. The red dashed line marks the end unsaturated fatty acid (C18:1) produced by *Saccharomyces cerevisiae*; the green dashed line the *Yarrowia lipolytica* end fatty acid (γ-linolenic acid - C18:2); the purple dashed line the *A. thaliana* end fatty acid (α-linoleic acid - C18:3); and the blue dashed lines, metabolites, encoded enzymes, and reactions, which demonstrate the *M. alpine* pathway into EPA

alpha, EC:2.3.1.86), **reduction** (3-oxoacyl-[acyl-carrier protein] reductase, EC:1.1.1.100 or fatty acid synthase subunit alpha, EC:2.3.1.86), **dehydration** (fatty acid synthase subunit beta, fungi type, EC:2.3.1.86), and **reduction** (enoyl-[acyl-carrier protein] reductase I, EC:1.3.1.9 or enoyl-[acyl-carrier protein] reductase II, EC:1.3.1.9 or enoyl-[acyl-carrier protein] reductase / trans-2-enoyl-CoA reductase (NAD+) EC:1.3.1.9 EC:1.3.1.44, or fatty acid synthase subunit beta, fungi type EC:2.3.1.86) (Fig. 7).

The biosynthesis of unsaturated fatty acids involves steps of desaturation and elongation (Fig. 7). In *S. cerevisiae*, the process takes place up to C16:1 and C18:1 by Δ-9 desaturase [58]. Wide-type *Y. lipolytica* can synthetize linoleic acid (C18:2) using a Δ-12 desaturase [34, 65]. *A. thaliana* synthetizes Gamma-linolenic acid (18:3, Omega-6) [66]. *M. alpina* can produce a low amount of EPA (C20:5) (Fig. 7, by its precursor, (5Z,8Z,11Z,14Z, 17Z)-Icosapentaenoyl-CoA), but in a possible non-efficient process, which involves sequential conversions from linoleoyl-CoA (Omega 6) to α-linolenoyl-COA (Omega 3) by Δ-15desaturase, from α-linolenoyl-COA to stearidonoyl-CoA by Δ-6 desaturase, Stearidonoyl-CoA into Eicosatetraenoyl-CoA by fatty acid elongase, finally using the Eicosateraenoyl-CoA as a precursor in its production by Δ-5 desaturase [37] (Fig. 7).

Concerning the ability of *P. irregulare* to produce unsaturated fatty acids, the annotation from *P. irregulare* predicted the presence of a gene that encodes the enzyme Δ17 desaturase (EC:1.14.19.-), which converts arachidonyl-CoA (precursor of arachidonic acid) into (5Z, 8Z,11Z,14Z,17Z)-icosapentaenoyl-CoA, and finally its conversion into eicosapentaenoic acid) [67–69], which has been reported previously in Oomycetes, such as *Pythium aphanidermatum*, *Phytophthora sojae*, and *Phytophthora ramorum* [67]. Other genes associated with elongation enzymes from palmitoyl-CoA to steaoryl-CoA Fig. 7) and β-oxidation (EC: 1.3.3.6, 4.2.1.17, 1.1.1.211, 2.3.1.16) from eicosapentaenoic acid to docosahexaenoic acid (DHA) were identified. However, there is no evidence of DHA production by *P. irregulare*, according to the review developed by Wu et al. [45].

Discussion

Table 4 summarizes the divergence in the functional annotation of *P. irregulare* compared with curated information for *A. thaliana*, *S. cerevisiae*, *Y. lipolytica*, and *M. alpina* [28, 31–37]. The metabolic annotation of *P. irregulare* showed high similarity with *A. thaliana*, mainly in carbohydrate metabolism, amino acids metabolism, and nitrogen assimilation, except by the photorespiratory pathways not expected in an Oomycete. Those results could be explained by the horizontal gene and chromosome transfer between *P. irregulare* and its host *A. thaliana* [70]. According to the metabolic annotation, the

fatty acid metabolism have high similarity with the fungi analyzed in this article, as producers of lipids with commercial biotechnological interest. However, for the assessed fatty acids, only *M. alpina* is able to produce EPA, but through a different pathway. EPA production with *P. irregulare*, by enzyme Δ-17 fatty acids' desaturase suggested by the metabolic annotation, provides a great commercial advantage, as this desaturase can use fatty acids both from the acyl-CoA fraction and the phospholipids fraction as substrates [67].

Finally this metabolic annotation can infer about *P. irregulare* metabolic capabilities, supplying information of great importance to future work.

Conclusions

This unprecedented functional annotation demonstrates the presence of relevant genes and is consistent with results described in literature. Genes associated with the amino acids production, consumption of carbon (glucose, sucrose, cellulose, fructose, glycerol, and others) and nitrogen sources (nitrate, nitrite, and ammonia), present in the wastewater (produced in large amounts around the world) provide great advantage in the production of value-added lipids using low cost carbon source and in an efficient way, for food and pharmaceutical industries. Several genes encode enzyme present in pathways able to maximize lipid production, notably, the enzyme Δ-17 fatty acid desaturase which can use not only fatty acids in acyl-CoA, but also fatty acids in the phospholipid fraction as substrates, providing a competitive advantage for EPA production [67].

This original functional annotation of *Pythium irregulare* can serve as the basis for the reconstruction of a genome scale metabolic model, which can be used to optimize biomass growth and EPA production.

Finally the metabolic annotation process developed in this article can be generalized to any strains and applied as an useful and straightforward tool in the metabolic engineering field.

Methods

Microorganism cultivation

P. irregulare strain CBS 494.86 was acquired from the CBS-KNAW Fungal Biodiversity Centre. It was inoculated on PDA plates (Potato Dextrose Agar) and incubated for 3 days at 29 °C, then 1 cm² of the grown microorganism was transferred and cultivated on YPDO medium (g/L: yeast extract 1.25, peptone 25, glucose 3, oatmeal 2) for 5 days at 29 °C in a shaker at 200 rpm.

Even though several articles have been published about *P. irregulare* DAOM BR 486 / CBS 250.28 [14, 18, 23, 71], mainly associated with its pathogenicity, with its genome sequences deposited at NCBI (Bioproject number: PRJNA169053), strain *Pythium irregulare* CBS 493.84 had not been explored yet.

Table 4 Key differences observed in *Pythium irregulare* metabolic annotation compared with *other microorganisms*[a]

Pathways	*Pyhtium irregulare* metabolic annotation (this study)	*Arabidopsis thaliana,*	*Saccharomyces cerevisiae*	*Yarrowia lipolytica*	*Mortierella alpina*
Carbohydrate metabolism					
Starch and Sucrose metabolism					
Cellulose consumption	√	√	–	–	–
Fatty acids metabolism					
Fatty acids biosynthesis					
Fatty acids synthase type I (IFAS)	√	√	√	√	√
Fatty acids synthase type II (IIFAS)	√	–	√	√	√
Unsaturated fatty acids					
EPA production	√ EC:1.14.19.-Δ17desaturase	–	–	–	√ EC: 1.14.19.-Δ5desaturase
Amino acids					
Lysine biosynthesis					
Diaminopimelate (DAP) pathway	√	√	–	–	–
Alphaaminoadipate (AAA) pathway	–	–	√	√	–
Serine biosynthesis					
Non-photorespiratory pathway	√	√	√	–	–
Photorespiratory pathway	–	√	–	–	–
Cysteine biosynthesis					
Cystathionine (CT) pathway	√	√	√	√	–
O-acetylserine (OAS) pathway	–	√			
Arginine biosynthesis					
Arginine reoxidation into Citrulline	√	√	–	–	–
Tyrosine biosynthesis					
Phenylalanine degradation	√	√	–	–	√
Nitrogen Metabolism					
Nitrate reduction	√	√	–	–	√
Nitrite reduction	√ EC:1.7.1.4	√ EC:1.7.7.1	–	–	√ EC:1.7.1.4

[a]Curated information

Species validation and genome sequencing

The DNA of the biomass was extracted following the de Graaff et al. protocol [72]. As the first step, to validate the species of this work, two conserved regions were chosen to be analyzed, as proposed by [71]. Cytochrome oxidase I (COI) and the Internal Transcribed Spacer Regions (ITS1 and ITS2) were amplified by PCR using the following set of primers: OomCoxI-Levup (5′-TCA WCWMGATGGCTTTTTTCAAC-3′) and Fm85mod (5′-RRHWACKTGACTDATRATACCAAA-3′), UN-up1 8S42 (5′-CGTAACAAGGTTTCCGTAGGTGAAC-3′) and UN-lo28S22 (5′-GTTTCTTTTCCTCCGCTTATT-GATATG-3′), respectively [73], and submitted for sequencing by Sanger method (Applied Biosystems), following the provider's protocol. Validating the working species, the genomic DNA libraries for Next Generation Sequencing were produced using the Nextera DNA library preparation kit (Illumina). Sequencing was carried out on a Illumina HiSeq 2500 instrument, using paired-

end chemistry (58.990.406 sequenced fragments 2*100 bp) at the NGS core facility of the Brazilian Bioethanol Science and Technology Laboratory (CTBE). K-mer statistics revealed an expected genome size of approx. 75 Mbp [74]. Genome assembly was carried out SPAdes using an ensemble of different k-mer values [75] and improved with Pilon [76]. Ploidy level analysis was carried out with ploidyNGS [77], which revealed that this organism is diploid, thus the inferred genome sequence was processed with Redundans to eliminate redundancies due to allelic polymorphisms. The genome of *P. irregulare* strain (CBS 494.86) is described at the GenBank sequence database provided by the National Center for Biotechnology Information (NCBI) in PRJNA371716.

Gene prediction and annotation
Functional annotation

The phenotypic potential of an organism is embedded in its genome sequence, and gene product

1 – – Perform Blast at EBI, and verify EC number score*:

If score > 0.5 : - - - - - - - Else:
  • Go to A   • If score < 0.2: - - - - - - - Else:
                       • Go to E   • Go to 2

2 – Hit from some Stamenopiles

If exists: - Else:
  • check status: • Go to 3
    • If entry is reviewed : - - - - Else:
      • Go to A   • Go to B

3 – Hit from _Arabidopsis thaliana_

If exists: - Else:
  • check status: • Go to 4
    • If entry is reviewed : - - - - Else:
      • Go to B   • Go to C

4 – Organism with the lowest e-value hit < 10E-10

If exists: - Else:
  • check status: • Go to E
    • If entry is reviewed : - - - - Else:
      • Go to C   • Go to D

A – annotate w/ Confidence Level Very High
B – annotate w/ Confidence Level High
C – annotate w/ Confidence Level Medium
D – annotate w/ Confidence Level Low
E – Discard classification

Fig. 8 Re-annotation pipeline for manual inspection of each candidate gene in _merlin_. Detailed legend: * The score was calculated using the first 100 priority reviewed homologues retrieved from the BLAST similarity, and the thresholds were manually curated by the authors, described in the methods section

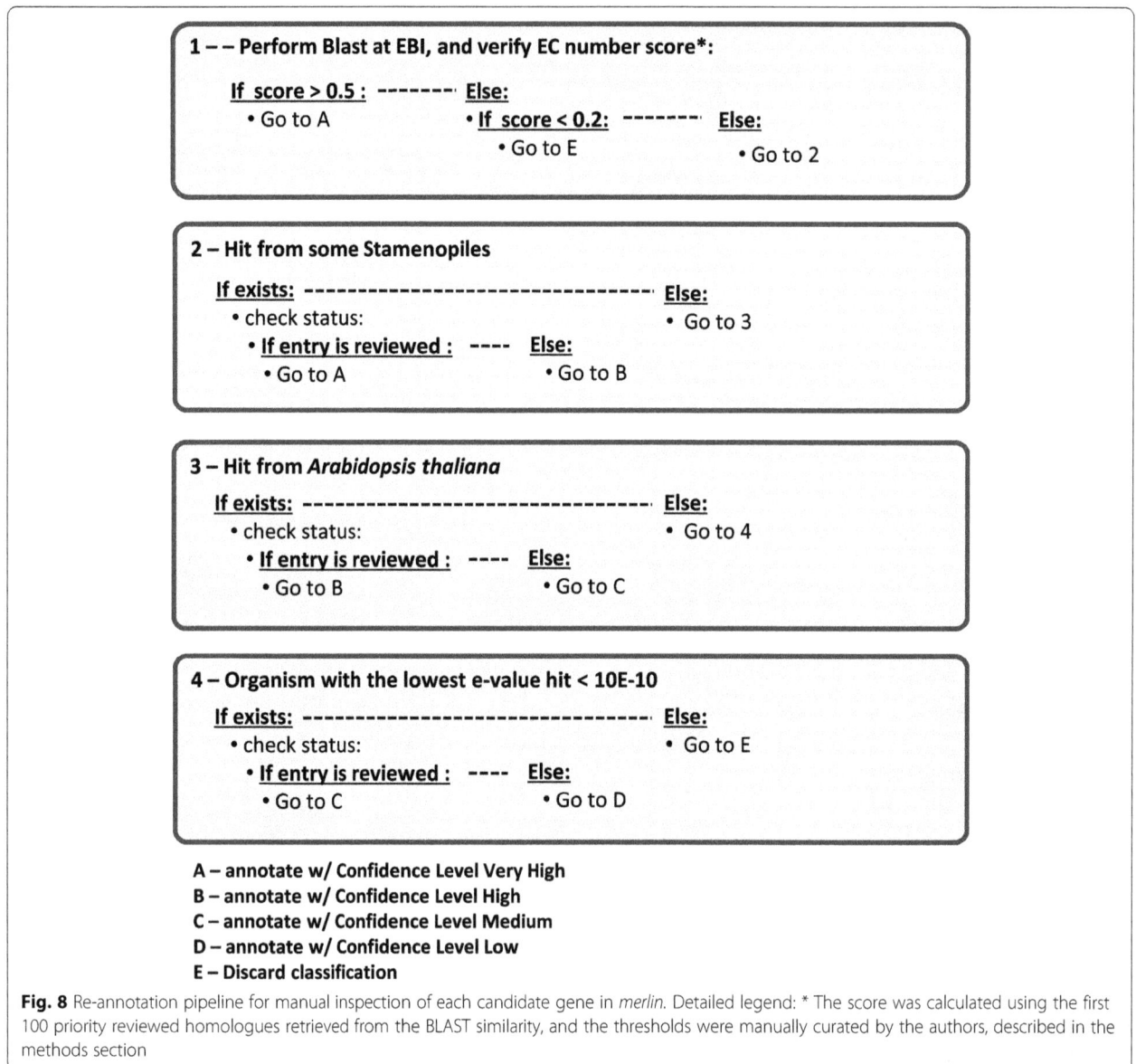

identification is compulsory in order to understand the occurring biological processes [78].

The annotation of a genome is the process of identifying and cataloging functional information of genes in a sequenced genome [79]. The important information retrieved from a genome annotation is gene name, assigned cellular functions, and Enzyme Commission (EC) number, for enzyme coding genes [80].

The _merlin_, a user-friendly software tool, was created to assist in the processes of annotation and reconstruction of genome-scale metabolic models, by performing automatic genome-wide functional annotations and providing a numeric score for each automatic assignment, taking into account the frequency and taxonomy within the annotation of all similar sequences [81].

The selection of the best threshold for automatic annotation _in merlin_ involved adjusting the alpha-value (a ratio of taxonomy and frequency score) using a set of random manually curated sequences and comparing these with the automatic annotation provided by the software (automatic classification and final metabolic annotation available in Additional files 3 and 4). In this process, the alpha-value in _merlin_ was set at 0.9, which emphasises the frequency score, due to the lack of reviewed information on the Stramenopiles lineage. The selected score threshold for automatically accepting annotation in _merlin_ was set to 0.5, meaning that any candidate genes with a score higher than 0.5 are automatically annotated. Candidate genes with a score below 0.2 were automatically discarded (Additional file 3). After analysing _merlin_'s similarity search output, an annotation workflow was developed to classify and curate the annotation. This workflow was developed to systematically analyze *merlin*'s classification and accept or reject it. The annotation workflow follows a series of simple steps to determine each gene's classification, together with the confidence level of such annotation. The confidence level was set by the authors. It starts by addressing specific situations with a high confidence level (A) (Fig. 8), extending then the search, covering a larger amount of possible annotations, whilst decreasing the confidence level. The EC number classification and the taxonomic distance of the results are taken into account, as well as reviewed information and literature on the studied gene.

A complete metabolic annotation involves identifying genes encoding enzymes and membrane transporters (Additional file 4).

The transporter candidate genes (TCGs) annotation of *Pythium irregulare* was performed in *merlin*'s TRIAGE (Transport Proteins Annotation and Reactions Generation) [24].

Initially, protein-encoding genes with transmembrane helices were identified using Phobius [82, 83]. Afterwards, *merlin* runs the Smith-Waterman (SW) algorithm [84] to compare the target TCGs' translated gene sequences with transmembrane helices with all protein sequences available in the TCDB database. Finally, the metabolites transported by each carrier are inferred from the annotations of the TCDB records that have similarities with thcarrier TCGs [24]. The assessment of the subcellular localization of the proteins was predicted usingLocTree3 [85]. The pipeline of annotating transporter candidate genes (TCGs) of *Pythium irregulare* and the genes associated are available in the Availability of data and materials (Additional files 5 and 6).

Additional files

Additional file 1: Figure S1. Wastewater composition and forecast for 2024 – Vinasse and Glycerol compostion and worldwild forecast production for 2024. (PDF 806 kb)

Additional file 2: Comparison *Pythium irregulare* – Genome comparison between *Pythium irregulare* strains. (XLSX 13 kb)

Additional file 3: Pipeline classification - Pipeline classification for annotation and reconstruction of genome-scale metabolic models established according dataset analysis. (XLSX 1790 kb)

Additional file 4: The metabolic annotation – The metabolic annotation of *Pythium irregulare* CBS 494.86 genome. (XLSX 4245 kb)

Additional file 5: The transporter annotation - Pipeline classification for transporter annotation established according dataset analysis. (XLSX 45 kb)

Additional file 6: Transporter candidate genes - The transporter annotation of *Pythium irregulare* CBS 494.86 genome. (XLSX 63 kb)

Abbreviations
AAA: Alphaaminoadipate; ACP: Acyl-carrier protein; CAGR: Compound annual growth rate; COI: Cytochrome oxidase I gene; DAP: Diaminopimelate; DHA: Docosahexaenoic acid; EC: Enzyme code; EPA: Eicosapentaenoic acid; FAs: Fatty acids; ITS: Internal transcribed spacer; NCBI: National Center for Biotechnology Information; OAS: O-acetylserine; TC: Transporter classification; TCA: Tricarboxylic acid

Acknowledgements
The authors are grateful to the Brazilian Bioethanol Science and Technology Laboratory - CTBE and the Institute for Biotechnology and Bioengineering, Centre of Biological Engineering, Universidade do Minho for the infrastructure. The authors also thank Dr. André Lévesque from Agriculture and Agri-Food Canada and Pedro Raposo from Universidade do Minho for valuable advice and discussion.

Authors' contributions
Experimental data: BSF performed all experiments coordinated by IR, OD, JGCP, and MZ. ITS, COI, and Genome sequencing: BSF, AAKN and JVCO performed sequencing. DMRP carried out genome assembly and gene prediction. Genome annotation: BSF, GC, OD, AAKN, TFCR, and DMRP. JVCO and DMRP coordinated the ITS, COI, and genome sequencing. BSF and GC performed the functional annotation with the support of OD and TFCR. BSF wrote the manuscript with support from AAKN, GC, TFCR, and DMRP. OD and IR supervised the functional annotation. BSF, IR, OD, MZ, and JGCP conceived and directed the study. All authors read and approved the final manuscript.

Funding
The genome sequencing, strain acquisition and preliminary experiments and data analysis were funded by the São Paulo Research Foundation (FAPESP) and Coordination for the Improvement of Higher Education Personnel (CAPES) (Grante: 2016/10562–4). The experiment performance, data collection, analysis and interpretation of data were supported by the Portuguese Foundation for Science and Technology (FCT) under the scope of the strategic funding of [UID/BIO/04469] unit and COMPETE 2020 [POCI-01-0145-FEDER-006684] and BioTecNorte operation [NORTE-01-0145-FEDER-000004] funded by the European Regional Development Fund under the scope of Norte2020 - Programa Operacional Regional do Norte. The authors thank the project DD-DeCaF - Bioinformatics Services for Data-Driven Design of Cell Factories and Communities, Ref. H2020-LEIT-BIO-2015-1 686070–1, funded by the European Commission.

Ethics approval and consent to participate
Not applicable.

Consent for publication
Not applicable.

Competing interests
The authors declare that they have no competing interests.

Author details
[1]Department of Civil and Environmental Engineering, Federal University of Pernambuco, Recife, PE, Brazil. [2]Centre of Biological Engineering, Universidade do Minho, Braga, Portugal. [3]Brazilian Bioethanol Science and Technology Laboratory (CTBE), Brazilian Centre of Research in Energy and Materials (CNPEM), Campinas, SP, Brazil. [4]Computational, Evolutionary and Systems Biology Laboratory, Center for Nuclear Energy in Agriculture, University of São Paulo, Piracicaba, São Paulo, Brazil. [5]Biological Processes Laboratory, Center for Research, Development and Innovation in Environmental Engineering, São Carlos School of Engineering (EESC), University of São Paulo, São Carlos, SP, Brazil. [6]PRBiotec Ltda, São José dos Campos, SP, Brazil.

References

1. Harvey PR, Butterworth PJ, Hawke BG, Pankhurst CE. Genetic variation among populations of Pythium irregulare in southern Australia. Plant Pathol. 2000;49(5):619–27.

2. Spies CFJ, Mazzola M, Botha WJ, Langenhoven SD, Mostert L, Mcleod A. Molecular analyses of Pythium irregulare isolates from grapevines in South Africa suggest a single variable species. Fungal Biol. 2011;115(12):1210–24.

3. de León IP, Montesano M. Activation of defense mechanisms against pathogens in mosses and flowering plants. Int J Mol Sci. 2013;14(2):3178–200.

4. Liang Y, Zhao X, Strait M, Wen Z. Use of dry-milling derived thin stillage for producing eicosapentaenoic acid (EPA) by the fungus Pythium irregulare. Bioresour Technol. 2012;111:404–9.

5. FAO Food and agriculture organization of the united nations. Fats and fatty acids in human nutrition report of an expert consultation. Rome: Food and agriculture organization of the united nations; 2010.

6. Albert BB, Derraik JGB, Cameron-Smith D, Hofman PL, Tumanov S, Villas-Boas SG, et al. Fish oil supplements in New Zealand are highly oxidised and do not meet label content of n-3 PUFA. Sci Rep. 2015;5(7928). https://doi.org/10.1038/srep07928.

7. Sitepu IR, Garay LA, Sestric R, Levin D, Block DE, German JB, et al. Oleaginous yeasts for biodiesel: current and future trends in biology and production. Biotechnol Adv. 2014;32(7):1336–60.

8. Grand View Research. Omega 3 Market Analysis And Segment Forecasts To 2020 [Internet]. 2014 [cited 2015 May 17]. Available from: http://www.grandviewresearch.com/industry-analysis/omega-3-market

9. Bajpai P, Bajpai PK. Eicosapentaenoic acid (EPA) production from microorganisms: a review. J Biotechnol. 1993;30:161–83.

10. Lee JM, Lee H, Kang SB, Park WJ. Fatty acid desaturases, polyunsaturated fatty acid regulation, and biotechnological advances. Nutrients. 2016;8(1):1–13.

11. Bajpai P, Bajpai P, Ward O. Eicosapentaenoic acid (EPA) formation_comparative studies with Mortierella strains and production by Mortierella elongata. Mycol Res. 1991;95(11):1294–8.

12. Abedi E, Sahari MA. Long-chain polyunsaturated fatty acid sources and evaluation of their nutritional and functional properties. Food Sci Nutr. 2014;2(5):443–63.

13. Athalye SK, Garcia RA, Wen Z. Use of biodiesel-derived crude glycerol for producing eicosapentaenoic acid (EPA) by the fungus pythium irregulare. J Agric Food Chem. 2009;57(7):2739–44.

14. Zerillo MM, Adhikari BN, Hamilton JP, Buell CR, Lévesque CA, Tisserat N. Carbohydrate-active enzymes in Pythium and their role in plant Cell Wall and storage polysaccharide degradation. PLoS One. 2013;8(9).

15. Seidl MF, Van Den Ackerveken G, Govers F, Snel B. Reconstruction of oomycete genome evolution identifies differences in evolutionary trajectories leading to present-day large gene families. Genome Biol Evol. 2012;4(3):199–211.

16. Cantarel BL, Korf I, Robb SMC, Parra G, Ross E, Moore B, et al. MAKER: an easy-to-use annotation pipeline designed for emerging model organism genomes. Genome Res. 2008;18(1):188–96.

17. Pythium Genome Database [Internet]. 2017 [cited 2017 Feb 10]. Available from: http://pythium.plantbiology.msu.edu

18. Adhikari BN, Hamilton JP, Zerillo MM, Tisserat N, Lévesque CA, Buell CR. Comparative genomics reveals insight into virulence strategies of plant pathogenic oomycetes. PLoS One. 2013;8(10).

19. Koonin EV, Galperin MY. Genome annotation and analysis. In: Sequence - evolution - function: computational approaches in comparative genomics. Boston: Kluwer Aca; 2003.

20. Altschul SF, Gish W, Miller W, Myers EW, Lipman DJ. Basic local alignment search tool. J Mol Biol. 1990;215:403–10.

21. Simão FA, Waterhouse RM, Ioannidis P, Kriventseva EV, Zdobnov EM. BUSCO: assessing genome assembly and annotation completeness with single-copy orthologs. Bioinformatics. 2015;31(19):3210–2.

22. Stanke M, Keller O, Gunduz I, Hayes A, Waack S, Morgenstern BAUGUSTUS. A b initio prediction of alternative transcripts. Nucleic Acids Res. 2006;34(WEB. SERV. ISS):435–9.

23. Lévesque CA, Brouwer H, Cano L, Hamilton JP, Holt C, Huitema E, et al. Genome sequence of the necrotrophic plant pathogen Pythium ultimum reveals original pathogenicity mechanisms and effector repertoire. Genome Biol. 2010;11(7):R73.

24. Dias O, Rocha M, Ferreira EC, Rocha I. Reconstructing genome-scale metabolic models with merlin. Nucleic Acids Res. 2015 Apr;43(8):3899–910.

25. Chen C, Huang H, Wu C. Protein bioinformatics databases and resources. Methods Mol Biol. 2017;1558:3–39.

26. Sierra R, Canas-Duarte SJ, Burki F, Schwelm A, Fogelqvist J, Dixelius C, et al. Evolutionary origins of rhizarian parasites. Mol Biol Evol. 2016;33(4):980–3.

27. Restrepo S, Enciso J, Tabima J, Riaño-Pachón DM. Evolutionary history of the group formerly known as protists using a phylogenomics approach. Rev la Acad Colomb Ciencias Exactas, Fis y Nat. 2016;40(154):147–60.

28. Plant Metabolic Network (PMN) [Internet]. 2016 [cited 2016 Feb 1]. Available from: www.plantcyc.org

29. Matsumoto C, Kageyama K, Suga H, Hyakumachi M. Intraspecific DNA polymorphisms of Pythium irregulare. Mycol Res. 2000;104(11):1333–41.

30. Dias O, Gomes DG, Vilaça P, Cardoso JJ, Rocha M, Ferreira EEC, et al. Genome-wide semi-automated annotation of transporter systems. IEEE/ACM Trans Comput Biol Bioinforma. 2014;XX(X):1–14.

31. Yu Y, Li T, Wu N, Ren L, Jiang L, Ji X, et al. Mechanism of arachidonic acid accumulation during aging in Mortierella alpina : a large-scale label-free comparative proteomics study. J Agric Food Chem. 2016;64(47):9124–34.

32. Santos S, Liu F., Costa C, Vilaça P, Rocha M, Rocha I. MIYeastK: The Metabolic Integrated Yeast Knowledgebase. In: SPB2016 - Book of Abstracts XIX National Congress of Biochemistry No O5/03. Guimarães, Portugal; 2016.

33. Ye C, Xu N, Chen H, Chen YQ, Chen W, Liu L. Reconstruction and analysis of a genome-scale metabolic model of the oleaginous fungus Mortierella alpina. BMC Syst Biol. 2015;9(1).

34. Pan P, Hua Q. Reconstruction and in silico analysis of metabolic Network for an oleaginous yeast, Yarrowia lipolytica. PLoS One. 2012;7(12):1–11.

35. Saccharomyces Genome Database (SGD) [Internet]. 2017 [cited 2017 Jan 10]. Available from: http://www.yeastgenome.org

36. Kyoto Encyclopedia of Genes and Genomes(KEGG) [Internet]. 2017 [cited 2017 Jan 10]. Available from: http://www.genome.jp/kegg/

37. Wang L, Chen W, Feng Y, Ren Y, Gu Z, Chen H, et al. Genome characterization of the oleaginous fungus mortierella alpina. PLoS One. 2011;6(12):e28319.

38. Hildebrand M, Abbriano RM, Polle JEW, Traller JC, Trentacoste EM, Smith SR, et al. Metabolic and cellular organization in evolutionarily diverse microalgae as related to biofuels production. Curr Opin Chem Biol. 2013;17(3):506–14.

39. Doelsch E, Masion A, Cazevieille P, Condom N. Spectroscopic characterization of organic matter of a soil and vinasse mixture during aerobic or anaerobic incubation. Waste Manag. 2009;29(6):1929–35.

40. Santos SC, Ferreira Rosa PR, Sakamoto IK, Amâncio Varesche MB, Silva EL. Continuous thermophilic hydrogen production and microbial community analysis from anaerobic digestion of diluted sugar cane stillage. Int J Hydrog Energy. 2014;39(17):9000–11.

41. Aditiya HB, Mahlia TMI, Chong WT, Nur H, Sebayang AH. Second generation bioethanol production: a critical review. Renew Sust Energ Rev. 2016;66:631–53.

42. Cheng M, Walher T, Hulbert G, Raman D. Fungal production of eicosapentaenoic and arachidonic acids from industrial waste streams and crude soybean oil. pdf Bioresour Technol. 1999;67:101–10.

43. Lio J, Wang T. Pythium irregulare fermentation to produce arachidonic acid (ARA) and Eicosapentaenoic acid (EPA) using soybean processing co-products as substrates. Appl Biochem Biotechnol. 2013;169(2):595–611.

44. Dong M, Walker TH. Production and recovery of polyunsaturated fatty acids-added lipids from fermented canola. Bioresour Technol. 2008;99(17):8504–6.

45. Wu L, Roe C, Wen Z. The safety assessment of Pythium irregulare as a producer of biomass and eicosapentaenoic acid for use in dietary supplements and food ingredients. Appl Microbiol Biotechnol. 2013;97:7579–85.

46. Caballero JRI, Tisserat NA. Transcriptome and secretome of two Pythiumspecies during infection and saprophytic growth. Physiol Mol Plant Pathol. 2017;99:41-54.

47. Bromke MA. Amino acid biosynthesis pathways in diatoms. Metabolites. 2013;3(2):294–311.

48. Tzin V, Galili G. The biosynthetic pathways for shikimate and aromatic amino acids in Arabidopsis thaliana. Arab book; Am Soc Plant Biol. 2010: e0132.

49. Braus GH. Aromatic amino acid biosynthesis in the yeast Saccharomyces cerevisiae: a model system for the regulation of a eukaryotic biosynthetic pathway. Microbiol Rev. 1991;55(3):349–70.

50. Wang H, Chen H, Hao G, Yang B, Feng Y, Wang Y, et al. Role of the phenylalanine-hydroxylating system in aromatic substance degradation and lipid metabolism in the oleaginous fungus mortierella alpina. Appl Environ Microbiol. 2013;79(10):3225–33.

51. Francisco JP, Folegatti MV, Silva LBD, Silva JBG, Diotto AV. Variations in the chemical composition of the solution extracted from a latosol under fertigation with vinasse. Rev Cienc Agron. 2016;47(2):229–39.

52. Winter G, Todd CD, Trovato M, Forlani G, Funck D. Physiological implications of arginine metabolism in plants. Front Plant Sci. 2015;6(July):1–14.

53. Danne JC, Gornik SG, MacRae JI, McConville MJ, Waller RF. Alveolate mitochondrial metabolic evolution: dinoflagellates force reassessment of the role of parasitism as a driver of change in apicomplexans. Mol Biol Evol. 2013;30(1):123–39.

54. Vorapreeda T, Thammarongtham C, Cheevadhanarak S, Laoteng K. Alternative routes of acetyl-CoA synthesis identified by comparative genomic analysis: involvement in the lipid production of oleaginous yeast and fungi. Microbiology. 2012;158(1):217–28.

55. Niu X. Gene regulatory machinery and proteomics of sexual reproduction in Phytophthora infestans. University of California Riverside; 2010.

56. Morris PF, Schlosser LR, Onasch KD, Wittenschlaeger T, Austin R, Provart N. Multiple horizontal gene transfer events and domain fusions have created novel regulatory and metabolic networks in the oomycete genome. PLoS One. 2009;4(7).

57. Koprivova A, Michael M, von Ballmoos P, Mandel T, Brunold C, Kopriva S. Assimilatory sulfate reduction in C3, C3-C4, and C4 species of Flaveria. Plant Physiol. 2001;127(2):543–50.

58. Tehlivets O, Scheuringer K, Kohlwein SD. Fatty acid synthesis and elongation in yeast. Biochim Biophys Acta - Mol Cell Biol Lipids. 2007;1771(3):255–70.

59. Griffiths RG, Dancer J, O'Neill E, Harwood JL. Effect of culture conditions on the lipid composition of Phytophthora infestans. New Phytol. 2003;158(2): 337–44.

60. Uttaro AD. Biosynthesis of polyunsaturated fatty acids in lower eukaryotes. IUBMB Life. 2006;58(10):563–71.

61. Klug L, Daum G. Yeast lipid metabolism at a glance. FEMS Yeast Res. 2014; 14(3):369–88.

62. Schweizer E, Hofmann J. Microbial type I fatty acid synthases (FAS): major players in a Network of cellular FAS systems. Microbiol Mol Biol Rev. 2004; 68(3):501–17.

63. Xia EH, Jiang JJ, Huang H, Zhang LP, Bin ZH, Gao LZ. Transcriptome analysis of the oil-rich tea plant, Camellia oleifera, reveals candidate genes related to lipid metabolism. PLoS One. 2014;9(8):e104150.

64. Shpilka T, Welter E, Borovsky N, Amar N, Shimron F, Peleg Y, et al. Fatty acid synthase is preferentially degraded by autophagy upon nitrogen starvation in yeast. Proc Natl Acad Sci U S A. 2015;112(5):1434–9.

65. Xie D, Jackson EN, Zhu Q. Sustainable source of omega-3 eicosapentaenoic acid from metabolically engineered Yarrowia lipolytica: from fundamental research to commercial production. Appl Microbiol Biotechnol. 2015;99(4): 1599–610.

66. Haslam RP, Sayanova O, Kim HJ, Cahoon EB, Napier JA. Synthetic redesign of plant lipid metabolism. Plant J. 2016;87(1):76–86.

67. Xue Z, He H, Hollerbach D, MacOol DJ, Yadav NS, Zhang H, et al. Identification and characterization of new Δ-17 fatty acid desaturases. Appl Microbiol Biotechnol. 2013;97(5):1973–85.

68. Hong H, Datla N, Reed DW, Covello PS, MacKenzie SL, Qiu X. High-level production of γ-linolenic acid in Brassica juncea using a Δ6 desaturase from Pythium irregulare. Plant Physiol. 2002;129(1):354–62.

69. Hong H, Datla N, MacKenzie S, Qiu X. Isolation and characterization of a delta5 FA desaturase from Pythium irregulare by heterologous expression in Saccharomyces cerevisiae and oilseed crops. Lipids. 2002;37(9):863–8.

70. Raffaele S, Kamoun S. Genome evolution in filamentous plant pathogens: why bigger can be better. Nat Rev Microbiol. 2012;10(6):417–30.

71. Lévesque CA, de Cock AWAM. Molecular phylogeny and taxonomy of the genus Pythium. Mycol Res. 2004;108(12):1363–83.

72. de Graaff L, van den Broek H, Visser J. Isolation and characterization of the Aspergillus nidulans pyruvate kinase gene. Curt Genet. 1988;13:315–21.

73. Robideau GP, De Cock AWAM, Coffey MD, Voglmayr H, Brouwer H, Bala K, et al. DNA barcoding of oomycetes with cytochrome c oxidase subunit I and internal transcribed spacer. Mol Ecol Resour. 2011;11(6):1002–11.

74. Chikhi R, Medvedev P. Informed and automated k-mer size selection for genome assembly. Bioinformatics. 2014;30(1):31–7.

75. Nurk S, Bankevich A, Antipov D, Gurevich AA, Korobeynikov A, Lapidus A, et al. Assembling single-cell genomes and mini-metagenomes from chimeric MDA products. J Comput Biol. 2013;20(10):714–37.

76. Walker BJ, Abeel T, Shea T, Priest M, Abouelliel A, Sakthikumar S, et al. Pilon: an integrated tool for comprehensive microbial variant detection and genome assembly improvement. PLoS One. 2014;9(11).

77. Corrêa dos Santos RA, Goldman GH, Riaño-Pachon DM. ploidyNGS: Visually exploring ploidy with next generation sequencing data. Not Peer-Reviewed. 2016;0–5.

78. Palsson B. Metabolic systems biology. FEBS Lett. 2009;583(24):3900–4.

79. Médigue C, Moszer I. Annotation, comparison and databases for hundreds of bacterial genomes. Res Microbiol. 2007;158(10):724–36.

80. Rocha I, Förster J, Nielsen J. Design and application of genome-scale reconstructed metabolic models. In: Inc HP, editor. Methods in molecular biology, vol 416: gene essentiality; 2007. p. 409–33.

81. Dias O, Gombert AK, Ferreira EC, Rocha I. Genome-wide metabolic (re-) annotation of Kluyveromyces lactis. BMC Genomics. 2012;13:517.

82. Käll L, Krogh A, Sonnhammer EL. L. a combined transmembrane topology and signal peptide prediction method. J Mol Biol. 2004 May;338(5):1027–36.

83. Kall L, Krogh A, Sonnhammer ELL. Advantages of combined transmembrane topology and signal peptide prediction--the Phobius web server. Nucleic Acids Res. 2007 May;35(Web Server):W429–32.

84. Smith TF, Waterman MS. Identification of common molecular subsequences. J Mol Biol. 1981;147(1):195–7.

85. Goldberg T, Hecht M, Hamp T, Karl T, Yachdav G, Ahmed N, et al. LocTree3 prediction of localization. Nucleic Acids Res. 2014;42(W1):350–5.

6

Nicotiana benthamiana is a suitable transient system for high-level expression of an active inhibitor of cotton boll weevil α-amylase

Guilherme Souza Prado[1,2], Pingdwende Kader Aziz Bamogo[3,4], Joel Antônio Cordeiro de Abreu[1,2], François-Xavier Gillet[1], Vanessa Olinto dos Santos[1], Maria Cristina Mattar Silva[1], Jean-Paul Brizard[3], Marcelo Porto Bemquerer[1], Martine Bangratz[3,4], Christophe Brugidou[3,4], Drissa Sérémé[4], Maria Fatima Grossi-de-Sa[1,2†] and Séverine Lacombe[3,4*†]

Abstract

Background: Insect resistance in crops represents a main challenge for agriculture. Transgenic approaches based on proteins displaying insect resistance properties are widely used as efficient breeding strategies. To extend the spectrum of targeted pathogens and overtake the development of resistance, molecular evolution strategies have been used on genes encoding these proteins to generate thousands of variants with new or improved functions. The cotton boll weevil (*Anthonomus grandis*) is one of the major pests of cotton in the Americas. An α-amylase inhibitor (α-AIC3) variant previously developed via molecular evolution strategy showed inhibitory activity against *A. grandis* α-amylase (AGA).

Results: We produced in a few days considerable amounts of α-AIC3 using an optimised transient heterologous expression system in *Nicotiana benthamiana*. This high α-AIC3 accumulation allowed its structural and functional characterizations. We demonstrated via MALDI-TOF MS/MS technique that the protein was processed as expected. It could inhibit up to 100% of AGA biological activity whereas it did not act on α-amylase of two non-pathogenic insects. These data confirmed that *N. benthamiana* is a suitable and simple system for high-level production of biologically active α-AIC3. Based on other benefits such as economic, health and environmental that need to be considerate, our data suggested that α-AIC3 could be a very promising candidate for the production of transgenic crops resistant to cotton boll weevil without lethal effect on at least two non-pathogenic insects.

Conclusions: We propose this expression system can be complementary to molecular evolution strategies to identify the most promising variants before starting long-lasting stable transgenic programs.

Keywords: Transient protein expression, α-amylase inhibitors, Gene silencing suppressors, Cotton boll weevil

* Correspondence: severine.lacombe@ird.fr
†Maria Fatima Grossi-de-Sa and Séverine Lacombe are contributed equally to this work.
³IRD, CIRAD, Université Montpellier, Interactions Plantes Microorganismes et Environnement (IPME), Montpellier, France
⁴INERA/LMI Patho-Bios, Institut de L'Environnement et de Recherches Agricoles (INERA), Laboratoire de Virologie et de Biotechnologies Végétales, Ouagadougou, Burkina Faso

Background

Biotic stresses such as insect pests induce dramatic damages in crops throughout the world, leading to significant losses for growers. To defend against these stresses, chemical treatments are largely used. However, due to health, environmental and cost concerns, for years attention has focused on genetic resistance, both in terms of conventional and transgenic applications [1, 2].

The most common transgenic plants displaying insect resistance (IR) carry genes encoding crystal toxins (Cry) from the soil bacterium *Bacillus thuringiensis* (*Bt*). Cry proteins solubilize in the insect midgut, where they become active and lead to cell lysis and insect death. Cry proteins are toxic to insects but not to humans or other vertebrates [3]. Despite a quite narrow range of control pathogens and low accumulation levels in plants, *Bt* IR crop plants represent one of the most successful achievements in plant transgene technology [2]. Currently, several *Bt* plants, including corn, cotton and soybean, grow under field conditions worldwide [4]. However, lack of high dose *cry* expression in plants still can lead to the selection of insect varieties that acquire resistance against the toxic effects of the Cry molecules via adaptation [5].

On the other hand, plants are equipped with natural defence systems against pests such as insects. These defences mainly involve antimetabolite proteins that induce alterations to the digestive system of insect pests. The transfer of proteinase inhibitor genes from one plant to another has been widely used to develop insect-resistant plants [6–8]. For example, when expressed in *Nicotiana benthamiana*, a beetroot gene encoding a proteinase inhibitor induces resistance to lepidopteran insect pests [9]. Lectins are plant carbohydrate-binding proteins that present a high toxicity to phytophagous insects [10]. Lectins have been used in genetic transformation to provide resistance against spider mite in papaya [11]. Chitinases are also plant-expressed proteins that can provide IR when expressed in a transgenic context [12, 13].

Alpha-amylase inhibitors (α-AI) produced in common bean (*Phaseolus vulgaris*) and other *Phaseolus* species act on α-amylase present in insect guts by inhibiting the processing of complex sugars and, consequently, the growth of insect larvae [14]. They exist as two isoforms, α-AI1 and α-AI2, that undergo proteolytic cleavage from a preprotein to two polypeptides: α- and β-subunits [15]. In addition, amino acid hydrolysis occurs at the C-terminal ends of both α- and β-subunits, giving rise to 10 and 15 kDa chains, respectively [16]. Even if the unprocessed and processed forms accumulated in plants, it has been shown that proteolysis is required for inhibitory activity [15]. Despite a relatively high similarity, α-AI1 and α-AI2 act on specific and distinct spectra of insect α-amylases [14]. Transgenic processes to express

bean α-AI have been widely used on several plant species for the improvement of IR [17–20].

Despite the efficiency of these IR strategies, the spectrum of insects controlled by any given protein is quite narrow. Moreover, whatever the controlling strategy is, it must face the development of resistant insects. Hence, to extend the spectrum of target pathogens and to overtake the development of insect resistance, molecular evolution strategies have been used on original IR proteins to generate thousands of variants with potentially new or improved functions [21, 22]. New resistances have been identified from these libraries for the cotton boll weevil (*A. grandis*), sugarcane giant borer (*Telchin licus licus*) and mustard aphid (*Acyrthosiphon pisum*) [23–26]. These findings highlight the importance of the variant libraries to create new IR to harmful insect pests that act on major crops worldwide. However, even with this important agricultural interest, a deep characterization of these proteins is required to demonstrate their economic interest and safety impact such as allergenic issues [27].

Systems allowing low-cost and rapid screening of these libraries are necessary to identify the most promising variants before starting long-term and costly transgenic programmes. Cry and trypsin inhibitor variants are expressed in phage systems before in vitro screening of inhibitory activity [23–25]. However, this phage-based system is not suitable for plant variants requiring post-translational modifications for their activities, such as α-AI. Moreover, the final goal is to express these variants in plants, implying that they would be processed by the plant cell machinery. Consequently, plant-based systems could be more convenient than phage- or prokaryote-based systems to screen these variants and select the most promising ones. The model plant *Arabidopsis thaliana* has been used to stably express α-AI variants. This system allowed the identification of a very promising variant, α-AIC3 that was able to inhibit 77% of the α-amylases from the insect *A. grandis*, whereas the original α-AI forms were ineffective. This variant differs from the original sequence by several amino acid changes induced by the molecular evolution strategy performed [26]. This outcome represents an important finding for the cotton culture in the Americas, where *A. grandis* is among the major insect pests. Consequently a deep characterization of this variant should be done before starting a promising transgenic cotton program. However, *A. thaliana* transgenic-based screenings may not be suitable for evaluating potentially interesting proteins from thousands of variant libraries. Therefore, in order to characterize accurately such protein variants, it is crucial to establish an alternative and robust plant-based expression system that allows the expression of recombinant proteins at high yield and with accuracy in terms of post-translational modifications.

In recent years, advances in biotechnology have led to the emergence of plants as bioreactors for the production of proteins of interest not only in stable transgenic systems but also in transient systems [28]. The first crucial advance was the use of transient expression systems relying on *Agrobacterium* as a vector to deliver DNA encoding proteins of interest directly into leaf cells by syringe infiltration – so-called agroinfiltration [29]. Moreover, protein production can be increased by the co-expression of viral proteins displaying suppression of gene silencing activity. Indeed, the presence of such viral proteins in transient expression systems allows overcoming the gene silencing triggered by the plant defence machinery to specifically degrade foreign nucleic acids. Consequently, the yield of the protein of interest is dramatically increased by 50 fold or more [30, 31].

Here, we describe a high-yielding, easier, quicker and cheaper system compared to the stable transformation of *A. thaliana*. This well-known system is based on the transient expression of the protein of interest in *N. benthamiana* leaves (see for review [32]). As previously described, a combination of three viral suppressors of gene silencing are used to improve the expression in terms of accumulation levels [31]. We focused on an α-AIC3 variant that was previously demonstrated to act on one of the most damaging insects to cotton culture in the Americas – the cotton boll weevil (*A. grandis*) [26]. We showed that these proteins that were transiently expressed in *N. benthamiana* leaves, accumulated at high levels and exhibited their expected post-translation maturation and in vitro function on the target insect enzyme. We proposed this system to be complementary to molecular evolution strategies to allow easy selection and characterization (within a few days) of the most promising variants from molecular evolution libraries before starting stable transgenic programs.

Results

α-AIC3 expression in *N. benthamiana* leaves

To optimize the accumulation of α-AIC3 in *N. benthamiana* leaves, the *aic3* gene was transiently co-expressed in 4-week-old wild-type *N. benthamiana* plants together with genes encoding three viral gene silencing suppressors. It has been previously demonstrated that these suppressors act synergistically by inhibiting three different steps of the gene silencing defence mechanism [31]. Agroinfiltrated leaf regions were collected at 5 dpi and weighted, after which protein was extracted. A total of 40 μg of soluble proteins representing approximately 10 mg of fresh leaves was separated by 15% (m/v) SDS-PAGE and blotted onto a nitrocellulose membrane. The Coomassie Blue-stained gel (Fig. 1a) showed additional

bands of lower molecular mass for samples 2 (pBIN61:α-AIC3) compared to samples 1 (pBIN61), suggesting that this difference was due to the *aic3* gene expression and protein accumulation in the leaves. This result was confirmed by Western blot (Fig. 1b) using a specific anti-α-AIC3 primary antibody; samples 1 did not show any visible band or signal, but samples 2 presented a pattern composed of three intense bands. The lower bands were very intense and referred to the processed α-AIC3 forms, which may correspond to α- and β-subunits of 12 kDa and 15 kDa, respectively. Moreover, bands of higher molecular weight also appeared that were approximately 28 kDa, strongly suggesting that they correspond to the unprocessed forms of α-AIC3; these bands had a less intense signal than the bands attributed to the processed subunits. The results here indicate that α-AIC3 was successfully expressed and mostly processed according to the expected proteolytic processing. Furthermore, the generated bands were not linear but dispersed. These patterns suggest several isoforms that could result from the expected post-translational maturation processes for these inhibitors including amino acid hydrolysis at the C-terminus ends of both subunits and glycosylation [16]. However, despite their accurate size, we cannot exclude that the observed bands were due to protein aggregation or degradation. The following structural and functional characterization were performed to exclude this possibility.

α-AIC3 yield and expression level analysis and protein purification

The yield of the protein of interest in the dialyzed samples was measured by indirect ELISA using a specific anti-α-AIC3 primary antibody. The expression level of α-AIC3 was estimated for the pBIN61:α-AIC3 samples, considering pBIN61 samples as negative controls. In the first experiment, 13.6 ng of α-AIC3 out of 40 ng of total protein were detected, indicating a yield of 34% of Total Soluble Proteins (TSP) for the heterologous protein. Based on the percentage of the specific expression of α-AIC3, this amount corresponded to a yield of 0.1 mg/g fresh weight (FW) tissue or 100 mg/kg FW. In another experiment, 70.4 ng of α-AIC3 of 160 ng of total protein was detected, indicating a yield of 44% TSP for the heterologous protein and corresponding to 0.15 mg/g FW or 150 mg/kg FW. For the protein purification, a dialyzed extract was used, and proteins were loaded on a gel-filtration column for performing size exclusion chromatography (SEC). A total of 90 fractions consisting of 2 mL each were obtained. The chromatograms showed different peaks for fractions 10–14, 16–20 and 30–58 (Fig. 2a). Hence, some fractions (12, 17, 18, 19, 26, 37, 40, and 42) from each peak were selected to perform electrophoresis and separate samples to further

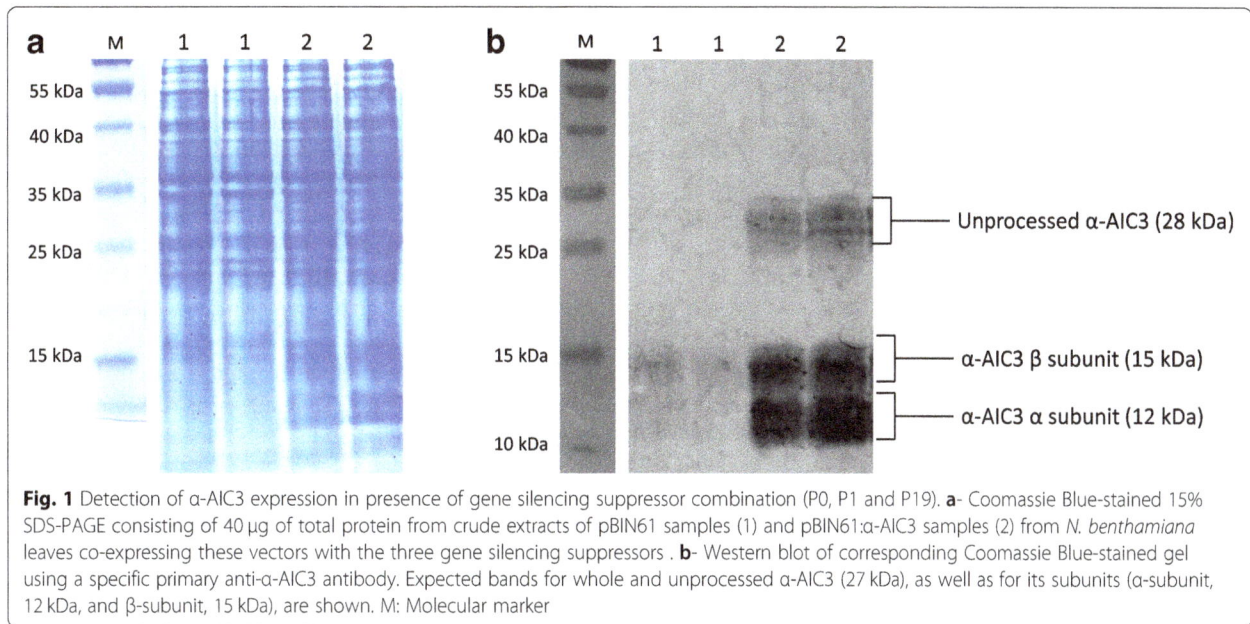

Fig. 1 Detection of α-AIC3 expression in presence of gene silencing suppressor combination (P0, P1 and P19). **a**- Coomassie Blue-stained 15% SDS-PAGE consisting of 40 µg of total protein from crude extracts of pBIN61 samples (1) and pBIN61:α-AIC3 samples (2) from *N. benthamiana* leaves co-expressing these vectors with the three gene silencing suppressors . **b**- Western blot of corresponding Coomassie Blue-stained gel using a specific primary anti-α-AIC3 antibody. Expected bands for whole and unprocessed α-AIC3 (27 kDa), as well as for its subunits (α-subunit, 12 kDa, and β-subunit, 15 kDa), are shown. M: Molecular marker

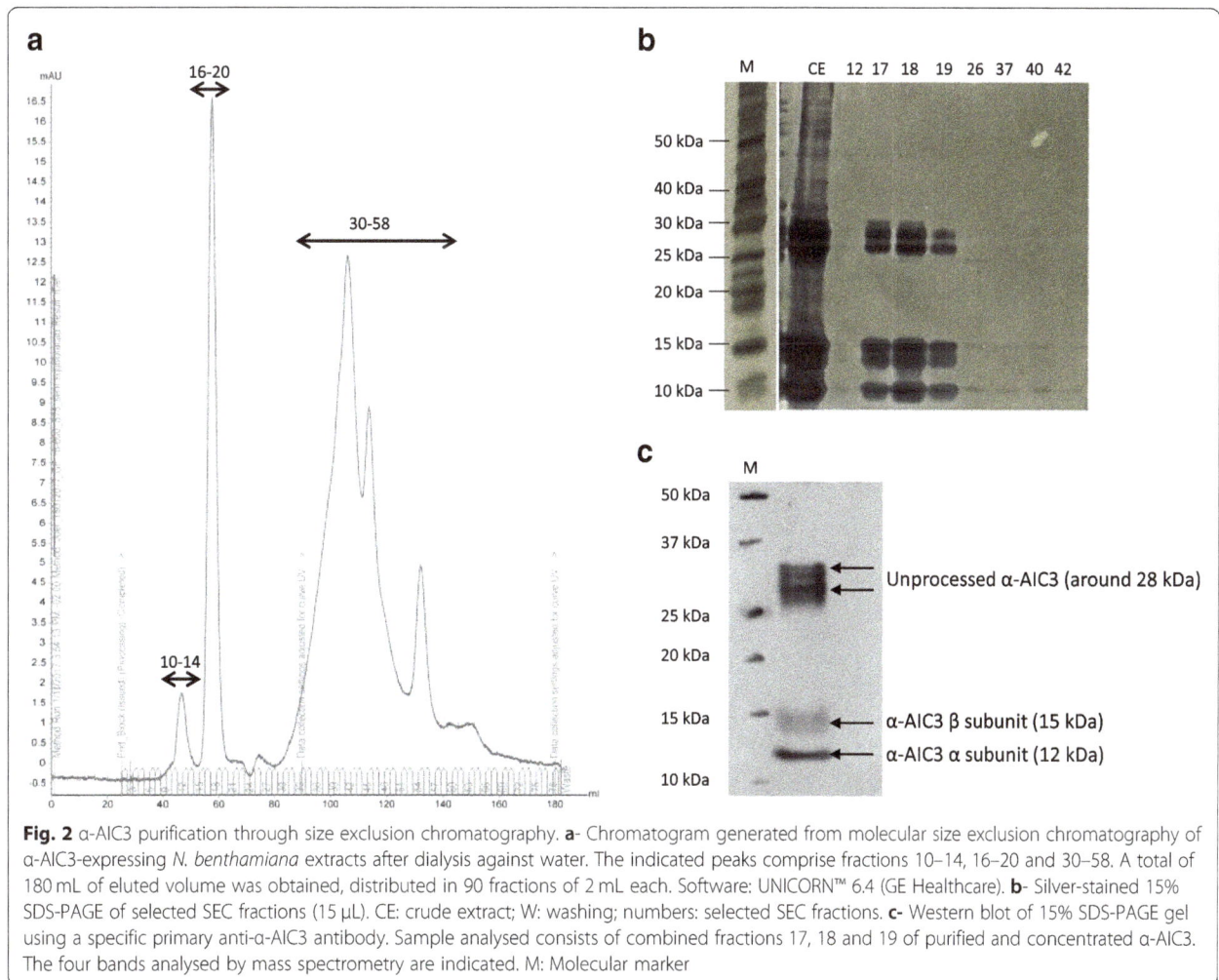

Fig. 2 α-AIC3 purification through size exclusion chromatography. **a**- Chromatogram generated from molecular size exclusion chromatography of α-AIC3-expressing *N. benthamiana* extracts after dialysis against water. The indicated peaks comprise fractions 10–14, 16–20 and 30–58. A total of 180 mL of eluted volume was obtained, distributed in 90 fractions of 2 mL each. Software: UNICORN™ 6.4 (GE Healthcare). **b**- Silver-stained 15% SDS-PAGE of selected SEC fractions (15 µL). CE: crude extract; W: washing; numbers: selected SEC fractions. **c**- Western blot of 15% SDS-PAGE gel using a specific primary anti-α-AIC3 antibody. Sample analysed consists of combined fractions 17, 18 and 19 of purified and concentrated α-AIC3. The four bands analysed by mass spectrometry are indicated. M: Molecular marker

identify presence of α-AIC3 subunits. Silver staining demonstrated that fractions 17, 18 and 19 (Fig. 2b) presented expected bands for α-AIC3. Indeed, these patterns were very similar to the one previously revealed via Western blotting using a specific anti-α-AIC3 primary antibody (Fig. 1b). Based on these results, these fractions were pooled concentrated and separated again for Coomassie Blue staining and western blotting. Four main bands were clearly detected by western blotting at the expected size for unprocessed (two bands around 28 kDa), β (15 kDa) and α subunits (12 kDa) (Fig. 2c). Corresponding bands visualized on Coomassie staining gel were excised to structurally characterize the proteins and confirm identity with the protein of interest.

Structural characterization

Spots were excised from the four bands and prepared for MALDI-TOF MS/MS analysis. Spectra of the generated peptides were fragmented, some of which are shown in Fig. 3 with respect to both α-AIC3 subunits. For the α-subunit, one of the four possible tryptic peptides was detected and confirmed after sequencing: AFYSAPIQIR. This finding indicates a coverage of 10 of 73 amino acid residues for the α-subunit, resulting in 14% coverage. For the β-subunit, five of the twelve possible tryptic peptides were detected and confirmed after sequencing: GDTVTVEFDTFLSR, SVPWDVHDYDGQNAEVR, ELDDWVR, VGFSAISGVHEYSFETR and DVLSWSFSSK. This finding indicates a coverage of 65 of 135 residues for the β-subunit, resulting in 48% coverage. In total, six peptides were detected, sequenced and confirmed, indicating a coverage of 75 of 221 residues for α-AIC3 or 34% coverage (Fig. 4). The peptide of the α-subunit was found in all samples corresponding to bands at 28 kDa, 25 kDa and 12 kDa. The five peptides of the β-subunit were found in the samples related to bands 28 kDa, 25 kDa and 15 kDa. This finding strongly supports that bands at 15 kDa and 12 kDa represent the β- and α-subunits, respectively, and

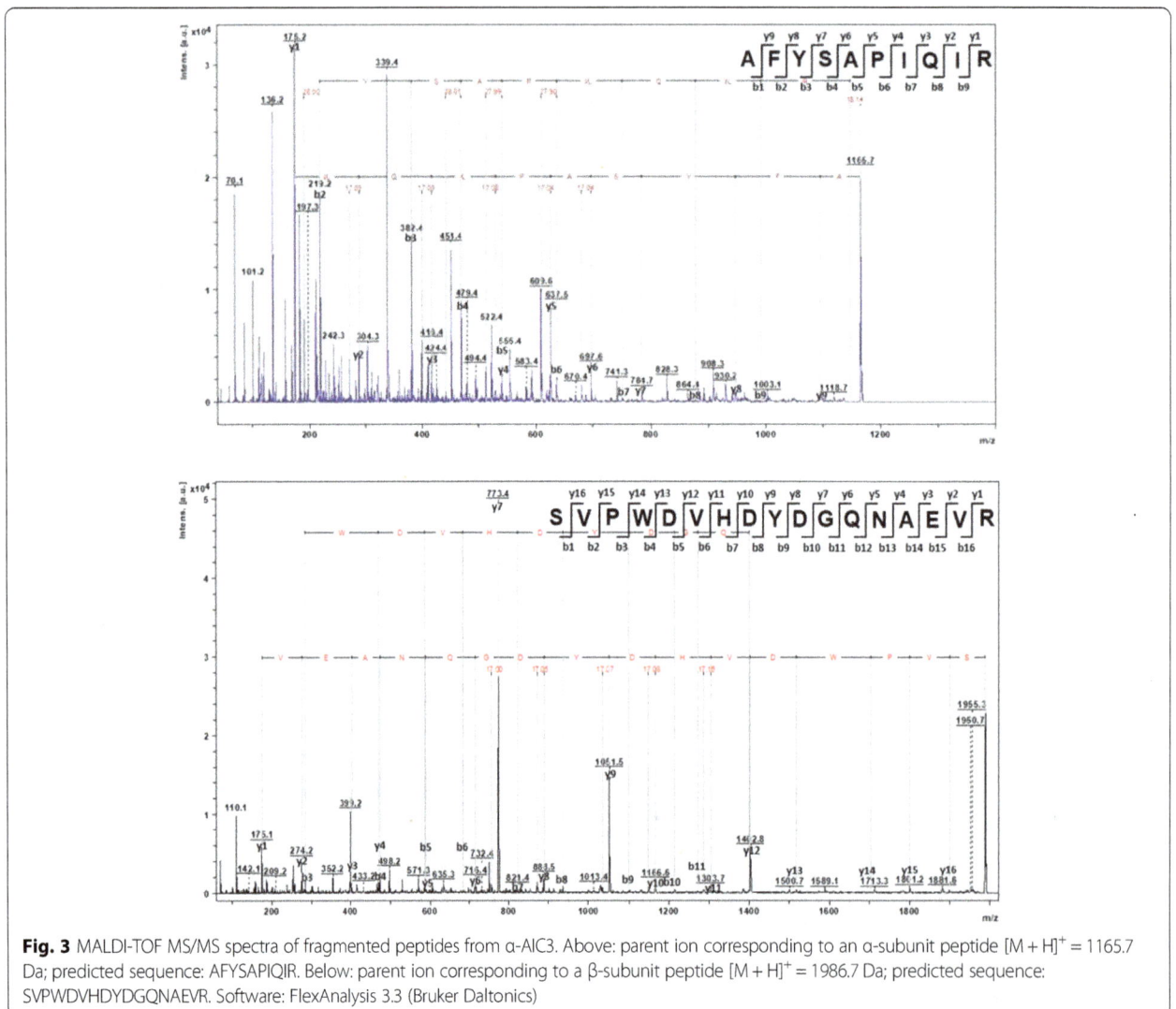

Fig. 3 MALDI-TOF MS/MS spectra of fragmented peptides from α-AIC3. Above: parent ion corresponding to an α-subunit peptide [M + H]⁺ = 1165.7 Da; predicted sequence: AFYSAPIQIR. Below: parent ion corresponding to a β-subunit peptide [M + H]⁺ = 1986.7 Da; predicted sequence: SVPWDVHDYDGQNAEVR. Software: FlexAnalysis 3.3 (Bruker Daltonics)

ATETSFIIDAFNKTNLILQGDATVSSNGNLQLSYNSYDSMSR AFYSAPIQIR DSTTGNVASFDTNFTMNIRTH**R**   —— **α-chain**

QANSAVGLDFVLVPVQPESK GDTVTVEFDTFLSR ISIDVNNNDIK SVPWDVHDYDGQNAEVR ITYNSSTKVLA

β-chain

VSLSNPSTGKSNEVSARMEVEK ELDDWVR VGFSAISGVHEYSFETR DVLSWSFSSK FSQHTTSERS<u>NILLNKIL</u>

Fig. 4 Total sequenced peptides from the α-AIC3 chain. α-AIC3 whole sequence, showing amino acid residues of the α- and β-subunits; the respective peptides were identified, fragmented and sequenced via MALDI-TOF MS/MS and are highlighted inside the rectangles. In total, six peptides were sequenced, one for the α-subunit and five for the β-subunit. C-terminal end peptides, which were cleaved off to yield the mature subunits, for both subunits are underlined in the figure

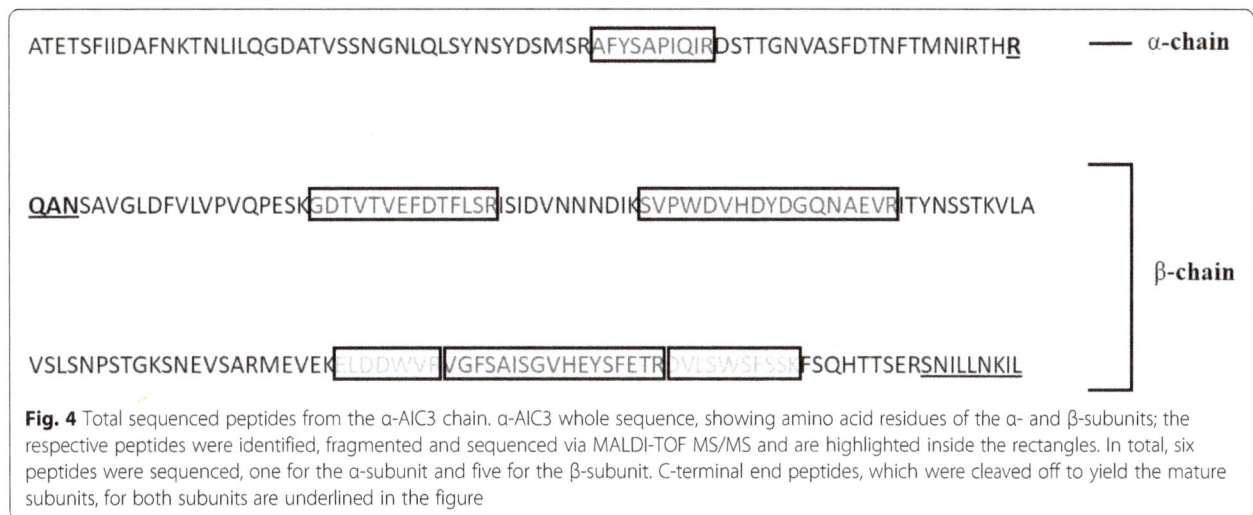

that bands 28 kDa and 25 kDa represent the whole unprocessed protein, since it contains sequences of both subunits. However, peptides detected do not cover both N- and C-terminus ends of each subunit. Consequently, we cannot exclude that subunits were not intact.

Functional characterization

Based on the DNS method, the α-AIC3-containing samples of *N. benthamiana* showed an average inhibition of 98.5% in *A.grandis* α-amylase (AGA) activity when using 1 unit of enzyme and 100 µg of soluble protein. This inhibition level was validated based on the inhibition obtained using the same amount of *A. thaliana*

α-AIC3-containing samples, which completely inhibited the AGA activity. The experiments were repeated for extracts from three different agroinfiltrations. These results showed that the AGA activity inhibition level varied from 96.7 to 100% (Fig. 5). The same extracts of the third agroinfiltration were simultaneously used to assess the inhibition level for AMA and SFA, and did not show any significant inhibition activity (Fig. 5), since absorbances were the same for reactions with or without α-AIC3 and containing active *Apis mellifera* amylase (AMA) and *Spodoptera frugiperda* amylase (SFA) enzymes. Hence, regardless of any assays using AMA and SFA specific inhibitor controls because of their currently

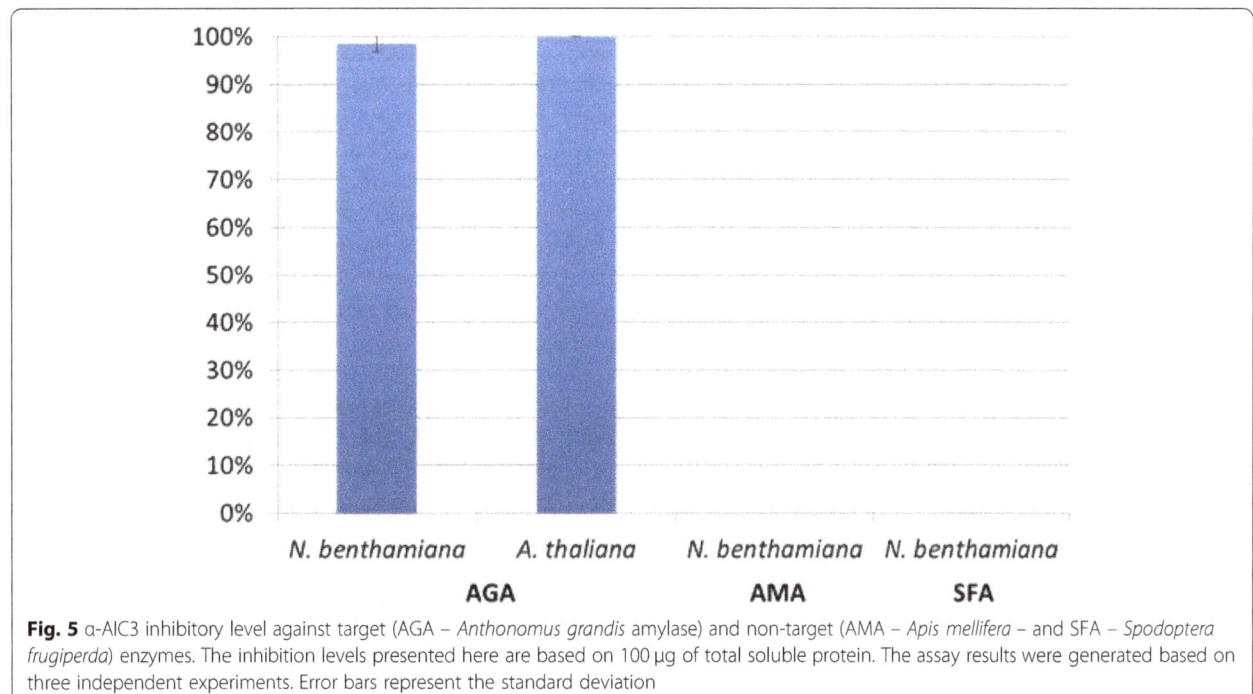

Fig. 5 α-AIC3 inhibitory level against target (AGA – *Anthonomus grandis* amylase) and non-target (AMA – *Apis mellifera* – and SFA – *Spodoptera frugiperda*) enzymes. The inhibition levels presented here are based on 100 µg of total soluble protein. The assay results were generated based on three independent experiments. Error bars represent the standard deviation

unavailability, these data suggested that α-AIC3 produced in *N. benthamiana* was unable to inhibit α-amylase from *A. mellifera* and *S. frugiperda*. Altogether, these results showed that the protein of interest exhibits its expected activity on AGA but exhibits no inhibitory activity against the amylases of these non-target insect species.

Discussion

The work presented here demonstrates that *N. benthamiana* coupled with the use of a cocktail of gene silencing suppressors is a suitable system for quickly and easily producing a quite high level of an α-AI variant, α-AIC3. The yield was estimated in the range of 100 to 150 mg/kg FW. Plant biosystem yield for proteins of interest is quite variable, reaching up to 2 g/kg of FW in the case of optimised viral based technology [33]. Here, we are in the upper range for non- viral based technology. Moreover, we demonstrated that all the expected α-subunit, β-subunit and unprocessed forms accumulated, mostly consisting of the processed α- and β-subunits. Finally, the expected protein functionality as an inhibitor of AGA was demonstrated, and α-AIC3 did not inhibit the α-amylases of two non-target insects tested (*A. mellifera* and *S. frugiperda*), preliminarily suggesting that α-AIC3 is biologically and environmentally safe. However, it is recommended to perform in vivo studies of α-AIC3, giving rise to an even more realistic result regarding the protein yield and safety,furnishing data concerning the feasibility to produce genetically modified cotton plants that could be resistant to *A. grandis* without triggering biosafety traits such as environmental imbalances or allergenicity.

Concerning the inhibitory assays, we could achieve up to 100% inhibition of AGA using 100 μg of *N. benthamiana* extracts, indicating that up to 44 μg of α-AIC3 were used to inhibit 1 U of AGA, while the complete inhibition of AGA was also achieved using the same amount of total protein from *A. thaliana* seeds. Silva et al. [26] demonstrated that the inhibitory activity was 77% when using 85 μg of total protein from *A. thaliana* leaves expressing this inhibitor at an expression level of 0.2% of TSP, which is very low compared to the level in transient expression obtained here. This means that, from the total protein, approximately 170 ng of α-AIC3 are necessary to inhibit 77% of the AGA activity, and in our study, we can estimate that approximately 200 ng of α-AIC3 are enough to inhibit 1 U of AGA if considering similar expression levels of α-AIC3 in seeds as in leaves for transgenic *Arabidopsis* as reported in the case of 35S transgenic constructs [34]. These reproducible *Arabidopsis* values validate the functional assay used here.

For *N. benthamiana* extracts, the amount of inhibitor used to inhibit completely the AGA activity was much higher, since 100 μg of soluble proteins contain up to 44 μg of α-AIC3 (considering an expression level of 44% TSP). As the kinetic parameters of this enzymatic assay were not known, we cannot exclude that this high amount of inhibitor present in *N. benthamiana* extract saturated the assay. Thus, fewer protein amounts could have triggered similar inhibition levels. Furthermore, we must also consider that the amount of useful α-AIC3, i.e., the amount that effectively participates in the enzyme inhibition was considerably lower than 44 μg. Indeed, based on the western (Fig. 1b), part of total α-AIC3 is composed of unprocessed chains, unable to inhibit the AGA activity as the post-translational processing is imperative for the acquisition of biological activity in α-AI proteins [15]. Moreover, from results from MALDI-TOF MS/MS analysis, we cannot determine the amount of α- and β-subunits produced that are fully active.

Regardless of this, the transient expression system remains a suitable alternative to stable expression systems because former exhibits practicality: it is simple as it does not require complex materials nor techniques and provides considerable amounts of protein without the need for regenerating plants and selecting transformants. Dias et al. [35] also used tobacco-based expression to produce an α-amylase inhibitor (αBIII) of *Secale cereale* in *Nicotiana tabacum* seeds via stable expression. This system also yielded low protein levels (0.1–0.29% TSP) that were similar to those in *A. thaliana* [26] and achieved a maximum inhibition of only 41% of AGA activity when using 250 mg of crude protein extract.

Altogether, these data suggest that this *N. benthamiana* transient expression system may be suitable for the rapid, easy and efficient production of α-AI variants obtained from molecular evolution strategies for preliminary functional screening and biosafety studies. The α-AIC3 variant analysed here was identified from a library that consisted of more than 8000 variants [26]. With such transient system, this library could be efficiently exploited to identify variants with new or improved IR functions against major pests as describe above for the potato gene encoding a disease resistant protein against a virus [36, 37].

Nicotiana-based transient expression systems have been widely used to express proteins of interest, such as those for vaccines and biopharmaceuticals [28]. Fewer examples have been described for proteins of agricultural interest. Farnham and Baulcombe [36] produced a variant library using random mutagenesis from a potato gene encoding a disease resistant protein (Rx) against a subset of potato virus X (PVX). Those authors used transient expression in *N. tabacum* to screen 1920 variants. Thirteen of those variants induced a cell-death response in the presence of the PVX coat protein, indicative of disease resistance [36]. The same potato gene encoding Rx resistant protein was also used to generate a library of 1500 mutants that were transiently expressed in *N. benthamiana* together with the gene encoding for the Poplar mosaic virus (PopMV) coat protein. This phenomenon allowed the identification of four variants inducing a cell-death response related to resistance to this new virus [37]. Similar to the results reported here, these studies indicate the interest in this dual variant library/*N. benthamiana* transient expression strategy for its easy, rapid, low-cost ability to identify new or improved disease resistance genes from thousands of variants.

Other proteins are known to be associated with IR and have been used in genetic transformation to bring new IR to important crops worldwide [2]. Molecular evolution libraries consisting of thousands of variants have been developed for Cry proteins and proteinase inhibitors. These variants have been screened using phage-based assays to identify variants with new IR functions not present in the original forms [23, 24]. Because the ultimate goal was to exploit these variants in transgenic plants, the *N. benthamiana* transient expression system presented here would be more accurate for the heterologous expression of the variants with the goal of performing functional tests in order to preview the responses of the transformed plant as a definite host. As such, functional tests must serve as a filtering step, as it is more difficult to regenerate several plants displaying a wide selection of candidate variants for additional sorting of promising proteins and events. Therefore, heterologous expression could save time and material and could reduce the complexity of the process for obtaining transformed and commercially feasible events. For this, it is important to gradually characterize the candidates, as was done in this study. An in vitro stage of characterization is needed to validate the proposed activity against the target. However, it is suitable to proceed with an in vivo and complementary stage of assays in which the insects are grown in the presence of the molecules, checking systemic effects in the insect. Once the biological activity is confirmed following this stepwise study, investigations on genetic transformation will be much more reliable since regenerated plants will display the same in vivo observed activity.

Plant breeding has been recently revolutionized with the advent of genome editing technologies allowing precise modifications in genomic sequences with the so-called genome engineering [38, 39]. Several economically important species, such as cotton, are suitable targets for these technologies [40, 41], especially concerning agronomic traits. These technologies have been successfully used in maize, soybean and rice to induce exact mutations in specific genes, leading to herbicide tolerance

[42–44]. Resistance development against biotic stresses can also benefit of these technologies as shown by the development of a genome-edited tomato displaying powdery mildew resistance [45] and an engineered cucumber showing broad virus resistance [46]. Based on that, we can speculate that the dual strategy variant library/*N. benthamiana* transient expression allowing identification of variants of interest could be followed by genome editing technologies to precisely induce modifications in the genome of crops. Results presented here suggest that α-AIC3 would be an ideal candidate to evaluate this hypothesis, as well as producing genome-edited plants displaying new or improved IR through α-AI specific modifications. Moreover, whatever the gene of interest, *N. benthamiana* system presented here could be a useful tool to rapidly and easily identify variants that could be integrated in plant genomes through genome editing strategies.

Conclusions

In this study, we reported successful transient expression of α-amylase inhibitors using *N. benthamiana*-based system with a recent established combination of gene silencing suppressors. We showed that this system is highly suitable for producing variants of mutant inhibitors, which were expressed not only at a very high yield but also with the correct, albeit incomplete, processing, preserving the expected biological function.

Methods
Expression vectors and silencing suppressors
The experiments were performed using *Agrobacterium tumefaciens* C58C1 strain harbouring pBIN61:α-AIC3 expression vector for producing the protein of interest or empty pBIN61 vector for negative control. Based on our previous work demonstrating the positive effect of the simultaneous expression of gene silencing suppressors on the accumulation of candidate protein by blocking the gene silencing defence mechanism [31], these additional gene silencing suppressor vectors were used for the co-expression with pBIN61 vectors. They encoded for P0 from Beet western yellow virus (pBIN61:P0 vector) [47], P1 from Rice yellow mottle virus (pCambia1300:P1Tz3 vector) [48] and P19 from Cymbidium ringspot virus (pBIN61:P19 vector) [49]. Each of them was cloned into expression vectors and transformed in *Agrobacterium tumefaciens* C58C1 strain.

Gene design, synthesis and cloning
The nucleotide sequence for the gene (*aic3*) encoding the α-AIC3 variant was obtained in silico via reverse translation and codon optimization of the α-AIC3 protein sequence [GenBank:AGB50990.1], as reported by Silva et al. [26]. Codon optimization was performed with

Gene Designer 2.0 software [50] based on the codon usage table for *N. benthamiana* species, available at Kazusa Codon Usage Database. The nucleotide sequence for the corresponding native signal peptide (MASSNLLSLALFLVLLTHANS) was also retrieved and codon-optimized. The final insert sequence was flanked by 5′-*Xba*I and 3′-*Bam*HI restriction sites, and a Kozak consensus sequence (GCCACC) was inserted immediately upstream of the start codon. No restriction sites for *Xba*I and *Bam*HI were detected within the CDS. SignalP 4.1 Server was used for signal peptide detection and validation. The sequence was synthesized de novo by Epoch Life Science® and cloned into the *Xba*I-*Bam*HI cloning sites of the pUC18 vector to generate pUC18:α-AIC3. The *aic3* gene was then excised from pUC18:α-AIC3 and cloned into the *Xba*I and *Bam*HI sites of the pBIN61 binary expression vector, which was previously described by Bendahmane et al. [51] under the control of the constitutive CaMV 35S promoter and terminator to generate pBIN61:α-AIC3 that was used to transform *A. tumefaciens* strain C58C1 via electroporation. The cloning into the pBIN61 vector was confirmed by sequencing using the M13 forward primer and carried out by Beckman Coulter Genomics®. Cells were also transformed with empty vectors; these cells served as negative controls.

Agroinfiltration, plant material and experimental conditions

Strains harbouring empty pBIN61, pBIN61:α-AIC3, pBIN61:P0, pCambia1300:P1Tz3 and pBIN61:P19 vectors were separately grown overnight from precultures at 28 °C and 200 rpm in an orbital shaker using LB medium containing rifampicin (100 µg/mL) and kanamycin (50 µg/mL). The cultures were pelleted by centrifugation for 10 min at 4000 g, after which the pellets were resuspended in 10 mM MgCl₂ to a final OD600 of 0.5. Acetosyringone (4-hydroxy-3,5-dimethoxyacetophenone) was added to each suspension to a final concentration of 100 µM for virulence induction, and the suspensions were incubated at 24 °C for 3 h. Agroinfiltration cocktails were prepared by combining cultures for co-infiltration: for the negative control, the pBIN61 culture was combined with the cultures of silencing suppressors (pBIN61:P0:P1:P19, 3:1:1:1, v/v:v/v), and the same procedure was employed for protein expression, in which the pBIN61:α-AIC3 culture was combined with the cultures of silencing suppressors (pBIN61:α-AIC3:P0:P1:P19, 3:1:1:1, v/v/v/v). Cocktails were infiltrated into the leaves of 4 weeks old wild-type *N. benthamiana* plants using syringes without needle. Four plants were used for the negative control per experiment, while twelve plants were used for α-AIC3 expression. The plants were placed in a growth chamber and

cultivated for 5 days before harvesting (12 h of light per day, 24 °C, 60% relative humidity). Three independent experiments were performed to generate three biological replicates for subsequent molecular analysis.

Protein extraction, dialysis and concentration

Infiltrated leaf tissues were harvested from the plants at 5 days post-infiltration (dpi). The fresh leaves were combined in their respective groups (negative control and α-AIC3 expression), weighted, frozen in liquid nitrogen and then ground using a mortar and pestle. Protein extraction was performed by adding 700 µL of extraction buffer (20 mM Tris-Cl, 100 mM NaCl, 10 mM Na₂EDTA·2H₂O, 25 mM D-glucose, 0.1% Triton X-100, 5 mM EGTA, 5% (v/v) glycerol, 5 mM dithiotreitol, and 1 mM phenylmethanesulfonyl fluoride, pH 7.4) to 300 mg of tissue powder. Crude extracts were incubated on ice for 20 min, strongly shaken for 20 min at 4 °C using a vortex and centrifuged at 14000 g for 30 min at 4 °C. The total soluble proteins (TSP) were recovered from the supernatants and dialyzed against water (1 mL of extract per 200 mL of distilled water) using Slide-A-Lyzer™ G2 Dialysis Cassettes (ThermoFisher Scientific) that had a 10 kDa molecular weight cut-off (MWCO). The dialyzed samples were clarified by centrifugation at 14000 g for 10 min and quantified by a Bio-Rad® Bradford protein assay [52] based on a bovine serum albumin (BSA) (Sigma Aldrich) standard curve.

SDS-PAGE and Western blot

A total of 40 µg of protein for each extract was subjected to low-pressure drying, resuspended in 15 µL of pure water and then diluted in protein loading buffer [53] with 2-mercaptoethanol. The samples were incubated at 95 °C for 5 min, loaded and then separated by 15% (m/v) SDS-PAGE. A mirror gel was also made for protein detection via immunoblotting. Proteins were stained with Coomassie Brilliant Blue G-250 or blotted onto a nitrocellulose membrane at 5 V for 20 min in a Trans-Blot® SD semi-dry system (Bio-Rad) after the membrane and gel were treated with blotting buffer (20 mM Tris base, 150 mM glycine, 20% methanol, pH 8.3) for 10 min. Western blot analysis proceeded by blocking the membrane with a 3% (m/v) solution of skimmed milk powder in TBS-T buffer (20 mM Tris base, 150 mM NaCl, 0.1% Tween 20, pH 7.5) for 2 h under shaking. The protein was probed by adding a primary specific anti-α-AIC3 rabbit IgG (GenScript) to the TBS-T buffer (1:2500 of antibody:buffer, or at 0.4 µg.mL⁻¹), after which the membrane was incubated for 2 h under shaking. After six five-minute rinses with TBS-T buffer, the bound antibodies were probed by adding an AP-conjugated secondary goat anti-rabbit IgG (Sigma Aldrich) to the TBS-T buffer (13,000 of antibody:buffer, or 0.3 µg.mL⁻¹), after which the membrane was incubated again for 1 h under shaking. Subsequent washing followed as described, and the proteins were detected using a colorimetric AP substrate reagent kit (Bio-Rad) according to the manufacturer's instructions.

ELISA: α-AIC3 quantification

Dialyzed samples were used to estimate the expression level of α-AIC3 in the total protein by indirect ELISAs. Assays were performed in triplicate by coating 96-well microplates with 40 ng or 160 ng of total protein. A standard curve of protein amount ($R^2 = 0.9948$) was constructed based on a gradient from 0.2 ng to 200 ng, in a total of 11 dilutions, of both the bacterial and purified β-subunits of α-AIC3 previously produced and kindly provided by Dr. Leonardo Macedo (Embrapa Genetic Resources and Biotechnology, Brasília, Brazil). The samples were diluted in coating buffer (50 mM sodium bicarbonate/carbonate, pH 9.6), and the coated plates were incubated at 4 °C for 18 h. Samples were incubated at 37 °C for 1 h and washed thrice with 200 μL of PBS-T buffer (136 mM NaCl, 3 mM KCl, 10 mM Na_2HPO_4, 2 mM KH_2PO_4, and 0.05% Tween 20, pH 7.4). The membrane was blocked by using 3% (m/v) gelatin in PBS-T buffer for 2 h at 37 °C. The samples were discarded, washed and incubated together with 100 μL of a primary anti-α-AIC3 antibody (GenScript) diluted in PBS-T buffer with 1% gelatin (1:1000 of antibody:buffer, v/v, or at 1 μg.mL^{-1}) for 2.5 h, at 37 °C. The samples were then washed and incubated together with 100 μL of an HRP-conjugated secondary goat anti-rabbit IgG H + L (Bio-Rad) in PBS-T buffer with 1% (m/v) gelatin (1:3000 antibody:buffer, v/v, or at 0.3 μg.mL^{-1}) for 1 h at 37 °C. The samples were detected with 100 μL of a revealing solution as peroxidase substrate consisting of 10 mL of phosphate-citrate buffer (24.3 mM citric acid, 51.4 mM Na_2HPO_4, and 0.06% H_2O_2, pH 5.0) and 1 mg of 3,3′,5,5′-tetramethylbenzidine (TMB) (Sigma Aldrich). The colour reaction was stopped after 15 min at room temperature with 100 μL of stop solution (3 M H_2SO_4). The absorbance values were read at 450 nm using a SpectraMax 190 microplate reader (Molecular Devices), and the samples were analysed according to the appropriate calculations using Excel 2007 software (Microsoft).

Protein purification

Dialyzed samples of the expressed α-amylase inhibitor were also used for purification via size exclusion chromatography (SEC) using a HiLoad 16/600 Superdex 75 pg (GE Healthcare) 120 mL column. As such, 15 mL of extract was completely dried under reduced pressure and resuspended in 1 mL of equilibration buffer (PBS 1X, 1 mM EDTA, 1 mM EGTA, and 1 mM dithiotreitol,

pH 7.4). Afterward, the column was washed with 120 mL of distilled water at a flow rate of 1 min/mL and then equilibrated with 240 mL of equilibration buffer at the same flow rate. The protein solution was loaded on the column, and 180 mL of equilibration buffer was injected at a continuous flow rate of 1 min/mL for elution; fractions were collected every 2 min. Chromatography was performed using an ÄKTAprime plus protein purification system (GE Healthcare), and chromatogram peaks at 280 nm were generated and analysed by UNICORN 6.4 software (GE Healthcare). Ninety fractions (2 mL each) were collected, and 15 μL of each fraction of the different peaks were separated by 15% (m/v) SDS-PAGE for silver staining according to the methods of Switzer et al. [54]. Fractions corresponding to the α-AIC3 peak were combined, lyophilized, resuspended in ultrapure water and quantified. Aliquots of 20 μg of proteins were separated by electrophoresis using 15% (m/v) SDS-PAGE.

In-gel digestion and mass spectrometry (MALDI-TOF) analysis

Spots of bands were excised from purified α-AIC3 corresponding bands, i.e., processed and unprocessed forms, and prepared for trypsin-based in-gel digestion. The samples were destained three times with 30% (v/v) ethanol under vigorous shaking for 20 min. Afterward, samples were dehydrated with a solution of 50% (v/v) acetonitrile (ACN) and 25 mM NH_4HCO_3 for 15 min, after which 200 μL of 100% (v/v) ACN was added to the recovered gel pieces, which were then shook for 10 min. The supernatant was discarded, the pieces were dried at room temperature and 15 μL of activated trypsin (Promega), which was prepared in digestion buffer according to the manufacturer's instructions, was added. The mixture was then incubated on ice for 30 min. Digestion proceeded by adding 25 μL of 50 mM NH_4HCO_3 to the samples, which were then incubated at 37 °C for 18 h. The hydrolysis products were collected, desalted, concentrated and purified using C18 resin ZipTip® pipette tips (Merck Millipore) according to the manufacturer's instructions, although peptides were eluted with 80% (v/v) aqueous ACN. The resulting peptides were dried under reduced pressure and resuspended in 10 μL of ultrapure and sterile water. Molecular mass analyses of α-AIC3 and its fragments were performed by MALDI-TOF MS/MS. A saturated α-cyano-4-hydroxycinnamic acid (CHCA, Sigma Aldrich) solution at 10 mg/mL was prepared in a 1:1 (v/v) aqueous acetonitrile solution containing 0.3% TFA. The solution of the hydrolysis products was mixed with CHCA solution (CHCA:sample, 3:1, v/v), spotted onto a MALDI target plate, and completely dried for crystallization at room temperature before analysis. Desorption/ionization, analysis and detection of peptides were performed using an Auto-flex™ Speed mass spectrometer (Bruker Daltonics), and ionization was carried out in positive reflection mode. Spectra were acquired based on external calibration using Protein Calibration Standard II (Bruker Daltonics) in accordance with the manufacturer's instructions. Peptide fragmentations were performed by using the LIFT™ method [55]. MS/MS spectra were manually interpreted, and the corresponding peptides were sequenced from the b/y series using FlexAnalysis 3.3 software (Bruker Daltonics). The peptide sequences were compared to the data from expected tryptic peptides generated by the theoretical tryptic digestion of α-AIC3 in ExPASy PeptideMass for confirming the already-known sequence and performing coverage analysis.

In vitro inhibitory assays
Activity validation of transiently expressed α-AIC3
The inhibitory activity of *N. benthamiana*-expressed α-AIC3 was first assessed and validated against cotton boll weevil amylase (AGA) based on the comparative inhibitory activity of α-AIC3 previously expressed in *A. thaliana* [26]. The colorimetric assay was performed by measuring the AGA activity using the 3,5-dinitrosalicylic acid (DNS) method adapted from Bernfeld [56] and using 2% (m/v) starch as substrate. Gut extracts as source of α-amylase were prepared by isolating gut from adults of *A. grandis* using a steel blade and mixing with AGA buffer (150 mM succinic acid, 20 mM CaCl$_2$, 60 mM NaCl, and 1 mM PMSF, pH 5.0) to a concentration of 0.5 g/mL. The assays were performed with a volume of gut extract containing one unit of α-amylase, which was defined as the amount of enzyme necessary to increase the absorbance (OD550) within 20 min to an amount between 0.11 and 0.15. Seed protein extracts from transgenic and non-transgenic *A. thaliana* were used as a control for α-AIC3 activity, whose transgenic one expressed α-AIC3 at a level of around 0.2% TSP [26]. These extracts were prepared by grinding seeds using a mortar and pestle, mixing each mg of powder with 7 μL of PBS-T buffer (10 mM sodium phosphate, 0.15 M NaCl, 0.05% (v/v) Tween-20, pH 7.5). Crude seed extracts were incubated on ice for 20 min, strongly shaken for 20 min at 4 °C using a vortex and centrifuged at 14000 g for 30 min at 4 °C. TSP were recovered from the supernatants and used for performing assays. Negative controls of digestion for all the samples were applied by inactivating the enzyme at 95 °C for five minutes before adding starch to the reaction system. Negative controls were used to prove that the enzyme was heat-inactivated and, thus, to give a background of inhibition to be used in calculations for inhibition level in digestion systems without heat-inactivation. *A. thaliana* seed extract controls were used to validate parameters of

the assay based on published data. This validation step allows conclusions concerning *N. benthamiana* extracts, such as the inhibition ability of AGA for the prepared extracts, and the comparison of inhibition level for each α-AIC3 against AGA. All of the reaction systems, i.e., digestions and negative controls, were performed in three technical replicates. We used 100 μg of dried protein resuspended in 75 μL of AGA buffer containing 1 unit of AGA as a source of plant material for each reaction. The absorbance values were recorded at 550 nm using a SpectraMax 190 microplate reader (Molecular Devices), and the samples were analysed using Excel 2007 software (Microsoft). Calculations were based on discounting the absorbance values for respective negative controls of digestion in each sample. Resulting values were used as following: absorbance values for samples containing α-AIC3 were discounted from the values for samples without α-AIC3, and the mean of triplicates indicated the level of activity remaining in each system.

Biosafety analysis: Non-target species enzymes
Once the inhibitory activity of the *N. benthamiana* extracts containing α-AIC3 was confirmed against AGA, these samples were used for assaying the inhibitory activity against enzymes of non-target species (*Apis mellifera* amylase – AMA – and *Spodoptera frugiperda* amylase – SFA). Samples at concentrations of 0.5 g/mL of ground whole insects were prepared using either AMA buffer (150 mM succinic acid, 20 mM CaCl$_2$, 60 mM NaCl, and 1 mM PMSF, pH 6.5) or SFA buffer (500 mM Tris-Cl, 20 mM CaCl$_2$, 60 mM NaCl, and 1 mM PMSF, pH 9.0) based on the recommended values of pH for enzyme activity according to the literature [57, 58]. The assays were performed following the same steps as those of the AGA test, as well as 100 μg of protein from the dialyzed *N. benthamiana* extracts was used. Since there are no specific amylase inhibitors developed, set and available for both insect species, comparison values relative to the absence of activity for AMA and SFA were exclusively derived from heat-inactivated enzyme systems, similarly to the negative control for AGA. The colour reactions were read at 550 nm, after which the appropriate calculations were used to analyse samples and inhibitory activities.

Abbreviations
ACN: Acetonitrile; AGA: *Anthonomus grandis* α-amylase; AMA: *Apis millefera* amylase; BSA: Bovine serum albumin; Bt: *Bacillus thuringiensis*; Cry: Crystal toxin; DNS: 3,5-dinitrosalicylic acid; FW: Fresh weight; IR: Insect resistance; MWCO: Molecular weight cut-off; PopMV: Poplar mosaic virus; PVX: Potato virus; SFA: *Spodoptera frugiperda* amylase; TSP: Total soluble proteins; X SEC: Size exclusion chromatography; α-AI: α-amylase inhibitor

Acknowledgments
Authors thank Dr. Leonardo Lima Pepino de Macedo (Embrapa Genetic Resources and Biotechnology) for the experimental support and the kind

provision of the bacterially derived β-subunit of the α-AIC3 for use as a standard molecule in the ELISAs. Authors thanks Pr Jacques Simporé for financial support.

Funding
GS was supported by the Coordination for the Improvement of Higher Education Personnel (CAPES, CfP AFCAPES 2012–03). The research was co-financed by the Agropolis Fondation under the reference ID 1203–005 through the "Investissements d'avenir" programme (Labex Agro: ANR-10-LABX-0001-01) and by the Embrapa Genetic Resources and Biotechnology lab.

Authors' contributions
GS, JPB, MP and SL designed the experiments. GS, PK, VO, JA, JPB, MB and SL performed the experiments and collected the data. FXG, MC, MP, CB, DS, SL and MF supervised the experiments. GS, PK and SL interpreted the data and wrote the article. MC, MP and MF supervised and complemented the writing. All authors have read and approved the manuscript.

Ethics approval and consent to participate
Not applicable.

Consent for publication
Not applicable.

Competing interests
The authors declare that they have no competing interests.

Author details
[1]Embrapa Genetic Resources and Biotechnology, Brasília, DF, Brazil. [2]Catholic University of Brasília, Brasília, DF, Brazil. [3]IRD, CIRAD, Université Montpellier, Interactions Plantes Microorganismes et Environnement (IPME), Montpellier, France. [4]INERA/LMI Patho-Bios, Institut de L'Environnement et de Recherches Agricoles (INERA), Laboratoire de Virologie et de Biotechnologies Végétales, Ouagadougou, Burkina Faso.

References
1. Ahmad P, Ashraf M, Younis M, Hu X, Kumar A, Akram NA, et al. Role of transgenic plants in agriculture and biopharming. Biotechnol Adv. 2012;30:524–40.
2. Lombardo L, Coppola G, Zelasco S. New technologies for insect-resistant and herbicide-tolerant plants. Trends Biotechnol. 2016;34:49–57.
3. Vachon V, Laprade R, Schwartz J. Current models of the mode of action of Bacillus thuringiensis insecticidal crystal proteins: a critical review. J Invertebr Pathol. 2012;111:1–12.
4. James C. ISAAA Brief 49–2014: Executive summary. Global status of commercialized biotech/GM crops: 2014. International Service for the Acquisition of Agri-Biotech Applications. 2014. http://www.isaaa.org/resources/publications/briefs/49/executivesummary/pdf/b49-execsum-english.pdf. Accessed 1 Aug 2018.
5. Gassmann AJ, Petzold-Maxwell JL, Clifton EH, Dunbar MW, Hoffmann AM, Ingber DA, et al. Field-evolved resistance by western corn rootworm to multiple Bacillus thuringiensis toxins in transgenic maize. Proc Natl Acad Sci U S A. 2014;111:5141–6.
6. Quilis J, López-García B, Meynard D, Guiderdoni E, San SB. Inducible expression of a fusion gene encoding two proteinase inhibitors leads to insect and pathogen resistance in transgenic rice. Plant Biotechnol J. 2014;12:367–77.
7. Chen PJ, Senthilkumar R, Jane WN, He Y, Tian Z, Yeh KW. Transplastomic Nicotiana benthamiana plants expressing multiple defence genes encoding protease inhibitors and chitinase display broad-spectrum resistance against insects, pathogens and abiotic stresses. Plant Biotechnol J. 2014;12:503–15.
8. Ma X, Zhu Z, Li Y, Yang G, Pei Y. Expressing a modified cowpea trypsin inhibitor gene to increase insect tolerance against Pieris rapae in Chinese cabbage. Hortic Environ Biotechnol. 2017;58:195–202.
9. Smigocki AC, Ivic-Haymes S, Li H, Savic J. Pest protection conferred by a Beta vulgaris serine proteinase inhibitor gene. PLoS One. 2013;8:2. https://doi.org/10.1371/journal.pone.0057303.
10. Vandenborre G, Smagghe G, Van Damme EJ. Plant lectins as defense proteins against phytophagous insects. Phytochemistry. 2011;72:1538–50.
11. McCafferty HRK, Moore PH, Zhu YJ. Papaya transformed with the Galanthus nivalis GNA gene produces a biologically active lectin with spider mite control activity. Plant Sci. 2008;175:385–93.
12. Wang J, Chen Z, Du J. Novel insect resistance in Brassica napus developed by transformation of chitinase and scorpion toxin genes. Plant Cell Rep. 2005;24:549–55.
13. McCafferty HR, Moore PH, Zhu YJ. Improved Carica papaya tolerance to carmine spider mite by the expression of Manduca sexta chitinase transgene. Transgenic Res. 2006;15:337–47.
14. Grossi-de-Sa MF, Mirkov TE, Ishimoto M, Colucci G, Bateman KS, Chrispeels MJ. Molecular characterization of a bean α-amylase inhibitor that inhibits the α-amylase of the Mexican bean weevil Zabrotes subfasciatus. Planta. 1997;203:295–303.
15. Pueyo JJ, Dale DC, Chrispeels MJ. Activation of bean (Phaseolus vulgaris) α-amylase inhibitor requires proteolytic process of the proprotein. Plant Physiol. 1993;101:1341–8.
16. Young NM, Thibault P, Watson DC, Chrispeels MJ. Post-translational processing of two α-amylase inhibitors and an arcelin from the common bean, Phaseolus vulgaris. FEBS Lett. 1999;446:203–6.
17. Morton RL, Schroeder HE, Bateman KS, Chrispeels MJ, Armstrong E, Higgins TJ. Bean α-amylase inhibitor 1 in transgenic peas (Pisum sativum) provides complete protection from pea weevil (Bruchus pisorum) under field conditions. Proc Natl Acad Sci U S A. 2000;97:3820–5.
18. Barbosa AEAD, Albuquerque EVS, Silva MCM, Souza DSL, Oliveira-Neto OB, Valencia A. α-amylase inhibitor-1 gene from Phaseolus vulgaris expressed in Coffea arabica plants inhibits α-amylases from the coffee berry borer pest. BMC Biotechnol. 2010;10:44. https://doi.org/10.1186/1472-6750-10-44.
19. Altabella T, Chrispeels MJ. Tobacco plants transformed with the bean α-ai gene express an inhibitor of insect α-amylase in their seeds. Plant Physiol. 1990;93:805–10.
20. Schroeder HE, Gollasch S, Moore A, Tabe LM, Craig S, Hardie DC. Bean α-amylase inhibitor confers resistance to the pea weevil (Bruchus pisorum) in transgenic peas (Pisum sativum L.). Plant Physiol. 1995;107:1233–9.
21. Lassner M, Bedbrook J. Directed molecular evolution in plant improvement. Curr Opin Plant Biol. 2001;4:152–6.
22. Yuan L, Kurek I, English J, Keenan R. Laboratory-directed protein evolution. Microbiol Mol Biol Rev. 2005;69:373–92.
23. Ceci LR, Volpicella M, Rahbé Y, Gallerani R, Beekwilder J, Jongsma MA. Selection by phage display of a variant mustard trypsin inhibitor toxic against aphids. Plant J. 2003;33:557–66.
24. Craveiro KI, Gomes Júnior JE, Silva MC, Macedo LL, Lucena WA, Silva MS, et al. Variant Cry1Ia toxins generated by DNA shuffling are active against sugarcane giant borer. J Biotechnol. 2010;145:215–21.
25. Oliveira GR, Silva MC, Lucena WA, Nakasu EY, Firmino AA, Beneventi MA, et al. Improving Cry8Ka toxin activity towards the cotton boll weevil (Anthonomus grandis). BMC Biotechnol. 2011;11:85. https://doi.org/10.1186/1472-6750-11-85.
26. Silva MCM, Del Sarto RP, Lucena WA, Rigden DJ, Teixeira FR, Bezerra CA, et al. Employing in vitro directed molecular evolution for the selection of α-amylase variant inhibitors with activity toward cotton boll weevil enzyme. J Biotechnol. 2013;167:377–85.
27. Rathinam M, Singh S, Pattanayak D, Sreevathsa R. Comprehensive in silico allergenicity assessment of novel protein engineered chimeric cry proteins for safe deployment in crops. BMC Biotechnol. 2017;17:64.
28. Daniell H, Singh ND, Mason H, Stratfield SJ. Plant-made vaccine antigens and biopharmaceuticals. Trends Plant Sci. 2009;14:669–79.
29. Fischer R, Vaquero-Martin C, Sack M, Drossard J, Emans N, Commandeur U. Towards molecular farming in the future: transient protein expression in plants. Biotechnol Appl Biochem. 1999;2:113–6.
30. Voinnet O, Rivas S, Mestre P, Baulcombe D. An enhanced transient expression system in plants based on suppression of gene silencing by the p19 protein of tomato bushy stunt virus. Plant J. 2003;33:949–56.

31. Lacombe S, Bangratz M, Brizard J, Petitdidier E, Pagniez J, Sérémé D, Lemesre J, Brugidou C. Optimized transitory ectopic expression of promastigote surface antigen protein in *Nicotiana benthamiana*, a potential anti-leishmaniasis vaccine candidate. J Biosci Bioeng. 2018;125:116–23.

32. Sainsbury F, Lomonossoff GP. Transient expressions of synthetic biology in plants. Curr Opin Plant Biol. 2014;19:1–7.

33. Peyret H, Lomonossoff GP. When plant virology met agrobacterium: the rise of the deconstructed clones. Plant Biotechnol J. 2015;13:1121–35.

34. Odell JT, Nagy F, Chua NH. Identification of DNA sequences required for activity of the cauliflower mosaic virus 35S promoter. Nature. 1985;6:810–2.

35. Dias SC, Silva MCM, Teixeira FR, Figueira ELZ, Oliveira-Neto OB, Lima LA, et al. Investigation of insecticidal activity of rye α-amylase inhibitor gene expressed in transgenic tobacco (*Nicotiana tabacum*) toward cotton boll weevil (*Anthonomus grandis*). Pestic Biochem Physiol. 2010;98:39–44.

36. Farnham G, Baulcombe DC. Artificial evolution extends the spectrum of viruses that are targeted by a disease-resistance gene from potato. Proc Natl Acad Sci U S A. 2006;103:18828–33.

37. Harris CJ, Slootweg EJ, Goverse A, Baulcombe DC. Stepwise artificial evolution of a plant disease resistance gene. Proc Natl Acad Sci U S A. 2013; 110:21189–94.

38. Ding Y, Li H, Chen L, Xie K. Recents advances in genome editing using CRISPR/Cas9. Front Plant Sci. 2016;7:703. https://doi.org/10.3389/fpls.2016.00703.

39. Quétier F. The CRISPR-Cas9 technology: closer to the ultimate toolkit for targeted genome editing. Plant Sci. 2016;242:65–76.

40. Liu X, Wu S, Xu J, Sui C, Wei J. Application of CRISPR/Cas9 in plant biology. Acta Pharm Sin B. 2017;7:292–302.

41. Wang P, Zhang J, Sun L, Ma Y, Xu J, Liang S, et al. High efficient multisites genome editing in allotetraploid cotton (*Gossypium hirsutum*) using CRISPR/Cas9 system. Plant Biotechnol J. 2018;16:137–50.

42. Li Z, Liu Z, Xing A, Moon BP, Koellhoffer JP, Huang L, et al. Cas9-guide RNA directed genome editing in soybean. Plant Physio. 2015;169:960–70.

43. Sun Y, Zhang X, Wu C, He Y, Ma Y, Hou H, et al. Engineering herbicide-resistant rice plants through CRISPR/Cas9-mediated homologous recombination of acetolactate synthase. Mol Plan. 2016;9:628–31.

44. Svitashev S, Young JK, Schwartz C, Gao H, Falco SC, Cigan AM. Targeted mutagenesis, precise gene editing, and site-specific gene insertion in maize using Cas9 and guide RNA. Plant Physiol. 2015;169:931–45.

45. Nekrasov V, Wang C, Win J, Lanz C, Weigel D, Kamoun S. Rapid generation of a transgene-free powdery mildew resistant tomato by genome deletion. Sci Rep. 2017;7:482.

46. Chandrasekaran J, Brumin M, Wolf D, Leibman D, Klap C, Pearlsman M, et al. Development of broad virus resistance in non-transgenic cucumber using CRISPR/Cas9 technology. Mol Plant Pathol. 2016;17:1140–53.

47. Baumberger N, Tsai CH, Lie M, Havecker E, Baulcombe DC. The Polerovirus silencing suppressor P0 targets ARGONAUTE proteins for degradation. Curr Biol. 2007;17:1609–14.

48. Siré C, Bangratz-Reyser M, Fargette D, Brugidou C. Genetic diversity and silencing suppression effects of *Rice yellow mottle virus* and the P1 protein. Virol J. 2008;5:55. https://doi.org/10.1186/1743-422X-5.

49. Hamilton A, Voinnet O, Chappell L, Baulcombe D. Two classes of short interfering RNA in RNA silencing. EMBO J. 2002;21:4671–9.

50. Villalobos A, Ness JE, Gustafsson C, Minshull J, Govindarajan S. Gene designer: a synthetic biology tool for constructing artificial DNA segments. BMC Bioinformatics. 2006;7:285. https://doi.org/10.1186/1471-2105-7-285.

51. Bendahmane A, Farnham G, Moffett P, Baulcombe DC. Constitutive gain-of-function mutants in a nucleotide binding site-leucine rich repeat protein encoded at the Rx locus of potato. Plant J. 2002;32:195–204.

52. Bradford MM. A rapid and sensitive method for the quantitation of microgram quantities of protein utilizing the principle of protein-dye binding. Anal Biochem. 1976;72:248–54.

53. Laemmli UK. Cleavage of structural proteins during the assembly of the head of bacteriophage T4. Nature. 1970;227:680–5.

54. Switzer RC, Merril CR, Shifrin S. A highly sensitive silver stain for detecting proteins and peptides in polyacrylamide gels. Anal Biochem. 1979;98:231–7.

55. Suckau D, Resemann A, Schuerenberg M, Hufnagel P, Franzen J, Holle A. A novel MALDI LIFT-TOF/TOF mass spectrometer for proteomics. Anal Bioanal Chem. 2003;376:952. https://doi.org/10.1007/s00216-003-2057-0.

56. Bernfeld P. Amylases, alpha and beta. In: Colowick SP, Kaplan NO, editors. Methods in enzymology. New York: Academic Press; 1955. p. 149–58.

57. Huber RE. The purification and study of a honey bee abdominal sucrase exhibiting unusual solubility and kinetic properties. Arch Biochem Biophys. 1975;168:198–209.

58. Ferreira C, Capella AN, Sitnik R, Terra WR. Properties of the digestive enzymes and the permeability of the peritrophic membrane of *Spodoptera frugiperda* (Lepidoptera) larvae. Comp Biochem Physiol. 1994;107:631–40.

7

Sustained transgene expression from sleeping beauty DNA transposons containing a core fragment of the *HNRPA2B1-CBX3* ubiquitous chromatin opening element (UCOE)

Kristian Alsbjerg Skipper[1], Anne Kruse Hollensen[1,2], Michael N. Antoniou[3] and Jacob Giehm Mikkelsen[1*] (iD)

Abstract

Background: DNA transposon-based vectors are effective nonviral tools for gene therapy and genetic engineering of cells. However, promoter DNA methylation and a near-random integration profile, which can result in transgene integration into heterochromatin, renders such vectors vulnerable to transcriptional repression. Therefore, to secure persistent transgene expression it may be necessary to protect transposon-embedded transgenes with anti-transcriptional silencing elements.

Results: We compare four different protective strategies in CHO-K1 cells. Our findings show robust protection from silencing of transgene cassettes mediated by the ubiquitous chromatin-opening element (UCOE) derived from the *HNRPA2B1-CBX3* locus. Using a bioinformatic approach, we define a shorter *HNRPA2B1-CBX3* UCOE core fragment and demonstrate that this can robustly maintain transgene expression after extended passaging of CHO-K1 cells carrying DNA transposon vectors equipped with this protective feature.

Conclusions: Our findings contribute to the understanding of the mechanism of *HNRPA2B1-CBX3* UCOE-based transgene protection and support the use of a correctly oriented core fragment of this UCOE for DNA transposon vector-based production of recombinant proteins in CHO-K1 cells.

Background

Genomic insertion of transgenes, leading to their stable expression, has been instrumental in studies of gene function and biomedical applications. Stable transgene expression is crucial for a wide range of in vitro experimental setups including disease modelling and production of recombinant proteins. Furthermore, some of the most successful genetic therapies rely on stably integrating and expressing correct copies of disease-causing gene variants [1–3]. Although precise genome editing using the CRISPR/Cas9 system is gaining increasing attention for introducing specific alterations in the genome, ways to achieve stable transgene expression in cell lines, patient cells, or tissues remain essential for many purposes including long-term therapeutic efficacy of cell and gene therapies.

In order to ensure reliable, long-lasting transgene expression, several *cis*-acting elements have been utilized and included in the design of gene transfer vectors [4]. These elements act by shielding the transgene cassette from position effect variegation (PEV), thereby avoiding the spread of heterochromatin and hypermethylation into the integrated transgene cassette. Until now, several different protective elements, including the 5'HS chicken β-globin (cHS4) insulator [5], the D4Z4 insulator [6] and Ubiquitous Chromatin Opening Elements (UCOE) [7, 8] have been exploited for protection of transgenes. The cHS4 insulator, widely used in the context of integrating viral and nonviral vectors [9–14], functions by blocking enhancer activity and, when flanking the transgene, by

* Correspondence: giehm@biomed.au.dk
[1]Department of Biomedicine, HEALTH, Aarhus University, DK- 8000 Aarhus C, Denmark

acting as a barrier against PEV [5, 15]. Interestingly, several DNA-binding proteins are recruited to the cHS4 insulator. Enhancer-blocking activity has been attributed to the CCCTC-binding factor (CTCF) [16, 17], whereas Upstream Stimulatory Factors 1 and 2 (USF 1 and 2) and Poly(ADP-ribose) Polymerase-1 (PARP-1) are thought to confer barrier activity of the insulator [18–20]. The UCOE sequence derived from a human CpG island encompassing the bidirectional promoters driving expression of Chromobox Protein Homolog 3 (CBX3) and Heterogeneous Nuclear Ribonucleoproteins A2/B1 (together referred to as the *HNRPA2B1-CBX3* locus), has also been extensively studied for protection against transcriptional silencing [21–24]. The element was initially shown to confer stable enhanced green fluorescent protein (eGFP) expression when integrated as part of the transgene expression cassette into centromeric heterochromatin [7]. Since then, several different fragments derived from the *HNRPA2B1-CBX3* locus have been utilized in various gene vehicles with a 1.5 kb fragment (1.5UCOE) as the most frequently used variant [21–23, 25–27]. Additional file 1: Figure S1 provides an overview of different UCOE fragments that have been used. In contrast to cHS4, for which insulating abilities have been attributed to the recruitment of several DNA-binding proteins, the UCOE mechanism of action is still poorly understood. The endogenous locus has been found to be hypomethylated in peripheral blood mononuclear cells [28], and different UCOE fragments in viral vectors have been shown to confer both hypomethylation and enrichment of the permissive histone H3K4 trimethylation (H3K4me3) mark [22, 27]. Furthermore, areas with high CpG density in the CBX3 region have recently been shown to be critical for conferring UCOE function and protection against transgene silencing [22, 29]. Nevertheless, whether a high CpG density in itself confers the anti-silencing function of the UCOE or whether this UCOE recruits DNA-binding proteins aiding in protecting transgenes, as has been described for both cHS4 and D4Z4 insulators [16, 18, 30], remains to be determined, although involvement of CTCF and CXXC finger protein (CFP1) has been suggested [27].

We have previously utilized the cHS4 insulator to shield gene cassettes in *sleeping beauty* (SB) transposon vectors to mediate protection against transgene silencing in F9 murine teratocarcinoma cells [31]. We also found increased vector mobilization when including protective *cis*-acting elements into DNA transposon vectors [32]. In the present study, we embarked on an investigation to assess UCOE-directed protection of transgene silencing in the context of SB DNA transposon vectors introduced into Chinese hamster ovary (CHO) K1 cells, which are widely used for the industrial production of recombinant proteins from chromosomally integrated transgenes [33]. Using an eGFP reporter gene, we showed rapid and robust repression of expression with

transgenes driven by the CMV promoter alone. In constructs where the CMV promoter was linked with either cHS4, D4Z4 or 1.5 kb *HNRPA2B1-CBX3* UCOE (1.5UCOE), we observed reduced silencing with both cHS4- and UCOE-protected DNA transposons. Based on a bioinformatic approach, we identified a core fragment within the 1.5UCOE and demonstrated that this element can confer effective protection from transcriptional silencing upon SB DNA transposon-based gene transfer.

Results

Comparison of protective elements in the context of sleeping beauty DNA transposon vectors

The SB DNA transposon system has become a powerful tool for mediating integration into the genome of target cells by means that do not involve virus-based gene transfer. As such, the system has been widely used for production of cell lines stably expressing transgenes of interest [34–36]. One of the key advantages of the SB transposon is its near random integration profile [37, 38], minimizing the risk of deleterious integrations. However, this increases the possibility of transgene integration into condensed (constitutive or facultative) heterochromatin, thereby leaving the integrated transposon cassette more vulnerable to transcriptional repression. In addition, even transgene integration events within transcriptionally permissive euchromatic regions can still be silenced by promoter DNA methylation. Reliable use of DNA transposon-based vectors may thus require inclusion of genetic elements that protect the inserted transgene against silencing. We therefore investigated the shielding of DNA transposon-embedded gene cassettes in CHO-K1 cells, by exploiting a model system for monitoring transgene stability following transposon-mediated integration. Similar to an approach that we previously described [32], we constructed a SB transposon vector, pT2/CGIP, carrying a CMV-driven eGFP-IRES-*pac* (CGIP) cassette conferring expression of both eGFP and puromycin N-acetyltransferase (pac) (Fig. 1a). This enabled us to not only monitor transgene expression over time by flow cytometry, but also to maintain stable expression by keeping CHO-K1 clones under continuous selection with puromycin. We co-transfected CHO-K1 cells with the pT2/CGIP vector together with a plasmid encoding either the hyperactive SB transposase SB100X or an inactive mutant variant (mSB). SB-mediated integration was confirmed by counting the number of puromycin-resistant colonies obtained with the SB100X and mSB transposase variants (Additional file 1: Figure S2). An overview of the experimental workflow is shown in Fig. 1a. A total of twelve clones harboring SB100X-directed transgene insertions were then passaged for 7 weeks in the absence of puromycin, and eGFP expression was monitored weekly by flow cytometry (Fig. 1b, Additional file 1: Figure S3). With six of the clones we observed a rapid decrease in eGFP expression even by 7 days

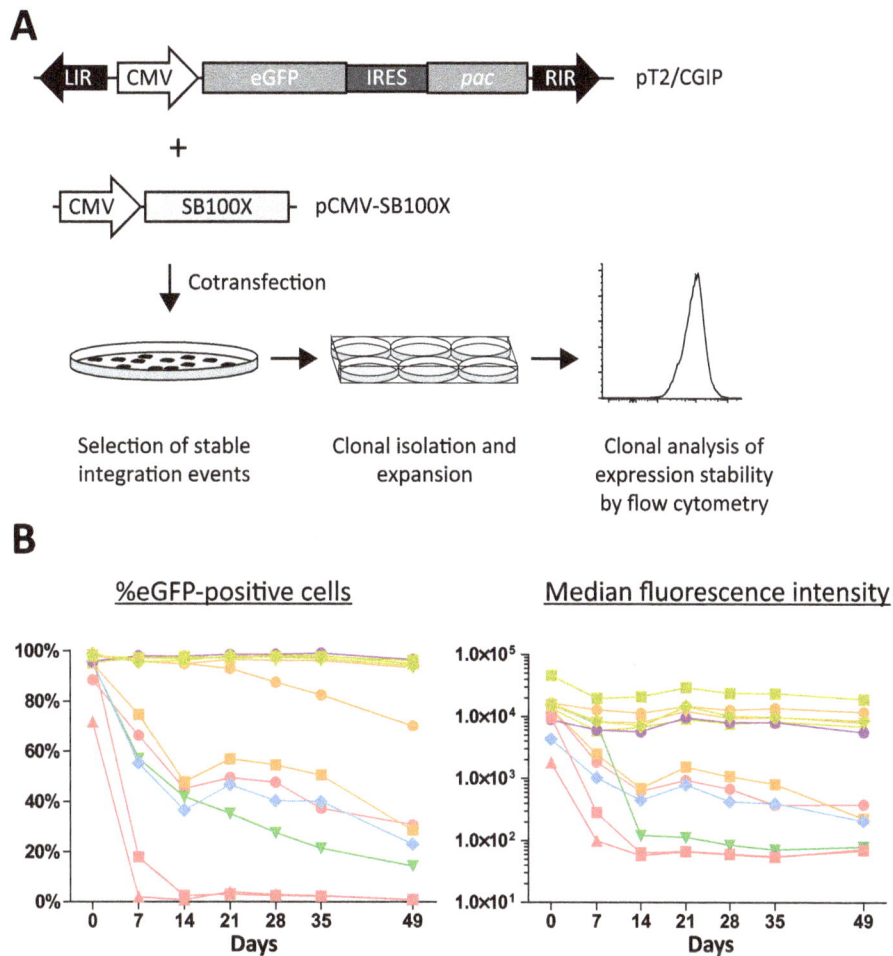

Fig. 1 Progressive silencing of *Sleeping Beauty* transposon vectors in CHO-K1 cells. **a** Illustration of the experimental procedure used in this study including a schematic representation of the SB vector construct. **b** Evaluation of stability of transgene expression in single cell CHO-K1 clones containing SB transposon vectors. On Day 0, puromycin was removed from the medium and eGFP expression monitored by flow cytometry at different time points for 7 weeks. Both the total percentage of eGFP-positive cells and the median fluorescence intensity is shown. Each line in the graphs represents the expression profile of a single clone over the course of the 7 weeks of continuous culture. LIR: Left inverted repeat, RIR: Right inverted repeat, CMV: Cytomegalovirus promoter, eGFP: enhanced Green fluorescent protein, IRES: Internal ribosomal entry site, pac: puromycin N-acetyl-transferase

following removal of puromycin selection pressure. In addition, in one clone, which initially appeared stable, we observed from day-21 to 49 a gradual decrease in the percentage of eGFP-positive cells (93 to 70%). Notably, the median fluorescence intensity (MFI) for this clone remained stable throughout the seven-week period, suggesting that expression of eGFP was only slowly silenced and that prolonged passage was required to measure an MFI decrease in this clone.

Having established that the CMV-driven eGFP cassette was frequently repressed in CHO-K1 cells as previously reported [21], we moved on to investigate the effect of including three different *cis*-acting elements, which have been reported to confer protection against epigenetic-mediated silencing, namely a 1.5 kb fragment of the *HNRPA2B1-CBX3* UCOE (1.5UCOE) [19–22], cHS4 [9, 11, 32] and D4Z4 [30]

in the SB transposon vector. The different elements were cloned into pT2/CGIP (Fig. 2a-d). The 1.5UCOE and D4Z4 elements were placed immediately upstream from the CMV promoter. As the 1.5UCOE has been shown to work both in the 5′-and 3′-orientation [21, 25] vectors carrying the element inserted in both orientations were constructed and tested. As the cHS4 insulator only displays anti-transcriptional silencing by flanking transgenes [9, 11, 32], we inserted copies of the 1.2-kb version of this insulator element both upstream and downstream from the expression cassette. The vectors were used to generate puromycin-resistant CHO-K1 clones using SB100X (Fig. 1a), and a total of 51 clones (14 with T2/5′1.5UCOE.CGIP, 16 with T2/3′ 1.5UCOE.CGIP, 14 with T2/cHS4.CGIP, and 7 with T2/ D4Z4.CGIP) were isolated and expanded. We first analyzed the level of eGFP expression in puromycin-selected clones

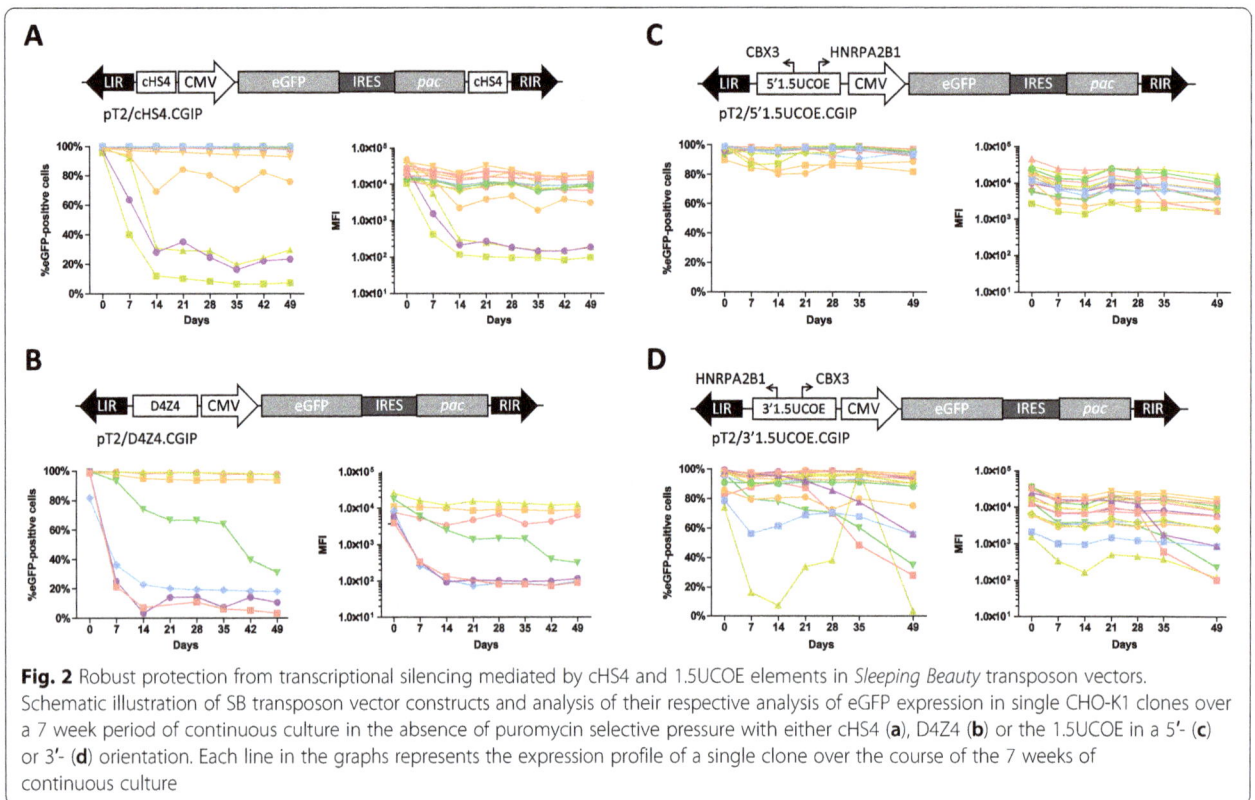

Fig. 2 Robust protection from transcriptional silencing mediated by cHS4 and 1.5UCOE elements in *Sleeping Beauty* transposon vectors. Schematic illustration of SB transposon vector constructs and analysis of their respective analysis of eGFP expression in single CHO-K1 clones over a 7 week period of continuous culture in the absence of puromycin selective pressure with either cHS4 (**a**), D4Z4 (**b**) or the 1.5UCOE in a 5'- (**c**) or 3'- (**d**) orientation. Each line in the graphs represents the expression profile of a single clone over the course of the 7 weeks of continuous culture

and as expected observed substantial interclonal differences (Additional file 1: Figure S4), most likely due to copy number variation and position effects affecting gene expression. Relative to clones carrying the unprotected CGIP cassette, the 1.5UCOE did not in either of the two orientations increase eGFP expression, in contrast to previous reports [7, 21], whereas cHS4 insulators flanking the CGIP cassette both increased mean expression levels and lowered the interclonal variation, in line with previous studies (Table 1, Additional file 1: Figure S4) [39, 40].

To evaluate the stability of expression over time, we then passaged the clones for 7 weeks in the absence of puromycin and monitored eGFP expression weekly by flow cytometry. Among the 14 clones harboring cHS4-flanked CGIP, only 4 clones gradually lost expression, validating the protective effect of cHS4 in CHO-K1 cells

(Fig. 2a, Additional file 1: Figure S3). The D4Z4 insulator, in contrast, did not appear to negate transcriptional silencing with 4 out of 7 clones being progressively silenced (Fig. 2b, Additional file 1: Figure S3). Notably, loss of expression was not observed in any of the 14 clones carrying T2/5′1.5UCOE.CGIP (Fig. 2c, Additional file 1: Figure S3), demonstrating a robust capability of the 1.5UCOE element to protect against silencing resulting in negligible interclonal variation. Among the 16 clones harboring CGIP with the 1.5UCOE in the 3′ orientation at least 6 clones gradually lost eGFP expression during passaging resulting in substantial variation between the clones (Fig. 2d, Additional file 1: Figure S3). Nevertheless, the majority of clones harboring T2/3′1.5UCOE.CGIP remained stable throughout the experiment, indicating that although minor direction-dependent differences were seen,

Table 1 Overview of the studied protective elements and their ability to sustain transgene expression in CHO-K1 cells

Group	Element size (kb)	No. of clones analyzed	Day 0 MFI[a]	Day 0 interclonal variation[a]	Day 0:Day 49 eGFP ratio[a]	Stable clones (%)[b]
CGIP	–	12	14,549 ± 0.3129	8.238 ± 1.772	0.56 ± 0.12	42
cHS4	2 × 1.2	14	24,838 ± 0.3454	2.559 ± 0.3559	0.79 ± 0.16	71
D4Z4	3.3	7	11,265 ± 0.2904	3.035 ± 0.7825	0.56 ± 0.10	43
5′1.5UCOE	1.5	14	17,014 ± 0.2988	6.358 ± 1.117	0.97 ± 0.01	100
3′1.5UCOE	1.5	16	17,085 ± 0.3009	10.97 ± 1.932	0.78 ± 0.08	63
5′UCOE-CORE	0.86	13	20,498 ± 0.5042	4.790 ± 1.178	0.46 ± 0.11	31
3′UCOE-CORE	0.86	13	20,256 ± 0.3585	4.454 ± 0.7882	0.87 ± 0.08	77

[a] mean ± SEM, [b] eGFP ratio ≥ 0.8

1.5UCOE efficiently protected against silencing in both orientations. Interestingly, this is in contrast to a previous study conducted in murine P19 embryonal carcinoma cells [26], in which UCOE function relied heavily on the orientation of the element with the 3′-orientation appearing to confer most robust protection. As the *HNRPA2B1-CBX3* UCOE encompasses two divergently transcribing promotors, it is plausible that transcriptional activation by these promoters is crucial for and even dictates the protective function by this element. In summary, our results demonstrate that both 1.5UCOE and the cHS4 insulator elements, but not the D4Z4 insulator, is able to shield SB transposon vectors and protect against transgene silencing in the CHO-K1 cell line. Due to protective properties and small size of 1.5UCOE, we focused on optimizing this element as a protective add-on to DNA transposon vectors used in CHO-K1 cells.

Bioinformatic analysis of the endogenous HNRPA2B1-CBX3 locus reveals extensive transcription factor binding

In order to further investigate the mechanism underpinning the transcriptional protective capability of the *HNRPA2B1-CBX3* UCOE, we utilized a bioinformatic approach to map transcription factor binding sites in this element. To identify potential key regulators of UCOE function, we combined data from the ENCODE project with an analysis of protein-protein interactions and conserved transcription factor binding motifs. A total of 89 proteins were found to be potentially associated with the endogenous *HNRPA2B1-CBX3* locus using ChIP-seq data from the ENCODE project (see Additional file 2 for a list of proteins with

identified peaks within the locus). In line with the status of both *CBX3* and *HNRPA2B1* as constitutively transcribed housekeeping genes, a large fraction of the identified proteins was associated with the RNA polymerase II transcription machinery and therefore not directly relevant for identifying key UCOE regulatory sites. Instead, based on reported roles in modulating chromatin structure and/or insulator function, we identified a subset of transcription factors (Fig. 3), for which enriched Gene Ontology (GO) terms are provided in Additional file 1: Figure S5. We identified physical protein-protein interactions by utilizing the BioGRID database together with esyN [41, 42] (Additional file 1: Figure S6) and identified a major cluster of potentially interacting transcription factors. E1A binding protein p300 (EP300) and SP1 had the highest number of interactions, six and five interactors, respectively, suggesting a role of these proteins in UCOE function. The two transcription factors, HMGN3 and CHD2, were not found to have any interactions within the cluster (Additional file 1: Figure S6). To strengthen our predictions, ConTra V3 [43] was utilized to identify predicted conserved transcription factor binding motifs, which were then cross-referenced with the ENCODE ChIP-seq data (Fig. 3). Several transcription factors showed significant overlap between the BioGRID database and the ChIP-seq data sets. Notably, SP1 and EP300 were again found to be prominent in this analysis emphasizing their potential recruitment to the locus and key role in UCOE function (Fig. 3, pink and yellow bars). Yin Yang 1 (YY1), a transcription factor described to be recruited to a methylation-sensitive insulator [44], was also identified. Intriguingly, together with SP1 binding motifs YY1 binding

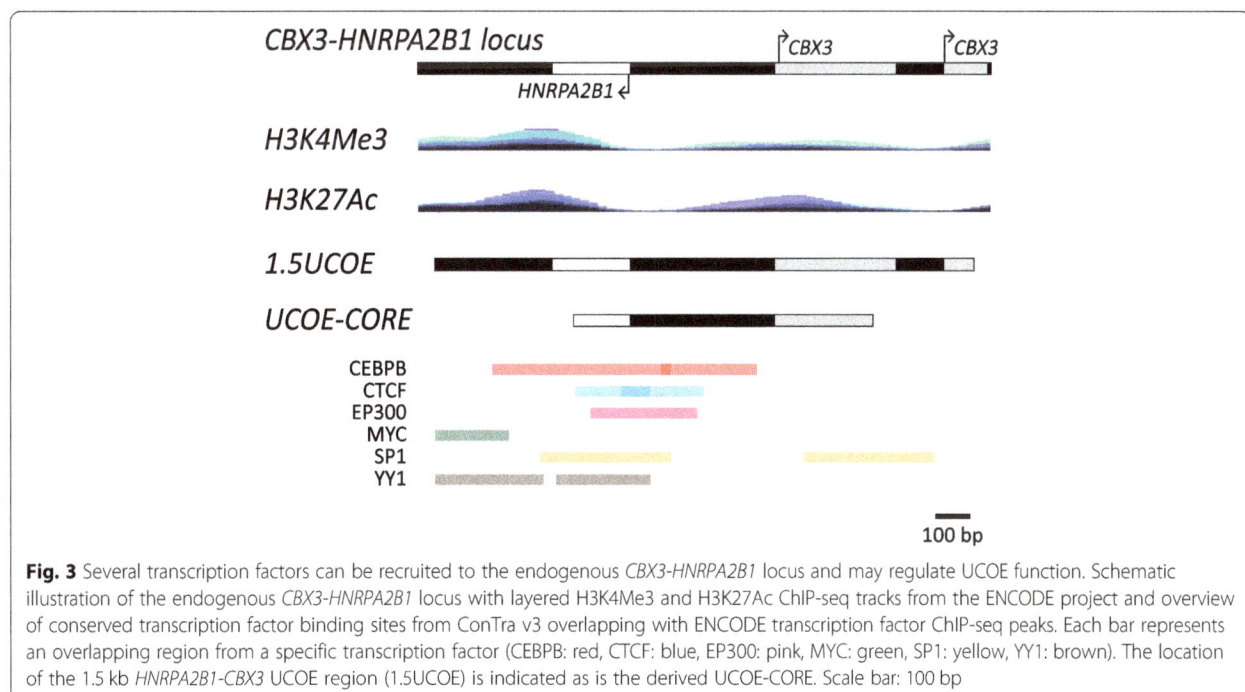

Fig. 3 Several transcription factors can be recruited to the endogenous *CBX3-HNRPA2B1* locus and may regulate UCOE function. Schematic illustration of the endogenous *CBX3-HNRPA2B1* locus with layered H3K4Me3 and H3K27Ac ChIP-seq tracks from the ENCODE project and overview of conserved transcription factor binding sites from ConTra v3 overlapping with ENCODE transcription factor ChIP-seq peaks. Each bar represents an overlapping region from a specific transcription factor (CEBPB: red, CTCF: blue, EP300: pink, MYC: green, SP1: yellow, YY1: brown). The location of the 1.5 kb *HNRPA2B1-CBX3* UCOE region (1.5UCOE) is indicated as is the derived UCOE-CORE. Scale bar: 100 bp

sites have been found to be enriched in bidirectional promoter loci compared to unidirectional promoter loci [45], supporting their potential recruitment to this UCOE locus. Furthermore, conserved binding motifs for CTCF were also found to overlap with ChIP-seq peaks. CTCF is associated with the function of methylation-determining regions [46] and is also responsible for mediating enhancer-blocking activity of the cHS4 insulator [16, 17]. CEBPB and MYC were also identified as potential players in UCOE function in this analysis. Based on these predictions, we deduced an 863-bp 'core' fragment within the *HNRPA2B1-CBX3* UCOE (UCOE-CORE) containing the motifs necessary to recruit and bind transcription factors to this UCOE and to maintain an open chromatin environment (Fig. 3).

Robust orientation-dependent protection by the UCOE-CORE element in DNA transposon vectors integrated into CHO-K1 cells

For expression vectors to accommodate promoters and genes of a considerable size, inclusion of silencing protective elements of restricted size is attractive and indeed may be necessary. In order to test the capability of the UCOE-CORE to protect a DNA-transposon-delivered CGIP cassette, the 863-bp fragment was inserted 5′ of the CMV promoter in the pT2/CGIP vector in both 5′-and 3′-orientations (Fig. 4a, b, pT2/5'UCOE-CORE.CGIP and pT2/3'UCOE-CORE.CGIP). These were then stably transposed into CHO-K1 cells using SB100X and a total of 26 clones isolated and eGFP expression followed over time. As in the case of the 1.5UCOE element, inclusion of the UCOE-CORE did not appear to increase initial eGFP expression levels (Table 1, Additional file 1: Figure S7). However, as predicted by the transcription factor binding site analysis (Fig. 3), the UCOE-CORE was capable of negating transcriptional repression in the CHO-K1 cell line. Notably, in contrast to the 1.5UCOE element, which showed little dependence on orientation for function (Fig. 2c, d), the UCOE-CORE element displayed a strong directional bias. We observed that 10 out of 13 clones with the 5'UCOE-CORE-CGIP cassette were progressively silenced after removal of puromycin during the period 7 weeks of continuous culture (Fig. 4a, Additional file 1: Figure S8). In marked contrast, in clones harboring constructs with the UCOE-CORE element placed in the 3′-orientation; that is, with the *CBX3* promoter in the sense direction, only 3 out of 13 clones showed a gradual loss in eGFP reporter gene expression over the same time course (Fig. 4b, Additional file 1: Figure S8). Thus, the 3'UCOE-CORE performed at a level comparable to that of the cHS4 insulator and the 1.5UCOE (Table 1). In addition, prolonged passage of the clones carrying the 3'UCOE-CORE element resulted in little change in eGFP expression levels over 100 days of continuous culture,

demonstrating the robustness of the element in conferring stability of function (Additional file 1: Figure S9).

Discussion

Although different *HNRPA2B1-CBX3* UCOE fragments have been extensively utilized for mediating stable gene expression, little is known about the mechanism of action of this element. *HNRPA2B1-CBX3* UCOE fragments have previously been shown to confer both hypomethylation and enrichment of H3K4me3 marks [22, 27]. In addition, the presence of an methylation-determining region [46] downstream of the *CBX3* promoter within the 2.2UCOE has been proposed [22]. The existence of several transcription factors that may potentially bind to the *HNRPA2B1-CBX3* UCOE and influence its function (Fig. 3), many of which are associated with chromatin remodeling, may suggest a possible link between the UCOE sequence and its ability to model chromatin. The location of the predicted transcription factor binding sites indicated the presence of a core region located between the transcriptional start sites of *HNRPA2B1* and *CBX3* generating an open chromatin environment. Indeed, when we tested this predicted core as a *cis*-acting element, we observed robust eGFP expression up to 100 days after removal of puromycin selective pressure (Fig. 4; Additional file 1: Figure S9).

Reports showing the anti-silencing capacity of several non-overlapping *HNRPA2B1-CBX3* UCOE fragments (Additional file 1: Figure S1) [21–23, 26, 27, 47], indicate that this element contains several functional subregions, all with the ability to generate an epigenetically, transcriptionally permissive environment. Whether these regions are completely independent or can work in concert to secure stable expression is not known. Of potential note is that the 3'UCOE-CORE element, which possesses both *HNRPA2B1* and *CBX3* promoters appears to provide somewhat greater stability of expression (Fig. 4b) than fragments encompassing just the *CBX3* half of the 1.5UCOE [22, 27, 48]. This suggests that the dual divergently transcribed promoter structure of the UCOE-CORE element may confer a more potent anti-transcriptional silencing capability.

In our study we found a bias related to the orientation of a *HNRPA2B1-CBX3* UCOE core segment that has not previously not been reported. Interestingly, this orientation bias appeared when parts of *HNRPA2B1* and *CBX3* were removed, indicating that one or more regulatory regions within either of these deleted fragments, necessary to ensure stable expression in the 5′-orientation, were lost in the UCOE-CORE element (Fig. 2c, d, 4). Indeed, the identification of functional elements located outside of the UCOE-CORE sequence that we defined in our studies, suggests that such additional regulatory sites exist [22, 47]. Interestingly, a direction-dependent function of the *HNRPA2B1-*

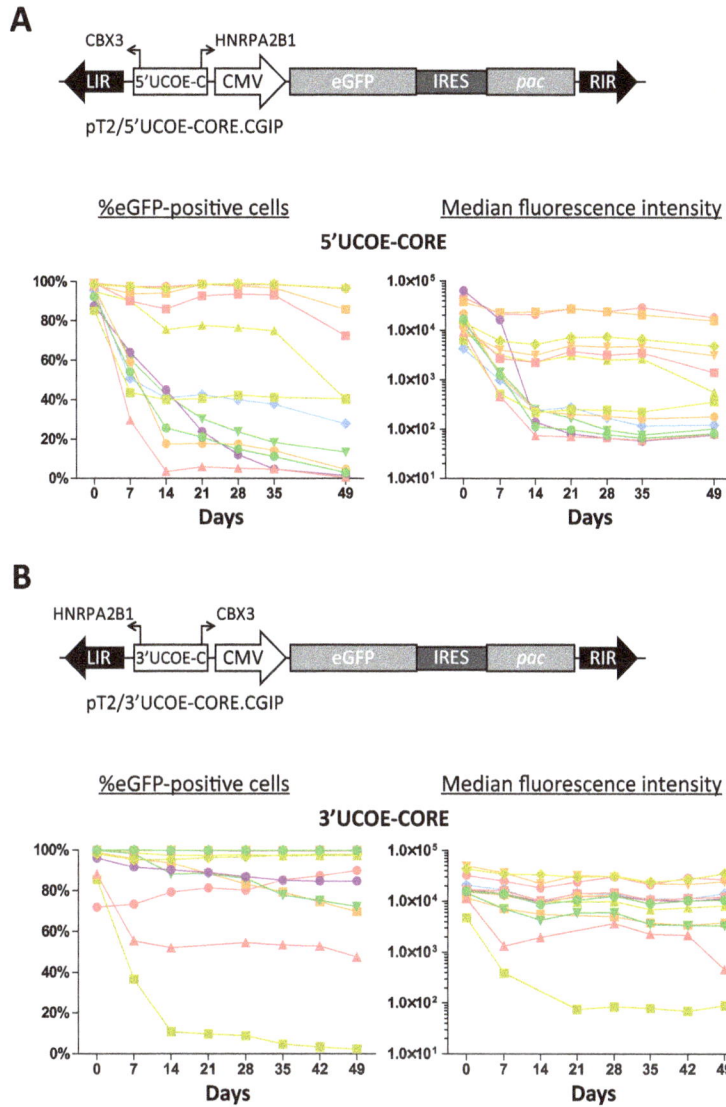

Fig. 4 UCOE-CORE element provides direction-dependent, robust protection against transgene repression. Schematic illustration of the SB transposon vectors carrying the UCOE-CORE element and eGFP expression profile of individual CHO-K1 clones harboring these vectors over a seven-week period of continuous culture in the absence of puromycin selection pressure with the UCOE-CORE in either the 5'-(**a**) or 3'- (**b**) orientation. Expression of eGFP was assessed by flow cytometry. Each line in the graphs represents the expression profile of a single clone over the course of the 7 weeks of continuous culture

CBX3 UCOE has been described in P19 cells for several different UCOE fragments [7, 22, 26]. The underlying cause for this apparent bias remains unknown. Nevertheless, it is possible that the overall structure of the *HNRPA2B1-CBX3* UCOE, encompassing two divergently transcribed promoters, plays a role in determining its function, as independent and differential expression from the *CBX3* and *HNRPA2B1* promoters may involve a direction-dependent function. Studies of *CBX5-HNRNPA1*, a 'sister' locus to *HNRPA2B1-CBX3*, have previously demonstrated independent transcription from two divergently transcribed promoters [49–51]. It is thus feasible that utilizing the *HNRPA2B1-CBX3* UCOE in a cell line with either low endogenous *CBX3* or *HNRPA2B1* expression can create a direction-dependent function, underlining the necessity of testing this element in both orientations when utilizing this UCOE in a new cellular context.

In summary, our studies compare the functionality of four different SB-based vectors containing protective elements for stable long-term expression in CHO-K1 cells. We found that the cHS4 insulator and 1.5UCOE possessed an equal anti-transcriptional silencing capability. However, unlike the cHS4 insulator that must be inserted at both 5′ and 3′ ends of a transgene to confer transcriptional stability, the 1.5UCOE and its smaller 863 bp 3'UCOE-CORE are fully functional as a single element placed upstream of the heterologous promoter driving expression of the gene of interest. Thus, the use of these UCOE elements, especially the 3'UCOE-CORE offers a significant space saving advantage over the cHS4 insulators that must be repeated at both ends of the transposon transgene cassette. In addition, the use of a single 3'UCOE-CORE also reduces propensity for transgene loss through recombination as may occur between insulators that that flank both ends of the transposon transgene. The data we present also lays the foundation for further investigations into the role of transcription factors associated with *HNRPA2B1-CBX3* UCOE-containing transgene cassettes. However, our initial attempts to generate CHO-K1 cells carrying CRISPR-derived knockout mutations restricting Sp1 and CTCF expression led to extensive cell death, suggesting that these factors are essential for cell growth and survival. Although this may complicate studies focusing on these genes, knockdown approaches based on CRISPRi or RNA interference may allow investigations of the impact of Sp1 and CTCF on UCOE function. Our data showed robustness of a the UCOE-CORE segment in the context of SB DNA transposon vectors, resulting in maintained transgene expression even after extended continuous cell culture. These findings define a short UCOE variant suitable for transgene expression purposes and provide a platform for functional analysis of UCOE action.

Conclusions

In conclusion, the results presented in this study contribute to the understanding of the mechanisms of action of the *HNRPA2B1-CBX3* UCOE and its use for protection against transgene silencing thereby laying the groundwork for improved vector design. Our data presented here furthermore stress the importance of including a protective *cis*-element, like the UCOE-CORE segment, in integrated DNA transposon vectors used for production of recombinant proteins in CHO-K1 and similar cell lines.

Methods

Plasmid construction

All plasmid constructions were done in pT2/CMV-eGFP(s).SV40-neo [52] containing a second-generation SB transposon backbone. Initially, the CMV-eGFP(s).SV40-neo fragment was replaced with a polylinker containing multiple restriction enzyme sites to expedite insertion of DNA fragments and cloning by HindIII digestion of the pT2 backbone and insertion of annealed double-stranded oligonucleotides with compatible overhangs; this created the construct pT2/Linker. This vector was then used to generate pT2/CMV-eGFP by insertion of the EcoRI-excised CMV-eGFP expression cassette from pT2/CMV-eGFP(s).SV40-neo into EcoRI-digested pT2/Linker. The

1.5UCOE element is a 1.5 kb Esp3I genomic fragment is derived from the *HNRPA2B1-CBX3* housekeeping gene region, extending over the transcriptional start sites of these two divergently transcribed genes [21]. The 1.5UCOE was isolated from a pBluescript subclone (MA895) of this region by PCR amplification and inserted into AvrII-digested and Klenow-treated pT2/CMV-eGFP by blunt-end ligation. Flanking the CMV-eGFP expression cassette with cHS4 was done by insertion of either AvrII -or AgeI-digested cHS4 from the pSBT/cHS4-PGK-Puro-cHS4 [31] upstream or downstream of the cassette, respectively. pT2/D4Z4.CMV-eGFP was created by insertion of a KpnI-digested D4Z4 fragment from C1X [6] into KpnI-digested pT2/CMV-eGFP. Finally, an IRES-*pac*-pA fragment was inserted into the pT2 vectors by PCR amplification from the pSBT/CMV-eGFP-IRES-*pac* and subsequent insertion into PacI -and AgeI-digested pT2 vectors to create pT2/CGIP, pT2/5'1.5UCOE.CGIP, pT2/3' 1.5UCOE.CGIP, pT2/cHS4.CGIP and pT2/D4Z4.CGIP. pT2/5'UCOE-CORE.CGIP and pT2/3'UCOE-CORE.CGIP were created by insertion of the PCR-amplified UCOE-CORE fragment from pT2/5'1.5UCOE.CGIP into a EcoRI-digested pT2/CGIP backbone. pCMV-SB100X has been previously described [53]. UCOE is an official registered trademark (Merck MilliporeSigma).

Cell culture and transfections

CHO-K1 were a kindly provided by professor Per Höllsberg, Department of Biomedicine, Aarhus University. The cells were cultured at 37 °C in a 5% (v/v) CO_2 atmosphere. The cells were maintained in Dulbecco's Modified Eagle's Medium (DMEM) (Sigma-Aldrich, St. Louis, MO, USA) supplemented with 5% fetal calf serum, 100 U/ml penicillin and 100 µg/ml streptomycin. For transfections, 2.5×10^5 cells were seeded in 6-well plates and the following day transfected with 250 ng of the hyperactive SB transposase-expressing plasmid pCMV-SB100X together with 250 ng of either pT2/CGIP, pT2/5'1.5UCOE.CGIP, pT2/3' 1.5UCOE.CGIP, pT2/cHS4.CGIP, pT2/D4Z4.CGIP, pT2/ 5'UCOE-CORE.CGIP or pT2/3'UCOE-CORE.CGIP in a 1:1 ratio using the TurboFect transfection reagent (Thermo Fisher Scientific, Waltham, MA, USA) according to manufacturer's instructions. For quantification of colony formation, the cells were then reseeded in P10 Petri dishes in appropriate dilutions and the medium was subsequently replaced with antibiotic selection medium containing 5 µg/ml puromycin. After 14 days of passage, the cells were fixed and stained with a 0.6% methylene blue solution, and colonies counted. For isolation and expansion of single clones, the cells were diluted 1:500 and reseeded in a 96-well plate the day after transfection. The cells were maintained in antibiotic selection medium as above, and resistant clones were expanded for analysis. All the resulting clones were kept in antibiotic selection medium unless otherwise specified.

Bioinformatics

Analysis of ChIP-seq data from the ENCODE project [54] was done using the built-in track on the UCSC Genome Browser, GRCh37/hg19 human genome assembly. All transcription factors showing a peak of binding activity within the 1.5UCOE region (chr7:26,239,852-26,241,407) were included. For analysis of physical protein-protein interactions, a subset of transcription factors were chosen based on current literature and analyzed using the esyN network builder [41] coupled with the BioGRID database v3.5.165 [42]. Finally, the ConTra v3 [43] web server was utilized to identify conserved transcription factor binding sites using all available databases and the following stringency settings; core = 0.95, similarity matrix = 0.85. These were then loaded into the UCSC Genome Browser to cross-reference with the ENCODE ChIP-seq data tracks.

Flow cytometry

CHO-K1 clones were trypsinized and centrifuged at 500 g for 3 min at 4 °C, washed in Dulbecco's Phosphate Buffered Saline (DPBS) and finally resuspended in 250 μl DPBS for analysis in 96-well plates. Flow cytometry analysis was carried out at the FACS Core Facility, Aarhus University, on a BD LSRFortessa Cell Analyzer (Becton Dickinson, Franklin Lakes, NJ, USA) with a BD High Throughput Sampler. Data analysis was performed using FlowJo (v.10.4, FlowJo, LLC, Ashland, OR, USA) and the gating strategy can be seen in, Additional file 1: Figure S10.

Supplementary information

Additional file 1: Figure S1. Schematic illustration of the endogenous CBX3-HNRPA2B1 locus with the different UCOE fragments indicated. **Figure S2.** Sleeping Beauty-mediated colony formation in CHO-K1 cells. Data is presented as mean ± SEM and $n = 3$. **Figure S3.** Flow cytometric analysis of representative clones within each group. **Figure S4.** Construct-dependent clonal variation in eGFP expression levels. eGFP-expression MFIs for the clones in each group were normalized to the lowest expressing clone in the individual groups. Boxes are displayed as Q2 + Q3 quantile, and whiskers show 10–90 percentile. **Figure S5.** Top ten enriched GO terms for the selected subset. **Figure S6.** esyN protein-protein interaction network for the selected subset of transcription factors. The size difference indicates the most central nodes as calculated by the betweenness centrality of each node. **Figure S7.** Clonal variation in eGFP expression levels in UCOE-CORE clones. eGFP-expression MFIs for the clones in each group were normalized to the lowest expressing clone in the individual groups. Boxes are displayed as Q2 + Q3 quantile, and whiskers show 10–90 percentile. **Figure S8.** Flow cytometric analysis of representative clones harbouring either the 5'UCOE-CORE or the 3'UCOE-CORE. **Figure S9.** Extended analysis of clones containing the 3'UCOE-CORE. Days 0–49 correspond to data presented in Fig. 4b. **Figure S10.** Flow cytometry gating strategy.

Additional file 2. List of proteins with identified ChIP-seq peaks from the ENCODE project within the CBX3-HNRPA2B1 locus.

Abbreviations

CBX3: Chromobox Protein Homolog 3; CFP1: CXXC finger protein; CHO: Chinese hamster ovary; cHS4: 5'HS chicken β-globin (cHS4) insulator; CTCF: CCCTC-binding factor; eGFP: Enhanced green fluorescent protein; EP300: E1A binding protein p300; GO: Gene ontology; H3K4me3: H3K4 trimethylation; HNRPA2B1: Heterogeneous Nuclear Ribonucleoproteins A2/B1; MFI: Median fluorescence intensity; PAC: Puromycin N-acetyltransferase; PARP-1: Poly(ADP-ribose) Polymerase-1; PEV: Position effect variegation; SB: Sleeping Beauty; UCOE: Ubiquitous chromatin-opening element; USF1: Upstream Stimulatory Factor 1; USF2: Upstream Stimulatory Factor 2; YY1: Yin Yang 1

Acknowledgements
Not applicable.

Authors' contributions
KAS, MNA, and JGM conceived the project and designed the experiments. KAS and AKH performed the experiments. KAS, MNA and JGM wrote the manuscript and assembled the figures. All authors read and approved the final manuscript.

Funding
This work was made possible through support of the Danish Council for Independent Research | Medical Sciences (grant DFF-4004-00220), The Lundbeck Foundation (grant R126–2012-12456), the Hørslev Foundation, Aase og Ejnar Danielsens Fond, Grosserer L. F. Foghts Fond, Agnes og Poul Friis Fond, Oda og Hans Svenningsens Fond, Snedkermester Sophus Jacobsen & Hustru Astrid Jacobsens Fond, and Familien Hede Nielsens Fond. None of the funding bodies influenced the study design, data collection, data analysis or manuscript drafting.

Ethics approval and consent to participate
Not applicable.

Consent for publication
Not applicable.

Competing interests
MNA has inventor status on patents covering the biotechnological applications of UCOE® elements and a paid consultancy with MilliporeSigma who hold the intellectual property and commercial rights to UCOE® technology. All other co-authors (KAS, AKH, and JGM) have no affiliation with MilliporeSigma and declare no conflicts of interest.

Author details
Department of Biomedicine, HEALTH, Aarhus University, DK- 8000 Aarhus C, Denmark. ²Department of Molecular Biology and Genetics, Science and Technology, Aarhus University, DK-8000 Aarhus C, Denmark. ³Gene Expression and Therapy Group, King's College London, Faculty of Life Sciences & Medicine, Department of Medical and Molecular Genetics, 8th Floor Tower Wing, Guy's Hospital, London SE1 9RT, UK.

References
1. Cavazzana-Calvo M, Payen E, Negre O, Wang G, Hehir K, Fusil F, et al. Transfusion independence and HMGA2 activation after gene therapy of human beta-thalassaemia. Nature. 2010;467(7313):318–22.
2. Aiuti A, Biasco L, Scaramuzza S, Ferrua F, Cicalese MP, Baricordi C, et al. Lentiviral hematopoietic stem cell gene therapy in patients with Wiskott-Aldrich syndrome. Science. 2013;341(6148):1233151.
3. Kebriaei P, Singh H, Huls MH, Figliola MJ, Bassett R, Olivares S, et al. Phase I trials using sleeping beauty to generate CD19-specific CAR T cells. J Clin Invest. 2016;126(9):3363–76.
4. Antoniou M, Skipper KA, Anakok O. Optimizing Retroviral Gene Expression for Effective Therapies. Hum Gene Ther. 2013;24(4):363–74.
5. Chung JH, Whiteley M, Felsenfeld G. A 5' element of the chicken beta-globin domain serves as an insulator in human erythroid cells and protects against position effect in drosophila. Cell. 1993;74(3):505–14.

6.   Ottaviani A, Rival-Gervier S, Boussouar A, Foerster AM, Rondier D, Sacconi S, et al. The D4Z4 macrosatellite repeat acts as a CTCF and A-type lamins-dependent insulator in facio-scapulo-humeral dystrophy. PLoS Genet. 2009; 5(2):e1000394.

7.   Antoniou M, Harland L, Mustoe T, Williams S, Holdstock J, Yague E, et al. Transgenes encompassing dual-promoter CpG islands from the human TBP and HNRPA2B1 loci are resistant to heterochromatin-mediated silencing. Genomics. 2003;82(3):269–79.

8.   Neville JJ, Orlando J, Mann K, McCloskey B, Antoniou MN. Ubiquitous chromatin-opening elements (UCOEs): applications in biomanufacturing and gene therapy. Biotechnol Adv. 2017;35(5):557–64.

9.   Arumugam PI, Scholes J, Perelman N, Xia P, Yee JK, Malik P. Improved human beta-globin expression from self-inactivating lentiviral vectors carrying the chicken hypersensitive site-4 (cHS4) insulator element. Mol Ther. 2007;15(10):1863–71.

10. Ramezani A, Hawley TS, Hawley RG. Performance- and safety-enhanced lentiviral vectors containing the human interferon-beta scaffold attachment region and the chicken beta-globin insulator. Blood. 2003;101(12):4717–24.

11. Li CL, Emery DW. The cHS4 chromatin insulator reduces gammaretroviral vector silencing by epigenetic modifications of integrated provirus. Gene Ther. 2008;15(1):49–53.

12. Ryu BY, Persons DA, Evans-Galea MV, Gray JT, Nienhuis AW. A chromatin insulator blocks interactions between globin regulatory elements and cellular promoters in erythroid cells. Blood Cells Mol Dis. 2007;39(3):221–8.

13. Evans-Galea MV, Wielgosz MM, Hanawa H, Srivastava DK, Nienhuis AW. Suppression of clonal dominance in cultured human lymphoid cells by addition of the cHS4 insulator to a lentiviral vector. Mol Ther. 2007; 15(4):801–9.

14. Li CL, Xiong D, Stamatoyannopoulos G, Emery DW. Genomic and functional assays demonstrate reduced gammaretroviral vector genotoxicity associated with use of the cHS4 chromatin insulator. Mol Ther. 2009;17(4):716–24.

15. Recillas-Targa F, Pikaart MJ, Burgess-Beusse B, Bell AC, Litt MD, West AG, et al. Position-effect protection and enhancer blocking by the chicken beta-globin insulator are separable activities. Proc Natl Acad Sci U S A. 2002; 99(10):6883–8.

16. Bell AC, West AG, Felsenfeld G. The protein CTCF is required for the enhancer blocking activity of vertebrate insulators. Cell. 1999;98(3):387–96.

17. Yao S, Osborne CS, Bharadwaj RR, Pasceri P, Sukonnik T, Pannell D, et al. Retrovirus silencer blocking by the cHS4 insulator is CTCF independent. Nucleic Acids Res. 2003;31(18):5317–23.

18. West AG, Huang S, Gaszner M, Litt MD, Felsenfeld G. Recruitment of histone modifications by USF proteins at a vertebrate barrier element. Mol Cell. 2004;16(3):453–63.

19. Huang S, Li X, Yusufzai TM, Qiu Y, Felsenfeld G. USF1 recruits histone modification complexes and is critical for maintenance of a chromatin barrier. Mol Cell Biol. 2007;27(22):7991–8002.

20. Aker M, Bomsztyk K, Emery DW. Poly(ADP-ribose) polymerase-1 (PARP-1) contributes to the barrier function of a vertebrate chromatin insulator. J Biol Chem. 2010;285(48):37589–97.

21. Williams S, Mustoe T, Mulcahy T, Griffiths M, Simpson D, Antoniou M, et al. CpG-island fragments from the HNRPA2B1/CBX3 genomic locus reduce silencing and enhance transgene expression from the hCMV promoter/enhancer in mammalian cells. BMC Biotechnol. 2005;5:17.

22. Zhang F, Santilli G, Thrasher AJ. Characterization of a core region in the A2UCOE that confers effective anti-silencing activity. Sci Rep. 2017;7(1): 10213.

23. Zhang F, Thornhill SI, Howe SJ, Ulaganathan M, Schambach A, Sinclair J, et al. Lentiviral vectors containing an enhancer-less ubiquitously acting chromatin opening element (UCOE) provide highly reproducible and stable transgene expression in hematopoietic cells. Blood. 2007;110(5):1448–57.

24. Dighe N, Khoury M, Mattar C, Chong M, Choolani M, Chen J, et al. Long-term reproducible expression in human fetal liver hematopoietic stem cells with a UCOE-based Lentiviral vector. PLoS One. 2014;9(8):e104805.

25. Brendel C, Muller-Kuller U, Schultze-Strasser S, Stein S, Chen-Wichmann L, Krattenmacher A, et al. Physiological regulation of transgene expression by a lentiviral vector containing the A2UCOE linked to a myeloid promoter. Gene Ther. 2012;19(10):1018–29.

26. Zhang F, Frost AR, Blundell MP, Bales O, Antoniou MN, Thrasher AJ. A ubiquitous chromatin opening element (UCOE) confers resistance to DNA methylation-mediated silencing of lentiviral vectors. Mol Ther. 2010;18(9):1640–9.

27. Muller-Kuller U, Ackermann M, Kolodziej S, Brendel C, Fritsch J, Lachmann N, et al. A minimal ubiquitous chromatin opening element (UCOE) effectively prevents silencing of juxtaposed heterologous promoters by epigenetic remodeling in multipotent and pluripotent stem cells. Nucleic Acids Res. 2015;43(3):1577–92.

28. Lindahl Allen M, Antoniou M. Correlation of DNA methylation with histone modifications across the HNRPA2B1-CBX3 ubiquitously-acting chromatin open element (UCOE). Epigenetics. 2007;2(4):227–36.

29. Kunkiel J, Godecke N, Ackermann M, Hoffmann D, Schambach A, Lachmann N, et al. The CpG-sites of the CBX3 ubiquitous chromatin opening element are critical structural determinants for the anti-silencing function. Sci Rep. 2017;7(1):7919.

30. Ottaviani A, Schluth-Bolard C, Gilson E, Magdinier F. D4Z4 as a prototype of CTCF and lamins-dependent insulator in human cells. Nucleus. 2010;1(1):30–6.

31. Dalsgaard T, Moldt B, Sharma N, Wolf G, Schmitz A, Pedersen FS, et al. Shielding of sleeping beauty DNA transposon-delivered transgene cassettes by heterologous insulators in early embryonal cells. Mol Ther. 2009;17(1): 121–30.

32. Sharma N, Hollensen AK, Bak RO, Staunstrup NH, Schroder LD, Mikkelsen JG. The impact of cHS4 insulators on DNA transposon vector mobilization and silencing in retinal pigment epithelium cells. PLoS One. 2012;7(10):e48421.

33. Kim JY, Kim YG, Lee GM. CHO cells in biotechnology for production of recombinant proteins: current state and further potential. Appl Microbiol Biotechnol. 2012;93(5):917–30.

34. Staunstrup NH, Stenderup K, Mortensen S, Primo MN, Rosada C, Steiniche T, et al. Psoriasiform skin disease in transgenic pigs with high-copy ectopic expression of human integrins alpha2 and beta1. Dis Model Mech. 2017; 10(7):869–80.

35. Holstein M, Mesa-Nunez C, Miskey C, Almarza E, Poletti V, Schmeer M, et al. Efficient non-viral gene delivery into human hematopoietic stem cells by Minicircle sleeping beauty transposon vectors. Mol Ther. 2018;26(4):1137–53.

36. Moldt B, Miskey C, Staunstrup NH, Gogol-Doring A, Bak RO, Sharma N, et al. Comparative genomic integration profiling of sleeping beauty transposons mobilized with high efficacy from Integrase-defective Lentiviral vectors in primary human cells. Mol Ther. 2011;19(8):1499–510.

37. Yant SR, Wu X, Huang Y, Garrison B, Burgess SM, Kay MA. High-resolution genome-wide mapping of transposon integration in mammals. Mol Cell Biol. 2005;25(6):2085–94.

38. Liu G, Geurts AM, Yae K, Srinivasan AR, Fahrenkrug SC, Largaespada DA, et al. Target-site preferences of sleeping beauty transposons. J Mol Biol. 2005;346(1):161–73.

39. Yannaki E, Tubb J, Aker M, Stamatoyannopoulos G, Emery DW. Topological constraints governing the use of the chicken HS4 chromatin insulator in oncoretrovirus vectors. Mol Ther. 2002;5(5 Pt 1):589–98.

40. Emery DW, Yannaki E, Tubb J, Nishino T, Li Q, Stamatoyannopoulos G. Development of virus vectors for gene therapy of beta chain hemoglobinopathies: flanking with a chromatin insulator reduces gamma-globin gene silencing in vivo. Blood. 2002;100(6):2012–9.

41. Bean DM, Heimbach J, Ficorella L, Micklem G, Oliver SG, Favrin G. esyN: network building, sharing and publishing. PLoS One. 2014;9(9):e106035.

42. Chatr-Aryamontri A, Oughtred R, Boucher L, Rust J, Chang C, Kolas NK, et al. The BioGRID interaction database: 2017 update. Nucleic Acids Res. 2017; 45(D1):D369–D79.

43. Kreft L, Soete A, Hulpiau P, Botzki A, Saeys Y, De Bleser P. ConTra v3: a tool to identify transcription factor binding sites across species, update 2017. Nucleic Acids Res. 2017;45(W1):W490–W4.

44. Kim J, Kollhoff A, Bergmann A, Stubbs L. Methylation-sensitive binding of transcription factor YY1 to an insulator sequence within the paternally expressed imprinted gene, Peg3. Hum Mol Genet. 2003;12(3):233–45.

45. Lin JM, Collins PJ, Trinklein ND, Fu Y, Xi H, Myers RM, et al. Transcription factor binding and modified histones in human bidirectional promoters. Genome Res. 2007;17(6):818–27.

46. Lienert F, Wirbelauer C, Som I, Dean A, Mohn F, Schubeler D. Identification of genetic elements that autonomously determine DNA methylation states. Nat Genet. 2011;43(11):1091–7.

47. Bandaranayake AD, Correnti C, Ryu BY, Brault M, Strong RK, Rawlings DJ. Daedalus: a robust, turnkey platform for rapid production of decigram quantities of active recombinant proteins in human cell lines using novel lentiviral vectors. Nucleic Acids Res. 2011;39(21):e143.

48. Cullmann K, Blokland KEC, Sebe A, Schenk F, Ivics Z, Heinz N, et al. Sustained and regulated gene expression by Tet-inducible "all-in-one"

retroviral vectors containing the HNRPA2B1-CBX3 UCOE((R)). Biomaterials. 2018;192:486–99.

49. Norwood LE, Grade SK, Cryderman DE, Hines KA, Furiasse N, Toro R, et al. Conserved properties of HP1(Hsalpha). Gene. 2004;336(1):37–46.

50. Kirschmann DA, Lininger RA, Gardner LM, Seftor EA, Odero VA, Ainsztein AM, et al. Down-regulation of HP1Hsalpha expression is associated with the metastatic phenotype in breast cancer. Cancer Res. 2000;60(13):3359–63.

51. Lieberthal JG, Kaminsky M, Parkhurst CN, Tanese N. The role of YY1 in reduced HP1alpha gene expression in invasive human breast cancer cells. Breast Cancer Res. 2009;11(3):R42.

52. Staunstrup NH, Madsen J, Primo MN, Li J, Liu Y, Kragh PM, et al. Development of transgenic cloned pig models of skin inflammation by DNA transposon-directed ectopic expression of human beta1 and alpha2 integrin. PLoS One. 2012;7(5):e36658.

53. Mates L, Chuah MK, Belay E, Jerchow B, Manoj N, Acosta-Sanchez A, et al. Molecular evolution of a novel hyperactive sleeping beauty transposase enables robust stable gene transfer in vertebrates. Nat Genet. 2009;41(6):753–61.

54. Consortium EP. An integrated encyclopedia of DNA elements in the human genome. Nature. 2012;489(7414):57–74.

8

Investigation of cell culture conditions for optimal foot-and-mouth disease virus production

Veronika Dill[1], Aline Zimmer[2], Martin Beer[1] and Michael Eschbaumer[1]* (iD)

Abstract

Background: Foot-and-mouth disease is a highly contagious and economically devastating disease with endemic occurrence in many parts of the world. Vaccination is the method of choice to eradicate the disease and to limit the viral spread. The vaccine production process is based on mammalian cell culture, in which the viral yield varies in dependence of the composition of the culture media. For foot-and-mouth disease virus (FMDV), very little is known about the culture media components that are necessary to grow the virus to high titers in cell culture.

Results: This study examined the influence of increasing concentrations of glucose, glutamine, ammonium chloride and different cell densities on the yield of FMDV. While an excess of glucose or glutamine does not affect the viral yield, increasing cell density reduces the viral titer by a $\log_{10}$ step at a cell density of 3×10^6 cells/mL. This can be mitigated by performing a 100% media exchange before infection of the cells.

Conclusions: The reasons for the diminished viral growth, if no complete media exchange has been performed prior to infection, remain unclear and further studies are necessary to investigate the causes more deeply. For now, the results argue for a vaccine production process with 100% media exchange to reliably obtain high viral titers.

Keywords: Foot-and-mouth disease virus, Vaccine, Glucose, Glutamine, Cell density, Suspension cells, Animal-component free

Background

Foot-and-mouth disease (FMD) is a viral disease of cloven-hoofed livestock with tremendous economic impact [1]. Every year, more than one billion doses of FMD vaccine are produced worldwide [2]. These vaccines are used for control programs in regions where FMD is endemic and for the emergency response to outbreaks in areas where the disease does not occur regularly [2, 3].

Mammalian cells are widely used for the propagation of viruses for vaccine production. The viral yield in cell culture varies greatly depending on the composition of the culture media [4]. For instance, the production of poliovirus in HeLa cells differs with the media composition, with salts, glucose and glutamine representing the only essential substrates for successful virus production [4, 5]. Glucose and glutamine are the main carbon sources for mammalian cells in culture. They are key nutrients to cover the cell's energy requirements [6]. The glycolysis and glutaminolysis pathways in the cell are utilized at high rates to metabolize these substrates, leading to the production of high amounts of waste products such as lactate and ammonium [7]. Many viral infections are characterized by an increase in the rate of glycolysis, e.g. poliomyelitis virus [8], feline leukemia virus [9], or herpes simplex virus [10] or an increase in glutamine uptake, e.g. vaccinia virus [11] or human cytomegalovirus [12]. In addition to these two important pathways, a viral infection of the cell leads to other changes in the cellular metabolism, such as fatty acid synthesis [13, 14].

For foot-and-mouth disease virus (FMDV), very little is known about the culture media components that are necessary to grow the virus to high titers in cell culture. An early study by Pledger et al. [15] named glucose as an important substrate for FMDV replication, while glutamine alone had no influence on the viral titer.

* Correspondence: michael.eschbaumer@fli.de
[1]Institute of Diagnostic Virology, Friedrich-Loeffler-Institut, Südufer 10, 17493 Greifswald, Insel Riems, Germany

However, the metabolism of glucose produces high amounts of lactate that are released into the culture media, decreasing its pH if the buffer capacity of the media is exceeded [13, 16]. Optimizing the glucose and glutamine content of the culture media can increase the viral harvest, while the use of increased cell densities allows a more efficient use of bioreactor capacity and decreased costs per dose of vaccine [16]. Using animal-component free (ACF) media for vaccine production can further reduce costs and minimize the risk of contamination through animal-derived raw materials [3, 17].

This study investigated the metabolism of baby hamster kidney (BHK) suspension cells infected with the recent FMDV isolates A IRN/8/2015 and O SAU/18/2015. The viral titer as well as the effect of different concentrations of glucose, glutamine and ammonium chloride in the medium were examined. Furthermore, infection at different cell densities in combination with a total or partial media exchange was compared.

Results

Abundant glucose and glutamine in the cell culture media does not increase the viral yield

Infection studies of BHK-2P suspension cells were performed using the recent FMDV isolates A IRN/8/2015 and O SAU/18/2015 and increasing concentrations of glucose and glutamine in the cell culture media. Independently of the glucose or glutamine concentration, stable virus titers of 7.7 ± 0.2 $\log_{10}$ $TCID_{50}$/mL across the different glucose concentrations and 7.9 ± 0.1 $\log_{10}$ $TCID_{50}$/mL across the different concentrations of glutamine were achieved for A IRN/08/2015, compared to 7.2 ± 0.1 $\log_{10}$ $TCID_{50}$/mL and 7.3 ± 0.1 $\log_{10}$ $TCID_{50}$/mL for O SAU/18/2015, respectively (Fig. 1).

No statistically significant differences in the content of glutamine and glucose in the cell culture media before and after viral infection were observed in any of the tested conditions for either virus isolate (Fig. 2).

Cell viability in the infected cultures dropped drastically after 20 hpi compared to the uninfected negative controls in all experiments (see Additional file 1). The amounts of lactate accumulated in the media of the infected cultures were 10 to 60% lower (depending on serotype and tested condition) in comparison to the negative control. Nevertheless, among the infected cultures, no statistically significant difference between the tested conditions was detected (Fig. 2).

In general, the cells consumed only a small fraction of the available carbohydrates when infected with FMDV based on the differences in the glucose and glutamine content of the media before and after viral infection independent of the total concentration of glucose or glutamine in the media. The difference in concentration of glutamine and glucose for the respective negative controls was nearly twice as much than for the infected cells (Table 1).

Calculations of the cell-specific glucose uptake rates revealed a positive trend of an increased glucose consumption with increased glucose availability when infected with FMDV serotype A that was not evident for cells infected with serotype O (see Additional file 3: Table S3). On the other hand, the data for the cell-specific glutamine uptake suggest an increased uptake of glutamine with increasing concentrations of glutamine available in the cell culture media when infected with serotype O, while no trend was evident for serotype-A-infected cells (see Additional file 3: Table S4).

The concentration of ammonium chloride in the cell culture media influences cell survival but not the viral yield

The ACF media was supplemented with increasing concentrations of ammonium chloride ranging from 0 mM to 12 mM and cells were infected with A IRN/

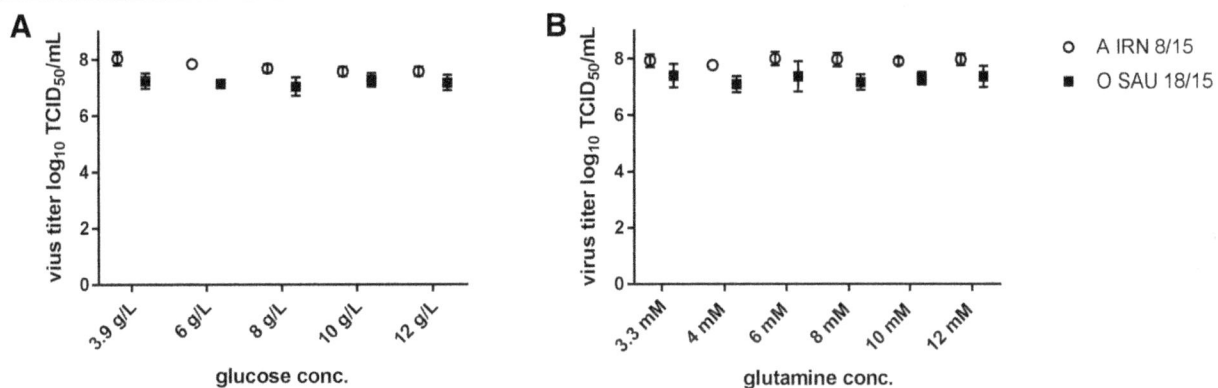

Fig. 1 Viral yield of BHK-2P cells maintained in ACFM with increasing concentrations of glucose (**a**) or glutamine (**b**). Cells were infected with FMDV A IRN/8/2015 (open circles) or O SAU/18/2015 (filled squares). No differences in viral titer were observed for any of the tested conditions. Experiments were performed three times independently and were titrated in duplicates

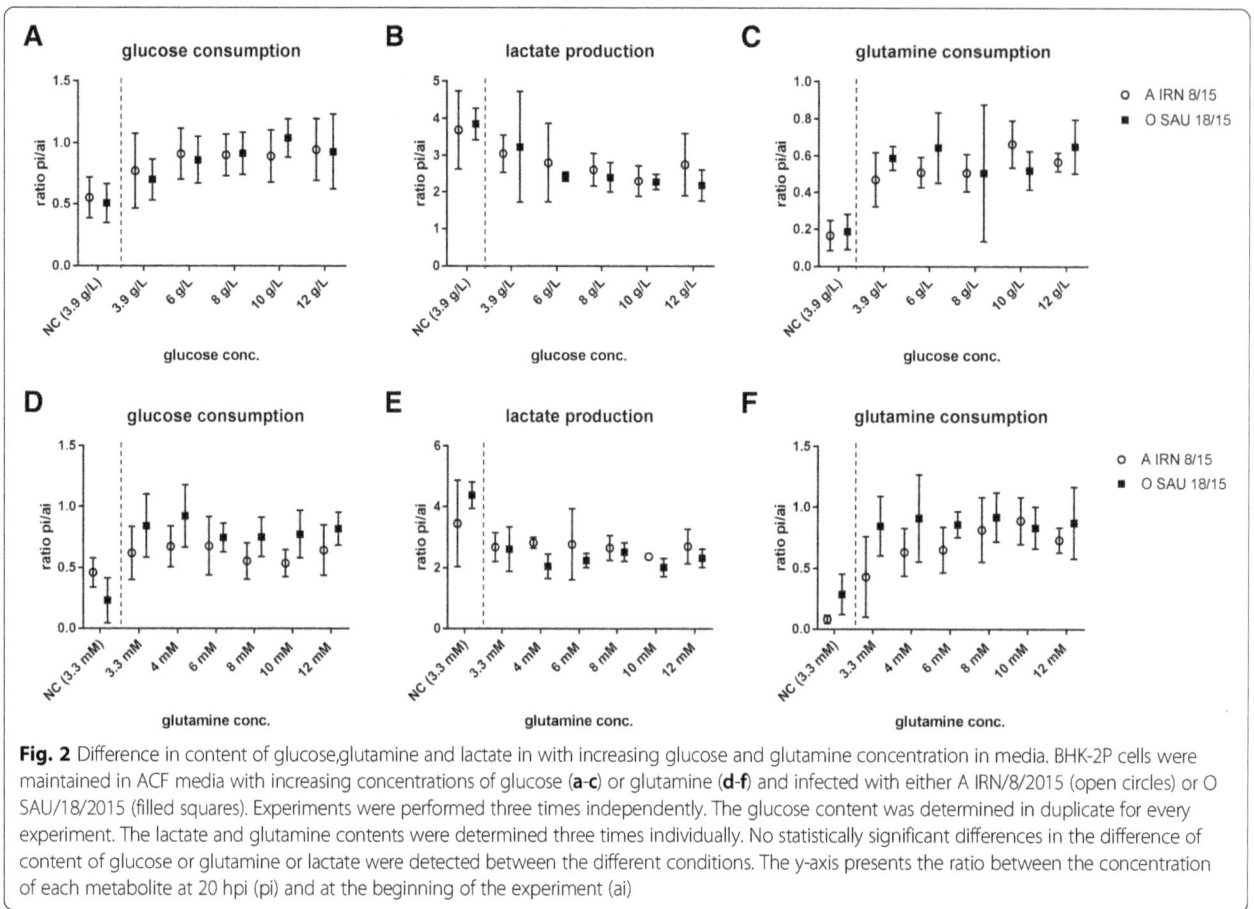

Fig. 2 Difference in content of glucose, glutamine and lactate in with increasing glucose and glutamine concentration in media. BHK-2P cells were maintained in ACF media with increasing concentrations of glucose (**a-c**) or glutamine (**d-f**) and infected with either A IRN/8/2015 (open circles) or O SAU/18/2015 (filled squares). Experiments were performed three times independently. The glucose content was determined in duplicate for every experiment. The lactate and glutamine contents were determined three times individually. No statistically significant differences in the difference of content of glucose or glutamine or lactate were detected between the different conditions. The y-axis presents the ratio between the concentration of each metabolite at 20 hpi (pi) and at the beginning of the experiment (ai)

8/2015 or O SAU/18/2015. Surprisingly, the viral yield was stable at a titer of 7.2 ± 0.1 $\log_{10}$ TCID$_{50}$/mL for A IRN 8/2015 and 6.9 ± 0.2 $\log_{10}$ TCID$_{50}$/mL for O SAU 18/2015, while cell viability at 20 hpi increased with increasing concentrations of ammonium chloride. That increase was statistically significant in the case of A IRN 8/15 (Fig. 3).

The prolonged survival of the infected BHK-2P cells at or above an ammonium chloride concentration of 6 mM is reflected in an increased content of glutamine and glucose as well as lactate in the cell culture media (Fig. 4). The difference in the content of glucose in the cell culture media before and after infection does not vary strongly between infected and uninfected cells but the

content of lactate is reduced in the cell culture media of FMDV-infected cells. The calculation of the cell-specific glucose and glutamine uptake as well as lactate production revealed an increasing trend for all three metabolites between a concentration of 2–10 mM NH$_4$Cl in the media for cells infected with the FMDV A/IRN/8/2015 isolate, while no such trend was evident for cells infected with FMDV O/SAU/18/2015 (see Additional file 3: Table S5).

Increased cell density leads to decreased viral yield

Cell densities of 1×10^6 cells/mL, 2×10^6 cells/mL and 3×10^6 cells/mL were infected with FMDV A IRN/8/2015 or O SAU/18/2015, either with a media exchange of 100% or with a media exchange of 30% before infection. Cell viability at 20 hpi was higher (Fig. 5) and viral titers were lower (Fig. 6) at higher cell densities when only 30% of the culture media was replaced with fresh media before infection. While differences in cell viability 20 hpi between 100 and 30% media exchange are already evident for A IRN/8/2015 at a cell density of 2×10^6 cells/mL, a statistically significant difference in cell death was only seen between the 30% media exchange preparations of 1×10^6 cells/mL and 3×10^6 cells/mL for O

Table 1 Mean glucose and glutamine content in the cell culture media of BHK-2P suspension cells when infected with FMDV

cells infected with	A IRN 8/15	O SAU 18/15	NC
glucose (g/L)	1.7 ± 0.5	1.0 ± 0.6	2.7 ± 1.1
glutamine (mM)	1.8 ± 0.7	1.5 ± 0.6	2.9 ± 0.4

Cells were either infected with FMDV A IRN/8/2015 or O SAU/18/2015 or uninfected (NC = negative control). The content was calculated by subtracting the measured concentration of the analyte at 20 hpi from the measured concentration before infection

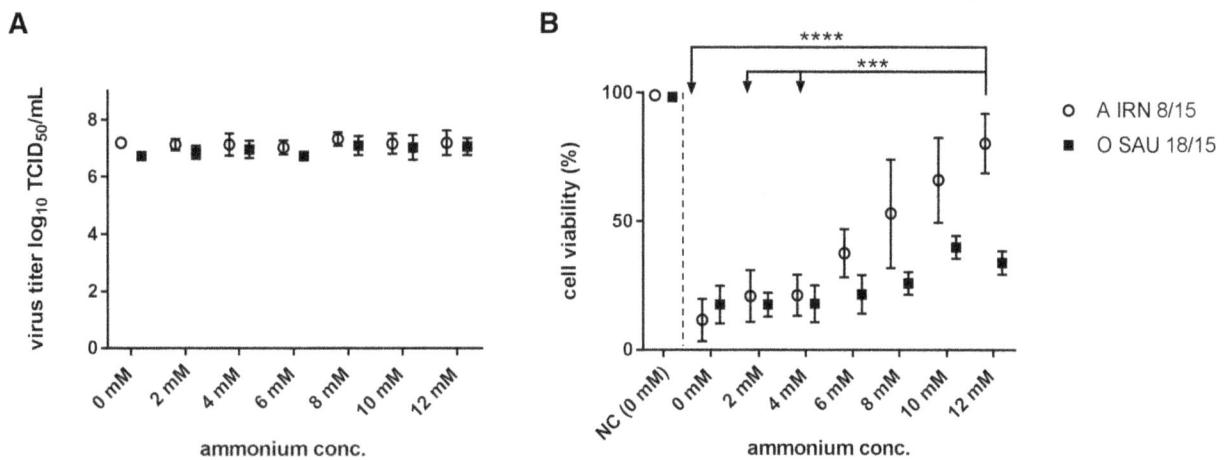

Fig. 3 Viral titers of BHK-2P cells maintained in ACF media with increasing concentrations of ammonium chloride. Cells were infected with FMDV A IRN/8/2015 (open circles) or O SAU/18/2015 (filled squares). No difference in viral titer (**a**) was observed for any of the tested conditions. Cell viability 20 hpi (**b**) increased with increasing concentration of ammonium chloride in the cell media. Experiments were performed three times independently and the viral yield was titrated in duplicate. Significance code: *** $p < 0.001$; **** $p < 0.0001$

SAU/18/2015 (Fig. 5). No differences in the cell viability were detected when the full volume of medium was exchanged, independently of the cell density.

Viral titers of 7.7 ± 0.3 $\log_{10}$ $TCID_{50}$/mL for A IRN/8/2015 and 7.5 ± 0.3 $\log_{10}$ $TCID_{50}$/mL for O SAU/18/2015 were achieved at all tested cell densities if a 100% media exchange was performed. A slight increase in viral titer with increasing cell density was seen, but it was not statistically significant (see Additional file 2: Table S1). With a 30% media exchange before infection, stable titers of 8.1 ± 0.0 $\log_{10}$ $TCID_{50}$/mL for A IRN/8/2015 and 7.3 ± 0.1 $\log_{10}$ $TCID_{50}$/mL for O SAU/18/2015 were reached for densities of 1×10^6 cells/mL and 2×10^6 cells/mL, while the viral titer significantly dropped by one $\log_{10}$ at a cell density of 3×10^6 cells/mL (serotype

A: 7.0 ± 0.6 $\log_{10}$ $TCID_{50}$/mL; serotype O: 6.7 ± 0.3 $\log_{10}$ $TCID_{50}$/mL) (Fig. 6).

The difference in the content of glucose, lactate and glutamine before and after viral infection was similar across the different cell densities when 100% of media were exchanged. Similar to the experiments with increasing glucose and glutamine concentrations in the media, only a small fraction of the provided carbohydrates was metabolized by the virus-infected cells. With an incomplete exchange of media (30%), the difference in the contents of glucose and lactate in the cell culture media of the virus-infected cell cultures were similar to the negative controls. High values of lactate (> 10 mM) and a decreased pH of the culture media were already evident at a cell density of 1×10^6 cells/mL (Table 2).

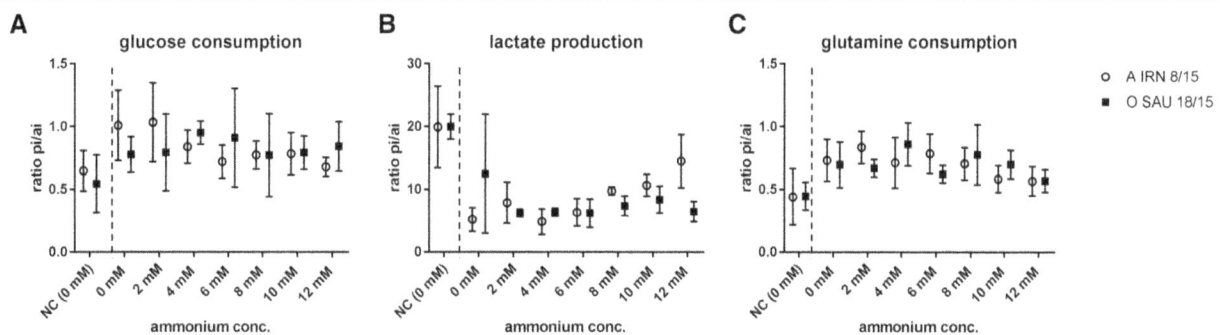

Fig. 4 Difference in content of glucose (**a**), lactate (**b**) and glutamine (**c**) in the cell culture media with increasing concentrations of ammonium chloride. BHK-2P cells were maintained in ACF media and infected with either A IRN/8/2015 (open circles) or O SAU/18/2015 (filled squares). Experiments were performed three times independently. The content of glucose was determined in duplicates for every experiment. The content of lactate and glutamine was determined three times individually. No statistically significant differences in the difference of content for glucose,glutamine or lactate could be detected between the conditions in infected cells. Nevertheless, an increase in the contents of glutamine and lactate is evident at ammonium chloride concentrations above 6 mM

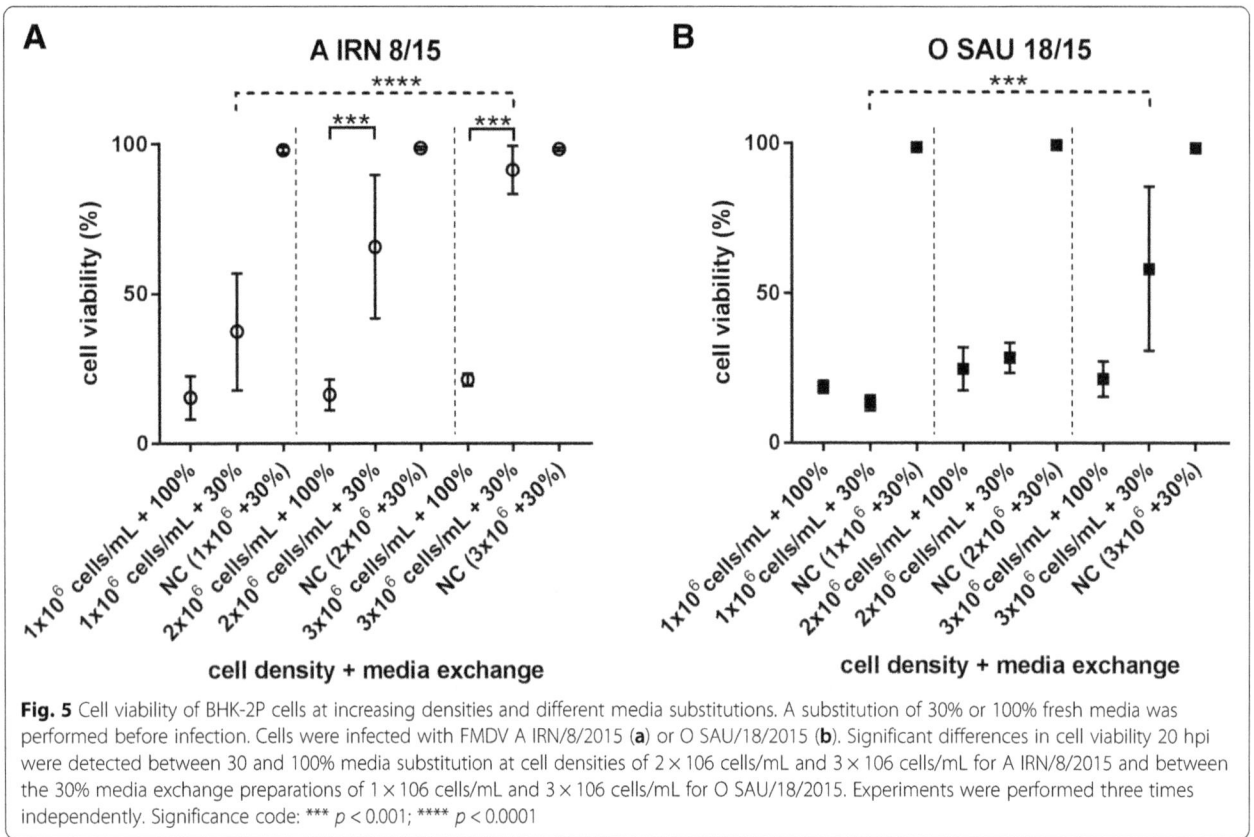

Fig. 5 Cell viability of BHK-2P cells at increasing densities and different media substitutions. A substitution of 30% or 100% fresh media was performed before infection. Cells were infected with FMDV A IRN/8/2015 (**a**) or O SAU/18/2015 (**b**). Significant differences in cell viability 20 hpi were detected between 30 and 100% media substitution at cell densities of 2×106 cells/mL and 3×106 cells/mL for A IRN/8/2015 and between the 30% media exchange preparations of 1×106 cells/mL and 3×106 cells/mL for O SAU/18/2015. Experiments were performed three times independently. Significance code: *** $p < 0.001$; **** $p < 0.0001$

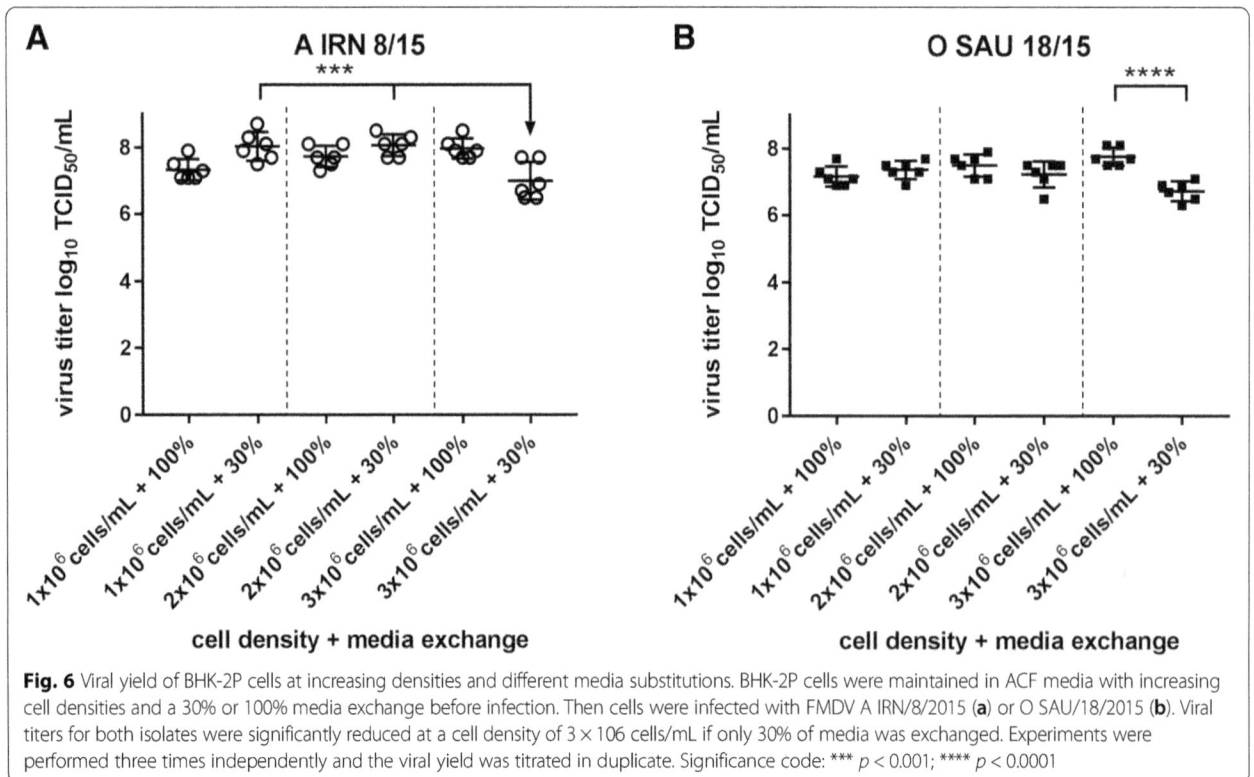

Fig. 6 Viral yield of BHK-2P cells at increasing densities and different media substitutions. BHK-2P cells were maintained in ACF media with increasing cell densities and a 30% or 100% media exchange before infection. Then cells were infected with FMDV A IRN/8/2015 (**a**) or O SAU/18/2015 (**b**). Viral titers for both isolates were significantly reduced at a cell density of 3×106 cells/mL if only 30% of media was exchanged. Experiments were performed three times independently and the viral yield was titrated in duplicate. Significance code: *** $p < 0.001$; **** $p < 0.0001$

Table 2 Glucose, lactate, and glutamine content and pH of the ACF media 20 h after viral infection

virus	A IRN/8/2015		NC	A IRN/8/2015		NC	A IRN/8/2015		NC
cell density	1×10^6 cells/mL			2×10^6 cells/mL			3×10^6 cells/mL		
media exchange	100%	30%	30%	100%	30%	30%	100%	30%	30%
pH	7.5 ± 0.1	7.3 ± 0.1	7.1 ± 0.2	7.3 ± 0.1	7.0 ± 0.1	7.0 ± 0.1	7.2 ± 0.1	7.0 ± 0.3	6.9 ± 0.1
glucose (g/L)	5.2 ± 1.2	3.4 ± 1.0	2.6 ± 0.3	3.7 ± 1.8	1.6 ± 0.4	1.4 ± 0.6	2.6 ± 1.0	1.1 ± 0.5	1.2 ± 0.6
lactate (mM)	8.2 ± 0.9	23.0 ± 6.1	26.9 ± 4.6	16.6 ± 3.1	28.5 ± 4.1	29.2 ± 3.5	18.5 ± 2.6	25.1 ± 5.3	33.6 ± 4.1
glutamine (mM)	2.5 ± 1.8	1.5 ± 1.0	0.4 ± 0.3	2.5 ± 0.9	0.6 ± 0.6	0.1 ± 0.1	1.6 ± 0.6	0.3 ± 0.3	0.7 ± 0.9
virus isolate	O SAU/18/2015		NC	O SAU/18/2015		NC	O SAU/18/2015		NC
cell density	1×10^6 cells/mL			2×10^6 cells/mL			3×10^6 cells/mL		
media exchange	100%	30%	30%	100%	30%	30%	100%	30%	30%
pH	7.4 ± 0.0	7.4 ± 0.0	7.1 ± 0.0	7.4 ± 0.0	7.2 ± 0.2	7.0 ± 0.2	7.2 ± 0.20	6.9 ± 0.2	6.9 ± 0.2
glucose (g/L)	7.0 ± 1.6	5.2 ± 1.0	3.5 ± 1.0	5.1 ± 0.7	3.1 ± 1.2	2.1 ± 0.8	4.1 ± 0.3	1.5 ± 0.9	1.6 ± 0.5
lactate (mM)	7.5 ± 4.1	11.5 ± 5.8	21.4 ± 10.7	14.9 ± 6.8	23.7 ± 11.1	27.9 ± 14.0	16.6 ± 8.4	29.2 ± 14.9	36.8 ± 19.1
glutamine (mM)	4.3 ± 1.5	3.1 ± 0.9	1.3 ± 0.3	4.0 ± 1.3	1.6 ± 0.7	0.5 ± 0.4	2.6 ± 0.5	0.7 ± 0.3	0.2 ± 0.1

The depletion of glutamine in the virus-infected cell cultures with 30% media exchange was slightly less compared to the negative controls of the same cell density (Table 2). In general, glutamine in the negative cultures was almost completely depleted. The comparison of the cell-specific uptake rates showed a negative trend for the mock-infected cells and cells infected in 30% fresh media with reduced glucose consumption at higher cell densities, whereas cells infected in 100% fresh media increased their glucose uptake (see Additional file 3: Table S6).

Discussion

Several studies have reported an increased uptake of glutamine and/or glucose in the course of a viral infection of cell cultures [13, 14]. Very little is known about the metabolic processes in an FMDV-infected cell culture or about the nutrient requirements for successful virus production. For poliomyelitis virus, another picornavirus, it is known that virus production depends on the media composition [4]. For that virus, it was hypothesized that glucose and glutamine are necessary as energy sources or for synthesis of the viral nucleic acid [5]. Of the two nutrients, glutamine was determined to be the more important factor, possibly due to the high inherent glutaminase activity of HeLa cells [4].

Conversely, for FMDV, an early study by Pledger and colleagues determined glucose as the only necessary factor for virus replication in primary bovine kidney cells [15]. The present study examined the possibility of increasing the FMD viral yield by providing more of these key nutrients in the culture media, but neither an increased concentration of glucose (up to 12 g/L) nor of glutamine (up to 12 mM) had a significant effect on the virus titer. In addition, no increase in glycolysis or glutaminolysis of the cell was observed. Darnell and colleagues have proposed that the optimal glucose and glutamine concentrations for virus production are the same as for cell growth [4]. Comparing the residual content of glucose and glutamine in the media between a mature infected culture and the corresponding negative control revealed that only a small fraction of the supplied nutrients had been metabolized. As previously proposed by Pledger et al., extracellular nutrient requirements seem to be of lesser importance for FMD virus particle production [15]. A possible explanation for this is the rapid growth of FMDV in infected cells. While one replication cycle of poliovirus takes 3–7.5 h (depending on the cell line), FMDV only needs 1.5 to 2.5 h to replicate in primary bovine kidney cells [15] and the first newly produced virus particles usually appear in the culture media after 4–6 h [18]. Due to the rapid progression of lytic infection, most cells in the culture vessel are dead before the available nutrients are exhausted. Consequently, metabolic waste products such as lactate were not sufficiently increased to lower the pH to a level that would negatively impact the viral yield. FMDV particles are highly acid-sensitive and the capsid dissociates at pH values slightly below neutrality (< pH 6.8) [19]. In the course of infection in this study, the pH of the media decreased from 7.5 to at most 7.2, which has no impact on the stability of virus particles.

To initiate infection, FMDV releases its genome into the host cell by dissociation of the capsid into pentameric subunits, which is triggered by the acidification of the endosome [20]. Ammonium ions ($NH4^+$), a waste product of the glutamine metabolism pathway, neutralize the acidic pH within endosomes and block the uncoating of the viral

RNA [21]. Surprisingly, we found no difference in the viral titers in cultures supplemented with increasing concentrations of ammonium chloride (up to 12 mM). At the same time, the cell viability in infected cultures increased dramatically at higher concentrations of ammonium chloride, particularly for the serotype A virus isolate.

Several studies have been performed to examine the effect of lysosomotropic agents on FMDV infection [20–23]. In most cases, only concentrations of 25 mM and 50 mM ammonium chloride were tested and these resulted in a drastically reduced viral titer at 25 mM and a complete block of infection at 50 mM NH_4Cl [20]. Carillo et al. tested the influence of lower concentrations of ammonium chloride (10, 20 and 30 mM), leading to a 95% reduction in the viral yield for 30 mM and a proportionally lower impact for 20 mM and 10 mM [21]. No data are available for batch cultures of BHK cells infected with FMDV, but glutamine metabolism and ammonium production during cell growth and virus infection in batch culture are well studied for MDCK cells and influenza virus [24]. Similar to the observations for conventional FMDV culture, a reduction in the yield of influenza virus was evident for NH_4Cl concentrations above 20 mM. However, this is not likely to have an effect under real production conditions. In theory, a maximum of 4 mM ammonium can be produced from the 4 mM glutamine available at the start of the process [24], but in the experiment, no more than 2 mM of ammonium were produced by the cells throughout the entire process [24]. Considering the rapid progression of FMDV infection, it is unlikely that the real ammonium concentrations produced during FMDV antigen production exceed these values and thus influence the yield. Nevertheless, in the absence of a satisfactory explanation for the increased cell viability observed at higher concentrations, production processes should be controlled in such a way that not more than 6 mM ammonium can accumulate in the culture media before viral infection.

Several studies have reported so-called "cell density effects" in different culture systems used for virus production [16, 25, 26]. For our studies, three different cell densities (1×10^6 cells/mL, 2×10^6 cells/mL and 3×10^6 cells/mL) with either 100% or 30% media exchange before infection were tested. No difference in viral titer was detected between the different cell densities if a full media exchange was performed. An exchange of 30% of the media led to a high variability in cell viability at the end of the process and together with a cell density of 3×10^6 cells/mL reduced the final viral titer by about one $\log_{10}$ step. This may be due to nutrient limitation or the accumulation of inhibitory factors [16, 25], but the exact cause of the specific "cell density effect" observed

in our system is unknown. The lactate content was elevated in cultures with 3×10^6 cells/mL and 30% media exchange, but it is unlikely that this directly influenced infectivity. The pH of the culture medium was adjusted to pH 7.5 before infection and high concentrations of lactate alone do not impair the growth of BHK cells [27]. With only 30% of media being replaced, glucose and especially glutamine levels are reduced compared to cultures with a full exchange, but the available nutrients should be still sufficient for virus production. Further experiments are needed to examine whether this effect is still visible if the concentrations of glucose and glutamine are increased at the time of infection.

The examination of cell growth rates at the different conditions could also offer valuable information about the ability of the cells to produce virus. An inhibited cell growth can be linked to reduced virus production. In addition, it is known that the impact of excessive cell densities on production depends on the culture medium [25]. It would be of interest if other culture media can mitigate the "cell density effect" and further improve FMDV vaccine production processes.

Conclusions

This study determined important culture parameters that influence FMDV titers in a small-scale bioreactor system. While an excess of glucose, glutamine and ammonia in the culture media does not directly influence the viral yield, the cell density seems to have the largest impact on the viral titers achieved in batch culture. Further experiments have to be performed to study the nature of this effect in greater detail. Based on the current state of knowledge, 100% media exchange is recommended for optimal yield and high reproducibility.

Methods

All experiments have been performed in a veterinary BSL-4 laboratory that meets the Minimum Biorisk Management Standards for Laboratories Working with FMDV [28].

Cells

The suspension cell line BHK21C13-2P (in short: BHK-2P; originally derived from the European Collection of Authenticated Cell Cultures specimen 84,111,301) was adapted to grow in Cellvento™ BHK-200 (Merck KGaA, Darmstadt, Germany) ACF medium in TubeSpin® bioreactors (TPP Techno Plastic Products AG, Trasadingen, Switzerland). The cells were maintained in a shaker incubator at 320 rpm (rpm) at 37 °C, 5% CO_2 and 80% relative humidity.

For virus titrations, the adherent BHK21 clone "Tübingen" cell line (in short: BHK164, specimen CCLV-RIE 164 in the Collection of Cell Lines in Veterinary Medicine, Friedrich-Loeffler-Institut [FLI], Greifswald, Germany) was cultured in Minimum Essential Medium Eagle, supplemented with Hanks' and Earle's salts (Sigma-Aldrich, St. Louis, USA) with 5% fetal bovine serum. The adherent cells were incubated at 37 °C and 5% CO_2.

Viruses and virus titrations

The FMDV isolates A IRN/8/2015 and O SAU/18/2015 were selected from archival stocks at the FLI. Their origin and passage history can be found in Additional file 2: Table S2 Viral titers were estimated by endpoint titration with the Spearman-Kärber method [29, 30] and expressed as 50% tissue culture infectious dose (TCID50) per milliliter. Virus titrations were performed on the adherent BHK164.

Small-scale bioreactor experiments

All experiments were performed in TubeSpin Bioreactors-50 (TPP) with a working volume of 30 ml. pH measurements before infection were performed with a potentiometric pH meter (Mettler Toledo, Gießen, Germany) and if necessary the pH was adjusted to 7.5 using 1 M sodium hydroxide. Cells were infected at a multiplicity of infection (MOI) of 0.1, and virus was harvested after 20 h of incubation. Virus samples were stored at − 70 °C, then thawed for further processing and cell debris were removed by centrifugation at 3200×g for 10 min at 4 °C. Samples to determine the content of glucose, lactate and glutamine were taken from the bioreactor immediately before infection and again 20 h post infection (hpi). These samples were centrifuged at 155×g for 5 min at 4 °C and the supernatant was stored at − 70 °C until further processing. pH measurements 20 hpi were performed using single-use pH test strips (range 6.5–10, Merck KGaA, Darmstadt, Germany) for biosafety reasons.

Glucose

TubeSpin cultures were seeded with 0.5×10^6 cells/ mL. The cells were pelleted at 290×g for 5 min and resuspended in 100% fresh media with different glucose concentrations (3.9 g/L, 6 g/L, 8 g/L, 10 g/L, 12 g/L). The cultures were placed in the shaker incubator overnight until a cell density of 1×10^6 cells/mL was reached. Then the cells were pelleted again and 30% of the culture media were replaced with fresh media supplemented with the respective glucose concentration. After the 30% media exchange, the pH was adjusted to 7.5 if necessary and the cells were infected with FMDV A IRN/8/2015 or O SAU/18/2015. An uninfected negative control with standard conditions (3.9 g/L glucose) was also included. The experiment was performed three times independently.

Glutamine

TubeSpin cultures were seeded with 0.5×10^6 cells/mL, cells were pelleted at 290×g for 5 min and resuspended in 100% fresh media with different glutamine concentrations (3.3 mM, 4 mM, 6 mM, 8 mM, 10 mM, 12 mM). The cultures were placed in the shaker incubator overnight until a cell density of 1×10^6 cells/mL was reached. Then the cells were pelleted again and 30% of the culture media were replaced with fresh media, supplemented with the respective glutamine concentration. After the 30% media exchange, the pH was adjusted to 7.5 if necessary and the cells were infected with FMDV A IRN/8/2015 or O SAU/18/2015. An uninfected negative control with standard conditions (3.3 mM glutamine) was also included. The experiment was performed three times independently.

Ammonium chloride

TubeSpin cultures were inoculated with 1×10^6 cells/ mL, the cells were pelleted at 290×g for 5 min and resuspended in 100% fresh media supplemented with different concentrations of ammonium chloride (0 mM, 2 mM, 4 mM, 6 mM, 8 mM, 10 mM, 12 mM). After resuspension of the cells, the pH was adjusted to 7.5 if necessary and the cells were infected with FMDV A IRN/8/2015 or O SAU/18/2015. An uninfected negative control with standard conditions (no ammonium chloride) was also included. The experiment was performed three times independently.

Cell density

TubeSpin cultures were inoculated with 0.5×10^6 cells/ mL, 1.0×10^6 cells/mL or 1.5×10^6 cells/mL. The cells were pelleted at 290×g for 5 min and resuspended in 100% fresh media. The cultures were placed in the shaker incubator overnight to reach cell densities of 1×10^6, 2×10^6 and 3×10^6 cells/mL, respectively. Then the cells were pelleted again and a media exchange of 100% or 30% was performed. After resuspension of the cells, the pH was adjusted to 7.5 if necessary and the cells were infected with FMDV A IRN/8/2015 or O SAU/18/2015. An uninfected negative control was also included. In total, three tubes per density were prepared: one tube with virus and 30% media exchange, one tube with virus and 100% media exchange, and one tube with no virus and 30% media exchange. The experiment was performed three times independently.

Assays

Cell concentration and viability were assessed using a trypan blue dye exclusion method with an automated cell counter (TC20™, Bio-Rad, Munich, Germany). Glucose, lactate and glutamine concentrations were determined using quantitative colorimetric assays (EnzyChrom™

EBGL-100, EGLN-100 or ECLC-100, BioAssay Systems, Hayward, USA) as directed by the manufacturer.

Statistical analysis

In all experiments, the differences between treatment groups were evaluated using one-way ANOVA, combined with Tukey's multiple comparisons test in GraphPad Prism 7 (www.graphpad.com). p-values of < 0.001 were considered significant.

Additional files

Additional file 1: Figure S1. Cell viability 20 hpi with FMDV in media with increasing concentrations of glucose (A) or glutamine (B). (TIF 770 kb)

Additional file 2: Table S1. Mean viral titers and standard deviation of cell density experiments. **Table S2**. Overview about the original virus isolates used in this study, their origin and passage history. (XLS 44 kb)

Additional file 3: Table S3. Cell-specific uptake and release rates of extracellular metabolites of infected BHK-2P cells and corresponding negative control with increasing extracellular glucose concentrations. **Table S4**. Cell-specific uptake and release rates of extracellular metabolites of infected BHK-2P cells and corresponding negative control with increasing extracellular glutamine concentrations. **Table S5**. Cell-specific uptake and release rates of extracellular metabolites of infected BHK-2P cells and corresponding negative control with increasing extracellular ammonium concentrations. **Table S6**. Cell-specific uptake and release rates of extracellular metabolites of infected BHK-2P cells and corresponding negative control at different cell densities and media exchange strategies. (XLSX 18 kb)

Abbreviations

ACF: animal-component free; ANOVA: analysis of variance; BHK: baby hamster kidney; BHK-2P: BHK21C13-2P; CCLV-RIE: Collection of Cell Lines in Veterinary Medicine; FLI: Friedrich-Loeffler-Institut; FMD(V): foot-and-mouth disease (virus); hpi: hours post infection; MDCK: Madin-Darby Canine Kidney; MOI: multiplicity of infection; NH_4Cl: ammonium chloride; RNA: ribonucleic acid; $TCID_{50}$: 50% tissue culture infectious dose

Acknowledgements

Not applicable.

Authors' contributions

VD performed the experiments and calculations, interpreted the data and drafted the manuscript. AZ contributed to the conception of the study and edited the manuscript. MB edited the manuscript. ME edited the manuscript and generally supervised the project. All authors read and approved the final manuscript.

Funding

The study was funded by Merck Life Science as part of the third-party project "MKS-Growth" (Ri-0367). Merck Life Science had no role in the collection, analysis, or interpretation of data and in the writing of the manuscript.

Ethics approval and consent to participate

Not applicable.

Consent for publication

Not applicable.

Competing interests

MB and ME declare that they have no competing interests. VD's position was funded by the project. AZ is an employee of Merck Life Science.

Author details

[1]Institute of Diagnostic Virology, Friedrich-Loeffler-Institut, Südufer 10, 17493 Greifswald, Insel Riems, Germany. [2]Merck KGaA, Merck Life Sciences, Upstream R&D, Frankfurter Straße, 250, 64293 Darmstadt, Germany.

References

1. Thompson D, Muriel P, Russell D, Osborne P, Bromley A, Rowland M, et al. Economic costs of the foot and mouth disease outbreak in the United Kingdom in 2001. Rev Sci Tech. 2002;21(3):675–87.
2. Spitteler MA, Fernandez I, Schabes E, Krimer A, Regulier EG, Guinzburg M, et al. Foot and mouth disease (FMD) virus: quantification of whole virus particles during the vaccine manufacturing process by size exclusion chromatography. Vaccine. 2011;29(41):7182–7.
3. Doel TR. FMD vaccines. Virus Res. 2003;91(1):81–99.
4. Darnell JE Jr, Eagle H. Glucose and glutamine in poliovirus production by HeLa cells. Virology. 1958;6(2):556–66.
5. Eagle H, Habel K. The nutritional requirements for the propagation of poliomyelitis virus by the HeLa cell. J Exp Med. 1956;104(2):271–87.
6. Cruz HJ, Moreira JL, Carrondo MJ. Metabolic shifts by nutrient manipulation in continuous cultures of BHK cells. Biotechnol Bioeng. 1999;66(2):104–13.
7. Baggetto LG. Deviant energetic metabolism of glycolytic cancer cells. Biochimie. 1992;74(11):959–74.
8. Levy HB, Baron S. The effect of animal viruses on host cell metabolism. II. Effect of poliomyelitis virus on glycolysis and uptake of glycine by monkey kidney tissue cultures. J Infect Dis. 1957;100(2):109–18.
9. Bardell D, Essex M. Glycolysis during early infection of feline and human cells with feline leukemia virus. Infect Immun. 1974;9(5):824–7.
10. Gray MA, James MH, Booth JC, Pasternak CA. Increased sugar transport in BHK cells infected with Semliki Forest virus or with herpes simplex virus. Arch Virol. 1986;87(1–2):37–48.
11. Fontaine KA, Camarda R, Lagunoff M. Vaccinia virus requires glutamine but not glucose for efficient replication. J Virol. 2014;88(8):4366–74.
12. Munger J, Bajad SU, Coller HA, Shenk T, Rabinowitz JD. Dynamics of the cellular metabolome during human cytomegalovirus infection. PLoS Pathog. 2006;2(12):e132.
13. Sanchez EL, Lagunoff M. Viral activation of cellular metabolism. Virology. 2015;479–480:609–18.
14. Maynard ND, Gutschow MV, Birch EW, Covert MW. The virus as metabolic engineer. Biotechnol J. 2010;5(7):686–94.
15. Pledger RA, Polatnick J. Defined medium for growth of foot-and-mouth disease virus. J Bacteriol. 1962;83:579–83.
16. Thomassen YE, Rubingh O, Wijffels RH, van der Pol LA, Bakker WA. Improved poliovirus D-antigen yields by application of different Vero cell cultivation methods. Vaccine. 2014;32(24):2782–8.
17. Genzel Y, Fischer M, Reichl U. Serum-free influenza virus production avoiding washing steps and medium exchange in large-scale microcarrier culture. Vaccine. 2006;24(16):3261–72.
18. Grubman MJ, Baxt B. Foot-and-mouth disease. Clin Microbiol Rev. 2004;17(2):465–93.
19. Caridi F, Vazquez-Calvo A, Sobrino F, Martin-Acebes MA. The pH stability of foot-and-mouth disease virus particles is modulated by residues located at the Pentameric Interface and in the N terminus of VP1. J Virol. 2015;89(10):5633–42.
20. Martin-Acebes MA, Gonzalez-Magaldi M, Sandvig K, Sobrino F, Armas-Portela R. Productive entry of type C foot-and-mouth disease virus into susceptible cultured cells requires clathrin and is dependent on the presence of plasma membrane cholesterol. Virology. 2007;369(1):105–18.
21. Carrillo EC, Giachetti C, Campos RH. Effect of lysosomotropic agents on the foot-and-mouth disease virus replication. Virology. 1984;135(2):542–5.
22. Martin-Acebes MA, Rincon V, Armas-Portela R, Mateu MG, Sobrino F. A single amino acid substitution in the capsid of foot-and-mouth disease virus can increase acid lability and confer resistance to acid-dependent uncoating inhibition. J Virol. 2010;84(6):2902–12.
23. Baxt B. Effect of lysosomotropic compounds on early events in foot-and-mouth disease virus replication. Virus Res. 1987;7(3):257–71.
24. Genzel Y, Behrendt I, Konig S, Sann H, Reichl U. Metabolism of MDCK cells during cell growth and influenza virus production in large-scale microcarrier culture. Vaccine. 2004;22(17–18):2202–8.
25. Kamen A, Henry O. Development and optimization of an adenovirus production process. J Gene Med. 2004;6(Suppl 1):S184–92.

26. Wood HA, Johnston LB, Burand JP. Inhibition of Autographa californica nuclear polyhedrosis virus replication in high-density Trichoplusia ni cell cultures. Virology. 1982;119(2):245–54.

27. Hassell T, Gleave S, Butler M. Growth inhibition in animal cell culture. The effect of lactate and ammonia. Appl Biochem Biotechnol. 1991;30(1):29–41.

28. European Commision for the Control of Foot-and-Mouth Disease. Minimum biorisk management standards for laboratories working with foot-and-mouth disease virus Italy 2013 Available from: http://www.fao.org/fileadmin/user_upload/eufmd/Lab_guidelines/FMD_Minimumstandards_2013_Final_version.pdf.

29. Spearman C. The method of "right and wrong cases" (constant stimuli) without Gauss's formula. Br J Psychol. 1908;2:227–42.

30. Kärber G. Beitrag zur kollektiven Behandlung pharmakologischer Reihenversuche. Archiv f experiment Pathol u Pharmakol. 1931;162:480–3.

9

A new method for quantitative detection of Lactobacillus casei based on casx gene and its application

Xiaoyang Pang[1,2], Ziyang Jia[2], Jing Lu[2], Shuwen Zhang[2], Cai Zhang[3], Min Zhang[1*] and Jiaping Lv[2*]

Abstract

Background: The traditional method of bacterial identification based on 16S rRNA is a widely used and very effective detection method, but this method still has some deficiencies, especially in the identification of closely related strains. A high homology with little differences is mostly observed in the 16S sequence of closely related bacteria, which results in difficulty to distinguish them by 16S rRNA-based detection method. In order to develop a rapid and accurate method of bacterial identification, we studied the possibility of identifying bacteria with other characteristic fragments without the use of 16S rRNA as detection targets.

Results: We analyzed the potential of using *cas* (CRISPR-associated proteins) gene as a target for bacteria detection. We found that certain fragment located in the *casx* gene was species-specific and could be used as a specific target gene. Based on these fragments, we established a TaqMan MGB Real-time PCR method for detecting bacteria. We found that the method used in this study had the advantages of high sensitivity and good specificity.

Conclusions: The *casx* gene-based method of bacterial identification could be used as a supplement to the conventional 16 s rRNA-based detection method. This method has an advantage over the 16 s rRNA-based detection method in distinguishing the genetic relationship between closely-related bacteria, such as subgroup bacteria, and can be used as a supplement to the 16 s rRNA-based detection method.

Keywords: TaqMan MGB RT-PCR, Lactobacillus casei, Cas, Rapid detection method, 16 s rRNA

Background

In recent years, many studies have confirmed that intestinal flora is associated with a variety of nutritional and metabolic diseases such as obesity [1, 2] and type 2 diabetes [3, 4]. In the field of scientific research, the study of intestinal flora has become of utmost importance. Billions of bacteria in the intestine live in symbiosis with each other for the host's nutritional and metabolic needs [5, 6]. The intestinal flora of the host have a close relationship with the storage and absorption of nutrition [7, 8], immunity [9], as well as the regulation of sRNA regulation [10]. Through their genes, intermediates and metabolites,

these florae affect the host's nutritional absorption, metabolism, weight, immunity, and several other aspects [11, 12]. Once the balance of intestinal flora is disrupted, a variety of nutritional and metabolic symptoms appears in the host [13, 14]. Although intestinal florae have been shown to be associated with many metabolic diseases, a lot of work still needs to be done in order to establish the differences between related and casual diseases.

Current research on intestinal flora are mostly based on Illumina's high-throughput sequencing technology; which has the advantages of high throughput, short time, and low cost [15, 16]. However, its low resolution characteristic is a big drawback, and most bacteria can be identified only at genus level. Consequently, at the species level only a few bacteria can be identified using the technology, with an inability to distinguish intestinal flora among sub-species or strains. In fact, the roles of different species of the same genus in a host are remarkably different. For instance, studies have shown that

* Correspondence: zmin@th.btbu.edu.cn; lvjiapingcaas@126.com
[1]Beijing Advanced Innovation Center for Food Nutrition and Human Health, Beijing Technology & Business University (BTBU), Beijing 100048, China
[2]Institute of Food Science and Technology, Chinese Academy of Agricultural Science, Beijing 100193, China
Full list of author information is available at the end of the article

different species of the same genus or family exhibit variations to increase or decrease during weight gain in high-fat-fed animals [17, 18]. Obviously, their relationships with the development of obesity cannot be fully elucidated. It is presumed that, while some of the intestinal bacteria are related to obesity, others are not. This suggests that it is necessary to establish a more suitable method with higher resolution to study the relationship between intestinal florae and their hosts. The search for specific gene fragments from the target bacterial genome, and the development of a corresponding detection method, could be the key factor to solve this fundamental problem.

The common strategy for the search of specific fragments of bacteria involves the analysis of bacterial 16 s rRNA sequencing, and then find the specific fragments from its variable area [19, 20]; However using this method, it is sometimes difficult to distinguish closely related bacteria such as *L. casei* and *L. rhamnosus*, because of the 99% similarity in the 16 s rRNA whole sequences (1540/1558). Due to the fact that a specific bacteria fragment from the 16 s rRNA sequence is difficult to find, it is necessary to search for new characteristic fragments from other areas of the bacterial genome. In this study, we found that some CRISPR-associated proteins (Cas) are strain-specific and could be used as target gene fragments for the identification of strains. The bacterial identification based on *casx* gene, could be used as an supplement to the conventional method based on 16 s rRNA.

CRISPR is a special-function DNA sequence that widely exists in bacteria and archaea genomes [21, 22]. The sequence covers one leader, multiple short and highly conserved repeats, as well as multiple spacers. CRISPR is considered to be the bacteria's immune system [23, 24]. After the bacteria are infected by a virus, the surviving bacteria can capture a characteristic DNA fragment from the virus and then integrate it into their genome CRISPR area. At a subsequent viral invasion, the bacteria can quickly identify them according to the CRISPR archive area and then activate the endonuclease to cut the invading virus; equivalently acting as immunity to the virus. Each time a new virus is encountered, the bacteria can capture its characteristic DNA fragments and insert them into their own CRISPR area. The above functions of bacteria are performed by a series of CAS proteins. Although some *cas* genes (such as the widely known *cas*9 gene), have great similarities in sequences among different bacteria, several others have low similarities. We selected all the *cas* genes annotated on the genome of *Lactobacillus casei* and then aligned them with their corresponding genes of ten *Lactobacillus* strains. The results showed that a *casx* gene in the flanking sequence of CRISPR had lower similarity with other *Lactobacillus* species. Primers and probes for fluorescence quantitative polymerase chain reaction (qPCR)

were designed according to the *casx* gene. Furthermore, the results also showed that *L. casei* from other intestinal microbes could be accurately distinguished with high sensitivity and reproducibility using this method. In this study, the bacteria from a large microbial flora were accurately identified and their abundance detected using fluorescence qPCR assay based on the *casx* gene of *L. casei*. The method is high sensitivity and repeatable. This study established the foundation for the study of the relationships between intestinal microbes and their host via species or subspecies.

Results

The acquisition of *Lactobacillus casei* specific gene fragments
The CRISPR sequences obtained from this study are shown in Table 1. We compared the CRISPR flanking sequence of *L. casei* with other strains of *Lactobacillus*, and found that one *casx* gene had a conserved region of ~ 270 bp (Fig. 1). The two *L. casei* strains in this region had an identical gene sequence (*L. casei* w56: 2325395–2,325,664; *L. casei* BL23: 2328749–2,329,018), and was quite different from other *Lactobacillus* species. Although *L. rhamnosus* is closely related to *L. casei*, the *casx* gene of *L. rhamnosus* is different from that of *L. casei*. Therefore, this region could be used as a candidate target gene for the detection of *L. casei*. In order to verify the specificity of this gene, the 270-bp *casx* gene fragment was obtained by Blast in the Genbank database. The results showed that the fragment had high similarity with the sequence of the six strains in the genome and Genbank database, and all six strains were *L. casei*; indicating the species specificity of the sequence.

Fragment-specific validation results
According to the specific fragment in this study, the primers for fluorescence qPCR were designed and named as 06232F and 06232R, while the probe match for the primers was designed and named as 06232P. The probe was linked to a luminophore FAM on the 5′ end and a quencher MGB-NFQ on the 3′ end. The details of the primers and probes are presented in Table 2.

The genetic relationship of 19 *Lactobacillus* strains was analyzed. The results show that *L. casei* was closely related to *L. brevis*, *L. plantarum*, *L. curvatus*, *L. coryniformis*, and *L. rhamnosus* (Fig. 2). Therefore, six strains of *Lactobacillus* (*L. casei* SY13, *L. plantarum* M15, *L. curvatus* znj160802, *L. coryniformis* znj160401, *L. rhamnosus* YL4, and *L. brevis* znj160202) were selected and their genomes extracted. The genomes of the six strains were amplified by PCR with 06232F and 06232R primers. As a result, the target fragment of about 90 bp was obtained from *L.casei* SY13 genome and no target fragment was obtained from the genomes of other bacteria (Fig. 3). This indicated that the specificity of the primers was good.

Table 1 CRISPR distribution of Lactobacillus

Strain	Genbank No.	DR consensus	Position
Lactobacillus acidophilus GCF_000934625	NZ_CP010432.1	GGATCACCTCCACATACGTGGAGAAAAT	1,541,318–1,543,298
Lactobacillus acidophilus NCFM	NC_006814	GGATCACCTCCACATACGTGGAGAAAAT	1,541,039–1,543,019
Lactobacillus backii GCF_001663655	NZ_CP014623	GATCTATTTTAGCTGAAAACTGAAGGAATCAATAGC	844,462–846,673
Lactobacillus brevis KB290	NC_020819	GTATTCCCCACACATGTGGGGGTGATCC	1,071,990–1,072,505
Lactobacillus casei W56	NC_018641	GCTCTTGAACTGATTGATTCGACATCTACCTGAGAC	2,323,692–2,325,115
Lactobacillus casei BL23	NC_010999	GCTCTTGAACTGATTGATTCGACATCTACCTGAGAC	2,327,048–2,328,469
Lactobacillus delbrueckii subsp. bulgaricus ATCC 11842	NC_008054	GTATTCCCCACGCAAGTGGGGGTGATCC	764,071–766,562
Lactobacillus fermentum F-6	NC_021235	GGATCACCCCCATATACATGGGGAGCAC	1,348,092–1,352,667
Lactobacillus helveticus GCF_001006025	NZ_CP011386	CTTTACATTTCTCTTAAGTTAAATAAAAAC	1,644,308–1,647,001
Lactobacillus plantarum ZJ316	NC_020229	GTCTTGAATAGTAGTCATATCAAACAGGTTTAGAAC	359,930–360,361
Lactobacillus rhamnosus GG	NC_013198	GCTCTTGAACTGATTGATCTGACATCTACCTGAGAC	2,265,855–2,267,474
Lactobacillus salivarius CECT 5713	NC_017481	GTTTCAGAAGTATGTTAAATCAATAAGGTTAAGACC	121,320–123,253

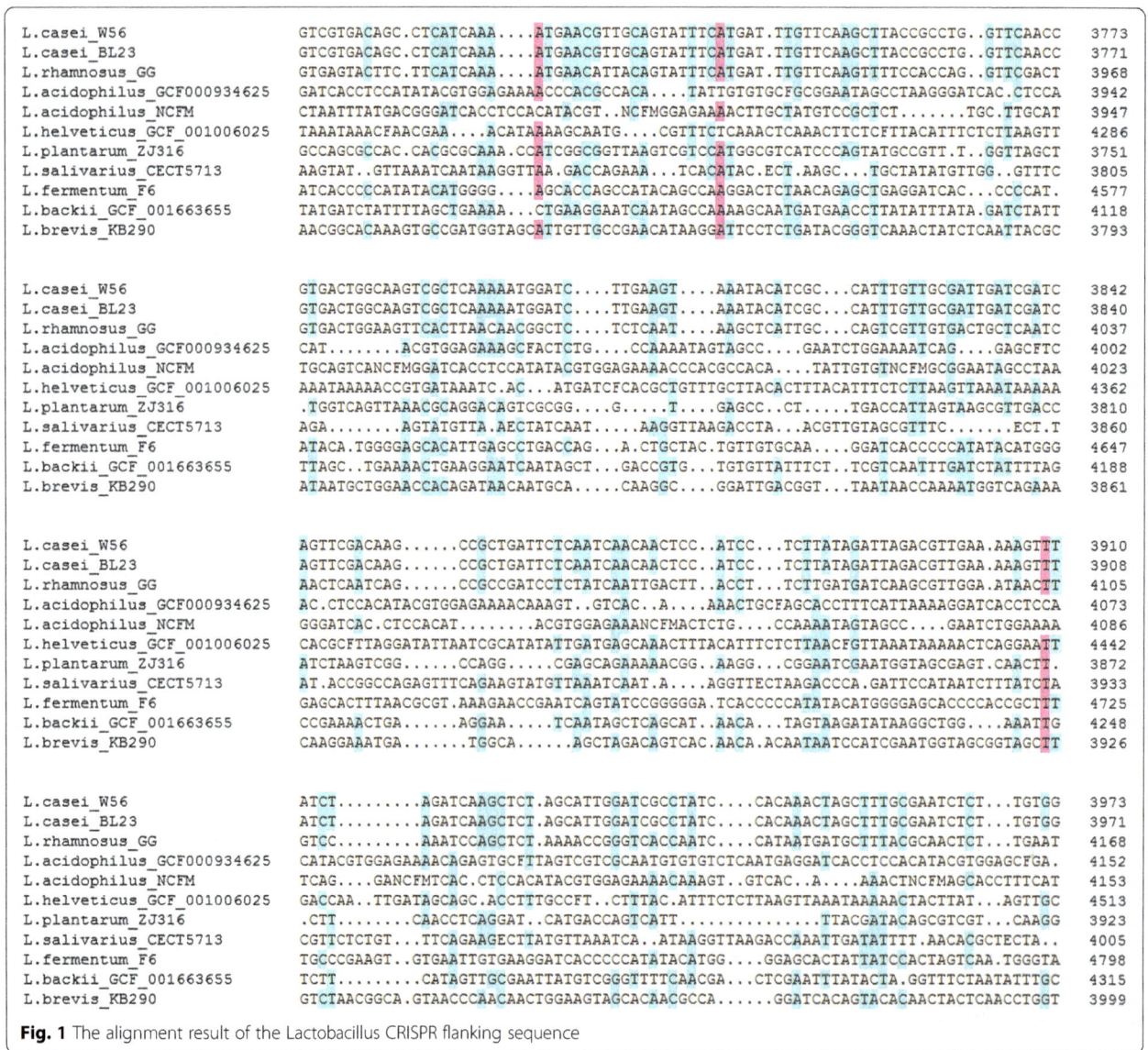

```
L.casei_W56                 GTCGTGACAGC.CTCATCAAA....ATGAACGTTGCAGTATTTCATGAT.TTGTTCAAGCTTACCGCCTG..GTTCAACC   3773
L.casei_BL23                GTCGTGACAGC.CTCATCAAA....ATGAACGTTGCAGTATTTCATGAT.TTGTTCAAGCTTACCGCCTG..GTTCAACC   3771
L.rhamnosus_GG              GTGAGTACTTC.TTCATCAAA....ATGAACATTACAGTATTTCATGAT.TTGTTCAAGTTTTCCACCAG..GTTCGACT   3968
L.acidophilus_GCF000934625  GATCACCTCCATATACGTGGAGAAAACCCACGCCACA....TATTGTGTGCFGCGGAATAGCCTAAGGGATCAC.CTCCA   3942
L.acidophilus_NCFM          CTAATTTATGCGGGATCACCTCCACATACGT.......NCFMGGAGAAAACTTGCTATGTCCGCTCT.......TGC.TTGCAT   3947
L.helveticus_GCF_001006025  TAAATAAACFAACGAA....ACATAAAAGCAATG....CGTTTCTCAAACTCAAACTTCTCFTTACATTTCTCTTAAGTT   4286
L.plantarum_ZJ316           GCCAGCGCCAC.CACGCGCAAA.CCATCGGCGGTTAAGTCGTCCATGGCGTCATCCCAGTATGCCGTT.T..GGTTAGCT   3751
L.salivarius_CECT5713       AAGTAT..GTTAAATCAATAAGGTTAA.GACCAGAAA...TCACATAC.ECT.AAGC...TGCTATATGTTGG..GTTTC   3805
L.fermentum_F6              ATCACCCCCATATACATGGGG....AGCACCAGCCATACAGCCAAGGACTCTAACAGAGCTGAGGATCAC...CCCCAT.   4577
L.backii_GCF_001663655      TATGATCTATTTTAGCTGAAAA...CTGAAGGAATCAATAGCCAAAAGCAATGATGAACCTTATATTTATA.GATCTATT   4118
L.brevis_KB290              AACGGCACAAAGTGCCGATGGTAGCATTGTTGCCGAACATAAGGATTCCTCTGATACGGGTCAAACTATCTCAATTACGC   3793

L.casei_W56                 GTGACTGGCAAGTCGCTCAAAAATGGATC....TTGAAGT....AAATACATCGC..CATTTGTTGCGATTGATCGATC   3842
L.casei_BL23                GTGACTGGCAAGTCGCTCAAAAATGGATC....TTGAAGT....AAATACATCGC...CATTTGTTGCGATTGATCGATC   3840
L.rhamnosus_GG              GTGACTGGAAGTTCACTTAACAACGGCTC....TCTCAAT....AAGCTCATTGC...CAGTCGTTGTGACTGCTCAATC   4037
L.acidophilus_GCF000934625  CAT........ACGTGGAGAAAGCFACTCTG....CCAAAATAGTAGCC....GAATCTGGAAAATCAG....GAGCFTC   4002
L.acidophilus_NCFM          TGCAGTCANCFMGGATCACCTCCATATACGTGGAGAAAACCCACGCCACA....TATTGTGTNCFMGCGGAATAGCCTAA   4023
L.helveticus_GCF_001006025  AAATAAAAACCGTGATAAATC.AC...ATGATCFCACGCTGTTTGCTTACACTTTACATTTCTCTTAAGTTAAATAAAAA   4362
L.plantarum_ZJ316           .TGGTCAGTTAAACGCAGGACAGTCGCGG....G.....T....GAGCC..CT.....TGACCATTAGTAAGCGTTGACC   3810
L.salivarius_CECT5713       AGA........AGTATGTTA.AECTATCAAT.....AAGGTTAAGACCTA...ACGTTGTAGCGTTTC.......ECT.T   3860
L.fermentum_F6              ATACA.TGGGGAGCACATTGAGCGTGACCAG...A.CGCTAC.TGTTGTGCAA....GGATCACCCCCATATACATGGG   4647
L.backii_GCF_001663655      TTAGC..TGAAAACTGAAGGAATCAATAGCT...GACCGTG...TGTGTTATTTCT..TCGTCAATTTGATCTATTTTAG   4188
L.brevis_KB290              ATAATGCTGGAACCACAGATAACAATGCA.....CAAGGC....GGATTGACGGT...TAATAACCAAAATGGTCAGAAA   3861

L.casei_W56                 AGTTCGACAAG......CCGCTGATTCTCAATCAACAACTCC..ATCC...TCTTATAGATTAGACGTTGAA.AAAGTTT   3910
L.casei_BL23                AGTTCGACAAG......CCGCTGATTCTCAATCAACAACTCC..ATCC...TCTTATAGATTAGACGTTGAA.AAAGTTT   3908
L.rhamnosus_GG              AACTCAATCAG......CCGCCGATCCTCTATCAATTGACTT..ACCT...TCTTGATGATCAAGCGTTGGA.ATAACTT   4105
L.acidophilus_GCF000934625  AC.CTCCACATACGTGGAGAAAACAAAGT..GTCAC..A....AAACTGCFAGCACCTTTCATTAAAAAGGATCACCTCCA   4073
L.acidophilus_NCFM          GGGGATCAC.CTCCACAT....ACGTGGAGAAANCFMACTCTG....CCAAAATAGTAGCC....GAATCTGGAAAA   4086
L.helveticus_GCF_001006025  CACGCFTTAGGATATTAATCGCATATATTGATGAGCAAACTTTACATTTCTCTTAACFGTTAAATAAAAACTCAGGAATT   4442
L.plantarum_ZJ316           ATCTAAGTCGG......CCAGG.....CGAGCAGAAAAACGG..AAGG....CGGAATCGAATGGTAGCGAGT.CAACTT.   3872
L.salivarius_CECT5713       AT.ACCGGCCAGAGTTTCAGAAGTATGTTAAATCAAT.A....AGGTTECTAAGACCCA.GATTCCATAATCTTTATCTA   3933
L.fermentum_F6              GAGCACTTTAACGCGT.AAAGAACCGAATCAGTATCCGGGGGA.TCACCCCCATATACATGGGGAGCACCCCACCGCTTT   4725
L.backii_GCF_001663655      CCGAAAACTGA......AGGAA.....TCAATAGCTCAGCAT..AACA...TAGTAAGATATAAGGCTGG....AAATTG   4248
L.brevis_KB290              CAAGGAAATGA.......TGGCA......AGCTAGACAGTCAC.AACA.ACAATAATCCATCGAATGGTAGCGGTAGCTT   3926

L.casei_W56                 ATCT........AGATCAAGCTCT.AGCATTGGATCGCCTATC....CACAAACTAGCTTTGCGAATCTCT...TGTGG   3973
L.casei_BL23                ATCT........AGATCAAGCTCT.AGCATTGGATCGCCTATC....CACAAACTAGCTTTGCGAATCTCT...TGTGG   3971
L.rhamnosus_GG              GTCC........AAATCAAGCTCT.AAAACCGGGTCACCAATC....CATAATGATGCTTTACGCAACTCT...TGAAT   4168
L.acidophilus_GCF000934625  CATACGTGGAGAAAACAGAGTGCFTTAGTCGTCGCAATGTGTGTCTCAATGAGGATCACCTCCACATACGTGGAGCFGA.   4152
L.acidophilus_NCFM          TCAG....GANCFMTCAC.CTCCACATACGTGGAGAAAACAAAGT..GTCAC..A....AAACTNCFMAGCACCTTTCAT   4153
L.helveticus_GCF_001006025  GACCAA..TTGATAGCAGC.ACCTTTGCCFT..CTTTAC.ATTTCTCTTAAGTTAAATAAAAAACTACTTAT...AGTTGC   4513
L.plantarum_ZJ316           .CTT........CAACCTCAGGAT..CATGACCAGTCATT...............TTACGATACAGCGTCGT..CAAGG   3923
L.salivarius_CECT5713       CGTTCTCTGT...TTCAGAAGECTTATGTTAAATCA..ATAAGGTTAAGACCAAATTGATATTTT.AACACGCTECTA..   4005
L.fermentum_F6              TGCCCGAAGT..GTGAATTGTGAAGGATCACCCCCATATACATGG...GGAGCACTATTATCCACTAGTCAA.TGGGTA   4798
L.backii_GCF_001663655      TCTT.......CATAGTTGCGAATTATGTCGGGTTTTCAACGA...CTCGAATTTATACTA.GGTTTCTAATATTTGC   4315
L.brevis_KB290              GTCTAACGGCA.GTAACCCAACAACTGGAAGTAGCACAACGCCA......GGATCACAGTACACAACTACTCAACCTGGT   3999
```

Fig. 1 The alignment result of the Lactobacillus CRISPR flanking sequence

Table 2 Amplification primers and Taqman–MGB probes used for specific detection

Primers or probes	Nucleotide sequence(5′ → 3′)	Amplicon size
Primers		
06232F	TCAACCGTGACTGGCAAGT	91
06232R	AGCGGCTTGTCGAACTGA	
M13–47	CGCCAGGGTTTTCCCAGTCACGAC	247
M13–48	AGCGGATAACAATTTCACACAGGA	
Probes		
06232P	FAM-CTCAAAAATGGATCTTG-MGB-NFQ	

Establishment of the fluorescent qPCR detection method

A 93 bp DNA fragment was amplified from the genome of *L. casei* SY13 using the 06232F and 06232R primers. The fragment was ligated with pMD19T plasmid and transformed into *E. coli* DH5α. Positive clones were screened on lysogeny broth plates which contained 50.0 µg/mL ampicillin, extracted and correctly sequenced. The target DNA of standard substance pMD19T-CS was assayed to determine its concentration using Spark 20 M Multiscan Spectrum. The concentration of pMD19T-CS was 30.05 ng/µL and the unit was converted to copies/µL according to formula:

$$\text{plasmid concentration(copies}/\mu L)$$
$$= \frac{\text{plasmid concentration(ng}/\mu L) \times 6.02 \times 10^{23}}{\text{pMD19T–CS length(bp)} \times 660 \text{g}/\text{mol}}$$

The DNA standard was diluted from 10^3 to 10^8 copies/µL and used to generate the standard curve (Fig. 4). The regression equation was:

$$Y = -3.53\lg C + 45.28$$

Where $R^2 = 0.998$, Y represents C_T, while C represents the concentration of standard DNA. The efficiency of amplification was 92.011% and the detection limit was 10^2 copies/µL.

Analysis of mice experiment

Balb/c mice were fed with *L. casei* SY13 for 7d, and then sacrificed at the end of the feeding trial. The content of the different parts of their intestines were analyzed to quantify *L. casei* SY13. The results showed that the target bacteria were not found in the intestinal tract of the negative control group, which implied that there were no endogenous *L. casei* in the intestines of the mice. However, the target bacteria were detected in the experimental group, and the highest numbers were found in the cecum. Interestingly, the target bacteria were not detected in the ileum (Table 3). The quantities of *L. casei* subsp. *casei* SY13 in the jejunum, cecum, and colon were 1.6×10^5 copies/g, 2.1×10^6 copies/g, and 1.7×10^6 copies/g respectively. The results indicated that the fluorescence qPCR method based on the *casx* gene could specifically detect *L. casei* from the intestinal microbial flora of mice.

Discussion

The effort to search for the specific gene fragments of bacteria had plagued the researchers of environmental microbiology for a long time. In the past, the conventional strategy was to search the 16 s rRNA sequence and then select the conserved region sequence as the

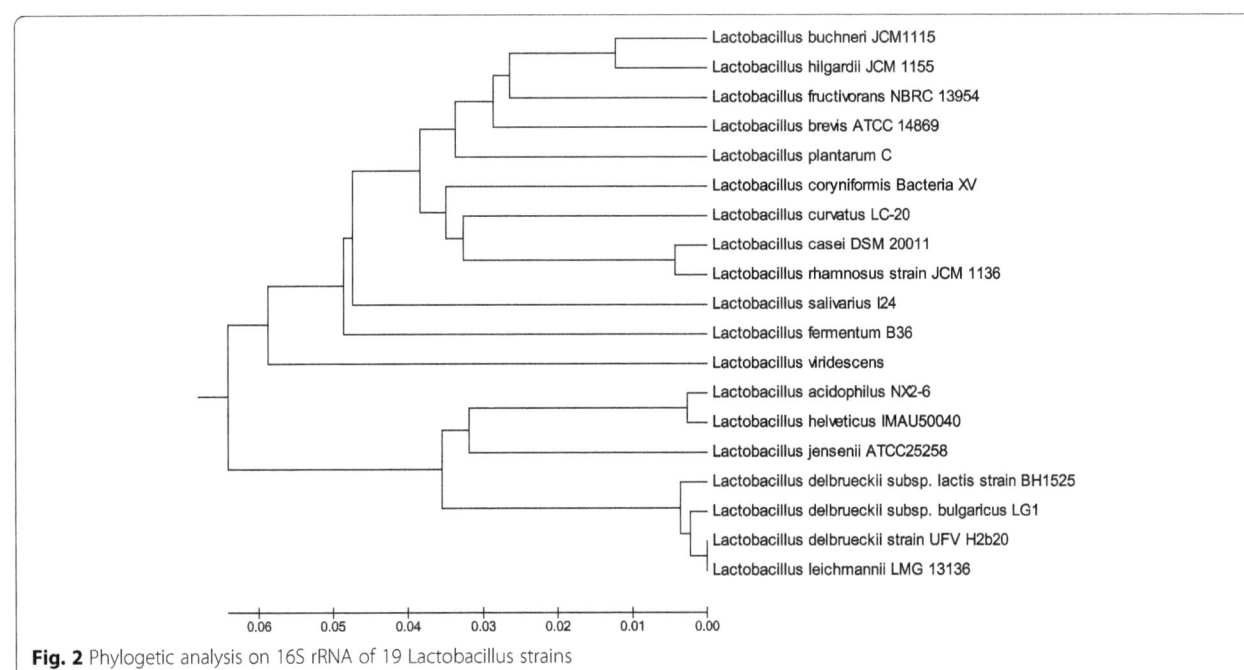

Fig. 2 Phylogetic analysis on 16S rRNA of 19 Lactobacillus strains

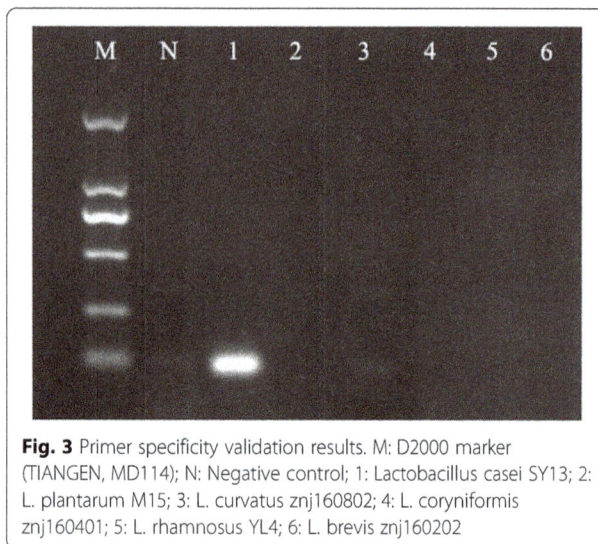

Fig. 3 Primer specificity validation results. M: D2000 marker (TIANGEN, MD114); N: Negative control; 1: Lactobacillus casei SY13; 2: L. plantarum M15; 3: L. curvatus znj160802; 4: L. coryniformis znj160401; 5: L. rhamnosus YL4; 6: L. brevis znj160202

combined with conventional microbiological cultivation. It is usually a difficult and labor-intensive procedure. FISH probe and hybridization were also used to detect *Lactobacillus*. Slides were made from intestinal-tract samples and examined using an Olympus BH2 epifluorescence microscope [31]. Without precision, the researchers visually recognized only the number of *Lactobacillus* that adhered to the intestinal tract. Thus, the development of a simple, highly efficient, and highly specific method was urgently required. We used the *casx* gene of *L. casei* and developed a method for the rapid and accurate detection of *L. casei* by fluorescence qPCR. The core of this method is to find the appropriate casx gene fragment on the flank of CRISPR. In order to verify the general applicability of this method, we tested it on *Legionella pneumophila*. *L. pneumophila* includes two main subspecies, one is *L. pneumophila* subsp. *fraseri*, the other is *L. pneumophila* subsp. *pneumophila*. Traditional methods based on 16S rRNA are difficult to distinguish these two kinds of bacteria. First of all, we searched for CRISPR region in the whole genome of *L. pneumophila* subsp. *fraseri* GCF_001886795_and *L.pneumophila* subsp. *pneumophila* GCF_001592705_respectively. The results are shown in Additional file 1: Table S1. Then, 3000 bp was taken from CRISPR's flank as candidate sequence, the extracted sequences are shown in Additional file 2: Table S2. Using ClustalX 2.0 to align the extracted flank sequence. Based on the sequence alignment results, a 256 bp specific sequence was found on *L.pneumophila* subsp. *pneumophila* GCF_001592705, which only existed in *L.pneumophila* subsp. *pneumophila* GCF_001592705 genome, but not in *L. pneumophila* subsp. *fraseri* GCF_001886795, the sequence information is shown in Additional file 3: Table S3. In order to verify the specificity of the sequence, we used blast tool to align the 256 bp

target fragment for detection [25–27]. However, this method is insufficient to distinguish closely related bacteria. The current microbiome technique involves the use of high-throughput sequencing technology based on the V3–V4 region of the 16s rRNA of the bacteria, to distinguish the different bacteria in the sample [28]. Although this method identified most of the microorganisms in the sample, however it cannot distinguish bacteria that are closely related in the same genus [29].

In order to identify bacteria more quickly and accurately, many researchers had explored several other options. For instance, multiplex PCR was used to detect *L. casei* [30] in such a way that two sets of primers were designed to ensure the specificity of *L. casei* ATCC 393 in the multiplex PCR system. However, this method was not able to quantify *L. casei* ATCC 393; thus it is usually

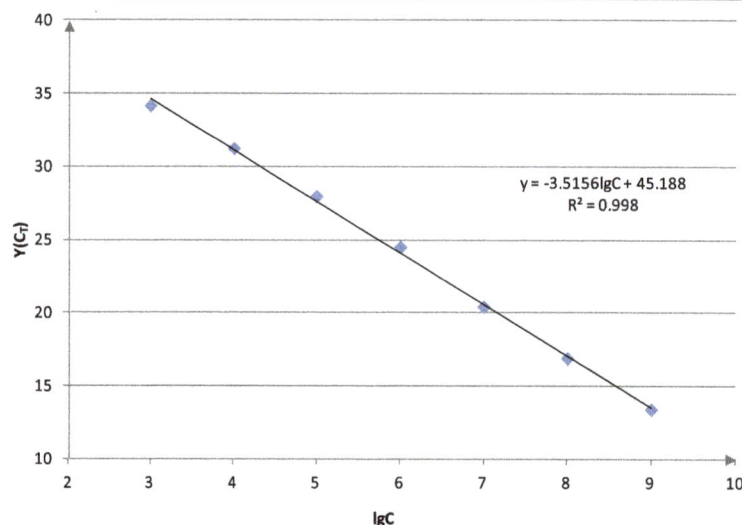

$$y = -3.5156 \lg C + 45.188$$
$$R^2 = 0.998$$

Fig. 4 Standard curve between concentrations of standard DNA and C_T value

Table 3 Quantity of target bacteria in intestinal tract

	jejunum	ileum	cecum	colon
	C_T			
Negative group	–	–	–	–
Experimental group	27.14 ± 0.79^a	–	23.21 ± 1.06^b	24.08 ± 1.08^b

ab Means in the same row different superscript letter differ significantly($P < 0.05$)

sequence in GenBank, the results are shown in Additional file 4: Figure S1. It can be seen from the results that the 256 bp fragment has good specificity and can distinguish *L.pneumophila* subsp. *pneumophila* from *L. pneumophila* subsp. *fraseri*. The above case in *L.pneumophila* can prove that the method provided in this study is not only applicable to *Lactobacillus casei*, but also applicable to other bacteria.

Conclusions

In this study, we used the *casx* gene of *L. casei* and developed a method for the rapid and accurate detection of *L. casei* by fluorescence qPCR. *L. casei* and *L. rhamnosus* were easily distinguished with the use of this method. There is an extremely high similarity between the two bacteria in 16 s rRNA sequences, therefore, it is difficult to distinguish them from each other based on the 16 s rRNA method. The *casx* gene-based method of identification developed in this study was able to rapidly and accurately distinguish the two bacteria. Finally, we validated the accuracy and sensitivity of the method using mouse experiments. This method has an advantage over the 16 s rRNA-based detection method in distinguishing the genetic relationship between closely-related bacteria, such as subgroup bacteria, and can be used as a supplement to the 16 s rRNA-based detection method.

Methods

Bacteria strains, plasmids, and mice

The bacteria and plasmids used in this study are shown in Table 4. *Lactobacillus* strains were statically cultured in MRS broth (Cat. No. CM187, Beijing Land Bridge Technology Co., Ltd., China) at 37 °C. *Escherichia coli* DH5α was grown at 37 °C in LB broth (1.0% peptone, 0.5% yeast extract powder, 1.0% NaCl; pH 7.4), SPF BALB/c mice were purchased from Beijing Vital River Laboratory Animal Technology Co., Ltd. (Beijing, China).

Acquisition and alignment of cas sequences of *Lactobacillus*

The CRISPR sequence of *Lactobacillus* was derived from the CRISPR database (http://crispr.i2bc.paris-saclay.fr). Due to the fact that some *cas* genes are not annotated in the genome of *Lactobacillus*, the CRISPR flank 3000 bp was selected as the analysis sequence to prevent the loss of some key *cas* genes. ClustalX 2.0 was used for the alignment of sequence.

Design of primers and TaqMan-MGB probes

According to the alignment results, we searched for the characteristic fragment of *L. casei*. The primers and TaqMan-MGB probe to detect *L. casei* were designed by Primer Express 3.0 based on the characteristic fragment. The syntheses of primers and probes were entrusted to the Beijing Genomics Institute (BGI). In this study, the specificity of the characteristic fragment was verified from two procedures. Firstly, the characteristic fragment was subjected to Blast in Genbank to examine whether the sequence matched the bacteria other than *L. casei*. Secondly, 19 *Lactobacillus* strains were used to reconstruct a phylogenetic tree based on their 16S rRNA

Table 4 Bacterial strains and plasmids used in this study

Strain or plasmid	Relevant genotype or description	Reference and/or source
Strains Lactobacillus casei SY13	Isolated from milk cheese	Preserved in our lab
Lactobacillus plantarum M15	Isolated from natural fermented yak milk	Preserved in our lab
Lactobacillus rhamnosus YL4	Isolated from milk cheese	Preserved in our lab
Lactobacillus curvatus znj160802	Isolated from traditional Chinese fermented dairy products	Dairy Processing Laboratory, Beijing Technology and Business University
Lactobacillus coryniformis znj160401	Isolated from a goat's milk cheese	Dairy Processing Laboratory, Beijing Technology and Business University
Lactobacillus brevis znj160202	Isolated from traditional Chinese fermented dairy products	Dairy Processing Laboratory, Beijing Technology and Business University
Escherichia coli DH5α	F⁻,φ80d lacZ ΔM15,Δ(lacZYA-argF)U169, deoR, recA1, endA1, hsdR17 (rk⁻,mk⁺), phoA, supE44, λ⁻, thi-1, gyrA96, relA1	TaKaRa
Plasmids		
pMD19T	clone vector; Amp^r	TaKaRa
pMD19T-CS	pMD19T derived plasmid, DNA standard for fluorescence quantitative PCR	this study

nucleotide sequences using MEGA 6.0. The five closely related strains to *L. casei* were selected to verify the specificity of the primers.

Establishment of the fluorescent qPCR detection method

Bacterial genome DNA was extracted and purified using a nucleic acid extraction kit (Cat. No. DP302, TIANGEN Biotech, Beijing, China) according to the manufacturer's instructions. Genome DNA was extracted from the intestinal contents and purified with the TIANamp Stool DNA Kit (Cat. No. DP328, TIANGEN Biotech, Beijing, China) according to the manufacturer's instructions.

PCR was performed to obtain target DNA fragments using primers 06232F and 06232R. Target DNA fragments were recovered from agarose gels using the TIAN Gel Purification Kit (Cat. No. DP209, TIANGEN Biotech, Beijing, China). The target fragment was then inserted into the pMD19T vector (Cat. No. 6013, Takara Biomedical Technology (Beijing) Co., Ltd. Beijing, China). The ligation products were transformed into DH5α competent cells using the heat shock method. Positive clones were chosen and inoculated into the LB broth containing 100 µg/mL ampicillin and then cultured overnight in an incubator at 37 °C and 200 rpm. The TIANprep Plasmid Kit (Cat. No. DP103, TIANGEN Biotech, Beijing, China) was used to extract plasmids from the cultured bacteria suspension. The extracted plasmids were sequenced to verify the inserted fragments. Sequencing was entrusted to BGI. Sequencing validates the correct transformants as the DNA standard. The concentration of the standard DNA was detected on Spark 20 M Multiscan Spectrum (Tecan Group Ltd., Switzerland).

The DNA standard was used to generate a standard curve and analyze assay sensitivity. The DNA diluted from 10^2 to 10^8 copies/µL was used as a template to perform RT-PCR. A TaqMan-MGB probe was used to detect the C_T value. A standard curve between the C_T value, dilution gradient and linear regression equation was generated automatically by ABI 7500 Real-Time PCR System. The coefficient of determination (R^2) value was also demonstrated. The DNA standard diluted from 10^0 to 10^4 copies/µL was tested to observe the detection limit.

Mice experiment

After the Balb/c mice were treated with *L. casei* for 7.0 d, we collected the cecum and colon contents and extracted the genomic DNA, to further test the specificity of the primers and probes in the intestinal flora. The number of *L. casei* in different parts of the intestinal tract was measured by fluorescence qPCR using the extracted genomic DNA as templates. The mice fed without *L. casei* were used as negative control. Six male mice were used in this experiment. They were randomly divided into two groups; three in each cage. They were left for 7.0 d to adapt to their environment, and water and basal diets were freely given. At the onset of the treatment, the experimental group was gavage-induced with 10^9 cfu *L. casei SY13*, while the negative group was administered sterile water. Mice were sacrificed 7 days after treatment. Carbon dioxide method was used to euthanize mice. The methods were referenced to previously published literature [32]. The sacrificed mice were dissected; the jejunum, ileum, cecum, and colon were extracted and preserved in liquid nitrogen. The genome DNA of the intestinal contents was also extracted. The quantity of the target bacteria was measured by RT-PCR. Data analysis was conducted using SPSS 20.0 (IBM Corporation, Armonk, NY, USA).

Abbreviations

Cas proteins: CRISPR-associated proteins; QPCR: Quantitative Polymerase Chain Reaction; RRNA: Ribosomal RNA; RT-PCR: Realtime Polymerase Chain Reaction

Acknowledgments

We thank China Agricultural University Laboratory Animal Service Center for its help with the animal experiments. We also showed our thanks to the anonymous reviewers for their helpful comments on the manuscript.

Authors' contributions

XP, MZ, JLv designed all the experiments; JLu, SZ, CZ performed the experiments. SZ, CZ analyzed the data. XP and ZJ wrote the paper. All authors read and approved the final manuscript.

Funding

This study was supported by National Key R&D Program of China (2017YFC1600903) and National Natural Science Foundation of China (31871833). These funds were used for collection of materials, analysis data, the interpretation and the writing/publication of the manuscript.

Ethics approval and consent to participate

Mice experiments were conducted in the China Agricultural University Laboratory Animal Service Center and were approved by the animal ethics committee of the university (No. CAU20161020–3, 20/10/2016). All the experiment processes were done under the supervision of the China Agricultural University Laboratory Animal Ethics Committee.

Consent for publication

Not applicable.

Competing interests

The authors declare that they have no competing interests.

Author details
[1]Beijing Advanced Innovation Center for Food Nutrition and Human Health, Beijing Technology & Business University (BTBU), Beijing 100048, China. [2]Institute of Food Science and Technology, Chinese Academy of Agricultural Science, Beijing 100193, China. [3]Laboratory of Environment and Livestock Products, Henan University of Science and Technology, Luoyang 471023, China.

References
1. Everard A, Belzer C, Geurts L, Ouwerkerk JP, Druart C, Bindels LB, Guiot Y, Derrien M, Muccioli GG, Delzenne NM, et al. Cross-talk between Akkermansia muciniphila and intestinal epithelium controls diet-induced obesity. P Natl Acad Sci USA. 2013;110(22):9066–71.
2. Bleau C, Karelis AD, St-Pierre DH, Lamontagne L. Crosstalk between intestinal microbiota, adipose tissue and skeletal muscle as an early event in systemic low-grade inflammation and the development of obesity and diabetes. Diabetes Metab Res Rev. 2015;31(6):545–61.
3. Sabatino A, Regolisti G, Cosola C, Gesualdo L, Fiaccadori E. Intestinal microbiota in type 2 diabetes and chronic kidney disease. Curr Diab Rep. 2017;17(3):16.
4. Wu T, Trahair LG, Little TJ, Bound MJ, Zhang X, Wu H, Sun Z, Horowitz M, Rayner CK, Jones KL. Effects of Vildagliptin and metformin on blood pressure and heart rate responses to small intestinal glucose in type 2 diabetes. Diabetes Care. 2017;40(5):702–5.
5. Gyang VP, Chuang TW, Liao CW, Lee YL, Akinwale OP, Orok A, Ajibaye O, Babasola AJ, Cheng PC, Chou CM, et al. Intestinal parasitic infections: current status and associated risk factors among school aged children in an archetypal African urban slum in Nigeria. J Microbiol Immunol Infect. 2019; 52(1): 106–113.
6. Marcial-Coba MS, Marshall IPG, Schreiber L, Nielsen DS. High-Quality Draft Genome Sequence of Lactobacillus casei Strain Z11, Isolated from a Human Adult Intestinal Biopsy Sample. Genome Announc. 2017;5(28):e00634–17.
7. Volynets V, Kuper MA, Strahl S, Maier IB, Spruss A, Wagnerberger S, Konigsrainer A, Bischoff SC, Bergheim I. Nutrition, intestinal permeability, and blood ethanol levels are altered in patients with nonalcoholic fatty liver disease (NAFLD). Dig Dis Sci. 2012;57(7):1932–41.
8. Kelly P. Nutrition, intestinal defence and the microbiome. Proc Nutr Soc. 2010;69(2):261–8.
9. Che C, Pang X, Hua X, Zhang B, Shen J, Zhu J, Wei H, Sun L, Chen P, Cui L, et al. Effects of human fecal flora on intestinal morphology and mucosal immunity in human flora-associated piglet. Scand J Immunol. 2009;69(3):223–33.
10. Sesto N, Koutero M, Cossart P. Bacterial and cellular RNAs at work during Listeria infection. Future Microbiol. 2014;9(9):1025–37.
11. Fernandes PC, Dolinger EJ, Abdallah VO, Resende DS, Gontijo Filho PP, Brito D. Late onset sepsis and intestinal bacterial colonization in very low birth weight infants receiving long-term parenteral nutrition. Rev Soc Bras Med Trop. 2011;44(4):447–50.
12. Levine A. Exclusive enteral nutrition: clues to the pathogenesis of Crohn's disease. Nestle Nutr Inst Workshop Ser. 2014;79:131–40.
13. Nicholson JK, Holmes E, Kinross J, Burcelin R, Gibson G, Jia W, Pettersson S. Host-gut microbiota metabolic interactions. Science. 2012;336(6086):1262–7.
14. Rastall RA, Gibson GR. Recent developments in prebiotics to selectively impact beneficial microbes and promote intestinal health. Curr Opin Biotechnol. 2015;32:42–6.
15. Li Y, Poroyko V, Yan Z, Pan L, Feng Y, Zhao P, Xie Z, Hong L. Characterization of intestinal microbiomes of Hirschsprung's disease patients with or without Enterocolitis using Illumina-MiSeq high-throughput sequencing. PLoS One. 2016;11(9):e0162079.
16. Moreau MM, Eades SC, Reinemeyer CR, Fugaro MN, Onishi JC. Illumina sequencing of the V4 hypervariable region 16S rRNA gene reveals extensive changes in bacterial communities in the cecum following carbohydrate oral infusion and development of early-stage acute laminitis in the horse. Vet Microbiol. 2014;168(2–4):436–41.
17. Barbosa A, Reiss A, Jackson B, Warren K, Paparini A, Gillespie G, Stokeld D, Irwin P, Ryan U. Prevalence, genetic diversity and potential clinical impact of blood-borne and enteric protozoan parasites in native mammals from northern Australia. Vet Parasitol. 2017;238:94–105.
18. Barrios C, Beaumont M, Pallister T, Villar J, Goodrich JK, Clark A, Pascual J, Ley RE, Spector TD, Bell JT, et al. Gut-microbiota-metabolite Axis in early renal function decline. PLoS One. 2015;10(8):e0134311.
19. Momose Y, Maruyama A, Iwasaki T, Miyamoto Y, Itoh K. 16S rRNA gene sequence-based analysis of clostridia related to conversion of germfree mice to the normal state. J Appl Microbiol. 2009;107(6):2088–97.
20. De Hertogh G, Aerssens J, de Hoogt R, Peeters P, Verhasselt P, Van Eyken P, Ectors N, Vermeire S, Rutgeerts P, Coulie B, et al. Validation of 16S rDNA sequencing in microdissected bowel biopsies from Crohn's disease patients to assess bacterial flora diversity. J Pathol. 2006;209(4):532–9.
21. Beneke T, Madden R, Makin L, Valli J, Sunter J, Gluenz E. A CRISPR Cas9 high-throughput genome editing toolkit for kinetoplastids. R Soc Open Sci. 2017;4(5):170095.
22. Comfort N. Genome editing: That's the way the CRISPR crumbles. Nature. 2017;546(7656):30–1.
23. Hryhorowicz M, Lipinski D, Zeyland J, Slomski R. CRISPR/Cas9 immune system as a tool for genome engineering. Arch Immunol Ther Exp. 2017; 65(3):233–40.
24. Rath D, Amlinger L, Rath A, Lundgren M. The CRISPR-Cas immune system: biology, mechanisms and applications. Biochimie. 2015;117:119–28.
25. Overmann J, Coolen MJ, Tuschak C. Specific detection of different phylogenetic groups of chemocline bacteria based on PCR and denaturing gradient gel electrophoresis of 16S rRNA gene fragments. Arch Microbiol. 1999;172(2):83–94.
26. Zhan XY, Hu CH, Zhu QY. Targeting single-nucleotide polymorphisms in the 16S rRNA gene to detect and differentiate legionella pneumophila and non-legionella pneumophila species. Arch Microbiol. 2016;198(6):591–4.
27. Ravasi DF, Peduzzi S, Guidi V, Peduzzi R, Wirth SB, Gilli A, Tonolla M. Development of a real-time PCR method for the detection of fossil 16S rDNA fragments of phototrophic sulfur bacteria in the sediments of Lake Cadagno. Geobiology. 2012;10(3):196–204.
28. Parikh HI, Koparde VN, Bradley SP, Buck GA, Sheth NU. MeFiT: merging and filtering tool for illumina paired-end reads for 16S rRNA amplicon sequencing. Bmc Bioinformatics. 2016;17(1):491.
29. Hua G, Cheng Y, Kong J, Li M, Zhao Z. High-throughput sequencing analysis of bacterial community spatiotemporal distribution in response to clogging in vertical flow constructed wetlands. Bioresour Technol. 2018;248:104–112.
30. Karapetsas A, Vavoulidis E, Galanis A, Sandaltzopoulos R, Kourkoutas Y. Rapid detection and identification of probiotic Lactobacillus casei ATCC 393 by multiplex PCR. J Mol Microb Biotech. 2010;18(3):156–61.
31. Wang B, Li JS, Li QR, Zhang HY, Li N. Isolation of adhesive strains and evaluation of the colonization and immune response by Lactobacillus plantarum L2 in the rat gastrointestinal tract. Int J Food Microbiol. 2009; 132(1):59–66.
32. Conlee KM, Stephens ML, Rowan AN, King LA. Carbon dioxide for euthanasia: concerns regarding pain and distress, with special reference to mice and rats. Lab Anim. 2005;39(2):137–61.

10

Mucosal delivery of *Lactococcus lactis* carrying an anti-TNF scFv expression vector ameliorates experimental colitis in mice

Maria José Chiabai[1], Juliana Franco Almeida[2], Mariana Gabriela Dantas de Azevedo[1], Suelen Soares Fernandes[1], Vanessa Bastos Pereira[3], Raffael Júnio Araújo de Castro[4], Márcio Sousa Jerônimo[4], Isabel Garcia Sousa[1], Leonora Maciel de Souza Vianna[5], Anderson Miyoshi[3], Anamelia Lorenzetti Bocca[4], Andrea Queiroz Maranhão[1,6] and Marcelo Macedo Brigido[1,6*]

Abstract

Background: Anti-Tumor Necrosis Factor-alpha therapy has become clinically important for treating inflammatory bowel disease. However, the use of conventional immunotherapy requires a systemic exposure of patients and collateral side effects. Lactic acid bacteria have been shown to be effective as mucosal delivering system for cytokine and single domain antibodies, and it is amenable to clinical purposes. Therefore, lactic acid bacteria may function as vehicles for delivery of therapeutic antibodies molecules to the gastrointestinal tract restricting the pharmacological effect towards the gut. Here, we use the mucosal delivery of *Lactococcus lactis* carrying an *anti-TNFα* scFv expression plasmid on a DSS-induced colitis model in mice.

Results: Experimental colitis was induced with DSS administered in drinking water. *L. lactis* carrying the scFv expression vector was introduced by gavage. After four days of treatment, animals showed a significant improvement in histological score and disease activity index compared to those of untreated animals. Moreover, treated mice display IL-6, IL17A, IL1β, IL10 and FOXP3 mRNA levels similar to health control mice. Therefore, morphological and molecular markers suggest amelioration of the experimentally induced colitis.

Conclusion: These results provide evidence for the use of this alternative system for delivering therapeutic biopharmaceuticals in loco for treating inflammatory bowel disease, paving the way for a novel low-cost and site-specific biotechnological route for the treatment of inflammatory disorders.

Keywords: *Lactococcus lactis*, Mucosal delivery, Anti-TNFα, scFv, Colitis, DSS

Background

Inflammatory bowel disease (IBD), which includes Crohn's disease (CD) and ulcerative colitis (UC), is characterized by chronic inflammation of the gastrointestinal tract (GIT) and a cryptogenic origin [1], with a global incidence of 0.3% [2]. The characteristic tissue damage of the disorder occurs due to the abnormal expression of anti-inflammatory and pro-inflammatory molecules from both the innate and adaptive responses [3]. Tumor

necrosis factor-α (TNFα) plays a crucial role in the pathogenesis of IBD, and indeed, monoclonal antibodies targeting TNFα are the most powerful treatment for IBD; however, the intravenous administration route causes immunogenic and systemic side effects [4, 5]. Therefore, the local delivery of this pharmaceutical would benefit patients restricting therapy towards the inflamed tissue [6].

Recently, using bacteria as a vehicle, novel approaches for the treatment of intestinal inflammation in IBD animal models have been proposed, showing promising anti-inflammatory results such as those described by Gomes-Santos et al. [7] and Luerce et al. [8]. The *Lactococcus lactis* subspecies cremoris MG1363 is one of the

* Correspondence: brigido@unb.br
[1]Laboratório de Imunologia Molecular, Departamento de Biologia Molecular, Universidade de Brasília, Brasília, Distrito Federal, Brazil
[6]Instituto Nacional de Investigação em Imunologia, INCTii, Brasília, Distrito Federal, Brazil

most explored bacteria; it is a noninvasive and non-pathogenic gram-positive species. It is the best characterized microorganism of the group named lactic acid bacteria (LAB) and is generally regarded as safe (GRAS) by the U. S. Food and Drug Administration (FDA) [9, 10]. Although these bacteria are used in the manufacture of dairy products such as cheese and yogurt [11], *L. lactis* subsp. cremoris MG1363 is considered a potential strategy for the treatment of IBD, once it has the ability to survive the gastric acid environment and is able to replicate and deliver therapeutic molecules locally to the GIT [12]. Moreover, the medical use of engineered bacteria to produce biopharmaceuticals will pave the way for a novel biotechnological route for the low-cost treatment of immune disorders.

The use of LAB as a drug delivery system has been proposed [9] as a substitute for the oral administration of biopharmaceuticals [6, 13, 14]. However, bacterial expression systems show a limited capacity for recombinant protein production. Complex heterologous eukaryotic protein production in bacteria is normally limited by the lack of specific chaperones and other modification enzymes. Therefore, the efficient delivery of monoclonal antibodies to the animal gut depends on novel strategies. In this work, we explored the ability of *L. lactis* MG1363 FnBPA+ [15] to locally deliver a single-chain fragment variable (scFv) of anti-TNFα antibody cloned in the eukaryotic expression plasmid pValac [16] for expression in the GIT lining. We use this delivery system in a dextran sulfate sodium (DSS)-induced colitis in mice and tested its effect on the inflammatory process. We showed that treating mice orally with *L. lactis* carrying pValac::*anti-TNFα* ameliorates disease indexes as well as immunological and molecular markers. The data support the use of this alternative delivery system for treating IBD.

Results

Construction of pValac::*anti-TNFα*

The synthetic *anti-TNFα* coding ORF was cloned into the pValac vector (Additional file 1: Figure S1A), forming an expression cassette for eukaryotic cells. The pValac::*anti-TNFα* construction was checked by restriction endonuclease digestion profile, PCR and sequencing to confirm ORF integrity (data not shown). To evaluate the ability of gene expression under the control of the CMV promoter, we transfected the plasmid pValac::*anti-TNFα* into the HEK-293 cell line. The cell culture supernatant was collected 48 h post-transfection, and soluble scFv was probed with anti-HA primary antibody by western blot. A reactive band at the expected size of 31 kDa was detected, showing the production and secretion of the expected antibody fragment (Additional file 1: Figure S1B).

Oral administration of the LL-FT strain ameliorates disease in DSS-induced colitis

The LL-F strain was transformed with the pValac::*anti-TNFα* plasmid by electroporation, and selected clones were checked by their restriction endonuclease digestion profile and PCR. The effects of LL-FT were evaluated in an animal model of acute colitis induced by DSS. An experimental protocol to mimic ulcerative colitis in humans was carried out. The mice received 2% DSS in drinking water for 4 days followed by a further 4 consecutive days of 2% DSS plus LL-F or LL-FT (Fig. 1a). As shown in Fig. 1b, body weight decreased with DSS ingestion, and there was no weight recovery in the mice that received LL-FT, despite clear clinical signs of improvement of inflammation, such as cessation of rectal bleeding and no signs of diarrhea. The colon length was examined after euthanasia. We found a shortening of the colons in the mice from the DSS group (average length of 3.2 ± 0.15 cm) in comparison with the group of mice that drank only saline (average length of 3.9 ± 0.05 cm). However, the group that received LL-FT showed a significant recovery of colon length (average length of 3.5 ± 0.13 cm) when compared with the LL-F group (average length of 2.9 ± 0.08 cm) (Fig. 1c). The effect of LL-FT on colitis in mice was evaluated using the disease activity index (DAI). This index reflects weight loss, diarrhea and rectal bleeding, parameters that had a milder occurrence in comparison with the LL-F group. This score was analyzed on day 9, before euthanasia. The mice treated with DSS alone without bacteria administration showed a score of 2.1 ± 0.13, and mice treated with LL-F showed a score of 2.0 ± 0.22. The mice that received LL-FT showed a significantly lower DAI of 1.1 ± 0.15 and thus a reduced inflammatory process compared to mice in the LL-F group (Fig. 1d). Moreover, CRP was measured (Fig. 1e). CRP levels increased among groups that received DSS, but the levels were significantly decreased in the LL-FT group, which reflects an anti-inflammatory effect.

LL-FT prevents colonic mucosal injury

DSS induces extensive injury in the mouse colon [17]. Thus, histological analysis was performed for all animals to evaluate histological damage caused by DSS and *L. lactis* treatment. We developed a histological score considering several histopathological parameters (Additional file 1: Table S1). The histological scores for the LL-FT-treated animals were statistically significantly lower (*p* = 0.0022) compared with the scores of the DSS and LL-F groups (Fig. 2a). The colonic samples of negative control animals remained intact, with no change in the normal histological architecture in the mucosa (Fig. 2b NC). When comparing the mice in the LL-F group (Fig. 2b LL-F) to the mice in the DSS group (Fig. 2b DSS), the mucosal and submucosal

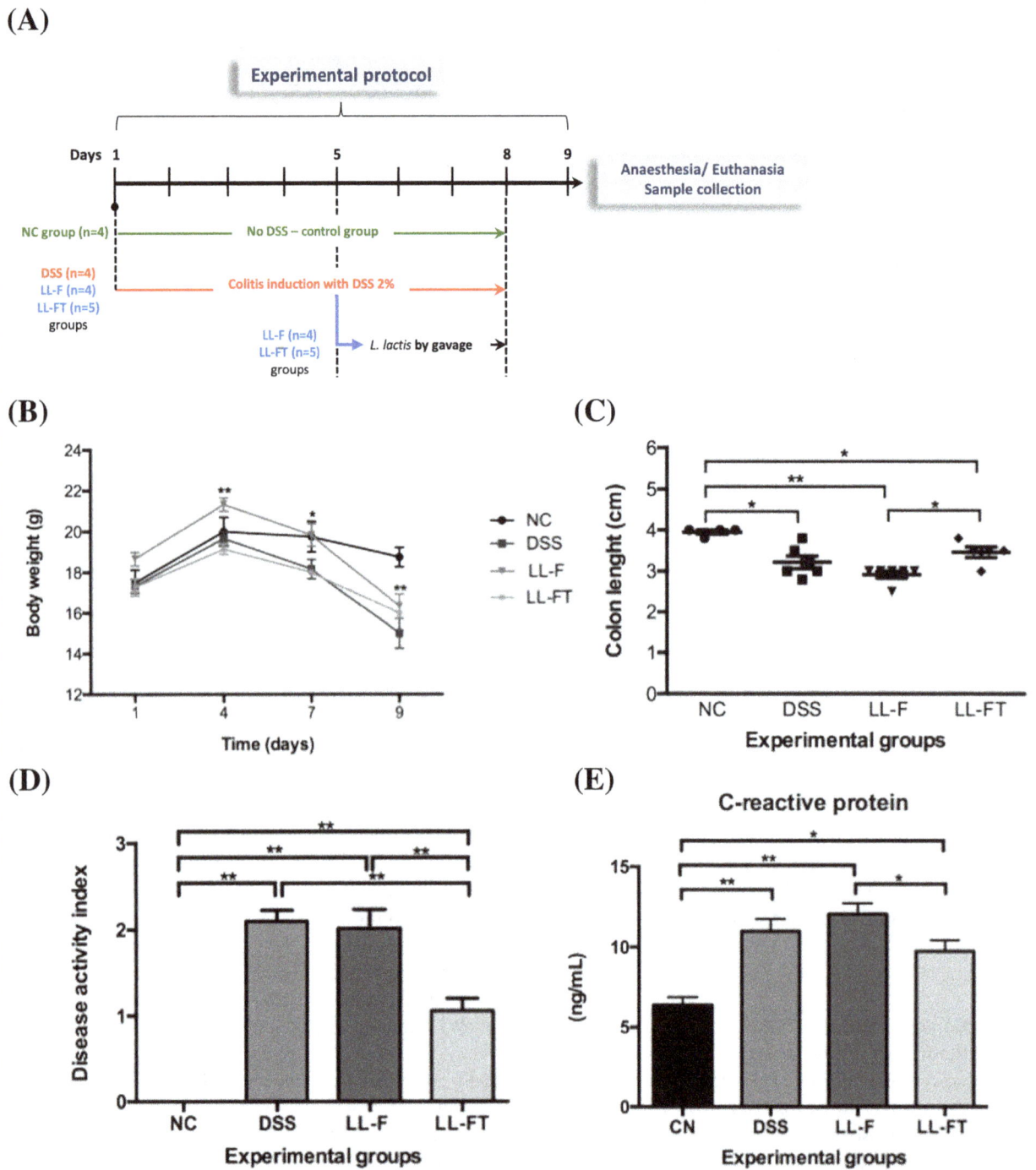

Fig. 1 Effects of oral administration of LL-FT in DSS-induced colitis. **a** Experimental protocol is as follows: C57BL/6 mice received 2% DSS for 8 days (day 1 to day 8). Recombinant *L. lactis* was administered for 4 days (day 5 to day 8) to LL-F and LL-FT groups. Euthanasia and samples collection occurred on day 9 (arrow). **b** Body weight was measured in grams from day 1 to 9. **c** Colon length was measured in centimeters after euthanasia. **d** Disease activity index (DAI) was evaluated on day 9 before euthanasia and included three major clinical signs related to weight loss, diarrhea and rectal bleeding. **e** Blood serum of mice was collected after euthanasia and measured by ELISA. Experimental groups are as follows: NC, negative control group; DSS, DSS group; LL-F, *L. lactis* MG1363 FnBPA+ group and LL-FT, *L. lactis* MG1363 FnBPA+ (pValac::*anti-TNFa*) group. Data are expressed as the means ± SEM from an experiment using 4–5 animals per group. Statistical analysis was performed using the Mann-Whitney test for charts and two-way ANOVA for curves; * $p < 0.05$ and ** $p < 0.01$

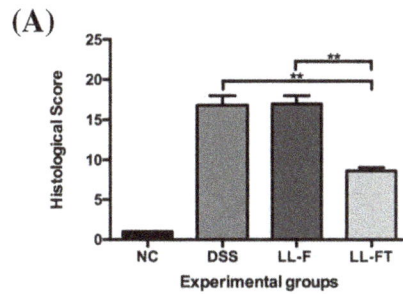

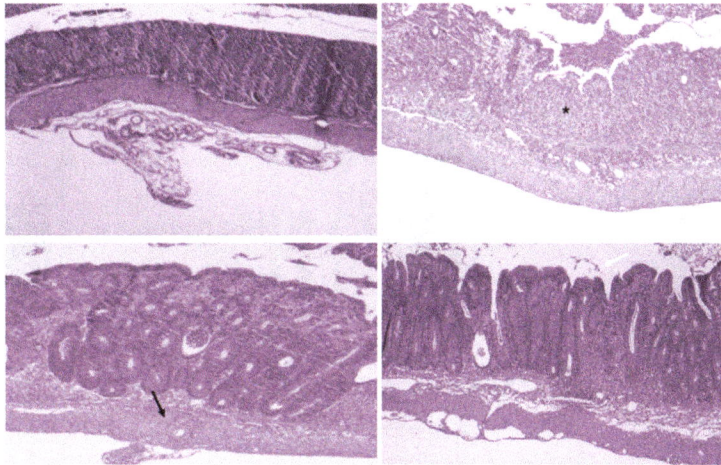

Fig. 2 Histopathological score and histopathology of colonic tissue. **a** Histopathological score was determined from colon samples that were photographed in paraffin sections by H&E staining of a representative distal colon from each group. **b** Representative photos from distal colon tissue of a mouse from each experimental group are shown: *Star* ulceration; *black arrow* depletion of goblet cells; *white arrow* intestinal wall with intact mucosa and discrete inflammatory infiltrate. Experimental groups: NC, negative control group; DSS, DSS group; LL-F, *L. lactis* MG1363 FnBPA+ group and LL-FT, *L. lactis* MG1363 FnBPA+ (pValac::*anti-TNFα*) group. Data are expressed as the means ± SEM from an experiment using 4–5 animals per group and representative of three independent experiments. Statistical analysis was performed using the Mann-Whitney test; * $p < 0.05$ and ** $p < 0.01$

inflammatory infiltrate ranged from moderate to severe. Furthermore, erosion with extensive ulceration, crypt abscesses, muscle herniation and depletion of goblet cells was observed. On the other hand, in the LL-FT group (Fig. 2b LL-FT), the mucosa, submucosa, muscular and serosal infiltrates were mild, with small or no erosion area and little gland inflammatory activity. In addition, mucosal ulceration or muscle thickening was not found, resembling the negative control of colitis. Hence, the presence of the *anti-TNFα* plasmid carried by *L. lactis* ameliorated the inflammatory symptoms, suggesting the participation of locally produced anti-TNFα.

Modulation of inflammatory gene markers suggests disease reversal

Cytokines and transcription factors involved in the mucosal immune response to colitis in mice were investigated by qPCR of the colonic total RNA. The mice treated with LL-FT were compared to control mice for the production of markers of T-cell and macrophage populations to investigate if anti-TNFα delivered by recombinant *L. lactis*

could evoke specific sets of immune cell populations. As shown in Fig. 3, we found that the mRNA levels of the pro-inflammatory cytokines IL-6, TNFα and IL-1β were induced in mice that received DSS but that they significantly decreased in mice treated with LL-FT towards mRNA levels found in healthy animals (Fig. 3a, b, c). Similarly, IL-17A levels increased in DSS and LL-F groups and decreased to healthy levels after LL-FT treatment (Fig. 3d). RORγt mRNA levels decreased compared to those in untreated animals after colitis induction or *L. lactis* treatment. However, no statistical significance was observed except for the LL-FT group (Fig. 3e). Similarly, TGF-β mRNA levels did not change significantly by DSS or *L. lactis* treatment despite a small increase after DSS treatment (Fig. 3f). The relative expression of the Th1 marker T-bet decreased significantly in all groups that received DSS (DSS, LL-F and LL-FT group) compared with that of the NC group (Fig. 3g). Additionally, STAT1 mRNA levels were significantly lower in the LL-FT group in comparison to those in the NC and LL-F groups (Fig. 3h) suggesting a marked effect of LL-FT on the expression of STAT1 in

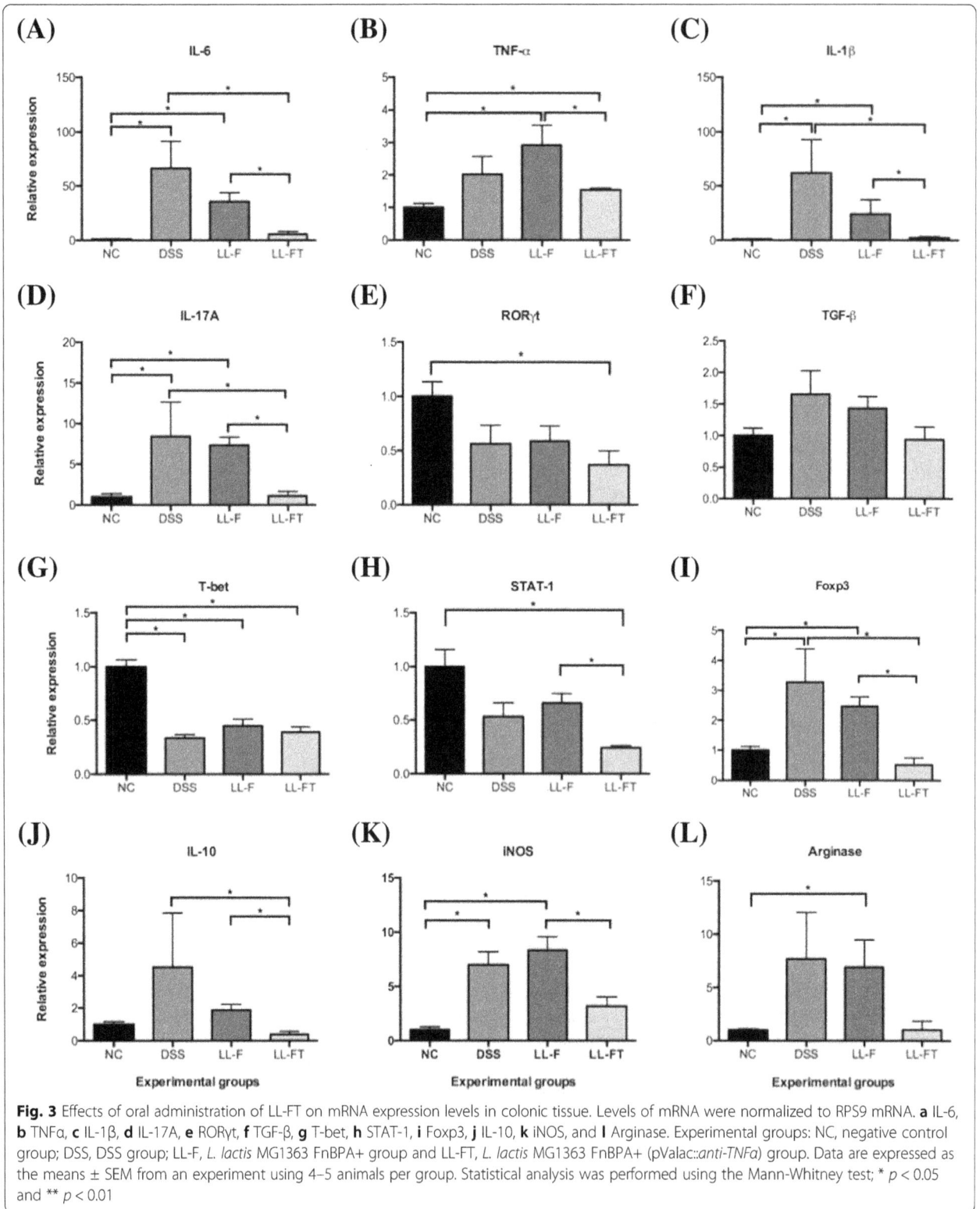

Fig. 3 Effects of oral administration of LL-FT on mRNA expression levels in colonic tissue. Levels of mRNA were normalized to RPS9 mRNA. **a** IL-6, **b** TNFα, **c** IL-1β, **d** IL-17A, **e** RORγt, **f** TGF-β, **g** T-bet, **h** STAT-1, **i** Foxp3, **j** IL-10, **k** iNOS, and **l** Arginase. Experimental groups: NC, negative control group; DSS, DSS group; LL-F, *L. lactis* MG1363 FnBPA+ group and LL-FT, *L. lactis* MG1363 FnBPA+ (pValac::*anti-TNFα*) group. Data are expressed as the means ± SEM from an experiment using 4–5 animals per group. Statistical analysis was performed using the Mann-Whitney test; * $p < 0.05$ and ** $p < 0.01$

the colon tissue. The Foxp3 expression increased in groups that received DSS and decreased significantly in mice treated with LL-FT (Fig. 3i), showing a profile similar to IL-17A. Likewise, the IL-10 anti-inflammatory cytokine

transcripts were increased after DSS but decreased in the LL-FT group, showing significant differences (Fig. 3j). iNOS expression increased in groups that received DSS or DSS plus LL-F but decreased significantly in the LL-FT group (Fig. 3k). The arginase mRNA levels were similar to iNOS levels, but statistically significant differences only occurred between the NC group and the LL-F group, and the LL-FT group which received the recombinant *L. lactis* with pValac::*anti-TNFα* seemed to recover the NC group levels (Fig. 3l). We also tested MUC-3 levels, but they were not affected by DSS or *L. lactis* treatment (Additional file 1: Figure S2).

In addition to the mRNA transcripts, we evaluated if LL-FT could alter the serum levels of cytokines in mice. The cytokines were quantified at the end of the experiments on day 9. Levels of IL-6, TNF and IL-10 increased with DSS administration followed by a decrease when mice were treated with LL-FT, repeating the tendency of cytokine mRNA levels (Additional file 1: Figure S3). However, only TNF levels in the DSS group were significantly different compared to those in the LL-FT group. Thus, corroborating with a return to homeostasis after treatment.

Because IgA on the mucosal surfaces is considered the first line of defense controlling the pro-inflammatory processes and preserving the integrity of the epithelial barrier [18, 19], we measured IgA levels in the fecal extracts of mice to verify the integrity of the mucosal tissue following treatment. We observed that mice receiving DSS presented higher IgA levels than the negative control group and that treatment with LL-F did not rescue untreated levels. However, no significance was observed at $p < 0.05$ (Additional file 1: Figure S4). On the other hand, LL-FT treatment showed a tendency to return IgA levels to untreated levels.

Discussion

In this work, we showed the construction of an anti-TNFα scFv eukaryotic expression vector to be delivered by oral administration to the gut of animals. This vector is based on a previously described vector (pValac) that allows the expression of a transgene in gut epithelial cells [16]. We showed that this vector induces the synthesis of a 31 kDa scFv in transfected HEK-293 cells, suggesting its competence to induce synthesis in the gut cells of animals. Therefore, we expected to delivery an anti-TNFα in the gut of mice via a genetically modified microorganism, *L. lactis*. We tested this hypothesis by treating mice suffering from ulcerative colitis induced by DSS, an experimental disease model known to be ameliorated by anti-TNF therapy. Furthermore, we tested the use of LAB as a delivery system for antibody fragments as an alternative strategy for the treatment of IBD.

Wild-type *L. lactis* are able to be internalized by eukaryotic cells and to deliver DNA efficiently, as shown previously both in vitro [20] and in vivo [21]. However, a recombinant *L. lactis* expressing a fibronectin-binding protein A (FnBPA) of *Staphylococcus aureus* has an improved ability to deliver DNA since FnBPA is an invasin that mediates invasivity in nonphagocytic host cells [22], facilitating intracellular spreading [15]. Therefore, for this study, we used a modified bacteria transformed with the plasmid pValac::*anti-TNFα* to adheres to eukaryotic cells of the intestinal mucosa [23]. Thus, LL-F transformed with the pValac::*anti-TNFα* (LL-FT) plasmid was administered by gavage to a DSS-induced colitis in mice.

DSS is a high molecular weight and irritating chemical known to induce colitis. At a concentration of 2%, it causes severe inflammation in the mucosa of the intestinal colon [24]. After receiving DSS, the mice presented distinguished weight loss, diarrhea and rectal bleeding. C57BL/6 mice were used in this work because this strain is highly susceptible to DSS colitis and because these mice do not evolve spontaneously to healing but rather develop a chronic state [25, 26]. In our model, mice exhibited the same disease characteristics, and after the treatment with LL-FT, we showed improvement in a variety of macroscopic parameters. The shortening of the colon length reflects the inflammatory process, and the treatment of mice with LL-FT improved the colon length compared to that of the LL-F group. Moreover, the same group LL-FT showed no significant difference compared to the untreated group. However, the macroscopic score (DAI), which reflects multiple macroscopic pathological parameters, showed that the mice that received the *anti-TNFa* transgene had a lower DAI than other control groups, suggesting a reversal of the disease.

The histopathological score as computed here gives an assessment of the microscopic aspect of the pathological process. The colon of the untreated animals appeared healthy, showing an integral histological architecture with visible microvillosities. This healthy aspect was lost after DSS treatment. Moreover, our experiments of inducing acute colitis by DSS was reproduced since, histologically, we found lesions with inflammatory infiltrate, focal crypt lesions and goblet cell loss [27]. The damage induced by DSS was not reduced after treatment with LL-F; however, a statistically significant benefit to tissue integrity was observed in mice receiving LL-FT. Therefore, the beneficial effect of LL-FT treatment most be the consequence of the anti-TNFα transgene expression after *L. lactis* treatment. The findings in the histopathology of the colon agreed with the macroscopic parameters and histopathological scores observed by other authors [26], and *L. lactis* harboring anti- TNFα treatment seemed to ameliorate colon tissue integrity. It is noteworthy that even though we are describing the results of a single experiment, these effects were reproducible by at least three independent experiments (data not shown).

To follow the inflammatory response of our model, we measured CRP, an acute-phase protein produced by the liver in acute inflammatory conditions and is a useful marker of IBD [28]. Although CRP is a nonspecific mucosal inflammatory marker, our data indicate that lower levels of this protein were associated with a lower active disease index in mice and could be used as a functional biomarker for the evaluation of intestinal inflammation. It is noteworthy that FnBPA acts as an inflammatory marker per se [29] that may counteract in part the beneficial effect of LL-F or LL-FT. Therefore, the administration of *L. lactis* delivering anti-TNFα after the fourth day of DSS treatment improved the overall aspect of disease, suggesting a reversal of the DSS-induced process.

DSS-induced colitis disrupts the epithelial barriers, allowing intestinal bacteria to invade the damaged mucosa and inducing excessive production of pro-inflammatory cytokines [30] that could be reduced with anti-TNF therapy [6]. In our model system, the mRNA levels of pro-inflammatory cytokines such as IL-6, TNFα and IL-1β increased after disease induction, as expected [31, 32] and decreased to levels similar to those of healthy controls after LL-FT treatment. IL-17A levels respond similarly, showing a clear upregulation after disease induction, supporting the Th17 axis in the experimental murine colitis model [32–35]. Coherently, T-bet, a hallmark of the Th1 response, was not affected by recombinant bacteria, corroborating an immune response associated with the Th17 phenotype. Regulatory T cells (Treg) are able to suppress abnormal immune responses, and they are involved in homeostasis of the intestinal mucosa [36]. These cells, marked by the expression of Foxp3, produce IL-10 and TGF-β and inhibit the effector function of T cells [3]. Treg markers are increased in the gut of CD and UC patients [37]. Our data suggest that mice treated with LL-FT recover to healthy levels of Foxp3 and IL-10, suggesting a reduction in Tregs, which possibly reflects the improvement in the disease index after treatment with anti-TNFα and a return to homeostasis.

Gobert et al. [38] showed an amelioration of colitis symptoms induced by DSS in iNOS-deficient mice, and our data showed a significant decrease in mRNA iNOS when mice were treated with LL-FT, suggesting that treatment exerted a beneficial effect. Arginase, an M2 macrophage-associated gene, also showed an increase when mice were exposed to DSS followed by a decrease with recombinant *L. lactis*. The association of increased arginase and increased IL-10 suggests that there is an accumulation of M2 macrophages in the colonic tissue of DSS-induced colitis as described by Lin et al. [39], and a lowering of these markers after LL-FT treatment leads to the resolution of the inflammatory process.

The homeostasis of the gastrointestinal tract is achieved by immune mechanisms such as secretion of mucus and IgA, which protects the intestinal epithelium against commensal and pathogenic microorganisms. MUC-3 codes for a structural protein are involved in the healthy epithelium. Some authors showed that MUC-3 is upregulated during the recovery of a damaged bowel after acute colitis [40]. In our model, MUC-3 seems to be downregulated, even though there was no statistical significance. IgA immunoglobulin is the most abundant isotype produced by the mucosa and constitutes an effective marker of inflammation within the microbiota, controlling and modulating it [41, 42]. In our model, the administration of LL-FT did not significantly increase IgA levels in the intestine. Surprisingly, those animals treated with DSS and received saline or LL-F without plasmid had high levels of IgA in comparison with the negative control group and treatment group. Souza et al. [13] also found no increase in IgA levels in mice that received doses of LL-F. On the other hand, a study showed that the use of this same strain carrying pValac::*il-10* was associated with high IgA levels when compared to those in the DSS group [14]. Despite the lack of statistical significance, our results suggest that the induction of disease shows a tendency to increase intestinal IgA levels compared to levels in healthy and anti-TNFα-treated groups. This increase may reflect disease activity directly or indirectly since IgA may be lacking from plasma due to colon tissue damage associated with DSS treatment.

Several cytokines are known to be involved in IBD [43, 44], therefore their serum concentrations were evaluated, and the results corroborated with local mRNA levels, validating the data achieved via qPCR. Even though we see scarce significant differences, the excessive secretion of pro-inflammatory cytokines is related to intestinal inflammation [3, 26, 37, 43], and this biological phenomenon seemed to be reproduced in our model system. The significative reduction of the systemic levels of TNFα could be reflecting a neutralization of this cytokine, hitherto a key player of the inflammatory process in human colitis. Infliximab and adalimumab [45], both anti-TNFα monoclonal antibodies, are currently used for treating Crohn's disease. Moreover, it lessens DSS induced colitis in mice [46, 47]. However, recent data suggests that neither infliximab binds TNFα [48], nor TNFα is necessary for colitis in TNFα$^{-/-}$ mice [49] contradicting current admitted mechanism for infliximab action on mouse model based on TNFα neutralization [6, 50, 51]. Despite of the mechanism, infliximab ameliorates DSS induced colitis that is corroborated by our results. It is possible that other mechanism, such as an apoptosis based one may underlie colitis mouse models response to anti-TNFα therapy [52–54], not TNFα neutralization. Consistently, the observed reduction of systemic TNFα observed in our experiment correlated to a decline in its mRNA levels, thus, TNFα seemed to decline due to a reduction of synthesis rather than from neutralization.

Anti-TNFα therapies have become popular for treating UC and CD in humans, and mucosal delivery of biopharmaceuticals may improve the outpatient's quality of life. However, the oral delivery of antibodies is hindered by the harsh conditions of the digestive tract. Thus, strategies based on mucosal delivery by microorganisms may overcome this obstacle by producing the antibody directly in the gut. We show here that genetically engineered *L. lactis* can be used to deliver an scFv anti-TNFα to the mammalian intestine. Vandenbroucke et al. [6] had previously reported the use of anti-TNF nanobodies secreting *L. lactis* in DSS-induced colitis in the IL-10$^{-/-}$ mouse, where they found the resolution of the inflammatory process. Their system was based on a single-domain camelid antibody fragment constitutively secreted from the bacteria. The expression system used here is based on the heterologous production of an scFv vector directly in the epithelium of the intestine in a eukaryotic expression system [16], restricting anti-TNFα to the gut milieu focusing the therapeutic intervention.

The mucosal delivery may help targeting immune modulation towards the gut, but some restriction could be pointed. The chemical instability of scFv could be partially overwhelmed by its in loco production drove by a mammalian expression system, but the amount of bioavailable pharmaceutical is still unpredictable. Therefore, finding an effective and reproducible dose of bacteria may be a challenging issue. Moreover, the use of recombinant proteins associated with symbiotic microbiota is only starting to be investigated [55], and variable results may be observed with non-conventional delivery, as observed with a bacterial membrane associated anti-TNF delivery [56].

In the present report, the scFv anti-TNFα was engineered inspired on the well-studied agent infliximab [57] and was delivered directly by cells in the gut. Hence, the delivery system proposed here may represent a more reliable model system for simulated anti-TNF treatment for UC in human subjects. Because this system uses a noninvasive route to carry the biopharmaceutical to the site of inflammation, it may represent an alternative for oral antibody therapies.

Conclusions

The use of LAB for delivering biopharmaceuticals may represent an alternative route for immunotherapy. The results reported here suggests that oral administration of *Lactococcus lactis* carrying the eukaryotic expression vector coding for an anti-TNFα induces a reduction of colitis associated inflammatory and histopathological markers suggesting an amelioration in disease. Novel therapeutic approach based on delivering recombinant

antibodies may soon substitute systemic immunotherapy for gut-associated diseases.

Methods
Bacterial strains, media and growth conditions
Escherichia coli XL1-Blue and *E. coli* TG1 were grown in Luria-Bertani (LB) medium with tetracycline (Tet; Sigma-Aldrich, St. Louis, MO, USA) at 30 μg/mL (only for XL1-Blue) at 37 °C and 250 rpm overnight. *L. lactis* MG1363 FnBPA+ were grown in M17 medium (Difco, Detroit, MI, USA) supplemented with 0.5% glucose, and, when necessary, chloramphenicol (Cm; Sigma-Aldrich, St. Louis, MO, USA) at 10 μg/mL and erythromycin (Ery; Sigma-Aldrich, St. Louis, MO, USA) at 5 μg/mL, at 30 °C without agitation for 18 h. The plasmids and bacterial strains used are described in Table 1 [15, 16].

Construction of pValac::*anti-TNFα* and development of LL-FT
An anti-TNFα scFv expression vector was constructed based on infliximab variable chain sequences (GenBank accession numbers: 471270577 and 471,270,576). A synthetic scFv was designed based on mouse codon usage and cloned in the pValac shuttle vector. The synthetic gene fragment was cloned into the pValac::*gfp* vector digested with *EcoR* I (Invitrogen, Carlsbad, CA, USA) and *Nhe* I (Invitrogen, Carlsbad, CA, USA), yielding pValac::*anti-TNFα*. An HA tag was included in the carboxy terminus of the scFv. This vector was used to transform *E. coli* TG1 as described by Sambrook and Russel [58]. Cloning was checked via the restriction endonuclease digestion profile, polymerase chain reactions (PCR) and sequencing [59], confirming the integrity of sequences. LL-F was transformed with pValac::*anti-TNFα* using electroporation as previously described [60], resulting in LL-FT. The presence of plasmid was confirmed by PCR and enzymatic digestion.

Transfection assays of mammalian HEK-293 cells with pValac::*anti-TNFα*
The pValac::*anti-TNFα* plasmid was tested for anti-TNFα protein expression by human embryonic kidney (HEK-293) cell line transfection. HEK-293 cells (ATCC number: CRL-1573™) obtained from Rio de Janeiro Cell Bank (Rio de Janeiro, Brazil), were cultured in Dulbecco's modified Eagle's medium (DMEM) (Gibco®, Glasgow, UK) supplemented with 10% fetal bovine serum (FBS) (Gibco®, Glasgow, UK) and Antibiotic-Antimycotic (100X) (Gibco®, Glasgow, UK) at 37 °C in 5% CO$_2$. Lipofectamine™ LTX reagent (Invitrogen, Carlsbad, CA, USA) was used for the transfection assay with 2 μg of pValac::*anti-TNFα* in wells containing 70–90% confluent cells according to the manufacturer's recommendations. After 48 h, samples were centrifuged at 10,000×g for 5

Table 1 Bacterial strains and plasmids used in this study

Plasmids	Characteristics	Reference
pValac	Eukaryotic expression vector (pCMV/Cm^r/RepA/RepC)	[16]
pValac::anti-TNFa	pValac containing anti-TNFa scFv ORF	This study.
Bacterial strain and plasmids	Characteristics	Reference
Escherichia coli XL1-Blue	E. coli; F'[Tn10 proAB$^+$ lacI^q Δ (lacZ)M15] hsdR17(r_K^- m_K^+)/ Tet^r	Stratagene® (Catalog n° #200249)
Escherichia coli TG1	E. coli, K-12-derived strain; F' [traD36 proAB$^+$ lacI^q lacZΔM15] supE thi-1 Δ (lac-proAB) Δ (mcrB-hsdSM)5, ($r_K^-m_K^-$)	Lucigen, Middleton, MI, USA (Catalog n° 60502-1)
Lactococcus lactis MG1363 FnBPA + (LL-F)	L. lactis MG1363 strain expressing FnBPA of S. aureus (Ery^r)	[15], obtained from Laboratory of Cellular and Molecular Genetics (LGCM), Federal University of Minas Gerais (UFMG), Brazil,
Lactococcus lactis MG1363 FnBPA + (pValac::anti-TNFa) (LL-FT)	L. lactis MG1363 FnBPA+ strain carrying the pValac::anti-TNFa plasmid (Cm^r/ Ery^r)	This study.

Cm^r chloramphenicol resistance; Tet^r tetracycline resistance; Ery^r erythromycin resistance

min, and the supernatant was stored at − 20 °C until anti-TNFα detection by western blot.

SDS-PAGE and Western blot analysis

Proteins from the supernatant of HEK-293 transfection were resolved by 12% SDS-PAGE using the Bio-Rad system (Bio-Rad, Hercules, CA, USA) for electrophoresis. Proteins were transferred to a nitrocellulose membrane (GE Healthcare, Uppsala, Sweden) that was blocked with PBST-milk (PBS buffer added 5% skim milk and 0.1% Tween 20) and then incubated with anti-HA probe (1:1000; Sigma-Aldrich, St. Louis, MO, USA). After PBST washing, the membrane was incubated with alkaline phosphatase-conjugated anti-rabbit IgG antibody (1:5000; Southern Biotechnology, Birmingham, AL, USA). After washing with PBST and APB, enzymatic activity was performed using a BCIP/ NBT chromogenic substrate (Invitrogen, Carlsbad, CA, USA).

Mice

Conventional female C57BL/6 mice [61] (10 week) were purchased from the CEMIB (Centro Multidisciplinar para Investigação Biológica) of Universidade Estadual de Campinas (Unicamp – Campinas, Brazil). Animals belonging to the same experimental group were housed in a single cage in a controlled temperature (25 °C) room with a 12:12-h light/dark cycle and ad libitum access to food and water. Sample number estimation in each experimental group and all animal procedures were performed following the rules of the Ethical Principles in Animal Experimentation adopted by the Ethics Committee on Animal Experimentation (CEUA/ICB-UnB/Brazil) and approved by CEUA (51,069/2015).

DSS-induced colitis and treatment with L. lactis MG1363 FnBPA+ (pValac::anti-TNFa)

Acute colitis was induced by adding 2% (w/v) DSS (MW 40–50 kDa; USB Affymetrix, Santa Clara, CA, USA) to the drinking water from day 1 to day 8 [24]. Experiments were carried out with 4 to 5 mice per group. The mice were divided into the following groups: i) a healthy negative control group (NC) in which the mice were gavaged with 100 µL of saline (0.9% NaCl) and allowed to ingest pure filtered water throughout the experiment, ii) a positive control group of colitis (DSS) in which the mice were gavaged with 100 µL of saline, iii) the LL-F group in which the mice received 100 µL of the corresponding bacterial strain as suspension without plasmid, and iv) the LL-FT group in which the mice received 100 µL of the corresponding bacterial strain as suspension. The DSS, LL-F and LL-FT groups ingested filtered water with 2% DSS added throughout the experiment. The recombinant strains were administered once daily by gavage from day 5 to day 8. Each dose corresponded to 100 µL of LL-FT suspension and contained $2.0–2.5 × 10^9$ colony forming units (CFU). Animals were euthanized on day 9 in a 5% carbon dioxide chamber with cervical dislocation to collect blood samples from the retro-orbital venous plexus; the colonic tissue was quickly removed and washed of feces. Animals that died during the experiment were not included in the analysis. The mean of the water-DSS intake per group was monitored, and each animal consumed 3–5 mL of water daily.

Disease activity index (DAI)

On day 9 (day of euthanasia), DSS-induced colitis was determined using the disease activity index (DAI) as described by Cooper et al. [62]. The DAI consisted of the combined scores for weight loss, stool consistency and rectal bleeding divided by 3. The features that were graded included the following: body weight loss (0, none; 1, 1–5% loss; 2, 5–10% loss; 3, 10–20% loss; and 4, > 20% loss), stool consistency (0, normal; 2, loose stools and 4, diarrhea) and rectal bleeding (0, absent; 2, moderate and 4, severe). Loss of body weight was defined by the difference between the initial and final weight. Stool consistency and rectal bleeding were confirmed by examination of the sectioned colon upon euthanasia.

Quantification of C-reactive protein in blood serum

C-Reactive Protein (CRP) was quantified in the blood serum of animals with a Mouse CRP ELISA Kit (Sigma-Aldrich, St. Louis, MO, USA) according to the manufacturer's instruction. After euthanasia, collected blood samples were kept at room temperature until coagulation and then centrifuged at 5000×g for 5 min. The serum samples (supernatant) were then transferred to new tubes and stored at − 20 °C. For ELISA, a dilution of 1:20,000 from each serum sample was used. The reading was performed at 450 nm on a VersaMax™ ELISA Microplate Reader (Molecular Devices, San Jose, CA, USA).

Histopathological score of colitis

On the day of euthanasia, the colon was removed, and its length was measured. Distal colon samples were sectioned into two fragments to be used for histological study and biochemical determinations. Colonic tissue was fixed in 10% formaldehyde, and H&E staining was performed. The histological analyses were carried out in a blind design and were based on the morphological findings regarding the presence of inflammatory infiltrate reaching the mucosa, submucosa, muscular and serosal layers; inflammatory activities in glands; abscesses of crypts; erosion or ulceration of the mucosa; thickening of the muscular layer; and depletion of goblet cells and herniation of the muscular layer. These findings were classified as mild, moderate or severe. Thus, the following scores were assigned: 0, mucosa with normal structures, without any alteration or with slight inflammatory infiltrate in the mucosa or submucosa; 1, the presence of mild-to-moderate inflammatory infiltrate with inflammatory activity in glands or abscesses of crypts with erosion but without ulceration; 2, the presence of all the previous findings associated with greater ulcerations in the mucosa but one or two ulcers; and 3, the presence of ulcerations compromising large areas of the mucosa. The score of each animal could range from 0 to 39.

RNA isolation and qPCR analysis

For gene expression analysis of the colonic samples by quantitative PCR (qPCR), the tissue was stored in RNAlater (Qiagen, Valencia, CA, USA) until total RNA extraction using a RNeasy Protect Mini Kit (Qiagen, Valencia, CA, USA) and TissueLyser LT (Qiagen, Valencia, CA, USA) to disrupt the samples. The samples were quantified using a NanoDrop™ One^C Microvolume UV-Vis Spectrophotometer (Thermo Scientific™, Waltham, MA, USA). For reverse transcription, RT^2 First Strand Kit (Qiagen, Valencia, CA, USA) was used. For elimination of genomic DNA during RNA purification, RNase-Free DNase Set (Qiagen, Valencia, CA, USA) was used. Amplification and detection were performed on optical 96-well plates (Applied Biosystem, Foster City, CA, USA) with the 7500 Fast Real Time PCR System (Applied Biosystem, Foster City, CA, USA) using a Fast SyBR Green Master Mix Kit (Applied Biosystem, Foster City, CA, USA). Levels of mRNA expression were normalized to ribosomal protein S9 (RPS9) mRNA, and RNA relative quantification was calculated using the method $2^{-\Delta Ct}$ [63]. The oligonucleotide primers are described in Additional file 1: Table S2.

Analysis of blood serum cytokines by flow cytometry

The sera of mice were collected to measure the levels of interleukin-6 (IL-6), TNF and IL-10 using a Cytometric Bead Array (CBA) Mouse Inflammation Kit (BD Biosciences, San Jose, CA, USA) as recommended by the manufacturer. Samples were acquired in an Accuri™ C6 flow cytometer (BD Biosciences, San Jose, CA, USA) and analyzed using FCAP Array™ Software version 3.0 (BD Biosciences, San Jose, CA, USA). At least 2100 events were acquired for each sample.

Measurement of fecal IgA

Levels of IgA were determined in fecal extract samples using a Mouse IgA ELISA Kit (Sigma-Aldrich, St. Louis, MO, USA) following the manufacturer's protocol. Before euthanasia, the mice were placed in clean cages and kept for 30 min. A pool of feces per group was collected, weighed, transferred to a Falcon tube containing 5 mL of PBS added to 100 mM PMSF 0.2% and kept on ice for 15 min. The extracts were homogenized via inversion of the tube and kept on ice for 15 min. Then, the tubes were centrifuged for 30 min at 3000 rpm at 4 °C. The supernatants were saved and stored at − 80 °C. For ELISA, dilutions of 1:10 and 1:100 from each pool were used. The reading was performed at 450 nm on a VersaMax™ ELISA Microplate Reader (Molecular Devices, San Jose, CA, USA).

Statistical analysis

The results of experiments are expressed as the means ± SEM. Statistical differences were determined by two-way ANOVA with a Bonferroni post hoc test for curves and by the Mann-Whitney test for charts. All statistical analyses were performed with Graph Pad Prism version 6.0 for Mac OS X (La Jolla, CA, USA). Statistical significance was considered at $p < 0.05$.

Additional files

Additional file 1: Figure S1. Structure of the eukaryotic expression vector pValac::anti-TNFα and scFv protein expression by HEK-293 on transfection assays. **Figure S2.** MUC-3 mRNA levels in colonic tissue. **Figure S3.** Effect of the treatment of colitis with LL-FT on systemic cytokines production. **Figure S4.** Changes in fecal IgA after oral administration of LL-FT. **Table S1.** Primer sequences used in qPCR assay. **Table S2.** Histological Score. (DOCX 684 kb)

Abbreviations
CBA: Cytometric bead array; CD: Crohn's disease; CFU: Colony forming units; CRP: C-reactive protein; DAI: Disease activity index; DSS: Dextran sulfate sodium; FBS: Fetal bovine serum; FnBPA: Fibronectin-binding protein A; GRAS: Generally regarded as safe; IBD: Inflammatory bowel disease; LAB: Lactic acid bacteria; LL-F: *Lactococcus lactis* MG1363 FnBPA+; LL-FT: *Lactococcus lactis* MG1363 FnBPA+ (pValac::*anti-TNFα*); ORF: Open read frame; PCR: Polymerase chain reactions; qPCR: Quantitative PCR; scFv: Single-chain fragment variable; TNFα: Tumor necrosis factor-α; UC: Ulcerative colitis

Acknowledgments
We are in debt with Christian Hoffmann for help with statistics and Fabiana Brandão for help with qPCR.

Funding
All experimentation was funded by Fundação de Apoio a Pesquisa do DF (grant application 193.000.834/2015). MJC and IGS received financial support of CAPES and Instituto Nacional de Investigação em Imunologia/CNPq. The funders played no role in the execution or design of this study.

Authors' contributions
MJC designed and performed all experiments and analysis and write manuscript; JFA designed experiments and strategies; MGDA and SSF performed animal experimentation and housing; VBP helped cloning expression vectors; RJAC and MSJ helped with animal experiments and collect histopathological data; IGS helped collecting data for molecular and immunological data; LMSV performed pathological analysis; AM helped design vector and transgene; ALB and AQM helped establishing experimental protocols; MMB design general strategies and experiments, performed analysis and write manuscript. All authors read and approved the final manuscript.

Ethics approval
All animal procedures were performed following the rules of the Ethical Principles in Animal Experimentation adopted by the Ethics Committee on Animal Experimentation (CEUA/ICB-UnB/Brazil) and approved by CEUA (51,069/2015).

Consent for publication
N/A

Competing interests
The authors declare that the research was conducted in the absence of any commercial or financial relationships that could be construed as a potential conflict of interest.

Author details
[1]Laboratório de Imunologia Molecular, Departamento de Biologia Molecular, Universidade de Brasília, Brasília, Distrito Federal, Brazil. [2]Centro de Biotecnologia, Departamento de Biologia Celular e Molecular, Universidade Federal da Paraíba, João Pessoa, Paraíba, Brazil. [3]Laboratório de Tecnologia Genética, Departamento de Biologia Geral, Universidade Federal de Minas Gerais, Belo Horizonte, Minas Gerais, Brazil. [4]Laboratório de Imunologia Aplicada, Departamento de Biologia Celular, Universidade de Brasília, Brasília, Distrito Federal, Brazil. [5]Departmento de Patologia, Escola de Medicina, Universidade de Brasília, Brasília, Distrito Federal, Brazil. [6]Instituto Nacional de Investigação em Imunologia, INCTii, Brasília, Distrito Federal, Brazil.

References
1. Podolsky DK. Inflammatory bowel disease. N Engl J Med. 2002;347:417–29.
2. Ng SC, Shi HY, Hamidi N, Underwood FE, Tang W, Benchimol EI, et al. Worldwide incidence and prevalence of inflammatory bowel disease in the 21st century: a systematic review of population-based studies. Lancet (London, England). 2018;390:2769–78 Available from: http://www.ncbi.nlm.nih.gov/pubmed/29050646.
3. Geremia A, Biancheri P, Allan P, Corazza GR, Di Sabatino A. Innate and adaptive immunity in inflammatory bowel disease. Autoimmun Rev. 2014;13(1):3–10.
4. Nielsen OH, Munck LK. Drug insight: Aminosalicylates for the treatment of IBD. Nat Clin Pract Gastroenterol Hepatol. 2007;4(3):160–70.
5. Scott Crowe J, Roberts KJ, Carlton TM, Maggiore L, Cubitt MF, Clare S, et al. Preclinical development of a novel, orally-administered anti-tumour necrosis factor domain antibody for the treatment of inflammatory bowel disease. Sci Rep. 2018;8(1):4941.
6. Vandenbroucke K, De Haard H, Beirnaert E, Dreier T, Lauwereys M, Huyck L, et al. Orally administered L. lactis secreting an anti-TNF Nanobody demonstrate efficacy in chronic colitis. Mucosal Immunol. 2010;3:49–56 Available from: https://doi.org/10.1038/mi.2009.116.
7. Gomes-Santos AC, de Oliveira RP, Moreira TG, Castro-Junior AB, Horta BC, Lemos L, et al. Hsp65-producing Lactococcus lactis prevents inflammatory intestinal disease in mice by IL-10- and TLR2-dependent pathways. Front Immunol. 2017;8:1–12.
8. Luerce TD, Gomes-Santos AC, Rocha CS, Moreira TG, Cruz DN, Lemos L, et al. Anti-inflammatory effects of Lactococcus lactis NCDO 2118 during the remission period of chemically induced colitis. Gut Pathog. 2014;6:1–11.
9. Mancha-Agresti P, Drumond MM, do CFLR, Santos MM, dos SJSC, Venanzi F, et al. A new broad range plasmid for DNA delivery in eukaryotic cells using lactic acid Bacteria: in vitro and in vivo assays. Mol Ther - Methods Clin Dev. 2017;4:83–91 Available from: http://linkinghub.elsevier.com/retrieve/pii/S2329050116301383.
10. Steidler L. Lactococcus lactis, a tool for the delivery of therapeutic proteins treatment of IBD. Sci World J. 2001;1:216–7 Available from: http://www.hindawi.com/journals/tswj/2001/137951/abs/.
11. Carr FJ, Chill D, Maida N. The lactic acid bacteria: a literature survey. Crit Rev Microbiol. 2002;28(4):281–370.
12. Vesa T, Pochart P, Marteau P. Pharmacokinetics of lactobacillus plantarum NCIMB 8826, lactobacillus fermentum KLD, and Lactococcus lactis MG 1363 in the human gastrointestinal tract. Aliment Pharmacol Ther. 2000;14:823–8.
13. Souza BM, Preisser TM, Pereira VB, Zurita-Turk M, Castro CP, Cunha VP, et al. Lactococcus lactis carrying the pValac eukaryotic expression vector coding for IL-4 reduces chemically-induced intestinal inflammation by increasing the levels of IL-10-producing regulatory cells. Microb Cell Fact BioMed Central. 2016;15:1–18.
14. Zurita-Turk M, del Carmen S, Santos ACG, Pereira VB, Cara DC, Leclercq SY, et al. Lactococcus lactis carrying the pValac DNA expression vector coding for IL-10 reduces inflammation in a murine model of experimental colitis. BMC Biotechnol. 2014;14:1–11.
15. Que YA, Francois P, Haefliger JA, Entenza JM, Vaudaux P, Moreillon P. Reassessing the role of Staphylococcus aureus clumping factor and fibronectin-binding protein by expression in Lactococcus lactis. Infect Immun. 2001;69:6296–302.
16. Guimarães V, Innocentin S, Chatel JM, Lefèvre F, Langella P, Azevedo V, et al. A new plasmid vector for DNA delivery using lactococci. Genet Vaccines Ther. 2009;7:1–7.
17. Okayasu I, Hatakeyama S, Yamada M, Ohkusa T, Inagaki Y, Nakaya R. A novel method in the induction of reliable experimental acute and chronic ulcerative colitis in mice. Gastroenterology. 1990;98:694–702 Available from: http://www.ncbi.nlm.nih.gov/pubmed/1688816.
18. Mathias A, Pais B, Favre L, Benyacoub J, Corthï¿½sy B. Role of secretory IgA in the mucosal sensing of commensal bacteria. Gut microbes; 2015. p. 688–95.
19. Zagato E, Mazzini E, Rescigno M. The variegated aspects of immunoglobulin a. Immunol Lett. 2016;178:45–9.
20. Guimarães VD, Innocentin S, Lefèvre F, Azevedo V, Wal JM, Langella P, et al. Use of native lactococci as vehicles for delivery of DNA into mammalian epithelial cells. Appl Environ Microbiol. 2006;72:7091–7.
21. Chatel JM, Pothelune L, Ah-Leung S, Corthier G, Wal JM, Langella P. In vivo transfer of plasmid from food-grade transiting lactococci to murine epithelial cells. Gene Ther. 2008;15:1184–90.
22. Sinha B, Francois P, Que YA, Hussain M, Heilmann C, Moreillon P, et al. Heterologously expressed staphylococcus aureus fibronectin-binding

proteins are sufficient for invasion of host cells. Infect Immun. 2000;68: 6871–8.

23. Innocentin S, Guimarães V, Miyoshi A, Azevedo V, Langella P, Chatel JM, et al. Lactococcus lactis expressing either Staphylococcus aureus fibronectin-binding protein a or listeria monocytogenes internalin a can efficiently internalize and deliver DNA in human epithelial cells. Appl Environ Microbiol. 2009;75:4870–8.

24. Wirtz S, Popp V, Kindermann M, Gerlach K, Weigmann B, Fichtner-Feigl S, et al. Chemically induced mouse models of acute and chronic intestinal inflammation. Nat Protoc. 2017;12:1295–309 Available from: https://doi.org/10.1038/nprot.2017.044.

25. Melgar S. Acute colitis induced by dextran sulfate sodium progresses to chronicity in C57BL/6 but not in BALB/c mice: correlation between symptoms and inflammation. AJP Gastrointest Liver Physiol. 2005;288: G1328–38 Available from: http://ajpgi.physiology.org/cgi/doi/10.1152/ajpgi.00467.2004.

26. Alex P, Zachos NC, Nguyen T, Gonzales L, Chen TE, Conklin LS, et al. Distinct cytokine patterns identified from multiplex profiles of murine DSS and TNBS-induced colitis. Inflamm Bowel Dis. 2009;15:341–52.

27. Chassaing B, Aitken JD, Malleshappa M, Vijay-Kumar M. Dextran sulfate sodium (DSS)-induced colitis in mice. Curr Protoc Immunol. 2014;104: Unit–15.25

28. Vermeire S, Van Assche G, Rutgeerts P. C-reactive protein as a marker for inflammatory bowel disease. Inflamm Bowel Dis. 2004;10(5):661–5.

29. Veloso TR, Chaouch A, Roger T, Giddey M, Vouillamoz J, Majcherczyk P, et al. Use of a human-like low-grade bacteremia model of experimental endocarditis to study the role of staphylococcus aureus adhesins and platelet aggregation in early endocarditis. Infect Immun. 2013;81(3):697–703.

30. Yan Y, Kolachala V, Dalmasso G, Nguyen H, Laroui H, Sitaraman SV, et al. Temporal and spatial analysis of clinical and molecular parameters in dextran sodium sulfate induced colitis. PLoS One. 2009;4(6):e6073.

31. Strober W, Fuss IJ. Proinflammatory cytokines in the pathogenesis of inflammatory bowel diseases. Gastroenterology. 2011;140(6):1756–67.

32. Kobayashi T, Okamoto S, Hisamatsu T, Kamada N, Chinen H, Saito R, et al. IL23 differentially regulates the Th1/Th17 balance in ulcerative colitis and Crohn's disease. Gut. 2008;57:1682–9.

33. Cao AT, Yao S, Gong B, Elson CO, Cong Y. Th17 cells upregulate polymeric Ig receptor and intestinal IgA and contribute to intestinal homeostasis. J Immunol. 2012;189(9):4666–73.

34. Nishikawa K, Seo N, Torii M, Ma N, Muraoka D, Tawara I, et al. Interleukin-17 induces an atypical M2-like macrophage subpopulation that regulates intestinal inflammation. PLoS One. 2014;9(9):e108494.

35. Huang XL, Zhang X, Fei XY, Chen ZG, Hao YP, Zhang S, et al. Faecalibacterium prausnitzii supernatant ameliorates dextran sulfate sodium induced colitis by regulating Th17 cell differentiation. World J Gastroenterol. 2016;22(22):5201–10.

36. Schmidt EGW, Larsen HL, Kristensen NN, Poulsen SS, Pedersen AML, Claesson MH, et al. TH17 cell induction and effects of IL-17A and IL-17F blockade in experimental colitis. Inflamm Bowel Dis. 2013;19(8):1567–76.

37. Cătană CS, Neagoe IB, Cozma V, Magdaş C, Tăbăran F, Dumitraşcu DL. Contribution of the IL-17/IL-23 axis to the pathogenesis of inflammatory bowel disease. World J Gastroenterol. 2015;21:5823–30.

38. Gobert AP, Cheng Y, Akhtar M, Mersey BD, Blumberg DR, Cross RK, et al. Protective role of arginase in a mouse model of colitis. J Immunol. 2004;173:2109–17.

39. Lin Y, Yang X, Yue W, Xu X, Li B, Zou L, et al. Chemerin aggravates DSS-induced colitis by suppressing M2 macrophage polarization. Cell Mol Immunol. 2014;11:355–66.

40. Garrido-Mesa J, Algieri F, Rodriguez-Nogales A, Utrilla MP, Rodriguez-Cabezas ME, Zarzuelo A, et al. A new therapeutic association to manage relapsing experimental colitis: Doxycycline plus Saccharomyces boulardii. Pharmacol Res. 2015;97:48–63 Available from: https://doi.org/10.1016/j.phrs.2015.04.005.

41. Macpherson AJ, Yilmaz B, Limenitakis JP, Ganal-Vonarburg SC. IgA function in relation to the intestinal microbiota. Annu Rev Immunol. 2018;36: annurev-immunol-042617-053238. Available from: http://www.annualreviews.org/doi/10.1146/annurev-immunol-042617-053238.

42. Corthésy B. Multi-faceted functions of secretory IgA at mucosal surfaces. Front Immunol. 2013;4:185.

43. Neurath MF. Cytokines in inflammatory bowel disease. Nat Rev Immunol. 2014;14(5):329–42.

44. Park JH, Peyrin-Biroulet L, Eisenhut M, Shin JI. IBD immunopathogenesis: a comprehensive review of inflammatory molecules. Autoimmun Rev. 2017; 16:416–26.

45. Steeland S, Libert C, Vandenbroucke RE. A new venue of TNF targeting. Int J Mol Sci. 2018;19(5):1442.

46. Lopetuso LR, Petito V, Cufino V, Arena V, Stigliano E, Gerardi V, et al. Locally injected infliximab ameliorates murine DSS colitis: differences in serum and intestinal levels of drug between healthy and colitic mice. Dig Liver Dis. 2013;45(12):1017–21.

47. Walldorf J, Hermann M, Porzner M, Pohl S, Metz H, Mäder K, et al. In-vivo monitoring of acute DSS-colitis using colonoscopy, high resolution ultrasound and bench-top magnetic resonance imaging in mice. Eur Radiol. 2015;25(10):2984–91.

48. Assas BM, Levison SE, Little M, England H, Battrick L, Bagnall J, et al. Anti-inflammatory effects of infliximab in mice are independent of tumour necrosis factor α neutralization. Clin Exp Immunol. 2017;187(2):225–33.

49. Naito Y, Takagi T, Handa O, Ishikawa T, Nakagawa S, Yamaguchi T, et al. Enhanced intestinal inflammation induced by dextran sulfate sodium in tumor necrosis factor-alpha deficient mice. J Gastroenterol Hepatol. 2003; 18(5):560–9.

50. Kojouharoff G, Hans W, Obermeier F, Männel DN, Andus T, Schölmerich J, et al. Neutralization of tumour necrosis factor (TNF) but not of IL-1 reduces inflammation in chronic dextran sulphate sodium-induced colitis in mice. Clin Exp Immunol. 1997;107:353–8 Available from: http://www.pubmedcentral.nih.gov/articlerender.fcgi?artid=1904573&tool=pmcentrez&rendertype=abstract.

51. Bhol KC, Tracey DE, Lemos BR, Lyng GD, Erlich EC, Keane DM, et al. AVX-470: a novel oral anti-TNF antibody with therapeutic potential in inflammatory bowel disease. Inflamm Bowel Dis. 2013;19(11):2273–81.

52. Fries W, Muja C, Crisafulli C, Costantino G, Longo G, Cuzzocrea S, et al. Infliximab and etanercept are equally effective in reducing enterocyte apoptosis in experimental colitis. Int J Med Sci. 2008;5(4):169–80.

53. Shen C, De Hertogh G, Bullens DMA, Van Assche G, Geboes K, Rutgeerts P, et al. Remission-inducing effect of anti-TNF monoclonal antibody in TNBS colitis: mechanisms beyond neutralization? Inflamm Bowel Dis. 2007;13(3): 308–16.

54. Qiu W, Wu B, Wang X, Buchanan ME, Regueiro MD, Hartman DJ, et al. PUMA-mediated intestinal epithelial apoptosis contributes to ulcerative colitis in humans and mice. J Clin Invest. 2011;121(5):1722–32.

55. Reardon S. Genetically modified bacteria enlisted in fight against disease. Nature. 2018;558:497–8 Available from: http://www.nature.com/articles/d41586-018-05476-4.

56. Berlec A, Perše M, Ravnikar M, Lunder M, Erman A, Cerar A, et al. Dextran sulphate sodium colitis in C57BL/6J mice is alleviated by Lactococcus lactis and worsened by the neutralization of tumor necrosis factor α. Int Immunopharmacol. 2017;43:219–26.

57. Tracey D, Klareskog L, Sasso EH, Salfeld JG, Tak PP. Tumor necrosis factor antagonist mechanisms of action: a comprehensive review. Pharmacol Ther. 2008;117:244–79.

58. Sambrook J, W Russell D. Molecular cloning: a laboratory manual. Cold Spring Harb Lab Press Cold Spring Harb, NY.

59. Sanger F, Nicklen S, Coulson AR. DNA sequencing with chain-terminating inhibitors. Proc Natl Acad Sci. 1977;74:5463–7.

60. Langella P, Le Loir Y, Ehrlich SD, Gruss A. Efficient plasmid mobilization by pIP501 in Lactococcus lactis subsp. lactis. J Bacteriol. 1993;175(18):5806–13.

61. Melgar S, Bjursell M, Gerdin A-K, Svensson L, Michaelsson E, Bohlooly-Y M. Mice with experimental colitis show an altered metabolism with decreased metabolic rate. Am J Physiol Gastrointest Liver Physiol. 2007;292(1):G165–72.

62. Cooper HS, Murthy SN, Shah RS, Sedergran DJ. Clinicopathologic study of dextran sulfate sodium experimental murine colitis. Lab Invest [Internet]. 1993;69:238–49 Available from: http://www.ncbi.nlm.nih.gov/pubmed/8350599.

63. Schmittgen TD, Livak KJ. Analyzing real-time PCR data by the comparative CT method. Nat Protoc. 2008;3:1101–8 Available from: http://www.nature.com/doifinder/10.1038/nprot.2008.73.

11

A new prokaryotic expression vector for the expression of antimicrobial peptide abaecin using SUMO fusion tag

Da Sol Kim[1], Seon Woong Kim[1], Jae Min Song[2], Soon Young Kim[1*] and Kwang-Chul Kwon[3*]⊕

Abstract

Background: Despite the growing demand for antimicrobial peptides (AMPs) for clinical use as an alternative approach against antibiotic-resistant bacteria, the manufacture of AMPs relies on expensive, small-scale chemical methods. The small ubiquitin-related modifier (SUMO) tag is industrially practical for increasing the yield of recombinant proteins by increasing solubility and preventing degradation in expression systems.

Results: A new vector system, pKSEC1, was designed to produce AMPs, which can work in prokaryotic systems such as *Escherichia coli* and plant chloroplasts. 6xHis was tagged to SUMO for purification of SUMO-fused AMPs. Abaecin, a 34-aa-long antimicrobial peptide from honeybees, was expressed in a fusion form to 6xHis-SUMO in a new vector system to evaluate the prokaryotic expression platform of the antimicrobial peptides. The fusion sequences were codon-optimized in three different combinations and expressed in *E. coli*. The combination of the native SUMO sequence with codon-optimized abaecin showed the highest expression level among the three combinations, and most of the expressed fusion proteins were detected in soluble fractions. Cleavage of the SUMO tag by sumoase produced a 29-aa-long abaecin derivative with a C-terminal deletion. However, this abaecin derivative still retained the binding sequence for its target protein, DnaK. Antibacterial activity of the 29-aa long abaecin was tested against *Bacillus subtilis* alone or in combination with cecropin B. The combined treatment of the abaecin derivative and cecropin B showed bacteriolytic activity 2 to 3 times greater than that of abaecin alone.

Conclusions: Using a SUMO-tag with an appropriate codon-optimization strategy could be an approach for the production of antimicrobial peptides in *E.coli* without affecting the viability of the host cell.

Keywords: Abaecin, Antimicrobial peptide, Small ubiquitin-related modifier (SUMO), Codon optimization, *Escherichia coli*

Background

The overuse of antibiotics for the last several decades and the prevalence of antibiotic-resistant bacterial infections present a threat to global health. About 30 million people are projected to be killed by antibiotic-resistance bacteria by 2050 [1]. Therefore, the discovery of novel antibiotic agents is of increasing medical importance. However, there is also a potential that over-use or abuse of a new antibiotic will give rise to more antibiotic-resistance bacteria. Growing difficulties from the discovery of new antibiotics to their clinical use have diverted attentions to a new therapeutic agent, antimicrobial peptides (AMPs) [2].

As an alternative approach to prevent the spread of bacteria which are resistant to antibiotics, antimicrobial peptides (AMPs) have been extensively studied since they have a broad spectrum of anti-infective activity against pathogenic bacteria with relatively low minimal inhibitory concentrations and with a property less capable of incurring resistance than conventional antibiotics, due to their nonspecific interaction with bacterial membranes, and their ability to work on multiple targets. In addition to the direct antimicrobial activities, many AMPs are associated with immunomodulatory properties as seen in host-defense peptides [3, 4].

* Correspondence: kimsy@anu.ac.kr; kwang-chul.kwon@microsynbiotix.com
[1]Department of Biological Sciences, Andong National University, Andong, South Korea
[3]MicroSynbiotiX Ltd, 11011 N Torrey Pines Rd Ste. #135, La Jolla, CA 92037, USA

To meet the demand of AMPs for clinical applications, the produciton of AMPs can be achieved by chemical synthesis. For example, phase peptide synthesis approaches can provide easy isolation of peptides with high purity and less use of solvents by avoiding chromatographic purification [5]. However, chemical synthesis has several disadvantages that prevent cost-effective, industrial-scale production of AMPs. In addition, the synthesis of longer peptides with more than 50 amino acids is not favored [6].

For high production of large peptides, biological systems such as bacteria and yeast have been preferably used as production platforms [7–13]. Although these biological systems don't need expensive active pharmaceutical ingredients (APIs) and toxic chemical solvents, there are still some issues remaining, such as the toxicity of the expressed AMPs to host cells, expensive purification, and low yield. As a solution to address these issues, plants can be a viable alternative expression platform [14–17]. In addition, they can be used as a delivery platform as well [18, 19]. Here, we report a new vector system for the production of AMPs, which can express the AMPs in both bacteria and plants.

The vector is designed to be operable in prokaryotic systems, and can be used to transform chloroplasts, a prokaryotic organelle of plants, for the large scale production of AMPs. The plant expression system can offer several other advantages over microbial expression systems, including no risk of endotoxin contamination, and oral delivery of bioencapsulated therapeutics using edible plants [18–21]. The expression cassettes of the vector were designed to be surrounded by flanking sequences, allowing the cassettes to be integrated into the chloroplast genome via double homologous recombination. Moreover, the copy number of the transgene in chloroplasts can be multiplied up to 10,000 per single plant cell [22, 23], leading to high expression of the transgene.

In addition, the vector was designed for AMPs to be expressed in a fused form using tags such as SUMO and 6xHis in order to eliminate the toxicity of AMPs to prokaryotic hosts, prevent protease degradation, increase solubility, and simplify purification and detection [24]. However, the tags need to be removed prior to clinical use. The final product needs to be an authentic or native amino acid sequence, so a cleavage recognition site should be placed between the therapeutic peptide and the tag. However, the removal of tag proteins at the cleavage site is often sterically hindered, and the cleavage action by proteases such as factor Xa, enterokinase, thrombin and tobacco etch virus protease, is often non-specific [24]. Unlike proteases that recognize linear amino acid sequences, the SUMO protease (sumoase) recognizes the tertiary structure of SUMO and prevents erroneous cleavage events in the target protein [24–26].

To evaluate our expression platform for AMPs in *E. coli*, abaecin from honeybee *Apis mellifera* was expressed using our new vector system. Abaecin is an important peptide in the bee innate immune system, and is found in a variety of bees including *Apis mellifera* [27], *Bombus pascuorum* [28] and *Bombus ignitus* [29]. As a proline-rich, non-glycosylated antimicrobial peptide [27], abaecin has a broad spectrum of bacteriolytic activity [30] and shows increased inhibitory effects on bacterial growth when treated with pore-forming peptides, such as cecropin A [13, 31], stomoxyn, and hymenoptaecin. The 34-aa-long cationic peptide contains 10 prolines (29%) with no cysteine residue, and the uniformed distribution of the proline residues through the entire peptide prevents the α-helical conformation [12, 27].

Here, we describe a new expression platform for the efficient produciton of AMPs, which can be potentially operable in both bacteria and plant chloroplast, and its use in an expression platform such as *E.coli* was first tested in this study. The new expression vector was built on the pUC19 backbone vector by assembling component DNA fragments, such as the promoter/5' UTR, 3' UTR and flanking sequences, which were derived from both plant chloroplast or bacterial genes, and further equipped with tagging systems such as SUMO and 6xHis.

Results

Expression vector design and its construction

A new expression vector was designed for use in prokaryotic systems such as bacteria and plant chloroplasts. The *psb*A promoter/5' UTR and 3' UTR were employed to drive the high expression of transgenes and stabilize the expressed transcripts, respectively [32, 33] (Fig. 1a). A second expression cassette, which can counteract spectinomycin, was constructed to select transformed cells. The antibiotic resistance gene, aminoglycoside-3″-adenylyltransferase (*aad*A), was put under the control of 16S rRNA promoter, and the transcripts were stabilized using the 3' UTR fragment of the *E.coli rrn*B operon [34, 35]. For chloroplast expression of transgenes, flanking sequences which can integrate the expression cassettes into the chloroplast genome at a specific location were added to both ends of the combined expression cassette fragments. The flanking sequences were derived from the *trn*I and *trn*A regions of the tobacco chloroplast genome. The integration region is transcriptionally active and has been widely used for high expression of transgenes [36, 37]. All of the DNA elements which were amplified by PCR or synthesized were combined into the pUC19 backbone vector to construct a new vector, pKSEC1 (Fig. 1a, b).

The vector was further modified by the addition of tagging systems. SUMO, derived from human SUMO1, was attached to abaecin to increase solubility and prevent toxicity of abaecin to host cells. Also, a 6xHis purification tag was added to the N-terminus of SUMO to facilitate the purification of the SUMO-fused abaecin, and to isolate abaecin from SUMO after cleavage by sumoase. Three different combinations of codon-optimized synthetic

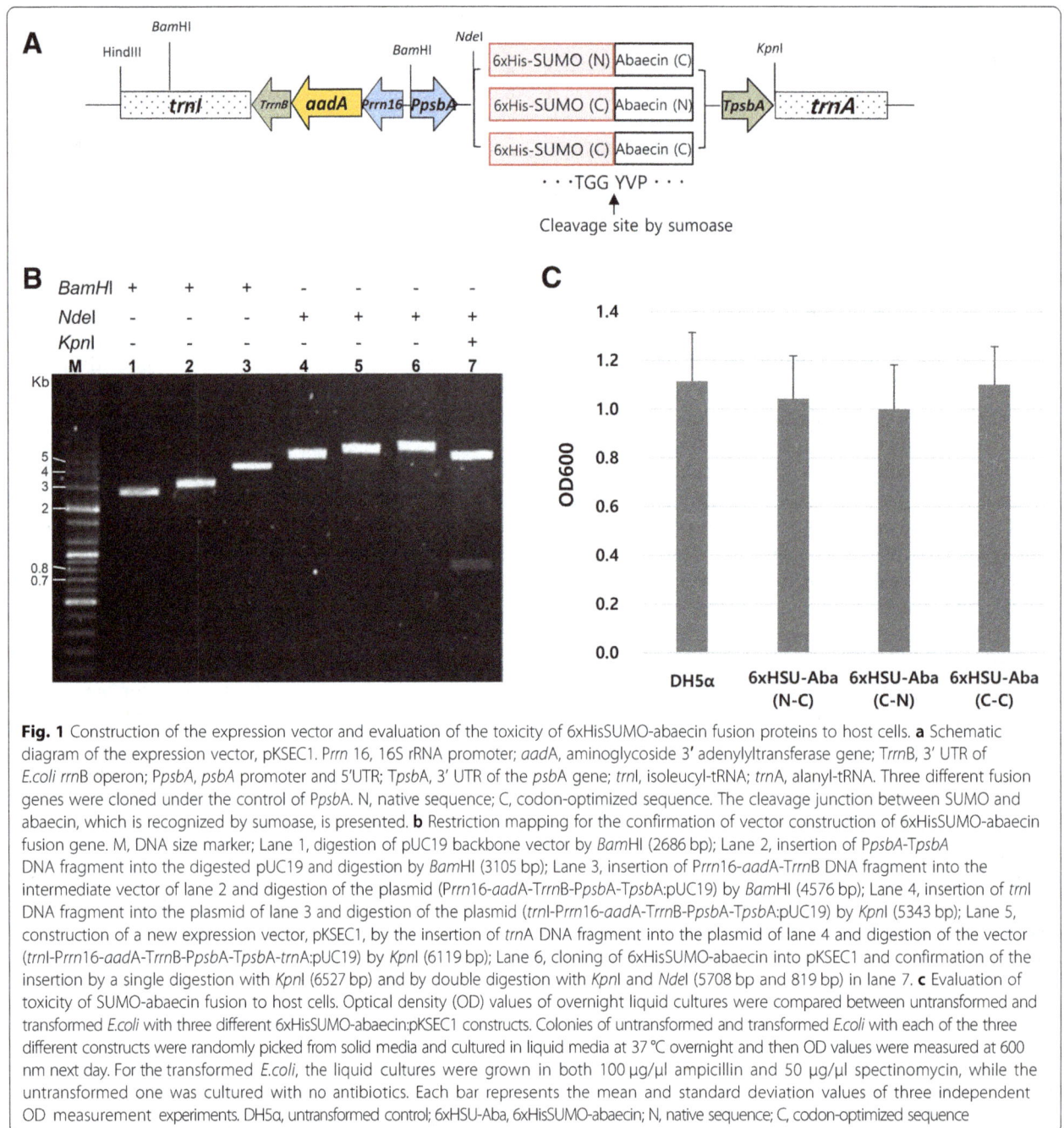

Fig. 1 Construction of the expression vector and evaluation of the toxicity of 6xHisSUMO-abaecin fusion proteins to host cells. **a** Schematic diagram of the expression vector, pKSEC1. Prrn 16, 16S rRNA promoter; *aad*A, aminoglycoside 3′ adenylyltransferase gene; TrrnB, 3′ UTR of *E.coli rrn*B operon; P*psb*A, *psb*A promoter and 5′UTR; T*psb*A, 3′ UTR of the *psb*A gene; *trn*I, isoleucyl-tRNA; *trn*A, alanyl-tRNA. Three different fusion genes were cloned under the control of P*psb*A. N, native sequence; C, codon-optimized sequence. The cleavage junction between SUMO and abaecin, which is recognized by sumoase, is presented. **b** Restriction mapping for the confirmation of vector construction of 6xHisSUMO-abaecin fusion gene. M, DNA size marker; Lane 1, digestion of pUC19 backbone vector by *Bam*HI (2686 bp); Lane 2, insertion of P*psb*A-T*psb*A DNA fragment into the digested pUC19 and digestion by *Bam*HI (3105 bp); Lane 3, insertion of Prrn16-*aad*A-TrrnB DNA fragment into the intermediate vector of lane 2 and digestion of the plasmid (Prrn16-*aad*A-TrrnB-P*psb*A-T*psb*A:pUC19) by *Bam*HI (4576 bp); Lane 4, insertion of *trn*I DNA fragment into the plasmid of lane 3 and digestion of the plasmid (*trn*I-Prrn16-*aad*A-TrrnB-P*psb*A-T*psb*A:pUC19) by *Kpn*I (5343 bp); Lane 5, construction of a new expression vector, pKSEC1, by the insertion of *trn*A DNA fragment into the plasmid of lane 4 and digestion of the vector (*trn*I-Prrn16-*aad*A-TrrnB-P*psb*A-T*psb*A-*trn*A:pUC19) by *Kpn*I (6119 bp); Lane 6, cloning of 6xHisSUMO-abaecin into pKSEC1 and confirmation of the insertion by a single digestion with *Kpn*I (6527 bp) and by double digestion with *Kpn*I and *Nde*I (5708 bp and 819 bp) in lane 7. **c** Evaluation of toxicity of SUMO-abaecin fusion to host cells. Optical density (OD) values of overnight liquid cultures were compared between untransformed and transformed *E.coli* with three different 6xHisSUMO-abaecin:pKSEC1 constructs. Colonies of untransformed and transformed *E.coli* with each of the three different constructs were randomly picked from solid media and cultured in liquid media at 37 °C overnight and then OD values were measured at 600 nm next day. For the transformed *E.coli*, the liquid cultures were grown in both 100 µg/µl ampicillin and 50 µg/µl spectinomycin, while the untransformed one was cultured with no antibiotics. Each bar represents the mean and standard deviation values of three independent OD measurement experiments. DH5α, untransformed control; 6xHSU-Aba, 6xHisSUMO-abaecin; N, native sequence; C, codon-optimized sequence

sequences of the 6xHisSUMO-abaecin were cloned into the new vector, and their expression in an *E. coli* system was investigated after the confirmation of their sequences (Fig. 1a, b and Additional file 1: Figure S1). For the codon optimization, codon adjustment was performed according to a previous study [33], in which a new algorithm for codon optimization was developed based on the codon usage hierarchy of chloroplast *psb*A genes, and the codon-optimized sequences under the control of *psb*A/5′ UTR showed increased expression levels over their respective native sequences in both *E. coli* and plant chloroplasts. The

expressions of all the three fusion constructs didn't show any toxicity to host cells. As shown in Fig. 1c, there was no significant difference of optical density (OD) values between untransformed and transformed cells with the three fusion constructs, which were grown overnight at 37 °C.

Evaluation of expression of His-tagged SUMO fused abaecin in *E. coli*

The three constructions of the 6xHisSUMO-abaecin fusion protein were evaluated using an *E. coli* expression system. Transformed *E. coli* cells with the three constructs were grown in liquid culture and the relative expression levels between the three fusion proteins were examined using an immunoblot assay with anti-His antibody. The expressed fusion proteins were detected around 20 kDa, not at 15.7 kDa which is a deduced molecular weight (Fig. 2a). This kind of discrepancy is very often observed in proline-rich proteins expressed in *E. coli*, due to presumably increased rigidity caused by high proline content, leading to slower migration than the same molecular weight protein [38–41]. Abaecin (3.9 kDa, 34 amino acids) contains 10 prolines in total of 34 amino acids, accounting for 34.9%, so it is assumed that the high content of proline residues affected the migration of all three fusion proteins (136 amino acids in length for each construct). Among three fusion proteins, 6xHisSUMO (N)-abaecin (C) (N and C, stand for "native sequence" and "codon-optimized sequence", respectively) showed the highest expression, with 2.8 or 3.5 fold

higher expression than 6xHisSUMO(C)-abaecin (N) or 6xHisSUMO(C)-abaecin(C), respectively (Fig. 2b). All three constructs showed that the expressed fusion proteins were detected at levels 1.4 to 2.1 times higher in the soluble fraction than in the insoluble one (Fig. 2b).

From these data, we confirmed that the newly designed vector is operable in a prokaryotic system and that the native sequence of the heterologous human SUMO gene improved the expression of the SUMO-fused abaecin in *E. coli* (Fig. 2a).

Cleavage of abaecin from purified His-tagged SUMO-abaecin and MALDI-TOF analysis

The cleavage efficiency of sumoase on the fusion protein was investigated by treating the soluble proteins extracted from *E. coli* with sumoase for 6 h and then detecting the treated proteins with Coomassie staining and

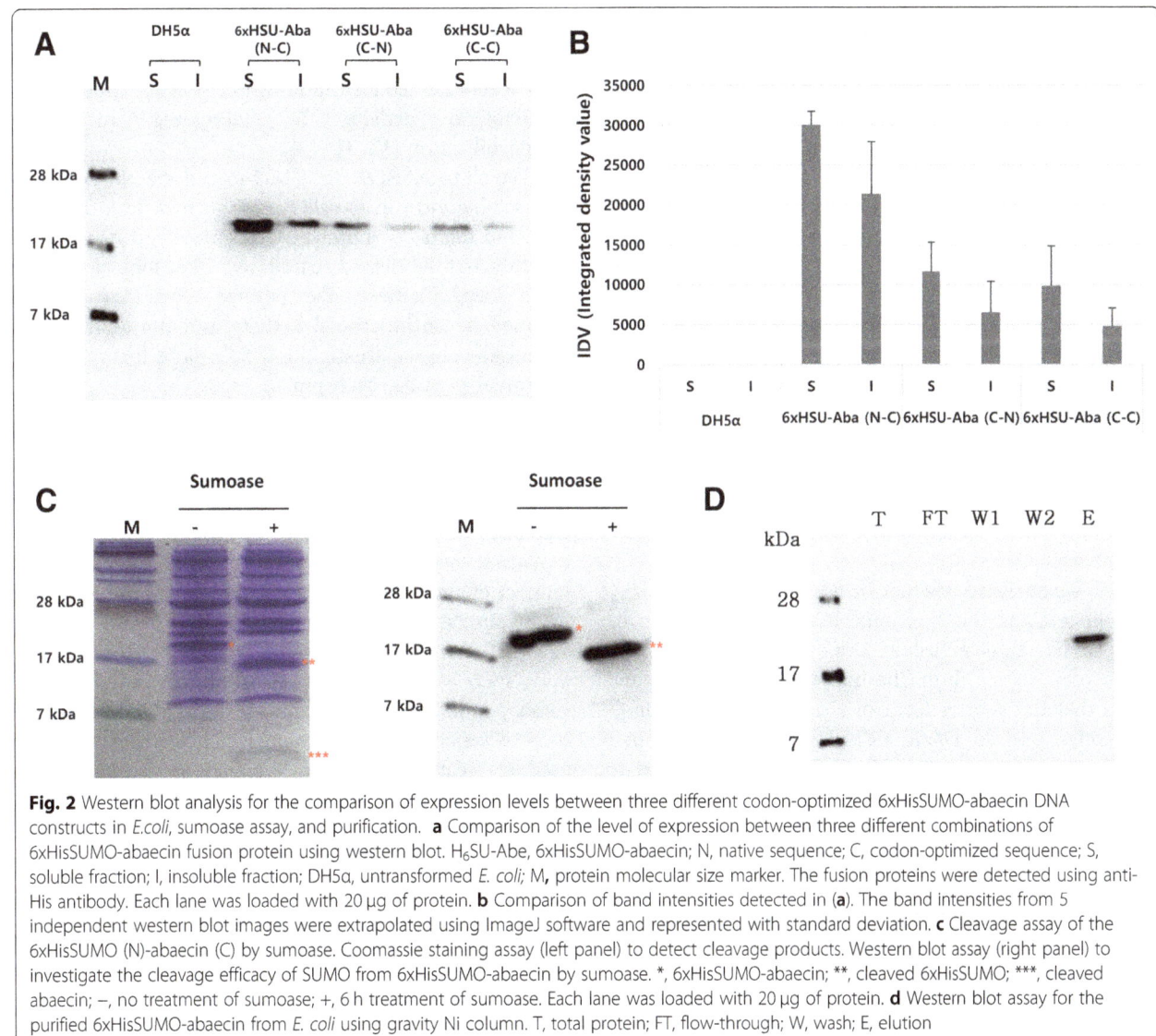

Fig. 2 Western blot analysis for the comparison of expression levels between three different codon-optimized 6xHisSUMO-abaecin DNA constructs in *E.coli*, sumoase assay, and purification. **a** Comparison of the level of expression between three different combinations of 6xHisSUMO-abaecin fusion protein using western blot. H₆SU-Abe, 6xHisSUMO-abaecin; N, native sequence; C, codon-optimized sequence; S, soluble fraction; I, insoluble fraction; DH5α, untransformed *E. coli*; M, protein molecular size marker. The fusion proteins were detected using anti-His antibody. Each lane was loaded with 20 μg of protein. **b** Comparison of band intensities detected in (**a**). The band intensities from 5 independent western blot images were extrapolated using ImageJ software and represented with standard deviation. **c** Cleavage assay of the 6xHisSUMO (N)-abaecin (C) by sumoase. Coomassie staining assay (left panel) to detect cleavage products. Western blot assay (right panel) to investigate the cleavage efficacy of SUMO from 6xHisSUMO-abaecin by sumoase. *, 6xHisSUMO-abaecin; **, cleaved 6xHisSUMO; ***, cleaved abaecin; –, no treatment of sumoase; +, 6 h treatment of sumoase. Each lane was loaded with 20 μg of protein. **d** Western blot assay for the purified 6xHisSUMO-abaecin from *E. coli* using gravity Ni column. T, total protein; FT, flow-through; W, wash; E, elution

western blot. As seen in Fig. 2c, the fusion proteins detected at 19 kDa before treatment of sumoase were cleaved and produced two products (Fig. 2c). To confirm the result, an immunoblot assay with a replicate gel was performed using anti-His antibody. The band for the cleaved-off 6xHisSUMO domain was detected around 16 kDa, along with a band for the fusion protein around 19 kDa (Fig. 2c). This data confirmed that the tertiary structure of SUMO was correctly formed in *E. coli* and recognized by sumoase (Fig. 2c).

With confirmation of the proper expression of 6xHis-SUMO-abaecin and cleavage of SUMO tag (Fig. 2c), the fusion proteins were purified for mass analysis of abaecin cleaved from the fusion protein (Fig. 2d). The abaecin cleaved by sumoase was examined by MALDI-TOF analysis to verify the precise cleavage between SUMO and abaecin. Singly protonated molecular ion (M + H) + of the cleaved-off abaecin was observed at m/z 3245.034 Da using reflector mode (Fig. 3a), while the theoretical molecular mass is 3877.049 Da (Fig. 3b). This discrepancy between the observed and theoretical values suggests the possibility of deletion of abaecin peptide at either N- or C-termini. To identify the peptide peak, we calculated the masses for a series of deleted abaecin peptides from either N- or C-termini and compared the theoretical masses to the observed mass (Fig. 3b) The best amino acid sequence corresponding to the observed mass at 3245.034 Da was deduced as YVPLPNVPQPGRRPFPTFPGQGPFNPKIK, which indicates the deletion of 5 amino acids from C-terminus of abaecin (Fig. 3b). Based on the results, although the proper cleavage between SUMO and abaecin occurred, undesired cleavage happened at the C-terminus.

Since the small cationic peptides are very susceptible to proteolytic degradation in *E.coli* [10], the positive charge distribution of the abaecin was analyzed (Fig. 3b) from which the positive charge patch created by two Lys (K) residues at C-terminus was likely targeted by proteolytic attack.

Next, we analyzed whether or not the 29-aa-long abaecin still retained the functional binding sequence to its intracellular target molecule, DnaK. For the analysis, we used limbo server (http://limbo.switchlab.org) [42] and found that the shorter form of abaecin still had a binding sequence to the DnaK; YVPLPNV, numbered from 1 to 7 (score 1.65) (Fig. 3c). Therefore, we continued to test an antimicrobial activity of the 29-aa-long abaecin derivative with or without cecropin B against *Bacillus subtilis*.

Antibacterial activity of purified abaecin against *Bacillus subtilis*

To evaluate the antimicrobial activity of 29-aa abaecin expressed in *E. coli*, the activity was tested against *B. subtilis* using an agar diffusion assay with the purified

abaecin derivative. We also tested whether the SUMO-fused abaecin is toxic to bacteria when treated externally, although we had already confirmed that there was no toxicity of the 6xHisSUMO-fused abaecin to host cells when expressed in the cells (Fig. 1c). The treatment of the 6xHisSUMO-abaecin didn't show bacteriolytic activity against *B. subtilis* (Fig. 4a, b). In the preliminary test, we found that anti-*B. subtilis* activity of 29-aa abaecin required an amount of 1.7 µg (Fig. 4a, b). The antimicrobial activity of 29-aa abaecin was enhanced in a combination treatment with cecropin B. No antimicrobial activity was observed by cecropin B when 0.125 µg of the peptide was treated (Fig. 4a, b), however, the antimicrobial activity of abaecin, at the dose of 1.7 µg, was increased by 40% when combined with the cecropin B (0.125 µg) (Fig. 4a, b). The antibacterial activity of abaecin was further enhanced by the increase of cecropin B quantity up to 0.25 µg (Fig. 4a, b). These results are consistent with the previous reports that the functional interaction of abaecin with other pore-forming peptides, such as cecropin A, stomoxyn and hymenoptaecin, at their sublethal concentrations, reciprocally potentiates bacteriolytic activity by increasing membrane permeabilization [13, 31, 43].

Taken all together, 6xHisSUMO-tagged abaecin expressed in a soluble form in *E. coli* caused no toxicity to the host cells, and the tag was properly recognized and removed by sumoase, but the cleaved abaecin was 29-aa long in size, not 34-aa long. However, the purified 29-aa-long abaecin retained its antimicrobial activity and the activity was potentiated by co-treatment with a pore-forming antimicrobial peptide, cecropin B.

Discussion

The limits of conventional antibiotics in treating pathogenic microbes and the increasing prevalence of antibiotic-resistant pathogens have led to the exploration of viable alternatives, including antimicrobial peptides. Only a limited number of antibiotics are available for clinical use, and they have similar activity spectrum and action mode [44, 45]. AMPs have distinct advantages over antibiotics such as remarkable structural and functional diversity, and immunomodulatory activity. Some AMPs show a broad range of action, which can be effectively used to treat multi-microbial infections including both Gram-positive and Gram-negative bacteria [45, 46].

Currently, no antibiotic peptides are available for clinical use. However, a number of AMPs are under clinical trials and development, and their applications are not limited to directly killing pathogens: e.g. pexiganan (for the treatment of bacterial infections, to diabetic foot ulcers), omiganan (catheter infections and rosacea), hLF-11 (bacterial and fungal infections in immunocompromised stem cell transplantation), novexatin (fungal infections), CZEN-002 (vaginal

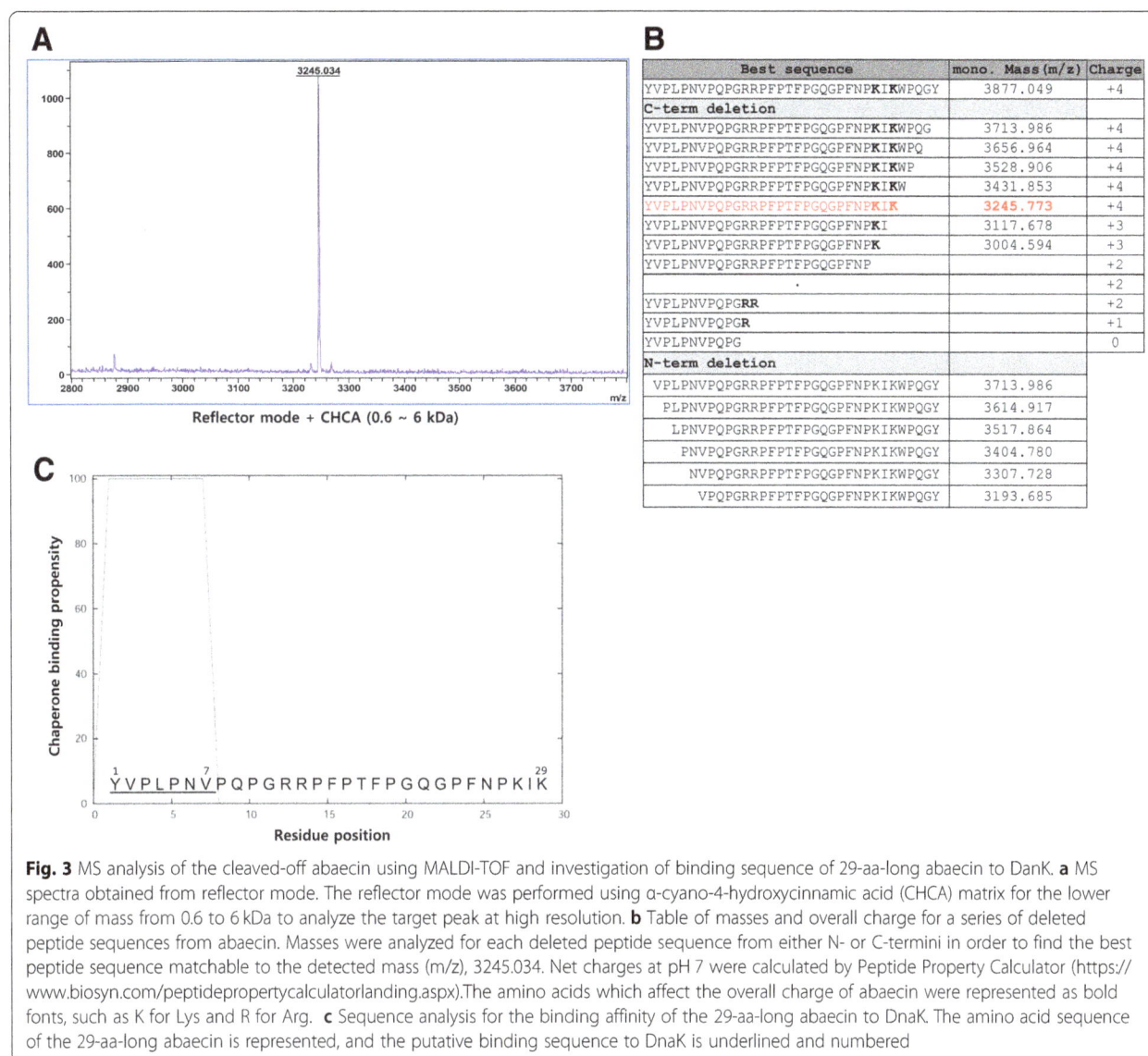

Fig. 3 MS analysis of the cleaved-off abaecin using MALDI-TOF and investigation of binding sequence of 29-aa-long abaecin to DanK. **a** MS spectra obtained from reflector mode. The reflector mode was performed using α-cyano-4-hydroxycinnamic acid (CHCA) matrix for the lower range of mass from 0.6 to 6 kDa to analyze the target peak at high resolution. **b** Table of masses and overall charge for a series of deleted peptide sequences from abaecin. Masses were analyzed for each deleted peptide sequence from either N- or C-termini in order to find the best peptide sequence matchable to the detected mass (m/z), 3245.034. Net charges at pH 7 were calculated by Peptide Property Calculator (https://www.biosyn.com/peptidepropertycalculatorlanding.aspx).The amino acids which affect the overall charge of abaecin were represented as bold fonts, such as K for Lys and R for Arg. **c** Sequence analysis for the binding affinity of the 29-aa-long abaecin to DnaK. The amino acid sequence of the 29-aa-long abaecin is represented, and the putative binding sequence to DnaK is underlined and numbered

candidiasis), LL-37 (wound healing), PXL01 (prevention of post-surgical adhesion formulation), Iseganan (oral mucositis), and PAC-113 (oral candidiasis) [45].

Despite their potential, the development and availability of AMPs for clinical use is met with several challenges. Primarily, the high production cost and low scalability of chemical peptide synthesis prevent the widespread development and adoption of AMPs as a viable clinical treatment [6]. Biological production using recombinant prokaryotic systems is a viable alternative to chemical synthesis, but issues such as toxicity to host cells, degradation of the product by protease, and low yield must be addressed [47].

In this study, a SUMO tagging system was used to prevent toxicity of the expressed antimicrobial peptide to host cells. The commonly used fusion carriers such as thioredoxin (12 kDa) and glutathione-S-transferase (GST, 26 kDa) have several advantages associated with increased solubility, promotion of proper folding and prevention of toxicity of AMPs [9, 46, 47]. However, GST increases the relative molecular weight ratio of carrier proteins to the peptides, which leads to low AMP yields. Also, several GST fused-AMPs expressed in *E.coli* showed proteolytic activity, resulting in inefficient or failed AMP productions. Thioredoxin is more favorably used for peptide production than the GST due to its small size, allowing high peptide yield attributed to the high peptide-to-carrier ratio [47]. However, the proteases used for the release of attached peptides from the carriers are more expensive and more sensitive to pH and chaotropes when compared to SUMO protease [47]. Moreover, AMPs tagged to aggregation-promoting carriers such as PurF fragment, PaP3:30 and ketosteroid isomerase have shown toxicity to host bacterial cells [9, 48].

A

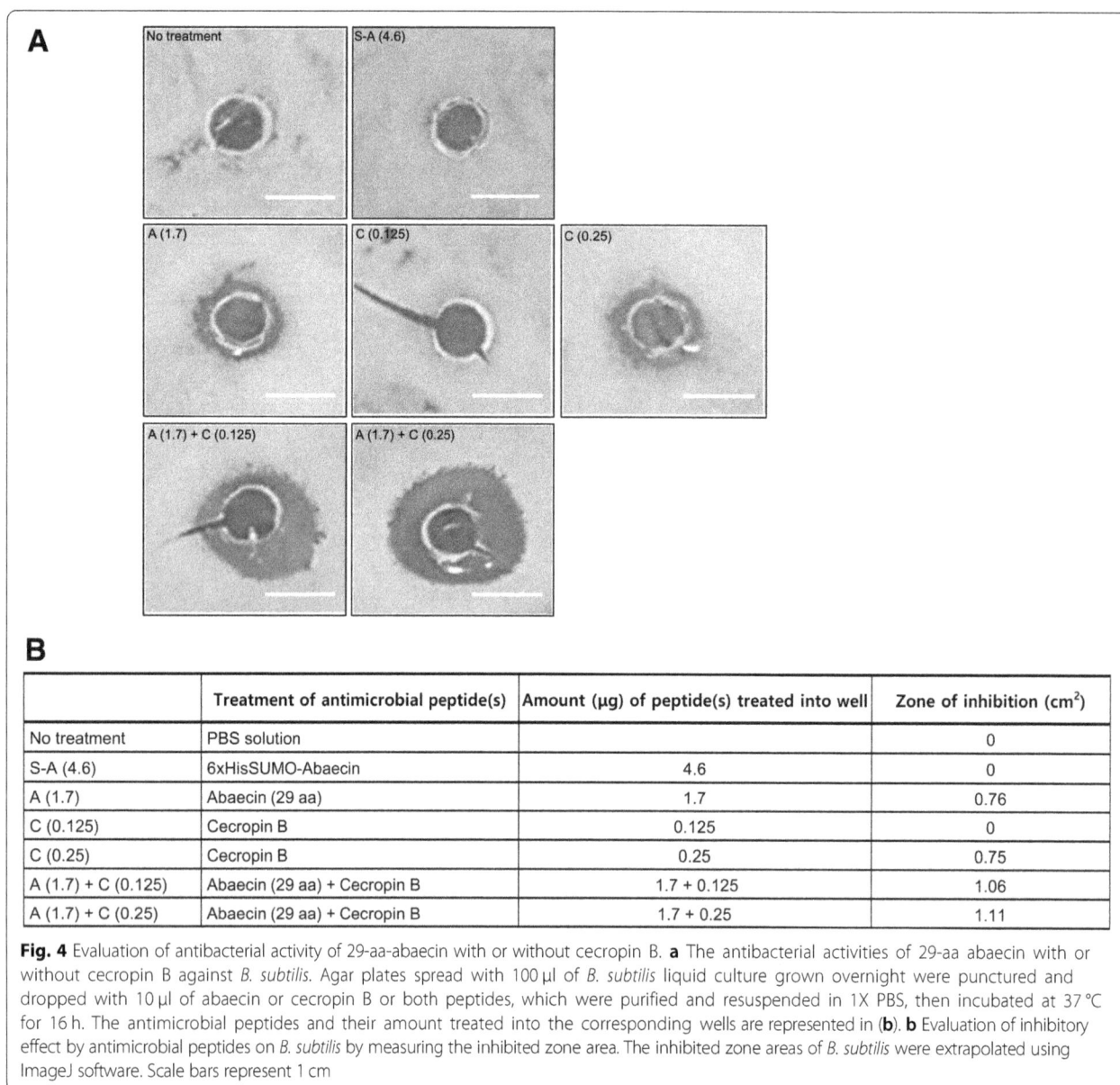

B

	Treatment of antimicrobial peptide(s)	Amount (µg) of peptide(s) treated into well	Zone of inhibition (cm²)
No treatment	PBS solution		0
S-A (4.6)	6xHisSUMO-Abaecin	4.6	0
A (1.7)	Abaecin (29 aa)	1.7	0.76
C (0.125)	Cecropin B	0.125	0
C (0.25)	Cecropin B	0.25	0.75
A (1.7) + C (0.125)	Abaecin (29 aa) + Cecropin B	1.7 + 0.125	1.06
A (1.7) + C (0.25)	Abaecin (29 aa) + Cecropin B	1.7 + 0.25	1.11

Fig. 4 Evaluation of antibacterial activity of 29-aa-abaecin with or without cecropin B. **a** The antibacterial activities of 29-aa abaecin with or without cecropin B against *B. subtilis*. Agar plates spread with 100 µl of *B. subtilis* liquid culture grown overnight were punctured and dropped with 10 µl of abaecin or cecropin B or both peptides, which were purified and resuspended in 1X PBS, then incubated at 37 °C for 16 h. The antimicrobial peptides and their amount treated into the corresponding wells are represented in (**b**). **b** Evaluation of inhibitory effect by antimicrobial peptides on *B. subtilis* by measuring the inhibited zone area. The inhibited zone areas of *B. subtilis* were extrapolated using ImageJ software. Scale bars represent 1 cm

The SUMO tag has many similar advantages to the thioredoxin or/and GST systems including high yield resulting from the high ratio of peptide-to-tagging protein, enhanced solubility, and no toxicity to host cells, but it also has unique advantages over other tags. Sumoase can recognize the tertiary structure of SUMO and cleave it from substrates with no attachment of unwanted amino acids to peptides [24–26]. The precise cleavage of target proteins from SUMO fusions by sumoase is confirmed by comparing the molecular mass and N-terminal amino acid sequence of the released peptides to the corresponding synthetic peptides using matrix-assisted laser desorption/ionization time-of-flight mass spectrometry (MALDI-TOF) and Edman degradation. More detailed information about the current tagging and expression systems for the

production of antimicrobial peptides in *E.coli* is included in Additional file 2: Table S1.

In this study, we confirmed the proper cleavage of abaecin from SUMO using MALDI-TOF, but there was 5-aa-long C-terminal deletion from the abaecin. In contrast to our expectation, the attached SUMO tag (overall charge − 5) didn't protect the abaecin (overall charge + 4, Fig. 3b) via electrostatic interaction from endogenous proteases in an *E.coli* system. Small size peptides with high cationic content are highly susceptible to proteolytic attack in *E.coli* as shown in Piers et al.'s study [10]. In the study, direct expression of small size cationic peptide such as human neutrophil peptide 1 (HNP-1, a kind of defensin) wasn't successful, while its transcript was detected. Even when the HNP-1 and other cationic peptide, synthetic cecropin/melittin hybrid (CEME), were fused to GST, the fusion proteins were proteolyzed. However, the proteolysis wasn't observed when anionic pre-pro defensin sequence was inserted between the GST and the cationic peptides. Based on our analysis, a positively charged patch created by two Arg (R) at 12th and 13th positions (Fig. 3b) seemed to not be recognized by endogenous proteases possibly due to charge neutralization effect by the electrostatic interaction with SUMO, while the second patch created by two Lys (K) at 28th and 30th positions (Fig. 3b) seemed subjected to proteolytic attack. Another possibility of the C-terminal deletion could be a consequence of excessive sonication, resulting in denaturation of the fusion protein and breakdown at the C-terminus. Also, it can't be ruled out that the sumoase could aberrantly cleave the abaecin. It is thought that sumoase is highly specific and active in a broad range of conditions; it cleaves effectively in a wide range of pH (5.5–10.5) and temperatures (4–37 °C), and even in detergents such as 2 M urea or 20 mM DTT or β-mercaptoethanol [25]. However, the cleavage study at the junction between SUMO and a partner protein showed that the cleavage didn't work when the first amino acid of the partner protein was Pro (P) [25]. Also, the introduction of a stretch of several Try (W) residues close to the cleavage site caused random cleavage within the fusion protein due to an inaccessibility of sumoase by steric hindrance, which, however, was relieved by addition of 1 M urea and led to the release of correct size peptide [26]. Considering the high content of proline in abaecin, the repeated kinks generated by 10 prolines in the abaecin could create a steric hindrance when attached to SUMO, which would lead an unexpected random cleavage. To investigate these issues mentioned above, SUMO could be further modified in a way which introduces additional glycine resides at the cleavage junction to release steric hindrance. In our unpublished study, cleavage of SUMO-cecropin B at the junction by sumoase happened only after introduction of additional glycine residues. Also, genetically modified SUMO which has more negative charge could provide a better protection to cationic peptides.

Although the C-terminus of abaecin sustained a 5-aa deletion after sumoase cleavage, the 29-aa-long abaecin showed antimicrobial activity against *B. subtilis* in the functionality study, while the SUMO-abaecin fusion protein had no bacteriolytic activity as shown in Fig. 4. Furthermore, the bacteriolytic activities were consistent with previous reports in which the antimicrobial activity of abaecin alone was further enhanced with combinatory treatment with other pore-forming AMPs [31, 43]. Although abaecin has high binding efficacy to DnaK like other proline-rich DnaK-binding AMPs including metalnikowins, metchnikowins, onocin Onc72, apidaecin

Api88, drosocin and pyrrhocoricin, the abaecin hasn't a conserved motif, YL/IPRP [43]. So we analyzed both abaecin amino acid sequences such as 34-aa and 29-aa-long abaecins for the functional binding sequence to DnaK using limbo server and found that the full length abaecin has two binding sequence sites: one is at N-terminus (^{1}YVPLPNV^7, score 1.7) and the other one is at C-terminus (^{25}NPKIKWP^{31}, score 2.8). So we assume that the reason why the 29-aa long abaecin still has antimicrobial activity (Fig. 4) is due to the presence of N-terminal binding sequence to DnaK (Fig. 3c).

The main purpose of this study was to introduce a new expression platform for the production of antimicrobial peptides in *E.coli* system in a way that does not harm the host cell. Abaecin was chosen as a reference AMP because it targets the intracellular molecule DnaK, prokaryotic heat shock protein 70 (Hsp70), and the binding of abaecin to DnaK can cause protein metabolism to be compromised in *E. coli*, resulting in host cell death [43]. The 6xHisSUMO-abaecin fusion protein, however, did not exhibit toxicity to *E. coli* host cells. The transformed *E. coli* cells grew in liquid culture at the similar growth rate to untransformed ones (Fig. 1c). Furthermore, the purified fusion protein showed no anti-bactericidal activity against *B. subtilis* (Fig. 4a, b). The normal growth of the transformed *E. coli* and the lack of antimicrobial activity of the fusion protein to *B. subtilis* proved that the lethal activity of the AMP to the host cells can be properly shielded by SUMO but its activity can be restored by the release from the tag. The antimicrobial activity of AMPs is generally determined by the hydrophobicity and the net positive charge. As reported in the previous study, the electrostatic interactions between the positive charges (between +4 and +6) of AMPs and negatively charge residues within SUMO (overall charge −5) seem to play a key role in neutralizing the bacteriolytic activity and protecting AMPs from degradation by endogenous proteases [26]. In this regard, although the protection of abaecin from proteolytic degradation by SUMO failed, the C-terminal deleted abaecin still had antimicrobial activity. Furthermore, the toxicity of abaecin was, in some way, successfully prevented by the SUMO tag via the presumed electrostatic interaction.

The SUMO tagging system is also advantageous because it has no disulfide bond. Due to the reducing environment of the *E. coli* cytoplasm, proteins that require disulfide bonds fail to achieve their active forms and ultimately form inclusion bodies or are degraded by proteases. To achieve and maintain their active forms, such proteins need to be redirected to periplasmic space or secreted, which are energy-consuming and could potentially lower target protein yield. However, the SUMO tagging system does not require the additional steps because it only contains one cysteine in its entire 96 amino acid sequence.

Overall, the SUMO-abaecin fusion proteins were detected more in soluble fractions than insoluble ones (Fig. 2a, b) with no toxicity to the host cells. There have been many reports that SUMO fusions increased the solubility of difficult-to-express proteins in *E.coli*, such as GFP, metallo-protease (MMP13) [25] and severe acute respiratory syndrome coronavirus (SARS CoV) proteins including 3CL protease, nucleocapsid protein and spike C protein [49, 50]. However, in contrast to antimicrobial peptides, those proteins are not toxic to the host cells. To conclusively distinguish the effect of SUMO on the solubility of the fusion protein, the expression of abaecin alone would need to be compared with SUMO-abaecin fusion protein, which, however, was not examined in this study due to technical issues such as possible low stability and potential toxicity to host cells. As mentioned above, the direct expression of the small size and high cationic content of AMPs are highly susceptible to endogenous proteolysis [10]. Furthermore, the expression of abaecin alone could be lethal to host cells because the peptide inhibits intracellular DnaK [43], which is a central organizer of the chaperone network in *E.coli*. The chaperone protein interacts with ~ 700 cytosolic proteins. Among them, ~ 180 proteins are relatively prone to be aggregation and rely extensively on DnaK during and after their initial folding [51].

The translation efficiency of codon-optimized heterologous sequences was evaluated in this study through partial or whole optimization in order to find the most accommodating nucleotide sequence in *E.coli* system. Codon usage was adjusted according to the codon usage preference of a gene, *psb*A, which is highly expressed in prokaryotic systems, and then three different combinations of codon-optimzed sequences were created. The codon adjustment was performed in a way to increase the compatibility between 5' UTR of the *psb*A promoter and the 5' coding region of the fusion gene, with an increase of AT content of the SUMO sequence to 63.5% from 59.5%. In contrast to our expectation, the codon optimization of N-terminal SUMO of the fusion protein failed to improve the expression level over the non-optimized counterpart. As seen in the Fig. 2a, b, the native sequence of SUMO performed better than its corresponding codon-optimized sequence for the expression of the fusion proteins. The expression level of 6xHisSUMO-abaecin (native – codon-optimized) was 2.8 or 3.5 times higher than that of 6xHisSUMO-abaecin (codon-optimized - native) or 6xHisSUMO-abaecin (codon-optimized – codon-optimized), respectively (Fig. 2b). This is likely due to the stability of the transcribed mRNA of the native SUMO sequence, which may be relatively higher than the stability of other codon-optimized sequences [52]. Another possible explanation is that the compatibility of the 5' UTR of the promoter with the 5' coding sequence of the codon-optimized SUMO could be compromised, resulting in an unstable or inefficient translational initiation complex [53, 54]. It is generally thought that translational efficiency is influenced by the efficiency of the formation of translational initiation complex and thus the first ~ 30–50 codons are considered more important than the rest of the sequence [55, 56]. Therefore, the marginal difference of expression level between codon-optimized SUMO fused native and codon-optimized abaecins could be a conseqeunce of the inefficient formation of the translational initiation complex.

Recent studies have found that the reciprocal functional interaction of abaecin with pore-forming peptides occurred not only with the peptides co-expressed in the same species, but with ones from other species [13, 31, 43]. Pores created in the membrane by the pore-forming peptide allow abaecin to access its intracellular target, DnaK. The inhibited heat shock proteins compromise protein metabolism, and the damaged pores remain unrepaired, thus allowing abaecin even greater access to its target. Abaecin's ability to increase membrane permeabilization consequently reduces the minimal inhibitory concentrations of both abaecin and other pore-forming peptides in a reciprocal manner [31, 43]. Likewise, the activity of the pore-forming peptide cecropin B from *Hyalophora cecropia* was potentiated in a combinatorial treatment with abaecin (Fig. 4), showing that abaecin can be used with diverse pore-forming peptides to inhibit bacteria which are renitent to conventional antibiotics.

One of the challenges that the application of AMPs presents is the demand for effective and patient-friendly delivery system, particularly, for the patients with chronic diseases. Currently, most AMPs under clinical development are designed to target local infections using a topical formulation. Only a few AMPs are being developed for systemic delivery [45]. Although an oral delivery system is most likely preferred due to straightforward administration, AMPs require special consideration because peptides are rapidly broken down in the gastrointestinal tracts due to the high concentration of proteases and high acidity. In this respect, edible plants can be used as a delivery platform, by which peptide based drugs bioencapsulated within the plant cells can be protected from the harsh environment of the gastrointestinal tract. But the drugs can be released into the intestine by the break-down of the cell walls by cellulolytic bacteria. The plant expression and delivery system can also eliminate the concern of endotoxin contamination, which causes fatal septic shock to recipients. Furthermore, the oral delivery of peptides by edible plant cells eliminates the need for expensive downstream purifications, reducing product cost and benefiting patients [18, 20, 21]. In our future study, we will evaluate the efficacy of the plant expression system as well.

Conclusions

Our SUMO tagging expression system showed an applicable method for the production of AMPs in *E. coli*, which although couldn't completely protect abaecin tested in this study from endogenous proteolytic degradation, did increase solubility and prevented toxicity of abaecin to host cells. Furthermore, the released AMP from the SUMO tag retained its antimicrobial activity and showed enhanced antimicrobial effects by combinatorial application with pore-forming peptides. Therefore, further studies for the enhancement of stability and translational efficiency of the SUMO-fused AMP will make the use of AMPs more affordable for clinics and patients.

Methods

DNA fragment amplification and synthesis

The recombinant protein/peptide coding sequences and DNA elements were designed referencing the deposited sequences in the National Center for Biotechnology Information (NCBI) or previous reports. The GenBank accession numbers for the sequences are as follows: Abaecin, NM_001011617.1; SUMO, NM_003352.4; Cecropin B, M34924.1. Codon-optimized nucleotide sequences such as abaecin and SUMO, and the native sequences of Prrn16, *aad*A and TrrnB deduced from a previous report ([57], AF327719.1) were synthesized by Macrogen (South Korea). The sequences of P*psb*A/5' UTR ([58], EU520589.1) and T*psb*A ([59], AY442171.1) were deduced from previous reports and amplified using tobacco chloroplast genomic DNA (Z00044.2) as a template by PCR. Codon optimizations were performed according to a previous study [33].

The primers used for PCR amplification of P*psb*A, T*psb*A, *trn*A and *trn*I DNA fragments are as follows: P*psb*A-F, 5'-CCCGGGCAACCCACTAGCATATC-3'; P*psb*A-R, 5'-C CTCCTATAGACTAGGCCAGGATCTAGATTACATATG AAAATCTTGGTTTATTTAATCATCAGGG-3'; T*psb*A-F, 5'-CCCTGATGATTAAATAAACCAAGATTTTCATATG TAATCTAGATCCTGGCCTAGTCTATAGGAGG-3' and T*psb*A-R, TCGAATATAGCTCTTCTTTCTTATTTCAATG ATATTATT-3'. *trn*A-F, 5'- GGGGAAGAATTCGGGGATA TAGCTCAGTTGGTAG-3'; *trn*A-R, 5'- GAAAAAGGT ACCTGGAGATAAGCGGACTCGAACC-3'; *trn*I-F, 5'- GG GGAAAAGCTTGGGCTATTAGCTCAGTGGTAG-3' and *trn*I-R, 5'- GAAAAAGTCGACTGGGCCATCC TGGACTTGAAC-3'.

Construction of expression vector

P*psb*A-T*psb*A DNA fragments amplified by overlapping PCR were treated by *Ase*I and *Sal*I and inserted into pUC19 backbone plasmid restricted with *Nde*I and *Sal*I. The ligated plasmids were further digested with *Bam*HI and *Eco*RI. The DNA fragments of Prrn16-*aad*A-TrrnB

were synthesized by Macrogen and digested with *Bam*HI and *Eco*RI, and were then ligated with the intermediate vector. The pUC19 vectors containing Prrn16-*aad*A-TrrnB-P*psb*A-T*psb*A were combined with *trn*I DNA fragments using *Hind*III and *Sal*I, and then *trn*A fragments were inserted using *Kpn*I and *Eco*RI. The newly constructed vector, pKSEC1, was used for the expression of recombinant fusion gene, 6xHisSUMO-abaecin, which was inserted into the vector by *Xba*I and *Nde*I.

Expression of 6xHisSUMO-abaecin in *E. coli*

The *E. coli* BL21 transformed with 6xHisSUMO-abaecin:pKSEC1 were grown in 4 L Terrific Broth (Sigma-Aldrich, USA) containing 100 μg/mL of ampicillin and 50 μg/mL of spectinomycin at 37 °C for 3 h at a speed of 200 rpm and the overnight culture was further grown at 18 °C overnight. The cells were collected by centrifugation (8000 rpm, 4 °C, 3 min) and resuspended in 140 mL of buffer including 20 mM Tris, 300 mM sodium chloride and 5 mM Imidazol. Then the cells were sonicated by a cycle of 5 s on and 15 s off (SONICS VC505, USA) for 40 m, and soluble and insoluble fractions were separated using centrifugation (13,000 rpm, 4 °C, 1 h) then quantified using Bradford (Sigma-Aldrich, USA). The quantified protein samples were mixed with 2X Laemmli sample buffer (Bio-Rad, USA) and heated at 95 °C for 5 min. The heated proteins were run on SDS-PAGE and the separated proteins were blotted onto PVDF membrane. The membrane was blocked with 1X TBS-T (0.1% Tween 20 and 5% skim milk) for 1 h at room temperature. After that, anti-His antibody (Santa Cruz, USA), diluted 1:1000 in the blocking solution, was incubated at 4 °C for 16 h. The membrane was washed with 1X TBS-T buffer for 5 min thrice. Secondary antibody (goat anti-rabbit IgG-HRP, Santa Cruz, USA), diluted 1: 5000 in the blocking solution, was incubated at room temperature for 1 h. The washed membrane was subject to ECL buffer for development using C-DiGit Blot Scanner (Li-Cor, USA).

Purification of the recombinant abaecin

The *E. coli* BL21 transformed with 6xHisSUMO-abaecin:pKSEC1 were grown and treated as described above. After sonication, the supernatant was filtered through filter paper (Advantec, Japan) and 0.45 μm syringe filter (Minisart syringe filter, Sartorius stedim biotech, Germany). The filtrate was incubated with His 60 Ni Suferflow resin (Takara, Japan) by inverting slowly for 1 h at 4 °C. After the binding incubation, the column was washed with 50 column volumes of wash buffer (20 mM Tris, 300 mM sodium chloride, 50 mM Imidazol, pH 8.0). 6xHisSUMO-abaecin fusion proteins bound to Ni resins were eluted approximately with 5 column volumes of elution buffer (20 mM Tris, 300 mM sodium chloride, 250 mM Imidazol, pH 8.0). Recombinant abaecin was isolated from 6xHisSUMO by treatment of SUMO protease

(Enzynomics, South Korea) at 30 °C for 6 h. To confirm the cleavage, SDS-PAGE was performed using NuPAGE™ 4–12% Bis-Tris Gel (Invitrogen, USA) with MES buffer.

Mass spectrometry

For MS analysis, the eluted samples were applied onto Anchor-Chip 600 targets (Bruker Daltonik, Germany) using α-cyano-4-hydroxycinnamic acid (CHCA) (Sigma, C2020) as matrix according to the manufacturer's recommendations. MS measurement was performed on an autoflex II TOF/TOF. Mass spectrum was acquired in the reflector mode using CHCA matrix. The mass analysis was carried out by Life Science Laboratory. Co (http://www.emass.co.kr, South Korea).

Antimicrobial activity

Agar diffusion assay was performed to evaluate the antimicrobial activity of abaecin. One hundred microliter of *B. subtilis* liquid cultured overnight was plated on LB agar medium and punctured using a tip then purified abaecin was dropped with or without cecropin B that was expressed and purified in the same way as described above in our lab. The plates were grown at 37 °C for 16 h and the zone areas, where the growth of *B. subtilis* was inhibited by abaecin, were measured using ImageJ software.

Additional files

Additional file 1: Figure S1. Sequences of three different codon-optimized 6xHisSUMO-abaecin. (PDF 163 kb)

Additional file 2: Table S1. Expression of antimicrobial peptides using *E.coli* and their subsequent release and purification. (PDF 73 kb)

Abbreviations

AMP: Antimicrobial peptide; Hsp70: Heat shock protein 70; SUMO: Small ubiquitin-related modifier 1

Acknowledgements

Authors thank Coleman Pinkerton and Antonio Lamb for English language editing of this manuscript.

Funding

This work was supported by a Research Grant of Andong National University.

Authors' contributions

KCK and SYK conceived the project, designed experiments, interpreted data and wrote the manuscript. DSK cloned 6xHisSUMO-abaecin into pKSEC1 vector, expressed the fusion protein, purified, performed antimicrobial activity and wrote the manuscript. SWK constructed prototype vector, pKSEC1, cloned cecropin B into the vector and verified the expression of cecropin B in transformed *E. coli* using western blot. JMS advised and supported antimicrobial activity against *B. subtilis*. All authors read and approved the final manuscript.

Ethics approval and consent to participate

Not applicable

Consent for publication

Not applicable

Competing interests

The authors declare that they have no competing interests.

Author details

[1]Department of Biological Sciences, Andong National University, Andong, South Korea. [2]Department of Global Medical Science, Health & Wellness College, Sungshin University, Seoul, South Korea. [3]MicroSynbiotiX Ltd, 11011 N Torrey Pines Rd Ste. #135, La Jolla, CA 92037, USA.

References

1. O'Neill J. Tackling drug-resistant infections globally: Final report and recommendations. 2016. HM Government and Welcome Trust: UK. 2018.
2. Lewis K. Platforms for antibiotic discovery. Nat Rev Drug Discov. 2013;12: 371–87.
3. Fjell CD, Hiss JA, Hancock REW, Schneider G. Designing antimicrobial peptides: form follows function. Nat Rev Drug Discov. 2011;11:37–51.
4. Li J, Koh J-J, Liu S, Lakshminarayanan R, Verma CS, Beuerman RW. Membrane Active Antimicrobial Peptides: Translating Mechanistic Insights to Design. Front Neurosci. 2017;11:73.
5. Latham PW. Therapeutic peptides revisited. Nat Biotechnol. 1999;17:755–7.
6. Pichereau C, Allary C. Therapeutic peptides under the spotlight. Eur Biopharm Rev. 2005;5:88–91.
7. Chen YQ, Zhang SQ, Li BC, Qiu W, Jiao B, Zhang J, et al. Expression of a cytotoxic cationic antibacterial peptide in *Escherichia coli* using two fusion partners. Protein Expr Purif. 2008;57:303–11.
8. Hsu K-H, Pei C, Yeh J-Y, Shih C-H, Chung Y-C, Hung L-T, et al. Production of bioactive human alpha-defensin 5 in Pichia pastoris. J Gen Appl Microbiol. 2009;55:395–401.
9. Li Y. Carrier proteins for fusion expression of antimicrobial peptides in Escherichia coli. Biotechnol Appl Biochem. 2009;54:1–9.
10. Piers KL, Brown MH, Hancock RE. Recombinant DNA procedures for producing small antimicrobial cationic peptides in bacteria. Gene. 1993; 134:7–13.
11. Zhang L, Falla T, Wu M, Fidai S, Burian J, Kay W, et al. Determinants of recombinant production of antimicrobial cationic peptides and creation of peptide variants in bacteria. Biochem Biophys Res Commun. 1998; 247:674–80.
12. Luiz DP, Almeida JF, Goulart LR, Nicolau-Junior N, Ueira-Vieira C. Heterologous expression of abaecin peptide from *Apis mellifera* in Pichia pastoris. Microb Cell Fact. 2017;16:76.
13. Li L, Mu L, Wang X, Yu J, Hu R, Li Z. A novel expression vector for the secretion of abaecin in Bacillus subtilis. Braz J Microbiol. 2017;48:809–14.
14. Lee S-B, Li B, Jin S, Daniell H. Expression and characterization of antimicrobial peptides Retrocyclin-101 and Protegrin-1 in chloroplasts to control viral and bacterial infections. Plant Biotechnol J. 2011;9:100–15.
15. Liu Y, Kamesh AC, Xiao Y, Sun V, Hayes M, Daniell H, et al. Topical delivery of low-cost protein drug candidates made in chloroplasts for biofilm disruption and uptake by oral epithelial cells. Biomaterials. 2016;105: 156–66.
16. Gupta K, Kotian A, Subramanian H, Daniell H, Ali H. Activation of human mast cells by retrocyclin and protegrin highlight their immunomodulatory and antimicrobial properties. Oncotarget. 2015;6:28573–87.
17. Maliga P, Bock R. Plastid biotechnology: food, fuel, and medicine for the 21st century. Plant Physiol. 2011;155:1501–10.
18. Kwon K-C, Daniell H. Oral Delivery of Protein Drugs Bioencapsulated in Plant Cells. Mol Ther. 2016;24:1342–50.
19. Xiao Y, Kwon K-C, Hoffman BE, Kamesh A, Jones NT, Herzog RW, et al. Low cost delivery of proteins bioencapsulated in plant cells to human nonimmune or immune modulatory cells. Biomaterials. 2016;80:68–79.
20. Jin S, Daniell H. The Engineered Chloroplast Genome Just Got Smarter. Trends Plant Sci. 2015;20:622–40.

21. Kwon K-C, Daniell H. Low-cost oral delivery of protein drugs bioencapsulated in plant cells. Plant Biotechnol J. 2015;13:1017–22.

22. Verma D, Daniell H. Chloroplast vector systems for biotechnology applications. Plant Physiol. 2007;145:1129–43.

23. Dhingra A, Daniell H. Chloroplast genetic engineering via organogenesis or somatic embryogenesis. Methods Mol Biol. 2006;323:245–62.

24. Lee C-D, Sun H-C, Hu S-M, Chiu C-F, Homhuan A, Liang S-M, et al. An improved SUMO fusion protein system for effective production of native proteins. Protein Sci. 2008;17:1241–8.

25. Malakhov MP, Mattern MR, Malakhova OA, Drinker M, Weeks SD, Butt TR. SUMO fusions and SUMO-specific protease for efficient expression and purification of proteins. J Struct Funct Genomics. 2004;5:75–86.

26. Bommarius B, Jenssen H, Elliott M, Kindrachuk J, Pasupuleti M, Gieren H, et al. Cost-effective expression and purification of antimicrobial and host defense peptides in Escherichia coli. Peptides. 2010;31:1957–65.

27. Casteels P, Ampe C, Riviere L, Van Damme J, Elicone C, Fleming M, et al. Isolation and characterization of abaecin, a major antibacterial response peptide in the honeybee (Apis mellifera). Eur J Biochem. 1990;187:381–6.

28. Rees JA, Moniatte M, Bulet P. Novel antibacterial peptides isolated from a European bumblebee, Bombus pascuorum (Hymenoptera, Apoidea). Insect Biochem Mol Biol. 1997;27:413–22.

29. Choi YS, Choo YM, Lee KS, Yoon HJ, Kim I, Je YH, et al. Cloning and expression profiling of four antibacterial peptide genes from the bumblebee Bombus ignitus. Comp Biochem Physiol B Biochem Mol Biol. 2008;150:141–6.

30. Otvos L Jr. The short proline-rich antibacterial peptide family. Cell Mol Life Sci. 2002;59:1138–50.

31. Rahnamaeian M, Cytryńska M, Zdybicka-Barabas A, Vilcinskas A. The functional interaction between abaecin and pore-forming peptides indicates a general mechanism of antibacterial potentiation. Peptides. 2016;78:17–23.

32. Daniell H, Ruiz G, Denes B, Sandberg L, Langridge W. Optimization of codon composition and regulatory elements for expression of human insulin like growth factor-1 in transgenic chloroplasts and evaluation of structural identity and function. BMC Biotechnol. 2009;9:33.

33. Kwon K-C, Chan H-T, León IR, Williams-Carrier R, Barkan A, Daniell H. Codon Optimization to Enhance Expression Yields Insights into Chloroplast Translation. Plant Physiol. 2016;172:62–77.

34. Tangphatsornruang S, Birch-Machin I, Newell CA, Gray JC. The effect of different 3′ untranslated regions on the accumulation and stability of transcripts of a gfp transgene in chloroplasts of transplastomic tobacco. Plant Mol Biol. 2011;76:385–96.

35. Bock R. Engineering chloroplasts for high-level foreign protein expression. Methods Mol Biol. 2014;1132:93–106.

36. Daniell H, Lin C-S, Yu M, Chang W-J. Chloroplast genomes: diversity, evolution, and applications in genetic engineering. Genome Biol. 2016; 17:134.

37. Verma D, Samson NP, Koya V, Daniell H. A protocol for expression of foreign genes in chloroplasts. Nat Protoc. 2008;3:739–58.

38. Laqueyrerie A, Militzer P, Romain F, Eiglmeier K, Cole S, Marchal G. Cloning, sequencing, and expression of the apa gene coding for the Mycobacterium tuberculosis 45/47-kilodalton secreted antigen complex. Infect Immun. 1995; 63:4003–10.

39. Ozaki LS, Svec P, Nussenzweig RS, Nussenzweig V, Godson GN. Structure of the plasmodium knowlesi gene coding for the circumsporozoite protein. Cell. 1983;34:815–22.

40. Staab JF, Ferrer CA, Sundstrom P. Developmental expression of a tandemly repeated, proline-and glutamine-rich amino acid motif on hyphal surfaces on Candida albicans. J Biol Chem. 1996;271:6298–305.

41. Kirkland TN, Finley F, Orsborn KI, Galgiani JN. Evaluation of the proline-rich antigen of Coccidioides immitis as a vaccine candidate in mice. Infect Immun. 1998;66:3519–22.

42. Van Durme J, Maurer-Stroh S, Gallardo R, Wilkinson H, Rousseau F, Schymkowitz J. Accurate prediction of DnaK-peptide binding via homology modelling and experimental data. PLoS Comput Biol. 2009;5:e1000475.

43. Rahnamaeian M, Cytryńska M, Zdybicka-Barabas A, Dobslaff K, Wiesner J, Twyman RM, et al. Insect antimicrobial peptides show potentiating functional interactions against Gram-negative bacteria. Proc Biol Sci. 2015;282:20150293.

44. Czaplewski L, Bax R, Clokie M, Dawson M, Fairhead H, Fischetti VA, et al. Alternatives to antibiotics-a pipeline portfolio review. Lancet Infect Dis. 2016;16:239–51.

45. Mahlapuu M, Håkansson J, Ringstad L, Björn C. Antimicrobial Peptides: An Emerging Category of Therapeutic Agents. Front Cell Infect Microbiol. 2016; 6:194.

46. Dryden MS. Complicated skin and soft tissue infection. J Antimicrob Chemother. 2010;65(Suppl 3):iii35–44.

47. Li Y. Recombinant production of antimicrobial peptides in Escherichia coli: a review. Protein Expr Purif. 2011;80:260–7.

48. Kim H-K, Chun D-S, Kim J-S, Yun C-H, Lee J-H, Hong S-K, et al. Expression of the cationic antimicrobial peptide lactoferricin fused with the anionic peptide in Escherichia coli. Appl Microbiol Biotechnol. 2006;72:330–8.

49. Zuo X, Mattern MR, Tan R, Li S, Hall J, Sterner DE, et al. Expression and purification of SARS coronavirus proteins using SUMO-fusions. Protein Expr Purif. 2005;42:100–10.

50. Butt TR, Edavettal SC, Hall JP, Mattern MR. SUMO fusion technology for difficult-to-express proteins. Protein Expr Purif. 2005;43:1–9.

51. Calloni G, Chen T, Schermann SM, Chang H-C, Genevaux P, Agostini F, et al. DnaK functions as a central hub in the E. coli chaperone network. Cell Rep. 2012;1:251–64.

52. Wu X, Jörnvall H, Berndt KD, Oppermann U. Codon optimization reveals critical factors for high level expression of two rare codon genes in Escherichia coli: RNA stability and secondary structure but not tRNA abundance. Biochem Biophys Res Commun. 2004;313:89–96.

53. Nakamura M, Hibi Y, Okamoto T, Sugiura M. Cooperation between the chloroplast psbA 5′-untranslated region and coding region is important for translational initiation: the chloroplast translation machinery cannot read a human viral gene coding region. Plant J. 2016;85:772–80.

54. Stenström CM, Isaksson LA. Influences on translation initiation and early elongation by the messenger RNA region flanking the initiation codon at the 3′ side. Gene. 2002;288:1–8.

55. Tuller T, Carmi A, Vestsigian K, Navon S, Dorfan Y, Zaborske J, et al. An evolutionarily conserved mechanism for controlling the efficiency of protein translation. Cell. 2010;141:344–54.

56. Chen GF, Inouye M. Suppression of the negative effect of minor arginine codons on gene expression; preferential usage of minor codons within the first 25 codons of the Escherichia coli genes. Nucleic Acids Res. 1990;18: 1465–73.

57. Marx CJ, Lidstrom ME. Development of improved versatile broad-host-range vectors for use in methylotrophs and other Gram-negative bacteria. Microbiology. 2001;147 Pt 8:2065–2075.

58. Farran I, Río-Manterola F, Iñiguez M, Gárate S, Prieto J, Mingo-Castel AM. High-density seedling expression system for the production of bioactive human cardiotrophin-1, a potential therapeutic cytokine, in transgenic tobacco chloroplasts. Plant Biotechnol J. 2008;6:516–27.

59. Lin C-H, Chen Y-Y, Tzeng C-C, Tsay H-S, Chen L-J. Expression of a Bacillus thuringiensis cry1C gene in plastid confers high insecticidal efficacy against tobacco cutworm - a Spodoptera insect. Bot Bull Acad Sinica. 2003;44:199–210.

12

Improvement and use of CRISPR/Cas9 to engineer a sperm-marking strain for the invasive fruit pest *Drosophila suzukii*

Hassan M. M. Ahmed[1,2], Luisa Hildebrand[1] and Ernst A. Wimmer[1*]

Abstract

Background: The invasive fruit pest *Drosophila suzukii* was reported for the first time in Europe and the USA in 2008 and has spread since then. The adoption of type II clustered regularly interspaced short palindromic repeats (CRISPR)/CRISPR-associated (Cas) as a tool for genome manipulation provides new ways to develop novel biotechnologically-based pest control approaches. Stage or tissue-specifically expressed genes are of particular importance in the field of insect biotechnology. The enhancer/promoter of the spermatogenesis-specific *beta-2-tubulin* (*β2t*) gene was used to drive the expression of fluorescent proteins or effector molecules in testes of agricultural pests and disease vectors for sexing, monitoring, and reproductive biology studies. Here, we demonstrate an improvement to CRISPR/Cas-based genome editing in *D. suzukii* and establish a sperm-marking system.

Results: To improve genome editing, we isolated and tested the *D. suzukii* endogenous promoters of the small nuclear RNA gene *U6* to drive the expression of a guide RNA and the *Ds heat shock protein 70* promoter to express *Cas9*. For comparison, we used recombinant Cas9 protein and in vitro transcribed gRNA as a preformed ribonucleoprotein. We demonstrate the homology-dependent repair (HDR)-based genome editing efficiency by applying a previously established transgenic line that expresses *DsRed* ubiquitously as a target platform. In addition, we isolated the *Ds_β2t* gene and used its promoter to drive the expression of a red fluorescence protein in the sperm. A transgenic sperm-marking strain was then established by the improved HDR-based genome editing.

Conclusion: The deployment of the endogenous promoters of the *D. suzukii U6* and *hsp70* genes to drive the expression of *gRNA* and *Cas9*, respectively, enabled the effective application of helper plasmid co-injections instead of preformed ribonucleoproteins used in previous reports for HDR-based genome editing. The sperm-marking system should help to monitor the success of pest control campaigns in the context of the Sterile Insect Technique and provides a tool for basic research in reproductive biology of this invasive pest. Furthermore, the promoter of the *β2t* gene can be used in developing novel transgenic pest control approaches and the CRISPR/Cas9 system as an additional tool for the modification of previously established transgenes.

Keywords: Cherry vinegar fly, Insect transgenesis, Molecular entomology, Pest management, Spotted Wing *Drosophila*

* Correspondence: ewimmer@gwdg.de
[1]Department of Developmental Biology,
Johann-Friedrich-Blumenbach-Institute of Zoology and Anthropology,
Göttingen Center for Molecular Biosciences, Georg-August-University
Göttingen, 37077 Göttingen, Germany
Full list of author information is available at the end of the article

Background

Native to East Asia [1], the cherry vinegar fly *D. suzukii*, also known as the Spotted Wing *Drosophila* (SWD) was reported for the first time in Europe, Spain and Italy, and the mainland USA in California in 2008 [1–3]. The pest has since then expanded its geographic distribution to include all of Europe as reported by the European Plant Protection Organization [2]. In the USA, the situation is as severe as in Europe. Four years after its first invasion in California, the SWD has been reported in more than 41 states [4]. By now, this invasive insect pest has also been reported further down in South America: for the first time between the years 2012 and 2013 in Brazil [5] and more recently also in Argentina in four localities [6].

The devastating fruit pest *D. suzukii* infests mainly soft-skinned as well as stone fruits with a wide host range spanning cultivated and wild plants [7]. In contrast to other *Drosophila* spp., the SWD is armoured with a sharp serrated ovipositor, which allows it to infest ripening and not only overripe or rotten fruits [8]. Earlier studies have shown that economic impact due to the infestation is in the order of millions of US dollar [9, 10]. Current control efforts mainly rely on heavy application of insecticides [11, 12], which is on the one hand not compatible with organic farming and prone to rapid emergence of insecticide resistance owning to the short generation time of this fly. And on the other hand, it is not safe, as the time between onset of infestation and harvest is very short and does not allow for a sufficiently long period post pesticide application. Other control strategies include the use of natural enemies such as parasitoids, predators, or pathogens [13], netting to cover the plants [14], and good cultural practices to minimise the source of infestation [15]. The sterile Insect technique (SIT) presents itself as an additional safe and effective pest management strategy. It provides a species-specific, environmentally sound pest control approach [16] and is compatible with other pest control strategies in integrated pest management (IPM) programs. The system has been proposed more than half a century ago and was used to successfully eradicate the tsetse fly from Zanzibar as well as the screw worm from Libya and the USA [17, 18]. It encompasses mass production of the target insect, removal of the females, and sterilization of males by ionizing radiation prior to release [16]. Using transposon-based germline transformation, many transgenic strategies have been developed to overcome some of the drawbacks of classical SIT. A transgene-based embryonic lethality system was developed for several dipterans including the model *D. melanogaster* and the cosmopolitan fruit pest *Ceratitis capitata* [19, 20]. The system relies on the ectopic expression of a pro-apoptotic gene during early embryonic stages, which leads to cell death and hence reproductive sterility [19]. The same system has

also been used for sexing, when the embryonic lethality was rendered female-specific by making use of the sex-specifically spliced intron of the *transformer* gene, which allows for elimination of females at the embryonic stage [20–22]. Furthermore, for monitoring the competitiveness of released males, sperm-marking systems were developed for a number of pest insects and diseases vectors by driving the expression of fluorescent protein during spermatogenesis [23–26].

Recently, a revolution in genome engineering was started by the application of the CRISPR/Cas system, which stands for type II clustered regularly interspaced short palindromic repeats CRISPR/CRISPR-associated. Respective sequences were first observed in bacterial genomes in 1987 [27]. Two decades later, researchers found an association between these repeated sequences and resistance of bacteria to bacteriophages [28] and showed that the bacteria use this system as an adaptive defence mechanism against invading DNA elements [29]. The system consists of the Cas9 effector endonuclease, the CRISPR RNA (*crRNA*), which confers specificity to Cas9, and the transactivating crRNA (*tracrRNA*), which facilitates maturation of *crRNAs* and the interaction with Cas9 protein for forming active RNP complexes [30, 31]. The *crRNA* and *tracrRNA* were fused together to generate a single chimeric gRNA that facilitated the use of the system [32]. The Cas9 endonuclease can easily be programmed to target and induce DNA double strands break (DSB) by replacing the 20 nucleotides (spacer) at the 5′ of the *crRNA* with 17–20 nucleotides (nt) complementary to the target of interest. The prerequisite for the RNP complex to unwind, bind, and induce DSB in the target DNA is a proto-spacer adjacent motif (PAM) immediately downstream of the 20 nt target sequence, which is NGG in the case of the most commonly used *Sp_Cas9* from *Streptococcus pyogenes* [31]. Similar to other programmable endonucleases such as Zinc finger nucleases (ZFNs) and Transcription activators like nucleases (TALENs), the role of Cas9 as a genome editing tool ends with the induction of a DSB. Repairing the genome - by either homology directed repair (HDR) or by non-homologous end joining (NHEJ) - is a function of the cell own DSB repair machinery, the stage of the cell at which the DSB is induced, and the availability of homologous DNA [32]. The system has rapidly been adopted as a genome engineering tool for many model and non-model organisms including zebrafish [33], mouse [34, 35], *Drosophila* [36], mosquitoes [37, 38], and human cell lines. The CRISPR/Cas9 system has also been used to induce chromosomal translocations in embryonic stem cells [39], and to engineer new balancer chromosomes in the nematode model *Caenorhabditis elegans* [40].

In the genetics power horse *D. melanogaster*, CRISPR/ Cas9 has been used and delivered in different forms: as helper plasmids, mRNA and gRNA, as well as a ribonucleoprotein complexes. Several promoters have been used to drive the expression of *Cas9* including germline-specific promoters of genes such as *nanos* and *vasa*, inducible promoters such as *heat shock protein 70 (hsp70)*, and promoters of ubiquitously expressed genes such as *Actin5C*. Systematic analysis of the three different promoters of the *small nuclear RNA (U6)* genes in *D. melanogaster* has shown that the *U6:3* promoter drives the strongest expression measured by gene editing events [41, 42].

In *Drosophila suzukii*, the CRISPR/Cas9 system has been used albeit with low efficiency to mutate the genes *white (w)* and *Sex lethal (Sxl)* using *D. melanogaster* promoters to drive the expression of *gRNA* and *Cas9* [43]. Another study reported on the use of pre-assembled a ribonucleoprotein complex (RNP) to induce mutations in the *white* gene [44]. The introduction of the mutations was in both studies based on NHEJ. The system has also been used to engineer by HDR a temperature sensitive mutation in the *Ds_transformer-2* gene (*Ds_ tra-2*) that leads to sex conversion. In this study a RNP complex in combination with RNA interference against the *Ds_lig4* gene was used and an HDR frequency of 7.3% was reported [45]. Furthermore, a RNP complex has also been used in a behavioural study of *D. suzukii* to knockout the gene that encodes the odorant receptor co-receptor (Orco) by HDR-mediated mutagenesis [46].

In applied insect biotechnology, CRISPR/Cas9 has become very popular particularly in the development of insect control strategies. One possible application for the system in SIT is the development of a reproductive sterility system that targets Cas9 to induce many DSBs at defined loci during spermatogenesis. This could mimic the desired effect of ionizing radiation in generating redundant sterility and at the same time overcome the random action of radiation affecting all organs, which reduces the overall fitness of the sterile males [47].

To restrict Cas9 activity to spermatogenesis, the isolation of a tissue-specific promoter is essential. The *Drosophila β2t* gene has been shown to code for a β-tubulin, which is expressed in a tissue-specific manner during spermatogenesis [48]. Its testes-specific expression makes it a good candidate for developmental studies related to reproductive biology and male germline development as well as pest control strategies. *Dm_β2t* is a TATA-less gene, which relies on an initiator element (Inr) as a core promoter with the testes-specific expression conferred by a 14 bp activator element called *β2 Upstream Element 1 (β2UE1)* [49]. Further elements required for the expression level are *β2UE2* at position − 25 and *β2DE1* at position + 60 [50]. Homologs of *Dm_*

β2t were identified in a number of insects including *Anopheles stephensi*, *Aedes aegypti*, *Ceratitis capitata*, *Anstrepha suspensa*, *Anastrepha ludens*, and *Bacterocera dorsalis* [23–26]. The upstream regulatory sequence has been used to drive the expression of fluorescent protein in the testes, which serves as a strategy for sex separation as well as for monitoring released males in SIT. In the major malaria vector *Anopheles gambiae*, the promoter of the *β2t* gene was used to drive the expression of the homing endonuclease *I-Ppol* during spermatogenesis. *I-Ppol* is a highly specific Homing Endonuclease Gene (HEG), which targets and cuts a conserved sequence within the *rDNA* on the X chromosome and thereby leads to X-chromosome shredding leaving mostly Y-chromosome bearing sperm functional, which results in sex-ratio distortion [51].

In this study, we present an improved CRISPR/Cas9-based genome engineering system for the invasive fruit pest *D. suzukii* and its application to edit a transgenic line generated using *piggyBac* germline transformation. Moreover, we report on the use of this editing system to generate a *D. suzukii* sperm marking line based on the *Ds_β2t* promoter driving the expression of *DsRed* in the testes.

Results

Improvement on CRISPR/Cas9 genome editing in *Drosophila suzukii*

In order to improve on the HDR-mediated genome editing based on CRISPR/Cas9-induced DSBs, we isolated endogenous polymerase II (*hsp70* gene) and polymerase III promoters (*U6* genes) from *D. suzukii* to drive *Cas9* or *gRNAs*, respectively. Searching for homologs of the *D. melanogaster heat shock protein 70 (hsp70)* gene, we identified the *D. suzukii Ds_hsp70* gene, cloned and sequenced 500 bp upstream of the ATG translation start codon and used this upstream sequence to drive the expression of *Cas9*.

First attempts using PCR to isolate the *U6* genes based on *D. suzukii* genome database sequences were not successful. The presence of three tandem copies obviously rendered the assembly inaccurate. Since *D. suzukii* is a close relative to *D. melanogaster*, we then tried to isolate the *U6* locus based on synteny cloning: we amplified and sequenced a 3.7 kbp fragment encompassing the *U6* locus. We identified three *U6* genes and refer to them in 5′ to 3′ direction as *U6a*, *U6b*, and *U6c* (Fig. 1a) to distinguish them from their *D. melanogaster* equivalents.

To test the efficiency of the endogenous *hsp70* and *U6* promoters in order to drive the expression of *Cas9* and *gRNA*, respectively, for mediating HDR-based genome editing, we used the embryonic line 06_F5M2 generated by *piggyBac* germline transformation as a target platform (Fig. 1b). This driver line can be used to express the

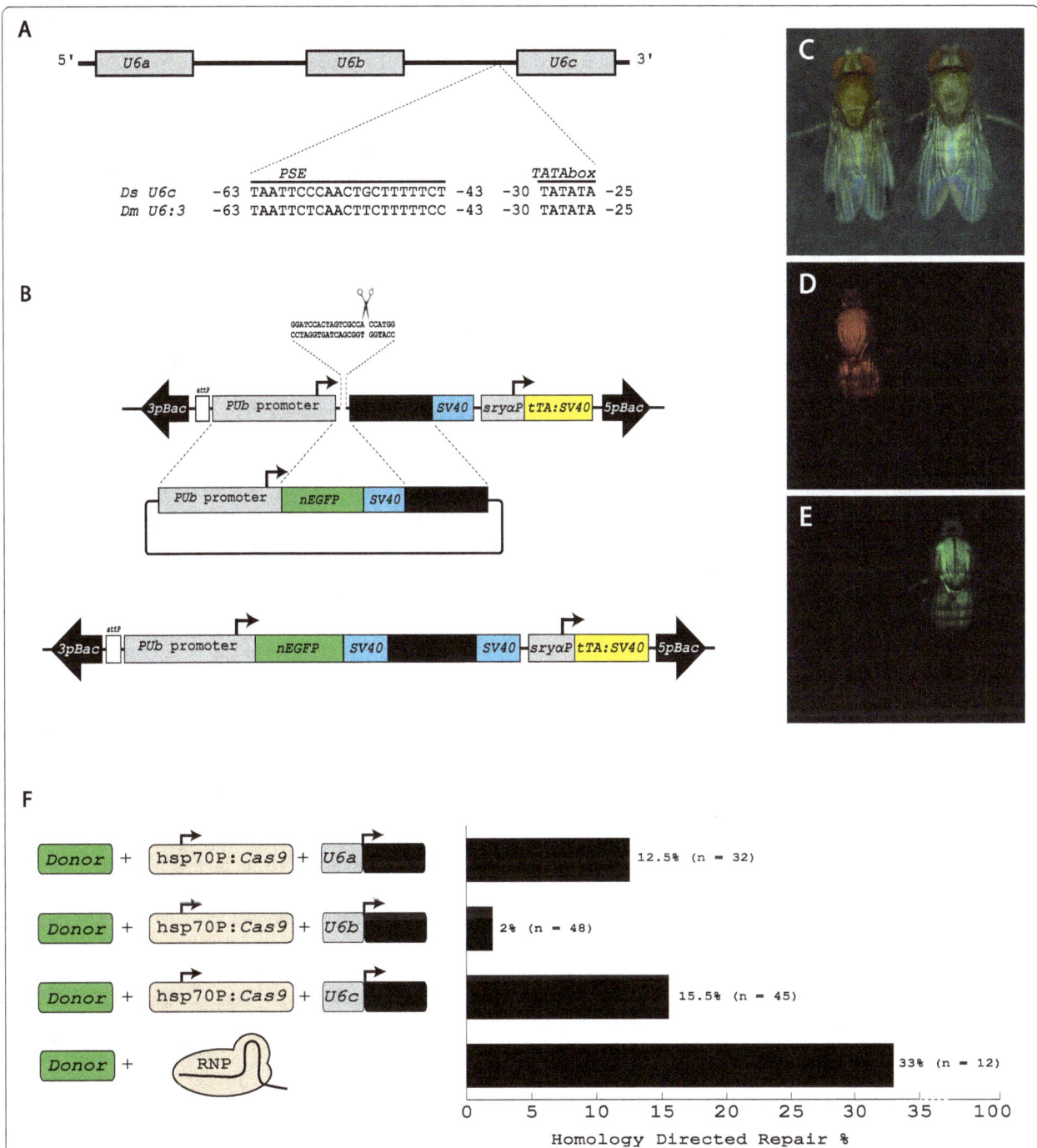

Fig. 1 Improvement of genome editing in *D. suzukii*. **a** Three copies of the *snRNA* gene *U6* in the genome of *D. suzkuii*. The transcription from *U6* genes by *RNA pol III* is directed by the proximal sequence element *PSE* which is highly conserved between *D. suzukii* and *D. melanogaster*. **b** Scheme for HDR-based genome editing at a transgenic target platform. Sequence of the target site in the transgenic strain showing the PAM sequence in red. The scissors indicate where Cas9 induces the DSB three nucleotides upstream of the PAM. **c-e** Fluorescent marker change as the result of the HDR knock-in: images of two male flies taken with cold light (**c**), RFP fliter (**d**), and EYFP filter (**e**). **f** Comparison of *Ds U6a, U6b, U6c* promoters as well as RNP in their efficiency to promote HDR-mediated knock-ins

heterologous tetracycline-controlled transactivator *tTA* gene specifically at early embryonic stages due to the use of the enhancer/promoter element of the cellularization gene *Ds_srya*. Such lines can be employed to establish conditional embryonic lethality for reproductive sterility [19, 20] or conditional female-specific embryonic lethality [21, 22, 52]. As a transgenic marker, this line expresses *DsRed* under the *D. melanogaster* promoter of the *polyubiquitin* (*PUb*) gene. Based on a T7EndoI assay, a functional guide targeting upstream of the *DsRed* translation start codon was identified (Fig. 1b). In a first attempt, in which donor (HMMA134), Cas9 (HMMA 056), and gRNA (HMMA104; *U6c*) plasmids were injected at concentrations of 350, 400, and 150 ng/µl, respectively, we obtained 9.5% homology directed repair (HDR) knock-in events, which we scored based on the change of the body marker from *DsRed* to *EGFP* (Fig. 1c-e). Sequencing of the knock-in junctions revealed faithful scar-less HDR events. The HDR was facilitated by the 1989 bp left homology arm (*PUb* promoter) and the 672 bp right homology arm (*DsRed*).

To compare the three promoters of the *DsU6* genes, we injected in a second attempt donor (HMMA134), *Cas9* (HMMA056), and either of the three gRNA plasmids HMMA102 (*U6a*), HMMA103 (*U6b*), or HMMA104 (*U6c*) at a concentration of 400, 400 and 250 ng/µl, respectively. This resulted in HDR events of 12.5, 2, and 15.5% for *U6a*, *U6b*, and *U6c*, respectively (Fig. 1f). Injection of a RNP complex resulted in 33% HDR events (Fig. 1f). This indicates, that at slightly higher concentrations of donor template and gRNA plasmids, we were able to obtain 15.5% knock-in events using the *U6c* promoter. The *U6b* showed the lowest performance with only 2% knock-in events, and *U6a* was intermediate with 12.5% efficiency (Fig. 1f). Interestingly, the tendency observed for the strength of the different promoters is in line with their *D. melanogaster* counterparts. The high HDR-rates of above 10% indicate that the use of the endogenous promoters allows for effective application of helper plamids instead of RNPs to induce HDR-dependent knock-ins, which represents an improvement for CRIPR/Cas9-based genome editing in *D. suzukii*.

Isolation of the *ß2 tubulin* gene from *Drosophila suzukii*

To be able to drive sperm-specific gene expression, we identified the *Ds_ß2t* gene by homology search in the *D. suzukii* genome database (www.spottedwingflybase.org) using the *Dm_ß2t* sequence as query. The open reading frame of the *Ds_ß2t* gene from the translation start codon to the stop codon is 1341 bp, which is interrupted by a 215 bp intron. The gene has a 5'UTR of 196 bp, which demarcates the transcription start site (Fig. 2a). Conceptual translation of the *Ds_ß2t* coding sequence gives rise to a protein of 446 amino acids.

To validate the testes-specific gene expression of the isolated *Ds_ß2t* gene, we performed whole mount in situ hybridization on the complete reproductive tract of 3–5 day old males using DIG-labelled antisense and sense RNA probes against the *Ds_ß2t* 5'UTR and exon I. These in situ hybridizations detected expression only in the testes with no expression at the apical part that

consists of stem cells (Fig. 2b). No transcription was detected in the rest of the reproductive tract (Fig. 2b) or with sense RNA probe as negative control (Fig. 2c).

Generation of a sperm-marking line of *Drosophila suzukii*

To identify the necessary upstream and downstream regulatory elements driving sperm-specific gene expression, we compared the *D. suzukii* *ß2t* sequence with the characterized counterpart in *D. melanogster*. The 14 bp upstream activator element *ß2tUE1* that confers testes specificity to the *ß2t* gene was found at the exact position − 51 to − 38 relative to the transcription start site with a C > G exchange at position − 41 and a T > A exchange at position − 39 (Fig. 3a). A second upstream regulatory element, *ß2tUE2*, which is not involved in specificity but its overall activity, was identified at position − 32 to − 25 with a G > T exchange at position − 32 and an A > C exchange at position − 28. Another element that functions as a TATAAA-box in TATA-less promoter is the 7 bp initiator sequence encompassing the transcription start, which was identified − 3 to + 4 with the first and last nucleotide differing from *D. melanogaster* (Fig. 3a). A further element involved in *ß2t* promoter function is the *ß2tDE1* element that is highly conserved and lies relative to the transcription start site at position + 51 to + 68 (Fig. 3a).

To examine whether the 51 bp upstream regulatory element plus 196 bp 5'UTR (− 51 to + 196) drives strong testes-specific gene expression, we fused this 247 bp enhancer/promoter fragment of the *Ds_ß2t* gene to *DsRed.T3* (Fig. 3b) and performed an HDR-based knock-in into the *D. suzukii* embryonic *piggyBac* line 06_F5M2, which we had used before as target platform (Fig. 3b). The repair template consisted in this case of *EGFP* fused to the *PUb* promoter followed by *SV40* 3'UTR and the 247 bp *Ds_ß2t* promoter fused to *DsRed.T3* (Fig. 3b). The HDR-based knock-in resulted with 13.3% efficiency. One of the resulting *D. suzukii* lines, 134M16M2, showing a ubiquitous green fluorescence and testes-specific red fluorescence (Fig. 3c-h), was molecular characterized to confirm the proper HDR event. In this line, red fluorescent sperm could be detected in the testes (Fig. 3i-l) and males of this line transferred red fluorescent sperm to the female spermatheca (Fig. 3m-p). This line 134M16M2 thus serves as a sperm-marking line for this invasive pest insect.

Discussion

The programmable genome editing system CRISPR/ Cas9 has enabled a series of new strategies of biotechnological engineering in model and non-model organisms. Based on the objective of the study, financial resources, and availability of functional promoters, researchers can chose the best strategy for delivery of CRISPR/Cas9

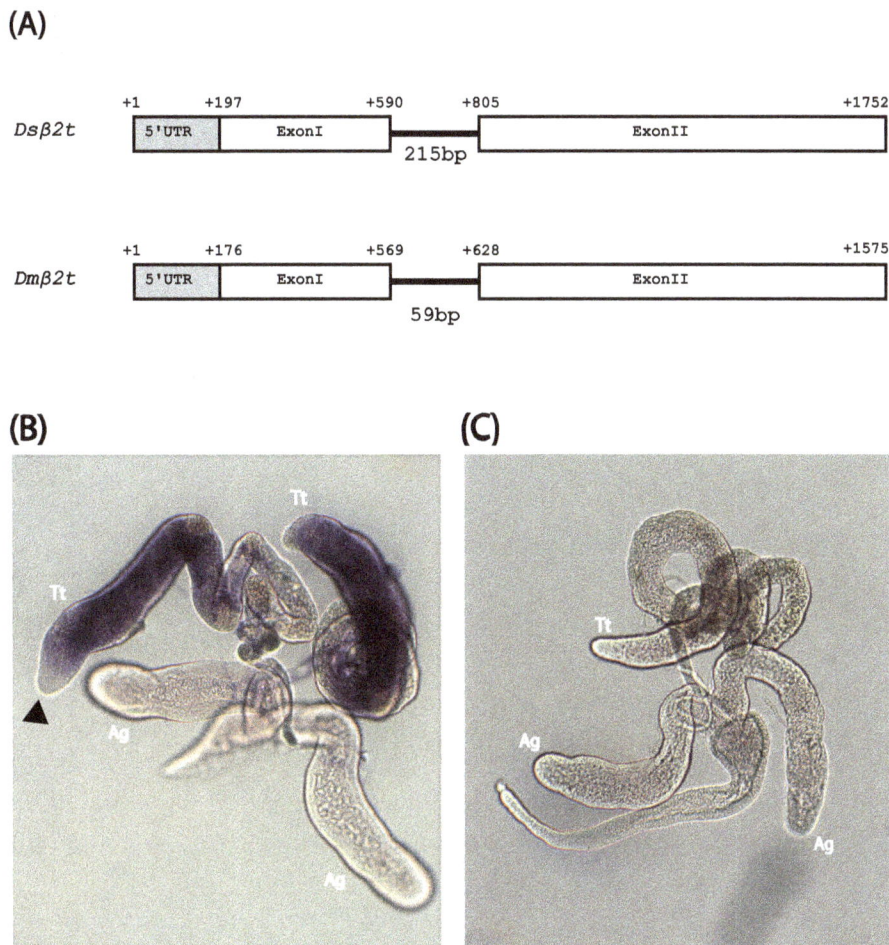

Fig. 2 *D. suzukii β2t* gene and its expression. **a** *Ds_β2t* gene has two exons and one intron similar to *D. melanogaster*. The gene is slightly longer in *D. suzukii* due to increase in the size of the 5'UTR and the intron. The numbers indicate the first nucleotide of the respective feature relative to the first transcribed nucleotide. **b** Testes whole mount in situ hybridization using DIG labeled RNA antisense probe against *Ds_β2t* 5'UTR and exon I detects strong and testes-specific expression. The gene is not expressed at the tip of the testes (black triangle) where stem cells reside. **c** Negative control using DIG labeled sense probe shows no signs of staining. The abbreviations Tt and Ag refer to testes or the accessory glands, respectively

components. From published literature, it can be concluded that the most efficient strategy is germline-specific transgenic expression of Cas9, followed by application of RNP-complexes, then mRNA and gRNA co-injection, and with the least efficiency helper plasmids co-injection [42, 53]. The latter, however, is the most convenient even though it requires the identification and characterization of suitable promoters.

CRISPR/Cas9 holds big promises in the field of insect biotechnology especially for the development of novel pest control strategies, such as reproductive sterility systems based on chromosome shredding [47]. To be able to engineer such strategies in *D. suzukii*, promoters that drive strong expression of gRNAs and other components are of particular importance. Inducible promoters of heat shock genes such as *D. melanogaster hsp70* and *Tribolium*

castaneum Tc_hsp68 have been used for a long time to conditionally express genes both transiently from a plasmid and as transgenes [54, 55].

Due to their defined transcription start site and transcription termination, the RNA *polIII* promoters of the small nuclear RNA genes (snRNA) *U6* have been widely used to express short hairpins to induce an RNA interference effect. With the development of the CRISPR/Cas9 genome editing system, such promoters gained even more popularity and have intensively been used to drive the expression of the chimeric gRNAs transiently and as transgene components from mammals to plants. *D. melanogaster* has three copies in tandem on the right arm of chromosome 3 and have the cytological map location 96A, based on which they were termed *U6:96Aa*, *U6:96Ab*, and *U6:96Ac*. The promoters of the three genes were

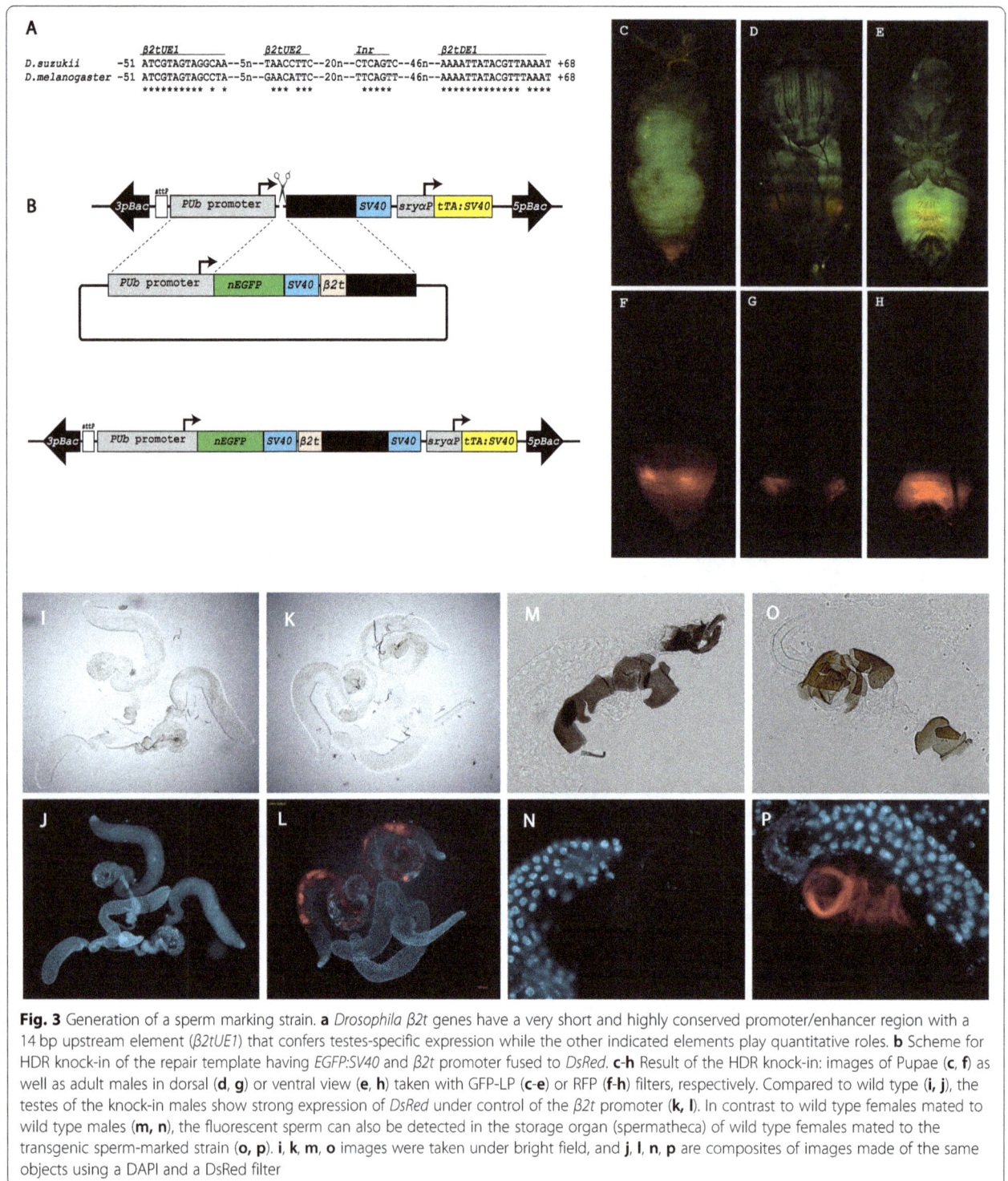

Fig. 3 Generation of a sperm marking strain. **a** *Drosophila β2t* genes have a very short and highly conserved promoter/enhancer region with a 14 bp upstream element (*β2tUE1*) that confers testes-specific expression while the other indicated elements play quantitative roles. **b** Scheme for HDR knock-in of the repair template having *EGFP:SV40* and *β2t* promoter fused to *DsRed*. **c-h** Result of the HDR knock-in: images of Pupae (**c, f**) as well as adult males in dorsal (**d, g**) or ventral view (**e, h**) taken with GFP-LP (**c-e**) or RFP (**f-h**) filters, respectively. Compared to wild type (**i, j**), the testes of the knock-in males show strong expression of *DsRed* under control of the *β2t* promoter (**k, l**). In contrast to wild type females mated to wild type males (**m, n**), the fluorescent sperm can also be detected in the storage organ (spermatheca) of wild type females mated to the transgenic sperm-marked strain (**o, p**). **i, k, m, o** images were taken under bright field, and **j, l, n, p** are composites of images made of the same objects using a DAPI and a DsRed filter

systematically tested and the promoter of the *U6:96Ac* gene (referred to also as *U6:3*) outperforms the other two, which made it the promoter of choice among Drosophilists. Our results are consistent in this respect, as also the *Ds_U6c* promoter has the highest effectivity (Fig. 1f).

Previous reports demonstrated the functionality of the promoters of *Dm-U6:3* and *vasa* genes to drive expression of *gRNA* and *Cas9*, respectively, to target and mutate *D. suzukii w* and *Sxl* by NHEJ but with low frequency. The authors argued that this low efficiency might be attributed to the use of plasmids to drive the expression of *Cas9* and *gRNA* or their bulk crossing scheme [43]. Another study demonstrated the feasibility of using RNP-complexes to induce mutations in *D. suzukii w* by NHEJ [44]. In a more recent study, researchers used RNP-complexes to induce DSBs and were able to knock-in by HDR a mutated temperature-sensitive version of *Ds-tra2* along with a transformation marker cassette. They reported on 7.3% HDR events even though they tried to shift the cell DSB repair machinery towards HDR by co-injection of dsRNA against the *Ds_lig4* gene [45]. In our hands, using RNP complex resulted in a four times higher rate of HDR-based knock-ins. However, no direct comparison with the previous studies is possible since the target itself is different. Anyway, also our helper plasmid co-injections yielded a two times higher rate of HDR-based knock-ins, which indicates that the isolated endogenous promoters allow for an efficient application of the CRISPR/Cas system with the more convenient use of plasmid helpers. However, if the objective is to manipulate the genome and recombinant Cas9 is available, the RNP approach is probably the best option, if no transgenic lines expressing *Cas9* in the germline are available. Studies in *D. melanogaster* and mosquitoes also showed that the use of RNP-complexes always leads to better editing results compared to injection of plasmids or mRNA and in vitro transcribed gRNA.

The use of the regulatory elements (enhancer/promoter) of sex-, tissue-, or stage-specifically expressed genes to drive effector molecules in a particular sex or developmental stage is not only useful in basic research to elucidate gene function, but also in applied insect biotechnology to develop transgene-based pest control strategies. The gene *β2t* has been identified in a number of insects to be testes-specific with its activity starting at the late larval instar. The gene in *D. melanogaster* is known to code for a 446aa protein. Here, we identified the *D. suzukii* homolog that shows at the amino acid level 100% identity but not at the nucleotide level. Interestingly, the transcript structure of the *Ds_β2t* gene revealed the presence of a 215 bp intron (Fig. 2a) compared to a highly conserved intron of 57 bp in *Aedes egypti* [24], 58 bp in *Anastrepha ludens*, 59 bp in *D. melanogaster*, 60 bp *Anstrepha suspensa*, and 67 bp in *Bacterocera dorsalis* [25]. Testes whole mount in situ hybridization identified a similar expression pattern as previously obtained in *D. melanogaster* with the apical part of the testes that contains the stem cells not expressing the gene. The testes specificity of the gene is conferred by a 14 bp activator element upstream of the transcription start site called upstream element *1 β2tUE1*, which is not only contextually conserved but also spatially relative to the transcription start site and other regulatory elements. This activator element was also identified in *D. suzukii*, which shares high similarity

to its *Dm_β2*t counterpart. The other elements that are quantitatively contributing to the expression of *β2t* were also identified in exactly the same positions as in *D. melanogaster* relative to each other and to the transcription start site.

The promoter of the *β2t* gene has been used to drive the expression of a fluorescent protein in mosquitoes and tephritid fruit flies [23, 24, 26], which serve as a sexing system to automate separation of males from females and also as a monitoring system for released males in the context of SIT programs. The generated sperm marking strain of *D. suzukii* proved that the 247 bp regulatory sequence made of 51 bp upstream sequence plus 196 bp leader immediately upstream of the translation start codon has the necessary elements to drive expression of effector molecules specifically in the sperm. The fluorescent sperm can also be identified stored in the spermathecae of wild type females mated to the transgenic sperm marked strain, which facilitates monitoring and allows assessment of the competitiveness of released sterile males compared to their wild type counterparts. The sperm marking system can also help in conducting reproductive biology studies that will enrich our understanding of the biology of this pest and allow us to better design pest control strategies. For example, the promoter of the *β2t* gene in *Anopheles* was used to drive the expression of an HEG that targets and shreds the X chromosome in the mosquito during spermatogenesis leading towards a Y sperm bias and as a consequence to sex ratio distortion, which eventually can lead to a population collapse [51].

Conclusion

We obtained improved usability of the CRISPR/Cas9 gene editing in *D. suzukii* compared to previous reports [43–45] by the employment of helper plasmids that contain endogenous promoters of the *U6* and *hsp70* genes to drive the expression of *gRNA* and *Cas9*, respectively. Moreover, we show that the CRISPR/Cas9 system can be used as an additional tool for the modification of previously established transgenes. The identification and cloning of the *β2t* promoter enabled us to generate a sperm-marking system in *D. suzukii*, which provides a tool for basic research in reproductive biology and should help to monitor the success of pest control campaigns in the context of SIT [23–26]. In addition, the *β2t* promoter can be used in developing novel transgenic pest control approaches [47] for this invasive pest insect.

Methods

Unless otherwise specified, all PCR amplifications were performed using Phusion DNA polymerase and Phusion-HF buffer (New England Biolabs GmbH, D-65926 Frankfurt am Main). Routine plasmid min-preps and PCR products were purified using NucleoSpin® Plasmid and NucleoSpin® Gel and PCR Clean-up kits (Macherey-Nagel GmbH & Co., 52,355 Dueren, Germany), respectively. Plasmid vectors for microinjections were prepared using NucleoSpin® Plasmid Transfection-grade (Macherey-Nagel) or QIAGEN Plasmid Plus Midi Kit (QIAGEN GmbH, 40,724 Hilden, Germany). Primers used are listed in Additional file 1: Table S1.

Fly strain and husbandry

All fly experiments were performed in our well-equipped safety level one (S1) laboratory, which is certified for generating and using genetically modified insects. Wild type *D. suzukii* from Italy (kindly provided by Prof. Marc F. Schetelig) as well as generated transgenic lines were reared on standard *Drosophila* food supplemented with baker yeast and kept at 25 °C throughout this study. For germline transformation, flies were transferred to *Drosophila* egg laying cages and allowed to lay eggs on apple juice agar plates with some yeast on top to increase egg laying.

Nucleic acid isolation

Genomic DNA was isolated from a mix of adult males and females of *D. suzukii* (Italian strain) using NucleoSpin® DNA Insect (Macherey-Nagel) according to the manufacturer instructions. To generate a testes-specific cDNA library, testes of 100 males (3–4 days old) were dissected in ice cold 1X PBS and used for total RNA preparation using ZR Tissue & Insect RNA MicroPrep (Zymo Research Europe, 79,110 Freiburg) according to manufacturer instructions.

Isolation of *DsU6* and *hsp70* genes

Based on synteny we identified *D. suzukii* the homologs of *D. melanogaster* genes *Esyt2* and *REPTOR* bordering the *U6* locus. Primer pair HM#137/138 was designed on the conserved parts of these genes and used to PCR amplify the sequence between them supposedly containing the *Ds_U6* locus, (initial denaturation temperature 98 °C 3 min followed by 35 cycles of 98 °C 30s, 72 °C 2 min 30 s). A 3.7 kbp fragment was obtained and sequenced.

To identify the *D. suzukii heat shock protein 70* (*Dshsp70*) gene, we BLASTed *D. melanogaster hsp70*Aa in the *D. suzukii* genome data base (www.spottedwingflybase.org) and compared the amino acid sequence as well as the corresponding DNA sequence individually to their *D. melanogaster* counterparts using the geneious program version 10.2.6 (Auckland, 1010, New Zealand).

Isolation of *Dsβ2t* gene and its 5′UTR

To isolate the spermatogenesis specific *beta-2-tubulin* (*β2t*) gene of *D. suzukii*, we searched in the www.

spottedwingflybase.org with the *D. melanogaster Dm_β2t* gene. A putative *Ds_β2t* gene sharing high homology to *Dm_β2t* was PCR amplified from genomic DNA using primer pair HM#25/26 and the PCR program 98 °C for 3 min followed by 35 cycles of 98 °C 30 s, 72 °C 1 min 40 s, and 7 min final elongation at 72 °C. The amplified fragment was purified, blunt cloned into pJet1.2 vector (Thermo Fisher Scientific, 64,293 Darmstadt, Germany), and sequenced using standard primers pJet1.2_fwd and pJet1.2_rev.

Since the 5′UTR of *β2t* has some regulatory elements, whose position relative to the transcription start site and the upstream regulatory elements is highly conserved and important for correct tissue specific expression, it was imperative to isolate the 5′UTR and to identify the transcription start site. To do so, 1.7 µg of testes total RNA were used to generate a 5′ RACE-ready cDNA library using the SMARTer™ RACE cDNA amplification kit (Takara Bio Europe SAS, 78100 Saint-Germain-en-Laye, France) according to manufacturer instructions. The 5′UTR was recovered by RACE PCR using gene specific primer HM#33 and universal primer (UPM) provided with the kit using Advantage2 DNA polymerase (Takara) with the following program: 94 °C 2 min, (94 °C 30 s, 72 °C 3 min) 5X, (94 °C 30 s, 70 °C 30 s, 72 °C 3 min) 5X, (94 °C 30 s, 68 °C 30 s, 72 °C 3 min) 30X. A single prominent band was recovered, purified, cloned into pCRII (Thermo Fisher Scientific) to generate pCRII_Dsb2t_5′UTR (HMMA24), and sequenced using a standard M13 primer.

Testes whole mount in situ hybridization

To generate DIG-labelled sense and antisense RNA probes of *Ds_β2t*, we prepared DNA templates for in vitro transcription by PCR amplification of the 5′RACE-fragment including the Sp6 or T7 promoters from pCRII_Ds β2t_5′UTR (HMMA24). Primer pairs HM#33/128and HM#41/127 were used respectively with the following PCR conditions: initial denaturation at 98 °C 3 min, followed by 35 cycles of 98 °C 30 s, 72 °C 50 s with a final elongation step of 7 min. RNA probes were synthesized using DIG-labelling kit (Thermo Fisher Scientific) according to manufacturer instructions using 200 ng of DNA as template in a total reaction mix of 10 µl. The reaction was allowed to proceed for 2 h at 37 °C followed by Turbo DNaseI treatment (Thermo Fisher Scientific) for 15 min to remove template DNA. Two microliter of 0.2 M EDTA was used to inactivate the reaction. Sense and antisense probes were precipitate and resuspended in 100 µl RNA resuspension buffer (5:3: 2 H2O: 20X SSC: formaldehyde) and stored at − 80 °C.

Testes of 3–5 days old males were dissected in ice cold 1X Phosphate buffered saline (PBS) and fixed in PBF-tween (4% formaldehyde and 0.1% tween 20 in 1X PBS) for 20 min at room temperature. In situ hybridization was performed according to an established protocol [56] with inclusion of dehydration steps according to Zimmerman et al. [57].

Plasmid construction

To generate plasmid HMMA006, 300 bp upstream of *Ds_srya* plus 50 bp 5'UTR sequence were PCR amplified using primer pair HM#23/24 introducing *AgeI/NheI* cut sites respectively and cloned into *AgeI/NheI* cut site of KNE007 [58] upstream of *tTA* CDS replacing the *Dm_β2t* promoter. Description of the *Ds_srya* gene and its cloning will be described elsewhere (Ahmed et al.)

To generate pSLaf_*T7-BbsI-BbsI-ChiRNA*_af (HMMA034) for in vitro transcription of gRNAs, annealed oligos HM#55/56 generating T7 promoter and 2X *BbsI* restriction sites were cloned into *BbsI/HindIII* digested plasmid pU6-chiRNA (Addgene: #45946) giving rise to HMMA033. Next, the *HindIII/SacI T7-BbsI-BbsI-chiRNA* fragment from HMMA033 was cloned into pSLaf1180af [59] *HindIII/SacI* cut sites.

To generate plasmids *pDsU6a-BbsI-BbsI-chiRNA-DSE* (HMMA091), *pDsU6b-BbsI-BbsI-ChiRNA DSE* (HMMA092), and *pDsU6c-BbsI-BbsI-chiRNA-DSE* (HMMA093) for transient expression of gRNAs, primer pairs HM#358/159, HM#104/158, and HM#360/160 were used to amplify the promoters of *snRNA* genes *U6a*, *U6b*, and *U6c*, respectively, with PCR condition 98 °C 3 min followed by 5 cycles of 98 °C 30 s, 66 °C 40 s, and 72 °C 1 min then 30 cycles of 98 °C 30 s, 72 °C 1 min 40 s with a final elongation 72 °C for 7 min. The promoters were then cloned into HMMA034 by megaprimer PCR cloning [60] using 30 ng of plasmid HMMA034 and 200 ng of the promoter as megaprimer in a 25 μl reaction with PCR (98 °C 3 min, [98 °C 30 s, 72 °C 2 min 30 s] 30X, 72 °C 7 min) generating plasmids HMMA088, HMMA089, and HMMA090. Finally, 250 bp of the sequence downstream of the *U6c* termination sequence was PCR amplified from genomic DNA using primer pair HM#186/187 with PCR (98 °C 3 min, [98 °C 30 s, 68 °C 30 s, 72 °C 20 s] 35X with a final elongation of 7 min at 72 °C). The amplified fragment was then cloned into HMMA088, HMMA089, and HMMA090 by megaprimer cloning as described above with annealing temperature at 68 °C.

For Cas9 recombinant protein expression, the plasmid *pET-T7-3XFlag-nls-Cas9-nls-6XHisTag* (HMMA101) was generated. The sumo part of the pET-SUMO expression vector was removed using *XhoI/NdeI* and the annealed oligos HM#152/153 were cloned introducing 2X *BsaI* sites giving rise to HMMA080. The 4.3Kb *BsaI/XbaI 3XFlag-nls-Cas9-nls* fragment was excised from HMMA066 and cloned into *BsaI* linearized HMMA080 to give rise to HMMA099. Finally, annealed oligos HM#180/181 introducing a *6XHisTag* were cloned into *FseI/BasI* digested plasmid HMMA099. Plasmid HMMA066 was generated by cloning *ClaI/HpaI* fragment *3XFlag-nls-Cas9-nls* from

HMMA039 into *ClaI/HpaI* cut #1215 [20] giving rise to HMMA065 followed by cloning of annealed self-complementary oligo HM#102 into the *ClaI* site of HMMA065 to introduce 2X *BbsI* restriction sites. Cas9 protein was expressed and purified according to Paix et al. [61], and frozen at − 20 °C until needed.

The plasmid *pSLaf_Dshsp70P-Cas9-SV40_af* (HMMA056) to express Cas9 transiently was generated by cloning of the 4.2Kb *ClaI/XbaI* fragment containing insect codon optimized *Cas9* CDS with N and C terminal nuclear localization signals from plasmid #46294 (Addgene) into *ClaI/XbaI* digested pCS2-Sp6-Cas9-SV40 (Addgene: #47322) replacing the mammalian codon optimized *Cas9* CDS giving rise to HMMA039. The *Ds_hsp70* promoter was PCR amplified from genomic DNA using primer pair HM#73/75 with PCR using the following condition: 98 °C 3 min [(98C°C 30 s, 66 °C 40 s, 72 °C 1 min) 5X, (98 °C 30 s, 72 °C 1 min 40 s) 35X with a final elongation step of 7 min at 72 °C. The fragment was purified and cloned into *EcoRI/ClaI* cut #1215 [20] to give rise to HMMA052. Finally, *Cas9-SV40* was excised from HMMA039 by *ClaI/HpaI* and cloned into *ClaI/HpaI* cut HMMA052 generating HMMA056.

To generate donor plasmid HMMA134, a 3.2Kb fragment containing *PUb-nls-EGFP-SV40* was excised from #1254 [20] using *SacI/AflII* and cloned into *SacI/AflII* cut pSLaf1108af [59] giving rise to plasmid HMMA094. *DsRed* CDS was PCR amplified from plasmid KNE007 [58] using primer pair (HM#37/167) with PCR (98 °C 3 min followed by 35 cycles of 98 °C 30 s, 72 °C 1 min and a final elongation of 7 min at 72 °C). The fragment was phosphorylated and ligated into blunted *AflII* cut HMMA095 generating HMMA096. To change the target PAM sequence in front of *EGFP* from TGG to TGA in the repair template (Fig. 1b), PCR mutagenesis using primer pair HM#221/222 was performed (98 °C 3 min followed by 30 cycles of 98 °C 30 s, 72 °C 4 min and final elongation of 7 min at 72 °C) to give rise to HMMA097, which results in changing the second amino acid of the EGFP from valine to methionine. Finally, the 247 bp *Ds_β2t* regulatory sequence spanning − 51 to + 196 was PCR amplified using primer pair HM#285/252 with PCR conditions 98 °C 3 min [(98 °C 30 s, 60 °C 30 s, 72 °C 20 s) 5X, (98 °C 30 s, 72 °C 1 min) 30X with a final elongation step of 7 min at 72 °C. The promoter was then cloned upstream of *DsRed* in HMMA097 by megaprimer PCR cloning as described previously with annealing at 61 °C.

Guide RNAs design, cloning, and validation

Guide RNAs were identified using the online target finder tool built by Wisconsin University (http://target-finder.flycrispr.neuro.brown.edu/). Identified potential targets were checked against *D. suzukii* database to exclude those with off-target sites. For each potential target, two oligos, a forward and reverse, were designed and the respective overhangs were added. Oligos were ordered as normal primers without phosphorylation. The two oligos for each target were annealed at a concentration of 10 µM in a total volume of 100 µl in a heat block. The gRNAs were validated using a T7EndoI assay [62, 63]. Each *gRNA* plasmid was mixed with *Cas9* plasmid HMMA056 at a concentration of 400/500 ng/µl, respectively, and injected into 50 pre-blastoderm embryos. Ten to fifteen hatching larvae were collected in 1.5 ml Eppendorf tubes and crushed by using a pipette tip against the tube wall. Two hundred microliter of squishing buffer [19] was added and mixed well. The tubes were then incubated at 55 °C for 1 h with occasional vortexing. Tubes were then centrifuged, and 5 µl of the supernatant was used as a template in 50 µl PCR reactions using primers HM#192/69. PCR products were gel purified, quantified, and 400 ng were mixed in 1X NEB 2.1 buffer in a total volume of 19 µl. DNA was denatured, rehybridized, 0.75 µl of T7 EndoI (NEB) were added, and incubated at 37 °C for 20 min. The reactions were stopped using 2 µl of 0.25 M EDTA and run in a 1.5% agarose gel. Only one guide showed obvious digest by T7 EndoI. Wild type un-injected larvae were used as control. To generate the plasmids expressing the functional guide RNA against the identified target upstream of *DsRed* (Fig. 1b), annealed oligos HM#161/162 and HM#169/162 were cloned by golden gate [64, 65] into gRNA vectors HMMA091, HMMA092, and HMMA093 to generate p*U6a_Red1chic* HMMA102, p*U6b_Red1chi* HMMA103, and p*U6c_Red1chi* HMMA104, respectively.

In vitro transcription of the gRNA
The functional gRNA was cloned by ligation of annealed oligos HM#162/215 into *BbsI* cut plasmid HMMA035, which was then used to generate the template for in vitro transcription by PCR using primer pair HM#84/128. In vitro transcription of *gRNA* was performed using MEGAscript® (Ambion) according to the manufacturer protocol. The reaction was allowed to proceed for 2 h at 37 °C followed by DNA template removal using 1 µl DNase I for 30 min. *gRNA* was purified using RNA clean and concentrator (Zymo Research) and the concentration was determined by nano-drop (Thermo Fisher Scientific) and stored at − 80 °C.

Germline transformation
All embryonic injections were performed using transfection grade plasmid preparations without further precipitation steps. To generate the embryonic driver line 06_F5M2 by random *piggyBac* integration, the transformation vector HMMA006 and the helper plasmid MK006 [58] were mixed at a final concentration of 400 and 200 ng/µl respectively. To validate that the transgene represents a single integration even, we performed inversePCR as described [58] using *XhoI and EcoRI* restriction enzymes. For both the 5 and 3′ junctions, we each obtained only a single fragment, whose sequences confirmed a single integration site in the second intron of a gene referred to as *Suppressor of Under Replication* (Additional File 2: *piggyBac* insertion in *D. suzukii* line 06_F5M2).

For the transgene editing experiments using CRISPR/Cas9, DNA was mixed at a concentration of 400, 150, and 350 ng/µl for *Cas9* (HMMA056), *gRNA* (HMMA102, HMMA103, or HMMA104), and donor plasmid HMMA097, respectively. Higher concentration was used at 400, 250, and 400 ng/µl, respectively. All DNA injection mixes were prepared in 1X injection buffer (5 mM KCl, 0.1 mM NaH$_2$PO$_4$, pH 6.8). For RNP injection, recombinant Cas9 endonuclease, gRNA, and donor plasmid HMMA097 were mixed together at a final concentration of 300 ng/µl, 150 ng/µl, and 400 ng/µl respectively, incubated at 37 °C for 10 min for the RNP-complex formation, and injected into 90 pre-blastoderm embryos.

Injection needles were prepared as previously described [58] .To inject in *D.suzukii* embryos, the eggs have to be squeezed out of the apple agar plates individually using home-made closed-tip glass pipettes. Embryos were then de-chorionated for 3 min using generic Clorox (DanKlorix, CP GABA GmbH, Hamburg, Germany) containing 2.5% sodium hypochlorite at final concentration of 1.25% sodium hypochlorite and washed in washing buffer (100 mM NaCl, 0.02% Triton X-100) followed by thorough wash with desalted water. Embryos were then aligned on apple agar blocks and transferred to double sticky tape on a coverslip and covered by Voltalef 10S oil (VWR International, Darmstadt, Germany). Injections were performed using a Femtojet (Eppendorf, Hamburg, Germany) and a manual micromanipulator. Excessive oil was drained and the injected embryos were incubated on apple agar plates at the room temperature until hatching. Larvae were manually transferred to fly food vials. Each emerging G$_0$ fly was out-crossed to 3–4 wild type individuals of the opposite sex.

Microscopy
Screening for transgenic flies and fluorescence imaging were performed using a Leica M205 FA fluorescence stereomicroscope equipped with camera Q imaging Micropublisher 5.0 RTV (Leica Mikrosysteme Vertrieb Gmb, Wetzlar, 35,578 Germany). Transgenic flies were screened using filter sets RFP (excitation: ET546/10x, emission: ET605/70 m) or GFP-LP (excitation: ET480/40, emission: ET510 LP), respectively, and imaged using cold light (Fig. 1c) or filter sets: RFP (Figs. 1d; Fig. 3 f-h), EYFP (excitation: ET500/20, emission: ET535/30) for Fig. 1e, or GFP-LP (Fig. 3c-e).

Epifluorescence microscopy was performed using a Zeiss Imager.Z2 equipped with two cameras, Axiocam

506 mono and Axiocam 305 colour (Zeiss, 73,447 Oberkochen, Germany). The testes or the spermathecae were dissected in ice-cold PBS, fixed for 10 min in 4% formaldehyde prepared in 0.1% PBS-tween 20, permeabilized for 10 min using 1% Triton X-100 in PBS, and nuclei were stained for 10 min using DAPI (4′,6-Diamidino-2-Phenylindole, Dihydrochloride) at a concentration of 1 µg/ml. Samples were mounted in 70% glycerol and the spermathecae were broken open using dissection needles. The tissues were imaged under bright field and to observe cell nuclei and expression of DsRed, images were taken with filters for DAPI (excitation: 335–383, emission: 420–470) or DsRed (excitation: 533–558, emission: 570–640), and composed in ZEN Blue (Zeiss).

Abbreviations

Cas9: CRISPR associated protein 9; CRISPR: Clustered regularly interspaced short palindromic repeat; crRNA: CRISPR RNA; DIG: Digoxigenin; *Ds_lig4*: *Drosophila suzukii ligase 4*; *Ds_srya*: *Drosophila suzukii serendipity alpha*; *Ds_tra2*: *Drosophila suzukii transformer 2*; DSB: Double strand break; DsRed: Discosoma Red; dsRNA: Double strand RNA; gRNA: Guide RNA; HDR: Homology directed repair; HEG: Homing endonuclease gene; Hsp70: Heat shock protein 70; mRNA: Messenger RNA; NHEJ: Non-homologous end joining; Orco: Odorant receptor co-receptor; PAM: Protospacer Adjacent Motif; PUb: Polyubiquitin gene; rDNA: Ribosomal deoxyribonucleic acid; RNApolIII: RNA polymerase III; RNP: Ribonucleoprotein; SIT: Sterile insect technique; snRNA: Small nuclear RNA gene; SWD: Spotted Wing *Drosophila*; *Sxl*: *Sex lethal*; TALENs: Transcription activator like endonucleases; Tc_hsp68: *Tribolium castaneum* heat shock protein 68 gene; TracrRNA: Transactivator RNA; TRE: tTA responsive element; tTA: Tetracycline controlled transactivator; ZFNs: Zinc finger nucleases; B2tUE1: Beta-2-tubulin Upstream Element 1

Acknowledgements

We would like to thank Prof. Dr. Marc F. Schetelig (Justus-Liebig-University Giessen) for providing wild type *Drosophila suzukii* (Italian strain) and Kolja N. Eckermann (University of Göttingen) for providing plasmid KNE007. Our thanks extend to our colleagues who made plasmids available through Addgene (#46294 and #47322). The project profited also from discussions at the International Atomic Energy Agency funded meetings of the Coordinated Research Project "Multifactorial reproductive sterility system to avoid resistance development against transgenic Sterile Insect Technique approaches". We acknowledge support by the Open Access Publication Funds of the Göttingen University.

Authors' contributions

EAW and HMMA conceived and designed the study; HMMA isolated the genes, designed the constructs and generated the transgenic lines; LH performed in situ hybridizations; EAW and HMMA wrote the manuscript; HMMA prepared the figures; all authors read and approved of the final manuscript.

Funding

H.M.M.A was supported by the German Academic Exchange Service (DAAD), which had no role in the design of the study, the collection, analysis, and interpretation of data, or in writing the manuscript.

Ethics approval and consent to participate

Not applicable.

Consent for publication

Not applicable.

Competing interests

The authors declare that they have no competing interests.

Author details

¹Department of Developmental Biology, Johann-Friedrich-Blumenbach-Institute of Zoology and Anthropology, Göttingen Center for Molecular Biosciences, Georg-August-University Göttingen, 37077 Göttingen, Germany. ²Department of Crop Protection, Faculty of Agriculture-University of Khartoum, P.O. Box 32, 13314 Khartoum, Khartoum North, Sudan.

References

1. Hauser M. A historic account of the invasion of Drosophila suzukii (Matsumura) (Diptera: Drosophilidae) in the continental United States, with remarks on their identification. Pest Manag Sci. 2011;67(11):1352–7.
2. Cini A, Ioriatti C, Anfora G. A review of the invasion of Drosophila suzukii in Europe and a draft research agenda for integrated pest management. Bull Insectol. 2012;65(1):12.
3. Walsh DB, Bolda MP, Goodhue RE, Dreves AJ, Lee J, Bruck DJ, et al. Drosophila suzukii (Diptera: Drosophilidae): invasive pest of ripening soft fruit expanding its geographic range and damage potential. J Integr Pest Manag. 2011;2(1):G1–7.
4. Asplen MK, Anfora G, Biondi A, Choi D-S, Chu D, Daane KM, et al. Invasion biology of spotted wing Drosophila (Drosophila suzukii): a global perspective and future priorities. J Pestic Sci. 2015;88(3):469–94.
5. Deprá M, Poppe JL, Schmitz HJ, De Toni DC, Valente VLS. The first records of the invasive pest Drosophila suzukii in the south American continent. J Pestic Sci. 2014;87(3):379–83.
6. Lavagnino NJ, Díaz BM, Cichón LI, De la Vega GJ, Garrido SA, Lago JD, et al. New records of the invasive pest Drosophila suzukii (Matsumura) (Diptera: Drosophilidae) in the south American continent. Rev Soc Entomológica Argent. 2018;77(1):27–31.
7. Kenis M, Tonina L, Eschen R, van der Sluis B, Sancassani M, Mori N, et al. Non-crop plants used as hosts by Drosophila suzukii in Europe. J Pestic Sci. 2016;89(3):735–48.
8. Lee JC, Bruck DJ, Curry H, Edwards D, Haviland DR, Van Steenwyk RA, et al. The susceptibility of small fruits and cherries to the spotted-wing drosophila, Drosophila suzukii. Pest Manag Sci. 2011;67(11):1358–67.
9. Lee JC, Bruck DJ, Dreves AJ, Ioriatti C, Vogt H, Baufeld P. In focus: spotted wing drosophila, Drosophila suzukii, across perspectives. Pest Manag Sci. 2011;67(11):1349–51.
10. Mazzi D, Bravin E, Meraner M, Finger R, Kuske S. Economic impact of the introduction and establishment of Drosophila suzukii on sweet cherry production in Switzerland. Insects. 2017;8(1):18.
11. Haviland DR, Beers EH. Chemical control programs for <I>Drosophila suzukii</I> that comply with international limitations on pesticide residues for exported sweet cherries. J Integr Pest Manag. 2012;3(2):1–6.
12. Diepenbrock LM, Rosensteel DO, Hardin JA, Sial AA, Burrack HJ. Season-long programs for control of Drosophila suzukii in southeastern U.S. blueberries. Crop Prot. 2016;81:76–84.
13. Lee JC, Wang X, Daane KM, Hoelmer KA, Isaacs R, Sial AA, et al. Biological control of spotted-wing Drosophila (Diptera: Drosophilidae)—current and pending tactics. J Integr Pest Manag. 2019;10(1):13.
14. Leach H, Van Timmeren S, Isaacs R. Exclusion netting delays and reduces Drosophila suzukii (Diptera: Drosophilidae) infestation in raspberries. J Econ Entomol. 2016;109(5):2151–8.
15. Rendon D, Hamby KA, Arsenault-Benoit AL, Taylor CM, Evans RK, Roubos CR, et al. Mulching as a cultural control strategy for Drosophila suzukii in blueberry. Pest Manag Sci. 2019;0(ja) [cited 2019 Jun 30]. Available from: https://onlinelibrary.wiley.com/doi/abs/10.1002/ps.5512.
16. Knipling EF. Possibilities of insect control or eradication through the use of sexually sterile Males1. J Econ Entomol. 1955;48(4):459–62.

17. Krafsur ES. Sterile insect technique for suppressing and eradicating insect population: 55 years and counting. J Agric Entomol. 1998;15(4):17.

18. Krafsur ES, Lindquist DA. Did the sterile insect technique or weather eradicate screwworms (Diptera: Calliphoridae) from Libya? J Med Entomol. 1996;33(6):877–87.

19. Horn C, Wimmer EA. A transgene-based, embryo-specific lethality system for insect pest management. Nat Biotechnol. 2003;21(1):64–70.

20. Schetelig MF, Caceres C, Zacharopoulou A, Franz G, Wimmer EA. Conditional embryonic lethality to improve the sterile insect technique in Ceratitis capitata(Diptera: Tephritidae). BMC Biol. 2009;7(1):4.

21. Ogaugwu CE, Schetelig MF, Wimmer EA. Transgenic sexing system for Ceratitis capitata (Diptera: Tephritidae) based on female-specific embryonic lethality. Insect Biochem Mol Biol. 2013;43(1):1–8.

22. Yan Y, Scott MJ. A transgenic embryonic sexing system for the Australian sheep blow fly Lucilia cuprina. Sci Rep. 2015;5(1):16090.

23. Scolari F, Schetelig MF, Bertin S, Malacrida AR, Gasperi G, Wimmer EA. Fluorescent sperm marking to improve the fight against the pest insect Ceratitis capitata (Wiedemann; Diptera: Tephritidae). N Biotechnol. 2008; 25(1):76–84.

24. Smith RC, Walter MF, Hice RH, O'Brochta DA, Atkinson PW. Testis-specific expression of the ?2 tubulin promoter of Aedes aegypti and its application as a genetic sex-separation marker. Insect Mol Biol. 2007;16(1):61–71.

25. Zimowska GJ, Nirmala X, Handler AM. The β2-tubulin gene from three tephritid fruit fly species and use of its promoter for sperm marking. Insect Biochem Mol Biol. 2009;39(8):508–15.

26. Catteruccia F, Benton JP, Crisanti A. An Anopheles transgenic sexing strain for vector control. Nat Biotechnol. 2005;23(11):1414–7.

27. Ishino Y, Shinagawa H, Makino K, Amemura M, Nakata A. Nucleotide sequence of the iap gene, responsible for alkaline phosphatase isozyme conversion in Escherichia coli, and identification of the gene product. J Bacteriol. 1987;169(12):5429–33.

28. Makarova KS, Grishin NV, Shabalina SA, Wolf YI, Koonin EV. No title found. Biol Direct. 2006;1(1):7.

29. Barrangou R, Fremaux C, Deveau H, Richards M, Boyaval P, Moineau S, et al. CRISPR provides acquired resistance against viruses in prokaryotes. Science. 2007;315(5819):1709–12.

30. Garneau JE, Dupuis M-È, Villion M, Romero DA, Barrangou R, Boyaval P, et al. The CRISPR/Cas bacterial immune system cleaves bacteriophage and plasmid DNA. Nature. 2010;468(7320):67–71.

31. Jinek M, Chylinski K, Fonfara I, Hauer M, Doudna JA, Charpentier E. A programmable dual-RNA-guided DNA endonuclease in adaptive bacterial immunity. Science. 2012;337(6096):816–21.

32. Bassett AR, Tibbit C, Ponting CP, Liu J-L. Highly efficient targeted mutagenesis of Drosophila with the CRISPR/Cas9 system. Cell Rep. 2013;4(1):220–8.

33. Hwang WY, Fu Y, Reyon D, Maeder ML, Tsai SQ, Sander JD, et al. Efficient genome editing in zebrafish using a CRISPR-Cas system. Nat Biotechnol. 2013;31(3):227–9.

34. Platt RJ, Chen S, Zhou Y, Yim MJ, Swiech L, Kempton HR, et al. CRISPR-Cas9 Knockin mice for genome editing and Cancer modeling. Cell. 2014;159(2): 440–55.

35. Hall B, Cho A, Limaye A, Cho K, Khillan J, Kulkarni AB. Genome editing in mice using CRISPR/Cas9 technology. Curr Protoc Cell Biol. 2018;81(1):e57.

36. Gratz SJ, Cummings AM, Nguyen JN, Hamm DC, Donohue LK, Harrison MM, et al. Genome engineering of Drosophila with the CRISPR RNA-guided Cas9 nuclease. Genetics. 2013;194(4):1029–35.

37. Li M, Bui M, Yang T, Bowman CS, White BJ, Akbari OS. Germline Cas9 expression yields highly efficient genome engineering in a major worldwide disease vector, Aedes aegypti. Proc Natl Acad Sci. 2017;114(49):E10540–9.

38. Li M, Akbari OS, White BJ. Highly Efficient Site-Specific Mutagenesis in Malaria Mosquitoes Using CRISPR. G3. 2018;8:653–8.

39. Jiang J, Zhang L, Zhou X, Chen X, Huang G, Li F, et al. Induction of site-specific chromosomal translocations in embryonic stem cells by CRISPR/Cas9. Sci Rep. 2016;6(1):21918.

40. Iwata S, Yoshina S, Suehiro Y, Hori S, Mitani S. Engineering new balancer chromosomes in C. elegans via CRISPR/Cas9. Sci Rep. 2016;6(1):33840.

41. Port F, Bullock SL. Augmenting CRISPR applications in Drosophila with tRNA-flanked sgRNAs. Nat Methods. 2016;13(10):852–4.

42. Port F, Chen H-M, Lee T, Bullock SL. Optimized CRISPR/Cas tools for efficient germline and somatic genome engineering in Drosophila. Proc Natl Acad Sci. 2014;111(29):E2967–76.

43. Li F, Scott MJ. CRISPR/Cas9-mediated mutagenesis of the white and sex lethal loci in the invasive pest, Drosophila suzukii. Biochem Biophys Res Commun. 2016;469(4):911–6.

44. Kalajdzic P, Schetelig MF. CRISPR/Cas-mediated gene editing using purified protein in Drosophila suzukii. Entomol Exp Appl. 2017;164(3):350–62.

45. Li J, Handler AM. Temperature-dependent sex-reversal by a transformer-2 gene-edited mutation in the spotted wing drosophila, Drosophila suzukii. Sci Rep. 2017;7(1):12363.

46. Karageorgi M, Bräcker LB, Lebreton S, Minervino C, Cavey M, Siju KP, et al. Evolution of multiple sensory systems drives novel egg-laying behavior in the fruit Pest Drosophila suzukii. Curr Biol. 2017;27(6):847–53.

47. Eckermann KN, Dippel S, KaramiNejadRanjbar M, Ahmed HM, Curril IM, Wimmer EA. Perspective on the combined use of an independent transgenic sexing and a multifactorial reproductive sterility system to avoid resistance development against transgenic sterile insect technique approaches. BMC Genet. 2014;15(Suppl 2):S17.

48. Kemphues J, Kaufman C, Raff A, Raff C. The testis-specific P-tubulin subunit in Drosophila melanogaster has multiple functions in spermatogenesis. Cell. 1982;31:655–70.

49. Michiels F, Gasch A, Kaltschmidt B, Renkawitz-Pohl R. A 14 bp promoter element directs the testis specificity of the Drosophila 32 tubulin gene. EMBO J. 1989;8(5):1559–65.

50. Michiels F, Wolk A, Renkawitz-Pohl R. Further sequence requirements for male germ cell-specific expression under the control of the 14 bp promoter element (B2UE1) of the Drosophila B2 tubulin gene. Nucleic Acids Res. 1991; 19(16):4515–21.

51. Galizi R, Doyle LA, Menichelli M, Bernardini F, Deredec A, Burt A, et al. A synthetic sex ratio distortion system for the control of the human malaria mosquito. Nat Commun. 2014;5(1):3977.

52. Schetelig MF, Handler AM. A transgenic embryonic sexing system for Anastrepha suspensa (Diptera: Tephritidae). Insect Biochem Mol Biol. 2012; 42(10):790–5.

53. Kondo S, Ueda R. Highly improved gene targeting by Germline-specific Cas9 expression in Drosophila. Genetics. 2013;195(3):715–21.

54. Shoji W, Sato-Maeda M. Application of heat shock promoter in transgenic zebrafish. Dev Growth Differ. 2008;50(6):401–6.

55. Schulte C, Leboulle G, Otte M, Grünewald B, Gehne N, Beye M. Honey bee promoter sequences for targeted gene expression: honey bee promoter analysis. Insect Mol Biol. 2013;22(4):399–410.

56. Lécuyer E. High Resolution Fluorescent In Situ Hybridization in Drosophila. In: Gerst JE, editor. RNA Detection and Visualization. Totowa: Humana Press; 2011. p. 31–47. [cited 2019 Jul 4]. Available from: http://link.springer.com/1 0.1007/978-1-61779-005-8_3.

57. Zimmerman SG, Peters NC, Altaras AE, Berg CA. Optimized RNA ISH, RNA FISH and protein-RNA double labeling (IF/FISH) in Drosophila ovaries. Nat Protoc. 2013;8(11):2158–79.

58. Eckermann KN, Ahmed HMM, KaramiNejadRanjbar M, Dippel S, Ogaugwu CE, Kitzmann P, et al. Hyperactive piggyBac transposase improves transformation efficiency in diverse insect species. Insect Biochem Mol Biol. 2018;98:16–24.

59. Horn C, Wimmer EA. A versatile vector set for animal transgenesis. Dev Genes Evol. 2000;210(12):630–7.

60. Ulrich A, Andersen KR, Schwartz TU. Exponential Megapriming PCR (EMP) Cloning—Seamless DNA Insertion into Any Target Plasmid without Sequence Constraints. Isalan M, editor. PLoS One. 2012;7(12):e53360.

61. Paix A, Folkmann A, Rasoloson D, Seydoux G. High efficiency, homology-directed genome editing in Caenorhabditis elegans using CRISPR-Cas9 Ribonucleoprotein complexes. Genetics. 2015;201(1):47–54.

62. Vouillot L, Thélie A, Pollet N. Comparison of T7E1 and Surveyor Mismatch Cleavage Assays to Detect Mutations Triggered by Engineered Nucleases. G3amp58 Gene Genom Genet. 2015;5(3):407–15.

63. Huang MC, Cheong WC, Lim LS, Li M-H. A simple, high sensitivity mutation screening using Ampligase mediated T7 endonuclease I and surveyor nuclease with microfluidic capillary electrophoresis. Electrophoresis. 2012; 33(5):788–96.

64. Engler C, Kandzia R, Marillonnet S. A one pot, one step, precision cloning method with high throughput capability. El-Shemy HA, editor. PLoS One. 2008;3(11):e3647.

65. Engler C, Gruetzner R, Kandzia R, Marillonnet S. Golden gate shuffling: A one-pot dna shuffling method based on type iis restriction enzymes. Peccoud J, editor. PLoS One. 2009;4(5):e5553.

13

Surface patches on recombinant erythropoietin predict protein solubility: engineering proteins to minimise aggregation

M. Alejandro Carballo-Amador[1], Edward A. McKenzie[2], Alan J. Dickson[3] and Jim Warwicker[2*]

Abstract

Background: Protein solubility characteristics are important determinants of success for recombinant proteins in relation to expression, purification, storage and administration. *Escherichia coli* offers a cost-efficient expression system. An important limitation, whether for biophysical studies or industrial-scale production, is the formation of insoluble protein aggregates in the cytoplasm. Several strategies have been implemented to improve soluble expression, ranging from modification of culture conditions to inclusion of solubility-enhancing tags.

Results: Surface patch analysis has been applied to predict amino acid changes that can alter the solubility of expressed recombinant human erythropoietin (rHuEPO) in *E. coli*, a factor that has importance for both yield and subsequent downstream processing of recombinant proteins. A set of rHuEPO proteins (rHuEPO E13K, F48D, R150D, and F48D/R150D) was designed (from the framework of wild-type protein, rHuEPO WT, via amino acid mutations) that varied in terms of positively-charged patches. A variant predicted to promote aggregation (rHuEPO E13K) decreased solubility significantly compared to rHuEPO WT. In contrast, variants predicted to diminish aggregation (rHuEPO F48D, R150D, and F48D/R150D) increased solubility up to 60% in relation to rHuEPO WT.

Conclusions: These findings are discussed in the wider context of biophysical calculations applied to the family of EPO orthologues, yielding a diverse range of calculated values. It is suggested that combining such calculations with naturally-occurring sequence variation, and 3D model generation, could lead to a valuable tool for protein solubility design.

Keywords: Protein solubility, Protein aggregates, Inclusion bodies, Erythropoietin, Solubility prediction, Protein expression

Background

Biological systems have evolved by orchestration of molecular interactions, with proteins as key elements. In the circulatory system 2.4 million red blood cells are replaced every second in human adults [1]. This requires stable and efficient regulation to fulfil this demand. Erythropoietin (EPO), the main glycoprotein behind this task, regulates the growth and proliferation of red blood cell progenitors [2]. EPO is one of the top-selling

therapeutics [3], providing therapy for millions of patients. There remains a continued demand to make EPO at large-scale and to increase the economic efficiency and, hence, availability to patients. Under this scheme, EPO has been a successful example of a biosimilar available in the market [3, 4]. Human erythropoietin (HuEPO) consists of 166 amino acid residues, in a structure that includes three N-linked glycosylation sites (N24, N38 and N83), an O-linked glycosylation (S126) and two disulphide bonds (C7-C161 and C29-C33) [5]. These complex post-translational modifications (PTMs) are the main challenge for expression of HuEPO in heterologous expression systems [6], and in the

* Correspondence: j.warwicker@manchester.ac.uk
[2]School of Chemistry, Manchester Institute of Biotechnology, University of Manchester, 131 Princess Street, Manchester M1 7DN, UK

cost-efficient *E. coli* system none of these PTMs are effectively incorporated in the cytoplasmic environment. However, the activity of a non-glycosylated version of HuEPO expressed in *E. coli* has been proved in in vitro cell proliferation assays [7]. Using bacteria to produce recombinant proteins efficiently entails a significant challenge since cysteine mispairing may lead to misfolding and low yields [8]. In order to overcome this challenge, engineered strains of *E. coli* have been developed such as the SHuffle system [9]. In this *E. coli* strain the cytoplasmic environment is altered by the overexpression of DsbC disulphide bond isomerase, and by deletion of two reductases (glutaredoxin [gor] and thioredoxin [trxB]).

Glycosylation of HuEPO contributes ~ 40% of the overall molecular mass, improving stability and solubility of the molecule as a therapeutic [10–12]. Lack of glycosylation in recombinant HuEPO (rHuEPO) derived from *E. coli* leads to aggregation during expression and, potentially, during subsequent purification, storage and delivery [13]. This protein aggregation phenomenon during expression in *E. coli* leads to incorporation into inclusion bodies (IBs), from the interactions of partially folded, misfolded or unfolded recombinant proteins in the cytoplasm [8]. Surface charge engineering has been illustrated by mutation of the three N-glycosylation sites on HuEPO to lysine (N24K, N38K and N83K), increasing net charge. This engineering decreased IB formation and facilitated the purification of protein to provide the rHuEPO crystal structure [13, 14].

Protein aggregation can involve chemical aggregation, such as disulphide bond formation, and/or physical aggregation, such as non-covalent interactions between hydrophobic surfaces [15]. Several approaches have been undertaken to diminish hydrophobic patches on the surface to prevent protein aggregation [16–23]. In this context, improved expression, stability and solubility of rHuEPO and granulocyte colony-stimulating factor (G-CSF) has been generated by application of the in vitro ribosome display technique, in combination with three parallel selection pressures (reducing agent, elevated temperature and hydrophobic interaction chromatography matrices) [24]. In the case of rHuEPO, a variant encoding four mutations resulted in a form that was less prone to aggregation [24]. Furthermore, the application of fusion tags to improve rHuEPO solubility has been successful [7, 25]. Some of these fusion partners, including NusA and maltose-binding protein (MBP), have large negatively-charged areas that may be involved in promotion of folding of the target protein by limiting protein aggregation [26]. Engineering of negatively-charged areas on protein surfaces is gaining strength as an approach for increasing solubility [27–31].

Here we report a novel experimental approach targeted at improvement of rHuEPO solubility for expression in *E. coli*, following the observation that soluble expression of proteins is inversely correlated with the size of the largest positively-charged patch on the protein surface [29, 32]. This result is based on data from cell-free expression [33], and here we test the hypothesis by mapping surface charge of rHuEPO, focusing on modulation of positively-charged patches through mutagenesis. A set of mutants has been generated, ranging from more (rHuEPO E13K) to less positively-charged surface patches (rHuEPO F48D, R150D and F48D/R150D), compared to natural (wild type) rHuEPO (rHuEPO WT). Experimental results support the prediction, i.e. largest positively-charged patch size correlated with the degree of protein aggregation in the cytoplasm of *E. coli*. Further application of this approach, particularly in the context of natural protein surface variation, could improve the rational design of proteins with enhanced solubility in cytoplasmic expression.

Results

Redesign of rHuEPO WT for altered charge surface

A published algorithm [29] was used to identify amino acids of rHuEPO for which mutation could be predicted to alter solubility (Table 1). The method is based on an observed correlation between positive charge patches and insolubility [29] for data derived in a cell-free expression system [33]. Design for improved solubility, therefore, involved identification and reduction of the larger positively-charged patch. For the protein variants shown in Table 1, the substitution R150D gave a lowered positive patch size and was predicted to be more soluble than rHuEPO WT. In contrast, substitution E13K had an increased positive patch compared with wild type and was predicted to generate a less soluble product. Both of these sites lie on the protein surface. A third site was chosen to introduce a negative charge (F48D), rather than make a charge swap, and decrease the size of the largest positive patch. It was recognized that although F48 was also on the surface, this mutation might present

Table 1 Predicted solubilities of recombinant human erythropoietin. Solubility profile was defined as described in Chan et al. [29]. Positive patch sizes are divided by that best separating soluble and insoluble datasets [33], above 1.0 implies predicted insolubility

Protein	Pos patch ratio to threshold	Prediction
rHuEPO wild-type	1.49	Insoluble
rHuEPO F48D	0.75	Soluble
rHuEPO R150D	0.61	Soluble
rHuEPO F48D/R150D	0.47	Soluble
rHuEPO E13K	2.47	Insoluble

a more challenging mutation structurally since the phenylalanine ring covers in part the hydrophobic side-chains of V46 and L155. Figure 1 shows the single site mutations and one double site mutation employed in this study, and their charge surfaces in comparison with wild type rHuEPO. Amino acid conservation analysis [34–36] showed that R150 is relatively conserved across evolution, but E13 and F48 showed less conservation (see Additional file 1: Figure S1). Previous site-directed mutagenesis of these residues had shown no alteration of the folded state of rHuEPO structure [37].

Soluble expression of rHuEPO is modulated in line with the predictions

The expression and solubility of rHuEPO variants were studied under low induction conditions (Fig. 2). The total expression of rHuEPO was approximately equivalent across WT and mutant constructs, particularly in the SHuffle system (Fig. 2d). Protein solubility, assessed as the ratio of EPO detected in soluble and total EPO fraction, agrees with the prediction for all 4 mutant constructs in the SHuffle system, but for only 2 of the 4 variants in the BL21 system, the exceptions being those involving mutation at F48 i.e. rHuEPO F48D and F48D/R150D (Fig. 2c).

That the two-way charge swap tests (R150D and E13K) match predictions in both expression systems used is encouraging in terms of applying a correlation [29] learned from a large dataset, albeit in cell-free expression [33]. With regard to the general mechanism through which charge may play a role in determination of solubility, there is an increasing body of work that indicates that negatively-charged residues (rather that positively-charged residues) are more favourable for protein solubility [27, 30, 38–41]. Whatever the molecular

mechanism by which the engineering of charged patches alters solubility, it may be associated with attainment of the native, or a near native state. This would be consistent with the observation here that charge-based predictions are matched for SHuffle but not with the BL21 system. Expression in the SHuffle system will, in relative terms, favour correct formation of the two disulphide bonds in the folded recombinant protein. The F48D mutation, apart from introducing negative charge, may expose more non-polar surface than it removes, due to the sidechains of V46 and L155 that lie beneath F48 (Fig. 3). It would be expected that the BL21 strain should be more susceptible than the SHuffle strain to this exposure, since it would be less able to refold partially unfolded protein. This difference may underpin the solubility data for F48D and F48D/R150D (Fig. 2c). Although the aspartic acid sidechain introduced in the F48D mutant could in principle alter pH-dependent properties, protein expression and measurements are made at neutral pH, for which a negative charge will be carried by the F48D carboxylate group.

Bioinformatics and solubility engineering

Having found that charge surface properties, and charge contributions to folded state stability, are factors that commonly arise in experimental studies aimed at improvement of EPO production, we assessed how these properties varied between EPO orthologues. The largest positively-charged patches, and the predicted contribution of ionisable group charge interactions to folded state stability at pH 7, were calculated for 115 EPO homologues found through a BLAST [42] search, and passed through a comparative modelling pipeline. Positive patches are distributed over a surprisingly large range (viewed as

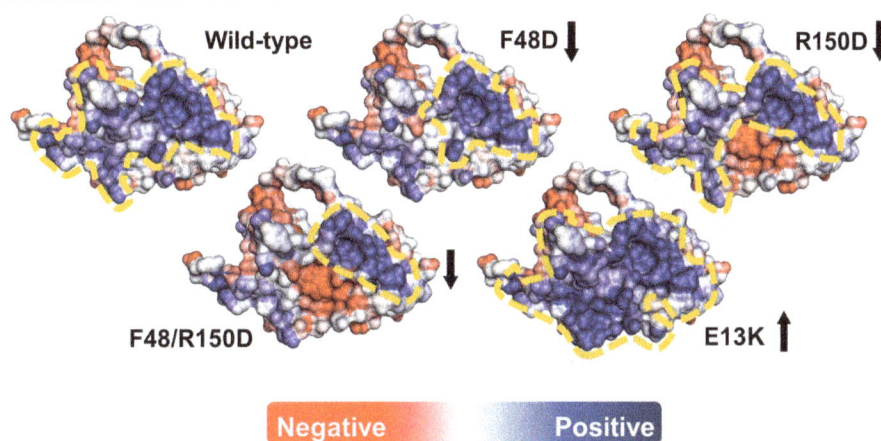

Fig. 1 HuEPO wild-type and variants surface illustration showing the electrostatic potential patches [29]. Amino acids in positive patches are represented by blue, non-charged patches by white and negatively charged by red colour, respectively, with dashed yellow contours drawn in to delineate the largest positive patches

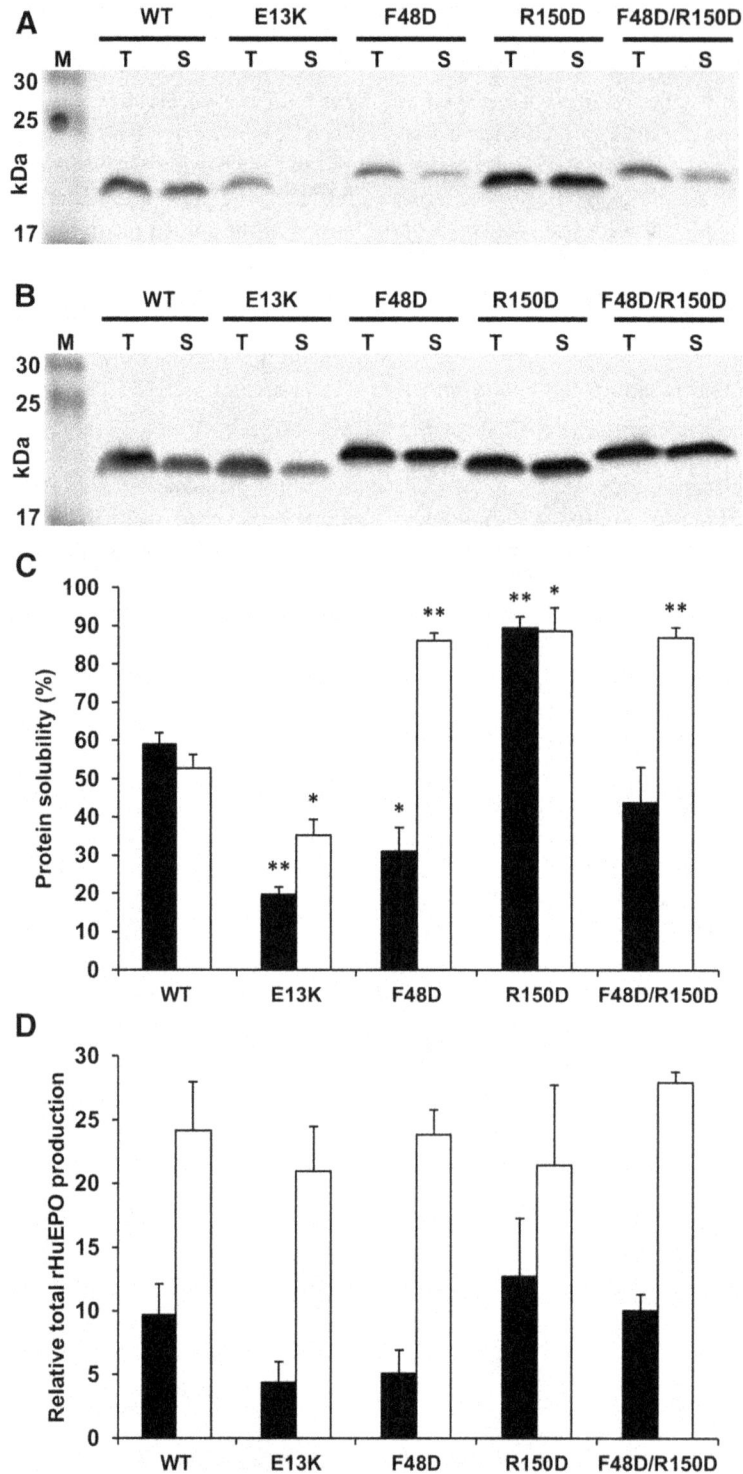

Fig. 2 Western blot of rHuEPO expression and solubility. (**a-b**) Equal volumes of total protein and soluble fraction were probed with Mouse anti-polyHis antibody and imaged with the Odyssey Imaging System for BL21 (DE3) pLysS (**a**) and SHuffle (**b**) strains. (**c**) Experimental solubility was determined by the distribution of rHuEPO between soluble and inclusion body fraction in *E. coli*. (**d**) Relative total rHuEPO production (arbitrary units). Error bars represent the + SEM for measurements in triplicate; statistical significance was calculated using a two-sided unpaired t-test (*P < 0.05, ** P < 0.01). BL21 (DE3) pLysS (■); SHuffle (□)

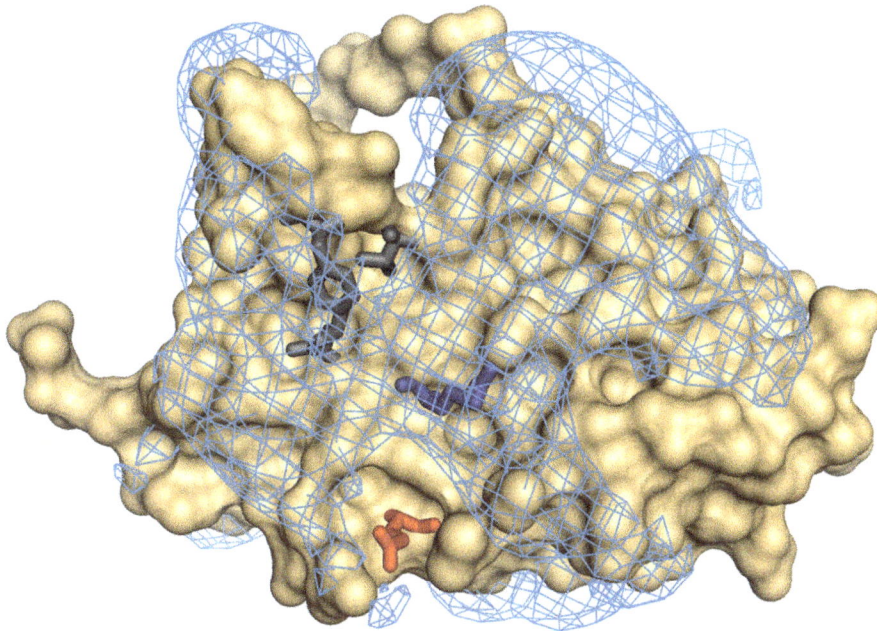

Fig. 3 Mutated residues on the surface of HuEPO. Molecular surface is shown (orange), with positive electrostatic field contoured at 30 mV (blue mesh). Residues E13 (red), F48 (grey) and R150 (blue) are shown in sticks, rather than surface representation. Also drawn are the non-polar residues V46 and L155 that are covered by F48 in the WT structure

a distribution or as individual examples, Fig. 4). Since it is not an evolutionarily conserved property, positive patch size presumably only becomes relevant for protein expression at the higher levels of over-expression, in comparison with expression in nature. As a general consideration, for over-expression, the range of values permissible naturally could present an important design feature for protein production (especially in relation to development of novel format species). A heat map, clustered according to pairwise sequence identity in Clustal [43], shows that positive patch sizes tend to cluster together, with occasional larger transitions (Fig. 4a). This indicates that a small number of mutations, or even a single mutation as in the current study, can significantly alter the charge distribution of a protein.

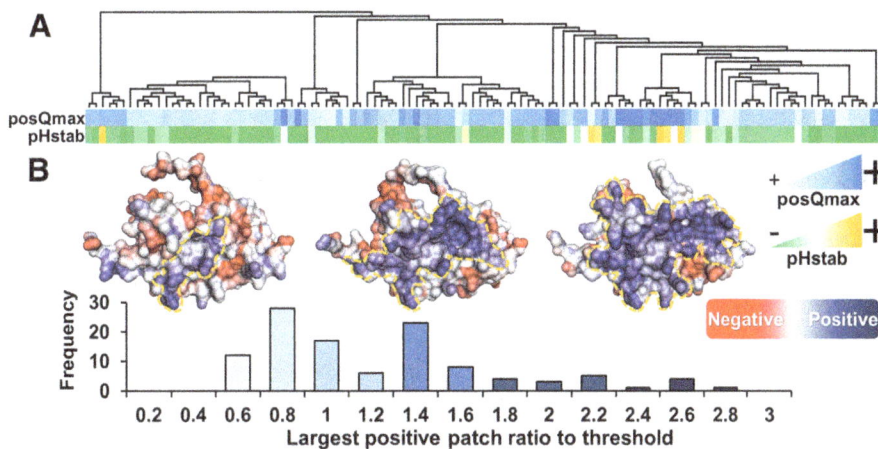

Fig. 4 Positively-charged patches and predicted stability in EPO orthologues. (**a**) A sequence-based phylogenetic tree is combined with positive patch and stability calculations. Colour coding for posQmax varies from lighter to darker blue as the calculated largest positive patch increases for an EPO orthologue (same colour code for histogram in panel **b**). Predicted pH-dependent contribution to folded state stability (pHstab) varies from yellow (positive, unfavourable) to green (negative, favourable). (**b**) Distribution of largest positive patch ratios to the threshold for the 115 EPO orthologues, showing that surface charge changes substantially. HuEPO is located roughly in the centre of the distribution. Frequency is the number of EPO orthologues having a largest positive patch ratio to threshold, in the given x-axis bin

Discussion

When expressed in *E. coli* rHuEPO WT tended to form insoluble protein aggregates in inclusion bodies [14]. The use of sub-optimal temperature and lower inducer concentrations has been argued to be a more appropriate approach to assess protein folding (and hence solubility) of recombinant proteins in *E. coli* [44]. Low induction conditions (decreased temperature, lower IPTG challenge) decreased cell growth and the elongation rate of translation [45] and induced chaperone activity and protein folding capability [46]. These responses led to better folding and less degradation [47] and offered a more refined system to investigate the consequences of variant structures on the expression and solubility of recombinant protein expression in *E. coli*.

Our understanding of IB formation and subsequent recovery of native protein is increasing [48, 49]. Important factors include the effects of a particular protein on metabolic burden in *E. coli* [50], chaperone and codon optimisation [51, 52], particular genetic loci [53], and solubility tag and medium engineering of the expression system [54]. However, there are still unknowns in what dictates IB formation for particular proteins. With regard to solubility in cell-free expression [33], a model was put forward [29] as one potential explanation of the correlation, involving charge interactions between the positive charge surface of a protein and large concentrations of a biological macro-anion i.e. mRNA. The current work does not address the underlying model. Other studies have highlighted the importance of charge properties in the production of EPO, from the incorporation of three lysines to remove the N-linked glycosylation sites and modify pI [14], to recent work in which the contribution of positively-charged patches was highlighted [55]. In the latter case, it was reasoned that improved solubility properties associated with increased ionic strength could be a result of lessening of repulsion within clusters of basic sidechains. The positively-charged region is the same as that studied here. However, the rationale for modification of aggregation properties is different, i.e. native state stabilization [55] compared with reduction of self-association (current work). The systems are quite different, purified protein [55] compared with cytoplasmic expression in *E. coli* (current work), and the suggested molecular mechanisms in each case remain unproven at this stage. It is intriguing that the same positively-charged region on EPO has featured in a predictive study (current work) and in a post-hoc rationalization [55]. Further, this patch has also been implicated in a selection mutagenesis screen for EPO variants with improved stability [24]. Ribosome display led to a variant that incorporated mutations at 4 sites, and exhibited a decreased aggregation in accelerated shelf-life studies. One of the mutations in that variant was G158E,

introducing a negative charge into the largest positively-charged patch [24].

The ionic strength dependence study of Banks (2015) implied that charge-charge interactions are delicately balanced in EPO (at pH 6.9), and thus predicted ionisable group contributions to folded state stability are also shown in the heat map format (at pH 7.0). It is relatively rare for proteins to have a predicted contribution (or indeed a measured contribution) that is net unfavourable as shown for a few examples here (Fig. 4a). This is likely to be correlated with the observation that increasing ionic strength improves the folded state stability for EPO, since unfavourable charge-charge interactions will be diminished [55]. Again, a relatively simple bioinformatics calculation shows molecular detail that is likely to underpin observed changes in EPO production, and furthermore it can guide design for improved production.

Importantly, no account has been taken of glycosylation in these interpretations, since it is difficult to model the conformations of these components with respect to the protein. The net negative charge carried by the glycosyl groups is emphasized by the change in pI of EPO from 9.2 to 4.4 upon glycosylation [10]. Interactions between sialic acid groups and positive charges on the protein are likely to improve the stabilizing component of the charge-charge balance. It should be emphasized that when charge-charge interactions contribute relatively little in net terms at neutral pH, this is generally the result of favourable and unfavourable terms cancelling, not due to a complete absence of ionisable group interactions. Such a situation lends itself to modification by relatively small changes in amino acid sequence (depending on location of those changes in the structure). In the heat map for EPO homologue stability contributions from charge-charge interactions (Fig. 4a), similar contributions generally cluster together (within the sequence-based phylogenetic tree), but there are also some abrupt changes, emphasizing the sequence-sensitivity. In general terms, the least stable contributions tend to group with the most positive patches (and vice versa). This may support the rationalization given for improved resistance to aggregation as ionic strength is increased for pure EPO [55], in that larger positively-charged patches are predicted to associate with less stable folded proteins.

Conclusions

Solubility prediction for proteins has been the focus of several groups in the last 25 years [38–41, 56–59]. Here, a method developed in our group [29] has been tested. Mutations of rHuEPO gave experimental results in line with predictions, excepting F48D in one of the two expression systems used. The F48D mutation stands apart from the other (charge swap) variants, in that it is likely to alter the hydrophobicity. It is therefore plausible that

F48D leads to reduced folding efficiency, a suggestion consistent with an improved solubility in the SHuffle system, compared with reduced solubility in the BL21 strain. Further purification might allow a better structural understanding of these mutations, however, native purification has not been possible [7, 11, 13, 14]. Finding that the region of EPO targeted here also appears in selection mutagenesis for improved stability [24] and a recent study of ionic strength effects [55], bioinformatics was employed to reveal the extent of variation in EPO homologues. Interestingly, both the largest positively-charged patch and the predicted contribution of ionisable group interactions to stability (pH 7) vary substantially through evolution, suggesting that the solubility of these EPO homologues, if over-expressed, would be divergent. Indeed, even small changes in amino acid sequence can lead to relatively large changes in solubility. Whilst it may not always be feasible to engineer a human protein across species (e.g. considering immunogenicity of a therapeutic protein), biophysical calculations for a set of homologues could guide design of protein with enhanced stability and solubility for large-scale expression, if incorporated at a sufficiently early point in the design cycle.

Methods

rHuEPO solubility profile and mutant design
The PyMOL Molecular Graphics System version 1.3 [60] and Swiss-PdbViewer 4.0.1 [61] were used to analyse rHuEPO structural and sequence features. Protein solubility predictions were calculated using an algorithm developed in our group [29] (see Additional file 2: Table S1 and Additional file 3: Table S2). The algorithm computes structured-based parameters, including the sizes of positively- and negatively-charged patches, when the electrostatic potential field is contoured at + 25 mV or – 25 mV. It also gives the size of the largest patch for which all points are between – 25 mV and + 25 mV (effectively non-charged). The principal prediction [29] is generated from the ratio of the largest positively-charged patch to a threshold, but a supplemental prediction from the combination of positive and non-charged patches is also made. Thresholds were calculated from a dataset of experimental solubilities determined for cell-free expression of *E. coli* proteins [33], as the value of a parameter that best separates less and more soluble proteins. Proteins with larger positive patches are predicted as less soluble, and have a ratio to threshold above 1.0. Where the ratio to threshold is below 1.0, a protein is predicted as soluble. Current work concentrates on positive patches (posQ), since this structure-based feature gives the best separation [29]. Substitutions to modify posQ, based on the protein data bank [62] file 1EER [63] were carried out in Swiss-PdbViewer, and the resulting structures analysed

for modification of the largest positive patch (Table 1). This in silico mutational screening gave the following candidate mutations: rHuEPO E13K, rHuEPO F48D, rHuEPO R150D and the double mutant rHuEPO F48D/R150D. Aspartic acid was selected for the introduction of negative charge, in preference to glutamic acid, due to its shorter side chain, lowering the possibility of non-specific interactions with surrounding side chains. Calculations were performed using a modified crystal structure of an analogue rHuEPO taken from the 1EER PDB entry in order to maintain consistency with our experimental and native rHuEPO cDNA (K24 N, K38 N, K83 N, N121P and S122P).

Bioinformatics analysis of rHuEPO surface and stability
Multiple sequence alignments and surface mapping coloured by residue conservation were performed using ConSurf with default parameters [34–36], using the UniRef90 database [64], which removes redundancy at 90% sequence identity (see Additional file 1: Figure S1). A separate structural study of EPO homologues was made with 115 models generated from a Clustal alignment [43] using a sidechain replacement method [65] for comparative modelling. Input to the Clustal alignment was from a search for EPO orthologues with BLAST [42], followed by manual checking of EPO annotation. Patch calculations were made based on the comparative models, to give a view of EPO variation over species.

The set of EPO comparative models was also used to estimate the contribution of ionisable groups (i.e. the pH-dependent stability term) to folded state stability at pH 7. A Debye-Hückel model for interactions between ionisable group charges, at 0.15 M ionic strength was used, with Monte Carlo sampling of protonation states to derive the pH-dependent term. This modelling follows previous methodology [66], and includes subtraction of an estimated unfolded state set of interactions to arrive at the predicted pH-dependent term for folded state compared with unfolded state.

Construction of rHuEPO mutants and expression vectors
Human erythropoietin cDNA was amplified from a pre-existing mammalian expression vector by applying primers containing the restriction sites 5′-*BamHI* and 3′-*EcoRI*. The PCR fragment (lacking signal peptide) was subcloned into a pHis vector. This plasmid is a modified version of the commercial pET-16b vector (Novagen). The gene sequence for each plasmid was as follows: 5′-6xHis-Thrombin cleavage site-*BamHI*-rHuEPO-*EcoRI*-3′. rHuEPO mutations were introduced using the GENEART Site-Directed Mutagenesis System with the enzyme AccuPrime *Pfx* (Invitrogen).

Protein expression and solubility assay

The bacterial cell lines used in this study were *Escherichia coli* BL21 (DE3) pLysS and SHuffle (New England BioLabs). Bacterial strains were transformed with the pHis-rHuEPO plasmids. Transformed cells were grown overnight in 5 ml working volume of Luria-Bertani (LB) medium (10 g tryptone, 5 g yeast extract, 5 g NaCl) containing 100 µg/ml ampicillin at 37 °C with shaking at 220 rpm. In addition, BL21 (DE3) pLysS were grown in the presence of chloramphenicol (50 µg/ml) in order to preserve the pLysS plasmid. On the following day, 1 ml of pre-culture was transferred to 50 ml 2% (v/v) LB supplemented with 2% (w/v) glucose with 100 µg/ml ampicillin in 250 ml shake flasks. Experiments were performed in triplicate. Shake flasks were incubated at a constant temperature of 25 °C, with shaking at 180 rpm. Bacteria were grown to an OD_{600} of approximately 0.6–0.8. Protein expression was induced by the addition of IPTG (0.05 mM, final). After 5 h, cultures were centrifuged at 6500 g for 15 min at 4 °C. Bacterial pellets were suspended in 5 ml of lysis buffer (25 mM Tris pH 7.5, 150 mM NaCl, 1% [v/v] Triton X-100) and were stored at − 20 °C until future use. The cells were disrupted by six sonication cycles of 30 s at 20% amplitude and then allowed to cool for 30 s on ice water bath. Separation of soluble and total fractions was performed by centrifugation at 18,000 g for 30 min at 4 °C of 1 ml of each sample from the whole cell lysate. The supernatants were collected and handled as the soluble fraction. An additional 1 ml of each lysate sample was processed as the total fraction, and rHuEPO solubility was calculated by densitometric ratio of protein detected in soluble and total fractions.

Proteins in soluble and total fractions were separated by sodium dodecyl sulphate-polyacrylamide gel electrophoresis (SDS-PAGE) with 12% (w/v) acrylamide using the Mini-PROTEAN Tetra Cell (BioRad). Samples containing equal volumes (20 µl) of protein were subjected to heat at 95 °C for 5 min in 6x denaturing buffer (375 mM Tris pH 6.8, 12% [w/v] SDS, 60% [v/v] glycerol, 0.06% [w/v] bromophenol blue, 5.5% [v/v] β-mercaptoethanol). Separated proteins were transferred to nitrocellulose membranes using a transblot semi-dry transfer cell (Bio-Rad) at 15 V for 45 min. Membranes were blocked overnight for non-specific binding in blocking buffer (5% [w/v] skimmed milk in TBS-Tween pH 7.4) at 4 °C with shaking. For detection of rHuEPO, a mouse anti-polyHis antibody (Sigma) was diluted 1:5000 in blocking buffer solution and the membrane was incubated in this solution for 2 h at room temperature with agitation. After three washes in TBS-Tween (5 min each time), samples were incubated with an IR-labelled secondary Donkey anti-Mouse IgG antibody (LI-COR) diluted 1:15000 in blocking buffer solution at room temperature for 45 min. Following

incubation, the secondary antibody was removed and the membrane was washed three times. For IR detection, blots were imaged with the Odyssey Imaging System. Bands were quantified in Image Studio Lite software (LI-COR).

Additional files

Additional file 1: Figure S1. Multiple alignment of HuEPO. (A) A surface map is coloured by residue conservation scores [34–36]. The image was rendered using PyMOL [60]. (B) Panel shows the same color-coding for conservation show in panel (A), but here applied to the amino acid sequence of rHuEPO. (PDF 1525 kb)

Additional file 2: Table S1. Positively-charged patches size profile of rHuEPO WT from the charged patch calculator. Complete screening of posQ ratio scores for the modified rHuEPO WT (PDB: 1EER) is shown. The largest positive patches are represented by blue (ratio > 1.0). Those proteins with ratio above 1.0 are predicted as insoluble and below 1.0 as soluble. The three targeted residues in this study are highlighted in red. Ratio: largest positively-charged patch (posQ) value from the charged patch calculator [29]. Charge patches: HYD, hydrophobic (non-charged); NEG, negatively-charged; POS, positively-charged. (PDF 478 kb)

Additional file 3: Table S2. Summary of the solubility screening of rHuEPO. Left column shows the complete mutational screening of all positive charge amino acids (i.e. arginine and lysine) within the largest positively-charged patch (posQ) for aspartic acid (D). Next two columns summarize a set of substitutions of any amino acid in the posQ for D. The column on the right shows all the negative charge residues (i.e. aspartic and glutamic acid) within the posQ for arginine or lysine. Those proteins with posQ ratio above 1.0 are predicted as insoluble and below 1.0 as soluble. Selected proteins for further site-directed mutagenesis are highlighted in red. (PDF 168 kb)

Abbreviations

EPO: Erythropoietin; HuEPO: Human erythropoietin; IBs: Inclusion bodies; posQ: positive patches; PTMs: Post-translational modifications; rHuEPO: Recombinant human erythropoietin; WT: Wild-type protein

Acknowledgements

We thank Professor Stephen High and Dr. E. Ceh-Pavia for inspiring discussions.

Funding

This work was supported by the Biotechnology and Biological Sciences Research Council (BBSRC) via a BioProNet award (BB/L013770/1), and a PhD studentship by CONACyT to MAC (309167). The funders had no role in the study design, or in the collection, analysis and interpretation of data, or in writing the manuscript.

Authors' contributions

MAC carried out the experiments, analysed all data, and wrote the first draft of the manuscript. EAM participated in its design and supervision. AJD and JW conceived the study, participated in its design and supervision, and finalised the manuscript with MAC. All authors read and approved the final manuscript.

Ethics approval and consent to participate

Not applicable.

Consent for publication

Not applicable.

Competing interests

The authors declare that they have no competing interests.

Author details

[1]Facultad de Ciencias, Universidad Autónoma de Baja California, Km. 103 Carretera Tijuana–Ensenada, Pedregal Playitas, 22860 Ensenada, Baja California, Mexico. [2]School of Chemistry, Manchester Institute of Biotechnology, University of Manchester, 131 Princess Street, Manchester M1 7DN, UK. [3]Faculty of Science and Engineering, Manchester Institute of Biotechnology, University of Manchester, 131 Princess Street, Manchester M1 7DN, UK.

References

1. Sackmann E: Biological membranes architecture and function. In: Handbook of Biological Physics. Edited by Lipowsky R, Sackmann E, vol. Volume 1. Amsterdam: Elsevier; 1995: 1–63.
2. Krantz SB. Erythropoietin. Blood. 1991;77(3):419–34.
3. Walsh G. Biopharmaceutical benchmarks 2018. Nat Biotechnol. 2018;36:1136.
4. Goldsmith D, Dellanna F, Schiestl M, Krendyukov A, Combe C. Epoetin Biosimilars in the treatment of renal Anemia: what have we learned from a decade of European experience? Clinical drug investigation. 2018;38(6):481–90.
5. Lai PH, Everett R, Wang FF, Arakawa T, Goldwasser E. Structural characterization of human erythropoietin. J Biol Chem. 1986;261(7):3116–21.
6. Skibeli V, Nissen-Lie G, Torjesen P. Sugar profiling proves that human serum erythropoietin differs from recombinant human erythropoietin. Blood. 2001; 98(13):3626–34.
7. Jeong TH, Son YJ, Ryu HB, Koo BK, Jeong SM, Hoang P, Do BH, Song JA, Chong SH, Robinson RC, et al. Soluble expression and partial purification of recombinant human erythropoietin from E. coli. Protein Expr Purif. 2014;95: 211–8.
8. Fink AL. Protein aggregation: folding aggregates, inclusion bodies and amyloid. Fold Des. 1998;3(1):9–23.
9. Lobstein J, Emrich C, Jeans C, Faulkner M, Riggs P, Berkmen M. SHuffle, a novel Escherichia coli protein expression strain capable of correctly folding disulfide bonded proteins in its cytoplasm. Microb Cell Factories. 2012;11(1): 56.
10. Davis JM, Arakawa T, Strickland TW, Yphantis DA. Characterization of recombinant human erythropoietin produced in Chinese hamster ovary cells. Biochemistry. 1987;26(9):2633–8.
11. Narhi LO, Arakawa T, Aoki KH, Elmore R, Rohde MF, Boone T, Strickland TW. The effect of carbohydrate on the structure and stability of erythropoietin. J Biol Chem. 1991;266(34):23022–6.
12. Banks DD. The effect of glycosylation on the folding kinetics of erythropoietin. J Mol Biol. 2011;412(3):536–50.
13. Cheetham JC, Smith DM, Aoki KH, Stevenson JL, Hoeffel TJ, Syed RS, Egrie J, Harvey TS. NMR structure of human erythropoietin and a comparison with its receptor bound conformation. Nat Struct Biol. 1998;5(10):861–6.
14. Narhi LO, Arakawa T, Aoki K, Wen J, Elliott S, Boone T, Cheetham J. Asn to Lys mutations at three sites which are N-glycosylated in the mammalian protein decrease the aggregation of Escherichia coli-derived erythropoietin. Protein Eng. 2001;14(2):135–40.
15. Wang W. Protein aggregation and its inhibition in biopharmaceutics. Int J Pharm. 2005;289(1–2):1–30.
16. Jenkins TM, Hickman AB, Dyda F, Ghirlando R, Davies DR, Craigie R. Catalytic domain of human immunodeficiency virus type 1 integrase: identification of a soluble mutant by systematic replacement of hydrophobic residues. Proc Natl Acad Sci U S A. 1995;92(13):6057–61.
17. Li Y, Yan Y, Zugay-Murphy J, Xu B, Cole JL, Witmer M, Felock P, Wolfe A, Hazuda D, Sardana MK, et al. Purification, solution properties and crystallization of SIV integrase containing a continuous core and C-terminal domain. Acta Crystallogr D Biol Crystallogr. 1999;55(Pt 11:1906–10.
18. Das D, Georgiadis MM. A directed approach to improving the solubility of Moloney murine leukemia virus reverse transcriptase. Protein Sci. 2001; 10(10):1936–41.
19. Slovic AM, Summa CM, Lear JD, DeGrado WF. Computational design of a water-soluble analog of phospholamban. Protein Sci. 2003;12(2):337–48.
20. Fan D, Li Q, Korando L, Jerome WG, Wang J. A monomeric human apolipoprotein E carboxyl-terminal domain. Biochemistry. 2004;43(17):5055–64.
21. Lawson AJ, Walker EA, White SA, Dafforn TR, Stewart PM, Ride JP. Mutations of key hydrophobic surface residues of 11 beta-hydroxysteroid dehydrogenase type 1 increase solubility and monodispersity in a bacterial expression system. Protein Sci. 2009;18(7):1552–63.
22. Andersen TCB, Lindsjø K, Hem CD, Koll L, Kristiansen PE, Skjeldal L, Andreotti AH, Spurkland A. Solubility of recombinant Src homology 2 domains expressed in E. coli can be predicted by TANGO. BMC Biotechnol. 2014;14(1):3.
23. Jetha A, Thorsteinson N, Jmeian Y, Jeganathan A, Giblin P, Fransson J. Homology modeling and structure-based design improve hydrophobic interaction chromatography behavior of integrin binding antibodies. mAbs. 2018;10(6):890–900.
24. Buchanan A, Ferraro F, Rust S, Sridharan S, Franks R, Dean G, McCourt M, Jermutus L, Minter R. Improved drug-like properties of therapeutic proteins by directed evolution. Protein Eng Des Sel. 2012;25(10):631–8.
25. Ahn JH, Keum JW, Kim DM. Expression screening of fusion partners from an E. coli genome for soluble expression of recombinant proteins in a cell-free protein synthesis system. PLoS One. 2011;6(11):e26875.
26. Zhang YB, Howitt J, McCorkle S, Lawrence P, Springer K, Freimuth P. Protein aggregation during overexpression limited by peptide extensions with large net negative charge. Protein Expr Purif. 2004;36(2):207–16.
27. Trevino SR, Scholtz JM, Pace CN. Amino acid contribution to protein solubility: asp, Glu, and Ser contribute more favorably than the other hydrophilic amino acids in RNase Sa. J Mol Biol. 2007;366(2):449–60.
28. Perchiacca JM, Ladiwala AR, Bhattacharya M, Tessier PM. Aggregation-resistant domain antibodies engineered with charged mutations near the edges of the complementarity-determining regions. Protein Eng Des Sel. 2012;25(10):591–601.
29. Chan P, Curtis RA, Warwicker J. Soluble expression of proteins correlates with a lack of positively-charged surface. Sci Rep. 2013;3:3333.
30. Chong SH, Ham S. Interaction with the surrounding water plays a key role in determining the aggregation propensity of proteins. Angew Chem Int Ed Engl. 2014;126(15):4042–5.
31. Laber JR, Dear BJ, Martins ML, Jackson DE, Divenere A, Gollihar JD, Ellington AD, Truskett TM, Johnston KP, Maynard JA. Charge shielding prevents aggregation of supercharged GFP variants at high protein concentration. Mol Pharm. 2017;14(10):3269–80.
32. Hussain H, Fisher DI, Roth RG, Mark Abbott W, Carballo-Amador MA, Warwicker J, Dickson AJ. A protein chimera strategy supports production of a model "difficult-to-express" recombinant target. FEBS Lett. 2018;592(14): 2499–511.
33. Niwa T, Ying BW, Saito K, Jin W, Takada S, Ueda T, Taguchi H. Bimodal protein solubility distribution revealed by an aggregation analysis of the entire ensemble of Escherichia coli proteins. Proc Natl Acad Sci U S A. 2009; 106(11):4201–6.
34. Ashkenazy H, Erez E, Martz E, Pupko T, Ben-Tal N. ConSurf 2010: calculating evolutionary conservation in sequence and structure of proteins and nucleic acids. Nucleic Acids Res. 2010;38(Web Server issue):W529–33.
35. Celniker G, Nimrod G, Ashkenazy H, Glaser F, Martz E, Mayrose I, Pupko T, Ben-Tal N. ConSurf: using evolutionary data to raise testable hypotheses about protein function. Isr J Chem. 2013;53(3–4):199–206.
36. Ashkenazy H, Abadi S, Martz E, Chay O, Mayrose I, Pupko T, Ben-Tal N. ConSurf 2016: an improved methodology to estimate and visualize evolutionary conservation in macromolecules. Nucleic Acids Res. 2016; 44(W1):W344–50.
37. Elliott S, Lorenzini T, Chang D, Barzilay J, Delorme E. Mapping of the active site of recombinant human erythropoietin. Blood. 1997;89(2):493–502.
38. Kuntz ID. Hydration of macromolecules. III. Hydration of polypeptides. JACS. 1971, 93(2):514–6.
39. Collins KD, Washabaugh MW. The Hofmeister effect and the behaviour of water at interfaces. Q Rev Biophys. 1985;18(4):323–422.
40. Collins KD. Charge density-dependent strength of hydration and biological structure. Biophys J. 1997;72(1):65–76.
41. Trevino SR, Scholtz JM, Pace CN. Measuring and increasing protein solubility. J Pharm Sci. 2008;97(10):4155–66.
42. Altschul SF, Gish W, Miller W, Myers EW, Lipman DJ. Basic local alignment search tool. J Mol Biol. 1990;215(3):403–10.
43. Larkin MA, Blackshields G, Brown NP, Chenna R, McGettigan PA, McWilliam H, Valentin F, Wallace IM, Wilm A, Lopez R, et al. Clustal W and Clustal X version 2.0. Bioinformatics. 2007;23(21):2947–8.
44. Sevastsyanovich Y, Alfasi S, Overton T, Hall R, Jones J, Hewitt C, Cole J. Exploitation of GFP fusion proteins and stress avoidance as a generic strategy for the production of high-quality recombinant proteins. FEMS Microbiol Lett. 2009;299(1):86–94.
45. Farewell A, Neidhardt FC. Effect of temperature on in vivo protein synthetic capacity in Escherichia coli. J Bacteriol. 1998;180(17):4704–10.

47. Chesshyre J, Hipkiss A. Low temperatures stabilize interferon α-2 against proteolysis in Methylophilus methylotrophus and Escherichia coli. Appl Microbiol Biotechnol. 1989;31(2):158–62.

48. Singh A, Upadhyay V, Upadhyay AK, Singh SM, Panda AK. Protein recovery from inclusion bodies of Escherichia coli using mild solubilization process. Microb Cell Factories. 2015;14(1):1–10.

49. Qi X, Sun Y, Xiong S. A single freeze-thawing cycle for highly efficient solubilization of inclusion body proteins and its refolding into bioactive form. Microb Cell Factories. 2015;14(1):1–12.

50. Rahmen N, Fulton A, Ihling N, Magni M, Jaeger K-E, Büchs J. Exchange of single amino acids at different positions of a recombinant protein affects metabolic burden in Escherichia coli. Microb Cell Factories. 2015;14(1):1–18.

51. Itkonen JM, Urtti A, Bird LE, Sarkhel S. Codon optimization and factorial screening for enhanced soluble expression of human ciliary neurotrophic factor in Escherichia coli. BMC Biotechnol. 2014;14(1):92.

52. Wang Y, Li Y-Z. Cultivation to improve in vivo solubility of overexpressed arginine deiminases in Escherichia coli and the enzyme characteristics. BMC Biotechnol. 2014;14(1):1–10.

53. Pandey N, Sachan A, Chen Q, Ruebling-Jass K, Bhalla R, Panguluri KK, Rouviere PE, Cheng Q. Screening and identification of genetic loci involved in producing more/denser inclusion bodies in Escherichia coli. Microb Cell Factories. 2013;12(1):1–12.

54. Zhou K, Zou R, Stephanopoulos G, Too H-P. Enhancing solubility of deoxyxylulose phosphate pathway enzymes for microbial isoprenoid production. Microb Cell Factories. 2012, 11(1):1–8.

55. Banks DD. Nonspecific shielding of unfavorable electrostatic intramolecular interactions in the erythropoietin native-state increase conformational stability and limit non-native aggregation. Protein Sci. 2015;24(5):803–11.

56. Sormanni P, Aprile FA, Vendruscolo M. The CamSol method of rational Design of Protein Mutants with enhanced solubility. J Mol Biol. 2015;427(2): 478–90.

57. Hebditch M, Carballo-Amador MA, Charonis S, Curtis R, Warwicker J. Protein–sol: a web tool for predicting protein solubility from sequence. Bioinformatics. 2017;33(19):3098–100.

58. Matsui D, Nakano S, Dadashipour M, Asano Y. Rational identification of aggregation hotspots based on secondary structure and amino acid hydrophobicity. Sci Rep. 2017;7(1):9558.

59. Wolf Pérez AM, Sormanni P, Andersen JS, Sakhnini LI, Rodriguez-Leon I, Bjelke JR, Gajhede AJ, De Maria L, Otzen DE, Vendruscolo M, Lorenzen N. In vitro and in silico assessment of the developability of a designed monoclonal antibody library. mAbs. 2018:1–13.

60. Schrödinger LLC: The PyMOL Molecular Graphics System, Version 1.3r1. In.; 2010.

61. Guex N, Peitsch MC. SWISS-MODEL and the Swiss-PdbViewer: an environment for comparative protein modeling. Electrophoresis. 1997; 18(15):2714–23.

62. Berman HM, Westbrook J, Feng Z, Gilliland G, Bhat TN, Weissig H, Shindyalov IN, Bourne PE. The Protein Data Bank. Nucleic Acids Res. 2000; 28(1):235–42.

63. Syed RS, Reid SW, Li C, Cheetham JC, Aoki KH, Liu B, Zhan H, Osslund TD, Chirino AJ, Zhang J, et al. Efficiency of signalling through cytokine receptors depends critically on receptor orientation. Nature. 1998;395(6701):511–6.

64. Suzek BE, Huang H, McGarvey P, Mazumder R, Wu CH. UniRef: comprehensive and non-redundant UniProt reference clusters. Bioinformatics. 2007;23(10):1282–8.

65. Cole C, Warwicker J. Side-chain conformational entropy at protein–protein interfaces. Protein Sci. 2002;11(12):2860–70.

66. Chan P, Warwicker J: Evidence for the adaptation of protein pH-dependence to subcellular pH. BMC Biol 2009, 7:69–69.

14

Formulation and evaluation of norcanthridin nanoemulsions against the *Plutella xylostella* (Lepidotera: Plutellidae)

Liya Zeng, Yongchang Liu, Jun Pan and Xiaowen Liu[*]

Abstract

Background: Norcantharidin (NCTD), a demethylated derivative of cantharidin (defensive toxin of blister beetles), has been reported to exhibit insecticidal activity against various types of agricultural pests. However, NCTD applications are limited by its poor water solubility and high dosage requirement. Nanoemulsions have attracted much attentions due to the transparent or translucence appearance, physical stability, high bioavailability and non-irritant in nature. In general, nanoemulsions with small droplet size can enhance the bioavailability of drugs, whereas this phenomenon is likely system dependent. In present study, NCTD nanoemulsions were developed and optimized to evaluate and improve the insecticidal activity of NCTD against *Plutella xylostella* (Lepidotera: Plutellidae) by a spontaneous emulsification method.

Results: Triacetin, Cremophor EL and butanol were selected as the constituents of NCTD nanoemulsions via solubility determination, emulsification efficiency and ternary phase diagram construction. Insecticidal activity of NCTD nanoemulsion was associated with the content of surfactant and cosurfactant: (1) Higher effective toxicity exhibited at Smix (surfactant to cosurfactant mass ratio) = 3:1 that may be associated with the changes in interfacial tension; (2) NCTD nanoemulsion at 3:7 < SOR (surfactant to oil mass ratio) < 6:4 was more effective at lower surfactant level, which was attributed to the relatively slow diffusion rate of NCTD hindering by excess surfactant. Interestingly, nanoemulsions with smaller droplets were not found to be more effective in our study.

Conclusions: The optimized NCTD nanoemulsion (triacetin/Cremophor EL/butanol (60/20/20, w/w)) exhibited effective insecticidal activity (LC_{50} 60.414 mg/l, LC_{90} 185.530 mg/l, 48 h) than the NCTD acetone solution (LC_{50} 175.602 mg/L, LC_{90} 303.050 mg/L, 48 h). Spontaneous emulsifying nanoemulsion employed to formulate this poor water-soluble pesticide is a potential system for agriculture application.

Keywords: Norcantharidin, Nanoemulsion, Insecticidal, *Plutella xylostella*, Spontaneous emulsification

Background

Cantharidin, the defensive toxin of the blister beetles (Colepotera: Meloidae) and some oedemerid beetles (Colepotera: Oedemeridae) [1], has been reported effective as an anticancer [2–4] and biopesticide [5–7]. However, it is difficult to synthesize cantharidin in the lab and to acquire from natural production. Norcantharidin (NCTD), an available analogue of cantharidin, has been approved for the treatment of multiple types of cancer [8–10]. Both cantharidin and NCTD have been shown

to be inhibitor of the highly conserved serine/threonine protein phosphatases (PPs) of eukaryote [11]. In addition, NCTD can act as a natural pesticide because of its effective oral toxicity [12, 13] and strong antifungal activity [14]. Nevertheless, applications of NCTD in agriculture are limited by its poor aqueous solubility and high dosage requirement. Therefore, appropriate formulation delivery systems need to be developed to utilize NCTD in agriculture widely. Formulation is a physical mixture of the one or more biologically active ingredient with a compatible inert (inactive) substances/ filler material in definite proportions. Conventional pesticide and their formulations has raised concerns about their residues in the food and the environment [15–18]. Designing of nanoformulation

* Correspondence: lxwforpaper@126.com
Key Laboratory of Comprehensive Utilization of Advantage Plants Resources in Hunan South, College of Chemistry and Bioengineering, Hunan University of Science and Engineering, Yongzhou, Hunan, China

of different pesticides has newly emerged field for controlling field pests with effective and low environmental risk.

Nanoemulsions-based delivery systems are particularly effective tools for encapsulating lipophilic compounds in pesticides [1, 19, 20]. Nanoemulsions are a class of emulsions with droplet size ranging from 20 to 200 nm and are kinetically stable. Due to their characteristic size, nanomulsions have been used for improving the bioactivity/bioavailability, stability and safety of some active substances [1, 20, 21]. Many methods have been reported on preparing nanoemulsions for agriculture. Recently, one of the most popular ideas is to prepare nanoemulsions use spontaneous emulsification methods because of its low-cost, simplicity of preparation and without any specialized homogenization equipment [22–24]. However, the main practical problem of spontaneous emulsification that confronts researchers is the requirement of high surfactant or organic solvent concentration [22, 23], which is undesirable on the basis of cost, safety and regulations. Consequently, there is a focus on the potential for NCTD-nanoemulsions that were prepared by the spontaneous emulsification method, stabilized by nonionic surfactant with low content and approved for insecticide use.

It has been proposed that the bioavailability of encapsulated compounds would increase as the droplet size of emulsions decreases, but this is likely to be system dependent. For instance, the antiradical efficiency and antioxidant activity of lycopene nanoemulsions with droplet sizes between 100 and 200 nm were higher than droplet size below 100 nm [21]. That is to say, there are other factors existing affect the bioactivity/bioavailability of the nanoemulsions containing active substances. Most studies undertaken so far have speculated that nanomulsions increase the lipid solubility of active materials which played an important role in improving bioactive [25]. Moreover, the larger surface area of nano-droplets improves the absorption and digestion of activity components [26]. However, the droplet size fabricated via spontaneous emulsification method is depended upon many factors such as the formulation compound types and content, system composition, interfacial tension, phase behavior, etc. [23, 27, 28]. Therefore, the bioactivity/bioavailability enhancement mechanism of nanomulsions is still not clear. In particular no study, to our knowledge, has considered the impact of processing conditions, physicochemical characteristics and/or droplet size on insecticidal activity of nanoemulsions formulated by spontaneous emulsification method.

The first objective of this study was to prepare and optimize the NCTD nanoemulsions using the spontaneous emulsification method. Second, the effect of modifying the self-emulsification processing conditions (surfactant and cosurfactant concentration) on the insecticidal activity of NCTD nanoemulsions against the

Plutella xylostella (Lepidotera: Plutellidae) was investigated. The insecticidal activity of NCTD nanoemulsions with different physicochemical characteristics (droplet size and size distribution) was also evaluated. Particularly, we were interested in investigating the relation between NCTD nanoemulsions characteristics and its bioactivity. The information obtained from this study would provide reference information for designing efficient pesticides for agriculture applications.

Results

Solubility determination in oils

NCTD has poor solubility both in the water and oil phase and the maximum solubility of NCTD in water is pH dependent [29]. Table 1 shows the variation of the NCTD saturated solubility across different oils. The highest solubility was observed in triacetin with 12.39 ± 0.21 mg/mL; the lower solubility was exhibited in tributyrin (3.06 ± 0.11 mg/mL). This difference may be attributed to their physical properties such as hydrophilcity, lipophicity or chemical polarity. The minimum solubility (0.44 ± 0.11 mg/mL) was obtained when olive oil was chosen as the oil phase, which could be attributed to its relatively higher viscosity affecting NCTD solubility. Although fatty acid esters have been well represented in many drug delivery systems, ethyl oleate and isopropyl myristate showed poor performance in our study. Numerous previously published studies showed that the addition of medium- and long-chain triglycerides did not reduce the droplet diameter but improved storage stability of nanoemulsions [30, 31]. The lethal concentration (LC_{50}) of cantharidin (12.37 mg/L) and NCTD (129.35 mg/L) against *P. xylostella* was reported [32]. Accordingly, triacetin was chosen as the optimized oil for the following studies.

The influence of surfactant

Two types of nonionic surfactants were used to stabilize the nanoemulsions including Cremophor EL, Cremophor RH 40, Tween 20 and Tween 80. The surfactant-to-oil mass ratio (SOR) was held constant at 1:1 and triacetin was used as the oil. The transmittance, droplet size, polydispersity index (PDI) and emulsification time of emulsions formed by spontaneous emulsification were significantly influenced by surfactant types (Table 2). No visible phase separation or creaming was observed in any of the formulation. The smallest droplet size with a bimodal size distribution was formed when Tween 80 was used ($r = 12.94 \pm 0.89$ nm, PDI $= 0.28 \pm 0.07$), but fine droplets could also be produced by Cremophor EL (17.03 ± 0.83 nm), Cremophor RH 40 (13.91 ± 0.55 nm) or Tween 20 (40.32 ± 18.08 nm) alone.

To further understand the emulsification ability of different surfactants, ternary phase diagrams of this

Table 1 Saturated solubility of NCTD in different oils

Oil Types	Saturated solubility (mg/mL)	Oil Types	Saturated solubility (mg/mL)
Ethyl Oleate	0.72 ± 0.01	Tributyrin	3.06 ± 0.11
Isopropyl Myristate	0.61 ± 0.01	Triacetin	12.38 ± 0.21
Soybean oil	0.47 ± 0.01	MCT	1.23 ± 0.02
Olive oil	0.44 ± 0.01		

Data are expressed as mean ± S.D. ($n = 3$)

three-components system containing water, surfactant and triacetin were constructed. No attempt was made to distinguish between oil-in-water microemulsion, bicontinuous phase, liquid crystalline phase and water-in-oil microemulsions. The oil-in-water nanoemulsions regions are identified in Fig. 1. It was observed that when Tween 20 and Tween 80 were used as surfactants, very low amount of oil (< 40% w/w) could be solubilized at a high surfactant concentration (> 60% w/w). With further increase the oil concentration to exceed 5:5 (SOR), no nanoemulsion was obtained and phase separation was observed quickly. However, in case of Cremophor EL and Cremophor RH 40, larger nanoemulsion areas were obtained compared to the Tween 20 or Tween 80 system. The maximum amount of triacetin that could be solubilized was increased to 70% w/w. This may indicate that Cremophor have better flexibility and greater ability to reduce the oil-water interfacial tension [33]. Although the nanoemulsions area scale of Cremophor EL formulations was very similar to the Cremophor RH 40, in consideration of size distribution, Cremophor EL was chosen for following study.

The influence of cosurfactant

The selection of a cosurfactant was executed by preparing a series of emulsions: 50% w/w triacetin, Cremophor EL (SOR = 1:1) and cosurfactant (ethanol, ethylene glycol, propanol, isopropanol, propylene glycol, butanol, glycerol, PEG 400, surfactant to cosurfactant mass ratio (Smix) = 3:2).

As shown in Table 3, the formation of nanoemulsions was affected by the addition of cosurfactant: the droplet diameter and particle size distribution were significantly decreased when compared to the cosurfactant-free systems (Table 2). In addition, with the cosurfactant presence, spontaneous emulsification occured conveniently

and smaller droplet size formed easily. Similar trends were also observed in relation to the size distribution. The smallest droplets with the narrowest distribution were produced for the system containing butanol ($r = 6.89 \pm 0.126$, PDI = 0.04 ± 0.064); the largest droplet size with widest distribution was exhibited for the system with PEG 400 ($r = 10.45 \pm 2.763$, PDI = 0.28 ± 0.088).

In order to accurately screen cosurfactants, the transparent temperature range, thermodynamics and dilution stability of nanoemulsions were measured (Additional file 1: Table S1). Most optional systems (except for systems with PEG 400 and ethylene glycol) exhibited thermodynamic and dilution stability. There was no phase separation, creaming or cracking between 4~75 °C in most nanoemulsions. When all of the above problems are considered together, butanol was selected as the cosurfactant to prepare NCTD nanoemulsions.

The effect of surfactant-to-cosurfactant

To develop a qualified pesticide formulation, the nanoemulsion region in the corresponding ternary phase diagram should be larger to maintain dilution stability and to avoid drug precipitation. Considering the economic efficiency and security, the SOR = 6: 4 was constant in this section. Figure 2 shows the relationship between Smix and the nanoemulsion formation (the NE means the nanoemulsions region, r < 100 nm and PDI < 0.2). The maximum concentration of oil that could be solubilized was 60% w/w. When butanol was incorporated along with the Cremophor EL at 1:2 and 1:3, the largest nanoemulsions region was appeared. The smallest nanoemulsions region was observed when the butanol concentration was decreased to Smix = 4:1.

Taken together, the NCTD nanoemulsions containing relatively small droplets and narrow size distribution

Table 2 Emulsification ability of various non-ionic surfactants to emulsify the Triacetin (triacetin:surfactant = 1:1, water 96 wt.%, 25 °C)

Surfactant	Molecular weight (g/mol)	HLB	Transmittance (%)	Droplet Size (r.nm)	PDI	Emulsification time (sec)
Cremophor EL	≈ 1630	12–14	100.08 ± 0.02	17.03 ± 0.83	0.33 ± 0.06	82.50 ± 3.12
Cremophor RH 40	≈ 2500	14–16	99.96 ± 0.05	13.91 ± 0.55	0.17 ± 0.08	84.5 ± 6.26
Tween 20	1228	16.7	100.36 ± 0.01	40.32 ± 18.08	0.52 ± 0.14	65.00 ± 7.07
Tween 80	1310	15	99.38 ± 0.03	12.94 ± 0.89	0.28 ± 0.07	171.00 ± 12.00

Data are expressed as mean ± S.D. ($n = 3$)

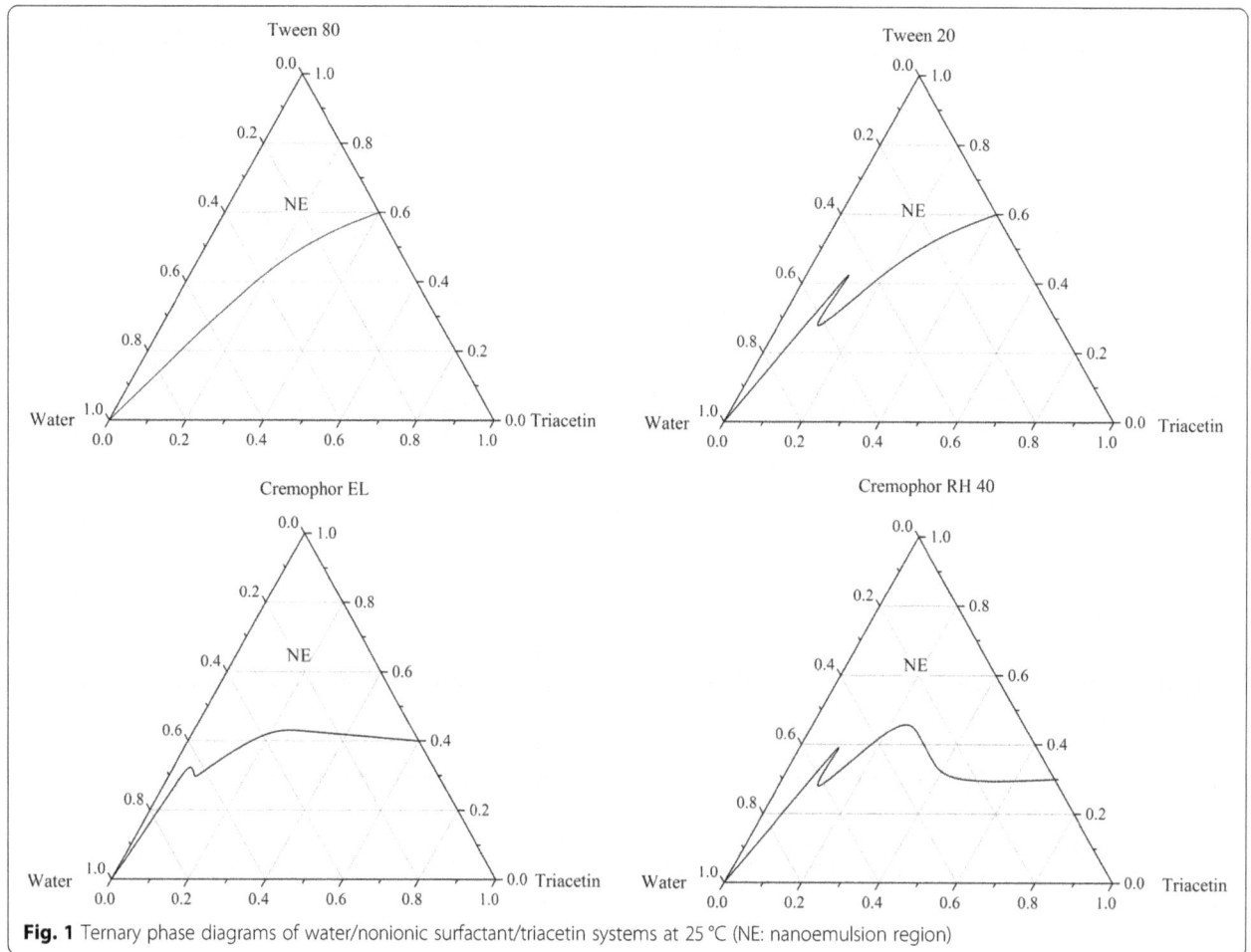

Fig. 1 Ternary phase diagrams of water/nonionic surfactant/triacetin systems at 25 °C (NE: nanoemulsion region)

could be formed with the triacetin/Cremophor EL/butanol/water system. From a commercial perspective, it is advantageous to have high content of triacetin in the nanoemulsions only with low Cremophor EL level were used in the subsequent insecticidal activity studies.

Table 3 Droplet size and size distribution of triacetin/Cremophor EL/cosurfactant/water systems at 25 °C (triacetin:Smix = 1:1, Cremophor EL:cosurfactant = 3:2, water = 96 wt.%)

Cosurfactant	Droplet size (r.nm)	Particle size distribution (PDI)
Ethanol	7.13 ± 0.20	0.09 ± 0.01
Ethylene glycol	7.07 ± 0.11	0.04 ± 0.02
Propanol	7.17 ± 0.12	0.06 ± 0.01
Isopropanol	7.38 ± 0.06	0.08 ± 0.06
Propylene glycol	7.76 ± 0.65	0.09 ± 0.03
Butanol	6.89 ± 0.13	0.04 ± 0.06
Glycerol	7.12 ± 0.12	0.05 ± 0.01
PEG 400	10.45 ± 2.76	0.28 ± 0.09

Data are expressed as mean ± S.D. (n = 3)

Insecticidal activity

The insecticidal activity of NCTD and NCTD nanoemulsions (triacetin, Cremophor EL, butanol and 95% w/w water) were evaluated. The impact of surfactant and cosurfactant concentration on the insecticidal activity was investigated via the variational Smix (1:3, 1:2, 1:1, 2:1, 3:1 and 4:1) and SOR (3:7, 4:6, 5:5 and 6:4). The insecticidal activity of NCTD-nanoemulsions with different physicochemical characteristics (droplet size and size distribution) was also evaluated.

NCTD insecticidal activity

Initially, the insecticidal activity of NCTD was determined against third-instar larvae of *P. xylostella*. Results on the mortality treated with increase NCTD concentrations (ranging from 50 to 250 mg/L) and different exposure time are shown in Table 4. Insecticidal activities of NCTD varied according to the exposure time and NCTD concentration. The insecticidal activity of NCTD at 50 mg/L was found to be relatively poor (48 h, mortality rate ≈ 20%), whereas there was 90% mortality at 250 mg/L concentration after 36 h. The fasting larvae rapidly

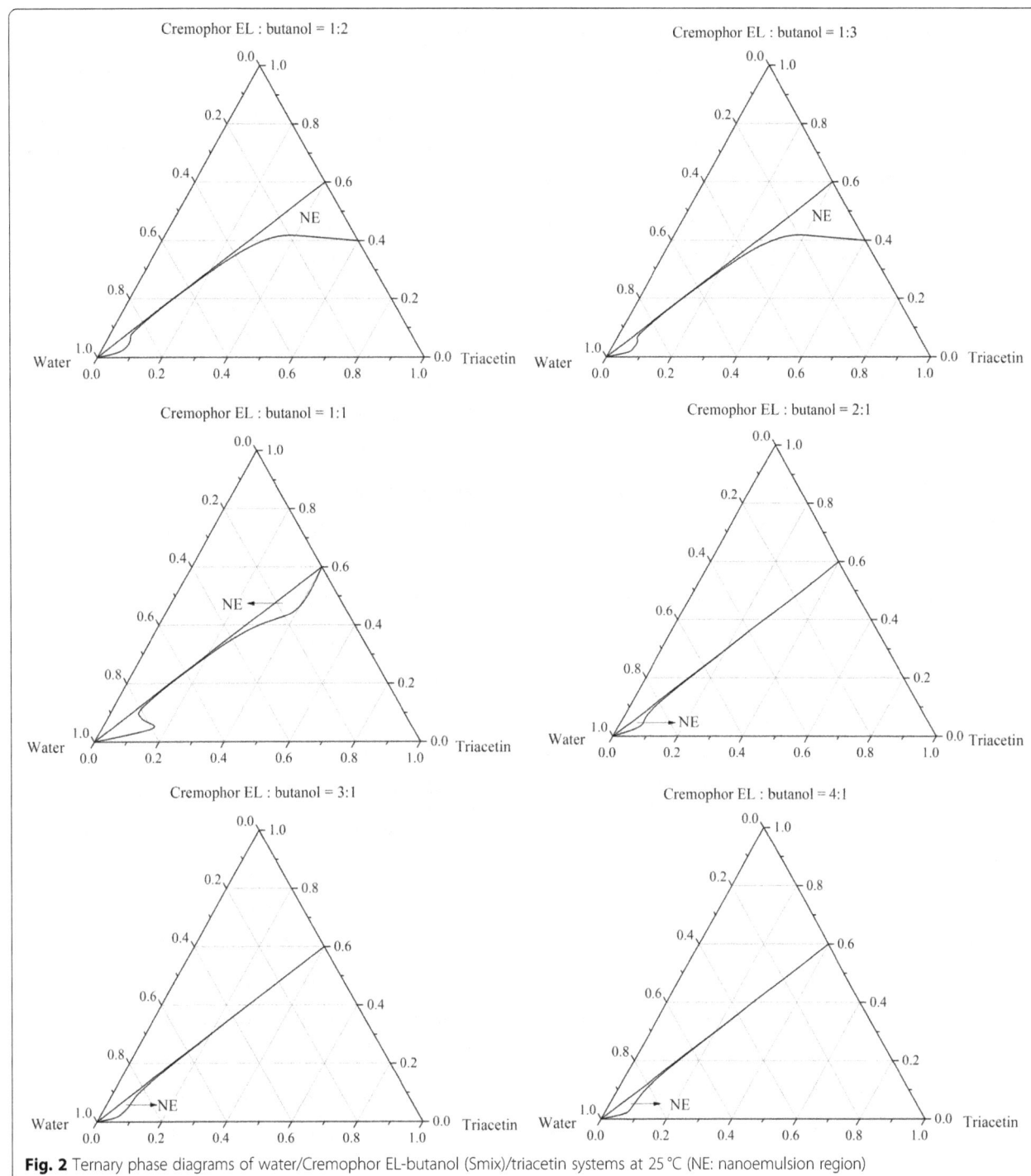

Fig. 2 Ternary phase diagrams of water/Cremophor EL-butanol (Smix)/triacetin systems at 25 °C (NE: nanoemulsion region)

approached to the seedlings with NCTD and fed. No restless movement was observed among the treatment as 90% mortality occured within 48 h. However, the bodies of the poisoned larvae were paralyzed and blacked with diarrhea symptoms. These phenomenons are consistent with earlier results [32, 34], where they investigated the oral toxicities of NCTD against *P. xylostella*. In addition, by increasing the NCTD concentrations (>

200 mg/L), an antifeedant effect was observed and the mortality was remained about the same.

The linear regression equation was determined between the NCTD concentration and mortality percentage (Additional file 1: Table S2). Oral toxicities showed that the LC_{50} and LC_{90} values decreased with increasing exposure time, from 247.010 mg/L to 175.602 mg/L and 414.479 mg/L to 303.050 mg/L, respectively. With regard

Table 4 Oral toxicity of NCTD against the 3rd-instar larvae of *P. xylostella*

Concentration (mg/L)	Mortality rate (%)			
	12 h	24 h	36 h	48 h
CK	0	0	12 ± 8.37	14 ± 8.94
50	0	0	8 ± 13.04	20 ± 15.81
80	0	4 ± 5.48	14 ± 15.17	28 ± 17.89
110	2 ± 4.47	10 ± 7.07	18 ± 8.37	24 ± 11.40
140	8 ± 13.04	12 ± 10.95	26 ± 11.40	26 ± 18.17
170	4 ± 8.94	24 ± 13.42	42 ± 22.80	56 ± 19.49
200	34 ± 34.35	46 ± 25.10	62 ± 24.90	64 ± 26.08
230	28 ± 17.89	48 ± 26.83	72 ± 8.37	72 ± 19.24
250	48 ± 13.04	72 ± 21.68	90 ± 7.07	90 ± 10.00

Data in this table were mean ± SE

to insecticidal activity, studies conducted by Liu Z [32] showed that NCTD was more active and the *P. xylostella* larvae were killed at lower concentration (LC_{50} = 129.35 mg/L, LC_{90} = 223.29 mg/L). The variation in the toxicity of NCTD against *P. xylostella* may be ascribed to the difference of the susceptible strain and NCTD purification. From the above results, we speculate that NCTD could be applied as a promising biopesticide against *P. xylostella*.

Effect of cosurfactant concentration

In present investigation, the insecticidal activity of various NCTD nanoemulsions were studied against third-instar *P. xylostella* larvae. In order to better understand the effect of cosurfactant concentration, droplet size and size distribution on the insecticidal activity of NCTD nanoemulsions, the concentration of NCTD was kept constant at 200 mg/L. The effect of butanol concentration on the mortality percentage of NCTD-nanoemulsions stabilized by Cremophor EL (SOR = 1:1) is shown in Table 5 (Additional file 1: Table S3-S5 are corresponded to SOR = 4:6, 6:4 and 3:7). The mortality rate increased with an

Table 5 Oral toxicity of different NCTD-nanoemulsions (SOR = 4:6, NCTD = 200 mg/L, water = 96% *w*/w) against the 3rd- instar larvae of *P. xylostella*

Smix	Mortality rate (%)			
	12 h	24 h	36 h	48 h
CK	0	10.00 ± 10.00	10.00 ± 10.00	10.00 ± 10.00
4:1	6.67 ± 5.77	23.33 ± 11.55	46.67 ± 15.28	53.33 ± 11.55
3:1	3.33 ± 5.78	53.33 ± 11.55	86.67 ± 11.55	93.33 ± 11.54
2:1	10.00 ± 10.00	36.67 ± 15.28	53.33 ± 15.28	66.67 ± 20.82
1:1	10.00 ± 10.00	46.67 ± 11.55	63.33 ± 15.28	90.00 ± 0
1:2	13.33 ± 5.78	36.67 ± 20.82	60.00 ± 0	90.00 ± 10.00
1:3	16.67 ± 20.82	36.67 ± 20.82	70.00 ± 17.32	76.67 ± 15.28

Data in this table were mean ± SE

increase in treatment time. The blank nanoemulsion containing Cremophor EL, triacetin or butanol exhibited week insecticidal activity. Nanoemulsion with an intermediate SOR value 5:5 showed 93.33% mortality at Smix = 3:1 after 48 h; while nanoemulsion formulations at Smix = 4:1 and 2:1, 53 and 67% mortality were observed after 48 h, respectively. To further understand the impact of butanol concentration on the insecticidal activity of NCTD nanoemulsions, the corresponding droplet size and size distribution are shown in Fig. 3. The NCTD nanoemulsion with the smallest droplet size (Smix = 2:1, r = 7.38 nm, mortality ≈ 66.67%) was found not to posses the highest mortality after 48 h exposure period.

Effect of surfactant concentration

In this section, we examined the influence of surfactant concentration on the insecticidal activity of NCTD nanoemulsions (Smix =1:1 and NCTD = 200 mg/L). As can be seen in Fig. 4, the higher mortality was observed at lower SOR in the first 12 h where larger droplets were formed [24, 31, 35]; a similar trend was observed in the next 12 h. The highest mortality (100%) was obtained from the nanoemulsions with SOR 4:6 after 36 h and a slight lower mortality (98%) was exhibited at SOR = 6:4 under the same time. Nanoemulsions with smaller droplets having better insecticidal activity was also not found in this experiment.

Optimized NCTD-nanoemulsion

On the basis of Section 3.5.2 and 3.5.3, the optimized NCTD-nanoemulsion formulation was consisted of triacetin/Cremophor EL/butanol (60/20/20, *w*/w) and the LC_{50} and LC_{90} of this formulation were determined in Table 6. A positive linear correlation was observed between the NCTD concentration and mortality, and the LC_{50} and LC_{90} value decreased as the exposure time increased. The LC_{50} (60.414 mg/l, 48 h) or LC_{90} (185.530 mg/l, 48 h) value of the optimizing formulation indicates that NCTD encapsulated in the nanoemulsion was more toxic than NCTD acetone solution (Additional file 1: Table S2, LC_{50} = 175.602 mg/L, LC_{90} = 303.050 mg/L, 48 h).

Discussion

The spontaneous emulsification method was used to fabricate nanoemulsions by adding an oil and hydrophilic surfactant mixture into the aqueous phase [35]. Studies have demonstrated that the rapid movement of hydrophilic surfactant from oil to aqueous phase is one of the most significant determinants affecting the formation of oil droplets by spontaneous emulsification [24, 28, 35]. Our transmittance values were not in accordance with earlier results [36], according to which the transmittance increased with decreasing droplet size. This effect can be

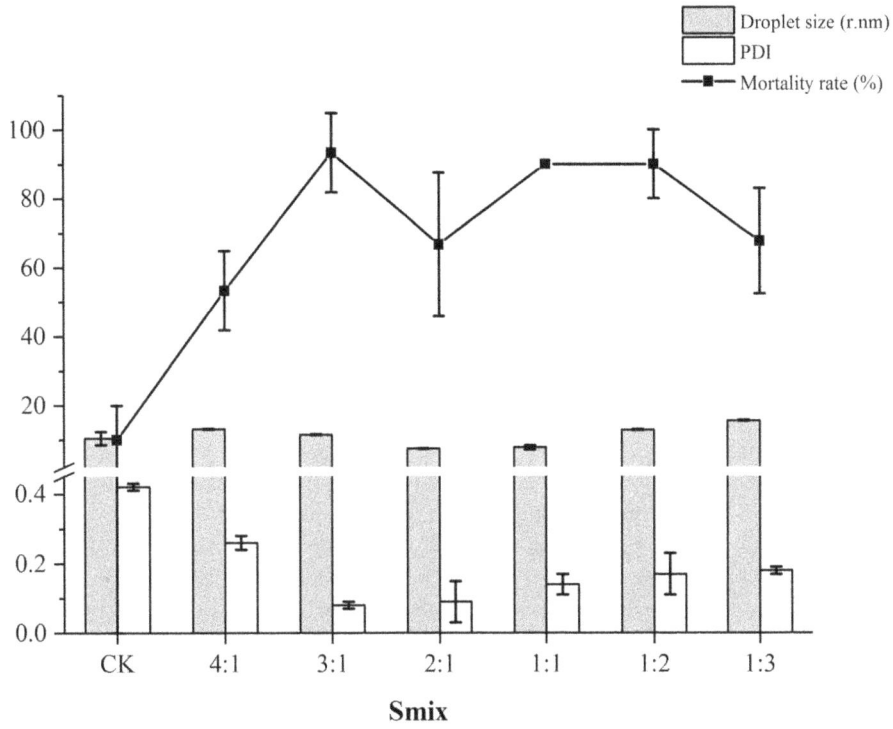

Fig. 3 Droplet size, size distribution and mortality rate (48 h) of triacetin/Cremophor EL/butanolt/water systems at 25 °C (triacetin: Smix = 5:5, water = 96 wt.%)

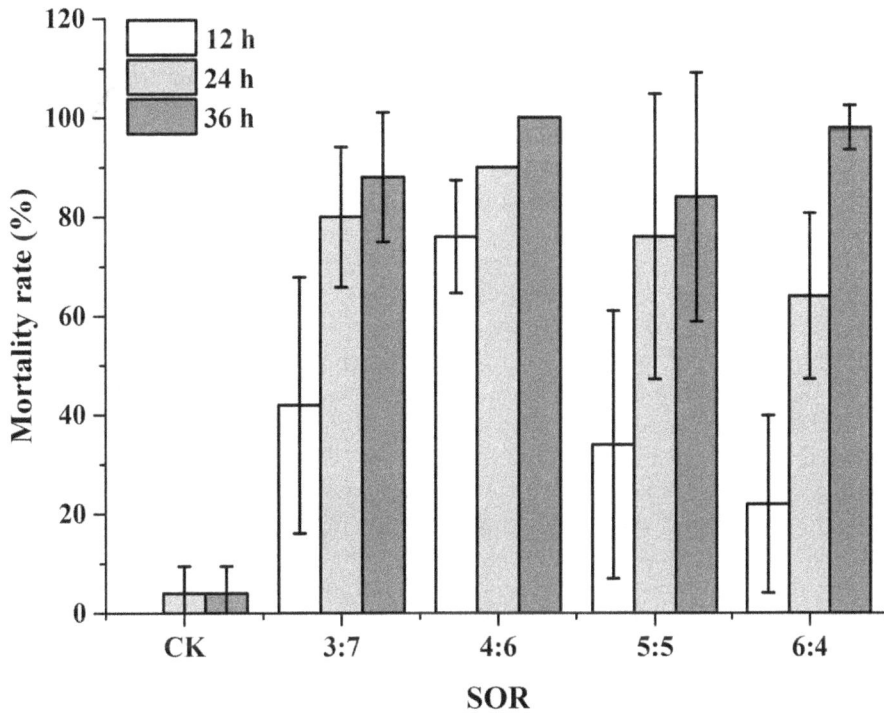

Fig. 4 The effect of different surfactant-to-oil mass ratio (SOR) on the oral toxicity of NCTD-nanoemulsion (Smix = 1:1, NCTD 200 mg/L) against the third-instar of *P. xylostella*

Table 6 Regression equation of NCTD against 3rd-instar larvae of *P. xylostella*

Time	Regression equation	R^2	χ^2	LC$_{50}$ (mg/L) (95% confidence limits)	LC$_{90}$ (mg/L) (95% confidence limits)
12 h	–	–	–	–	–
24 h	$y = 6.379x - 14.881$	0.921	0.462	150.090 (114.135~310.777)	312.188 (203.841~10,200.727)
36 h	$y = 5.637x - 12.727$	0.992	0.156	75.992 (30.489~108.489)	191.903 (130.619~791.051)
48 h	$y = 2.630x - 4.684$	0.985	0.264	60.414 (23.136~92.727)	185.530 (117.860~693.284)

LC$_{50}$ = Lethal concentration at which 50% of the larvae showed mortality
LC$_{90}$ = Lethal concentration at which 90% of the larvae showed mortality
x = log concentration
y = percentage mortality

attributed to the droplets multimodal distribution: the bigger droplets scattered light strongly though the formulation had a translucent appearance. The affinity of surfactant for the hydrophobic or hydrophilic phase is a significantly important factor for the nanoemulsions formation [24, 35]. The HLB value of four surfactants were around 15 in our study while, different emulsification abilities were exhibited. That indicates that there were other factors responsible for spontaneous emulsification, such as the molecular geometry of surfactant.

It is well known that the main drawback of spontaneous emulsification is the requirement of relatively high amounts of synthetics surfactants to maintain stability. Consequently, there is interest in developing stable nanoemulsions with small droplets and low surfactant concentration. Cosurfactant addition is an appealing idea for assisting a surfactant in reducing interfacial tension, fluidifying interfacial film, controlling droplet size and improving nanoemulsion stability [37, 38]. What is notewothy is that the smallest droplets with the narrowest distribution were produced for the system containing butanol, whilt the largest droplet size with widest distribution was exhibited with PEG 400. This phenomenon was agreed with our previous hypothesis postulating that the oil-water interfacial may be more flexible during spontaneous emulsification in cosurfactant-presence system, due to the hydrophilicity and viscosity of cosurfactant [33]. It is worth noting that some cosurfactants are volatile, a factor making it impractical for transportation and storage.

The concentration of the butanol had a notable impact on the spontaneous emulsification of emulsions. Additionally, spontaneous emulsification became easier and faster which was attributed to the combination of Cremophor EL and butanol decreased the oil-water interfacial tension and increased the flexibility of the surfactant layer. Decreasing the butanol content, the nanoemulsions areas were reduced appreciably. However, the opposite observation was found in previous report where the nanomeulsion region was increased with decreased cosurfactant concentration [38]. This difference could be explained by the effect of triacetin that acted as an emulsifier and co-emulsifier simultaneously in our systems.

In this study, the cosurfactant concentration significantly influenced the insecticidal activity of NCTD nanoemulsions prepared by spontaneous emulsification; this has not yet been reported, and the mortality was in the following order: 3:1 > 1:1 = 1:2 > 1:3 > 2:1 > 4:1(48 h). As above, the mortality rate of NCTD-nanoemulsions in the first 24 h exhibited a similar order. This might be associated with the presence of 5–10% *w/w* alcohol which could give rise to a change in the interfacial tension [39, 40]. In addition, there was a nonlinear correlation between droplet size or size distribution and the mortality rate of *P. xylostella*. This finding does not agree with previous research indicating that nanoemulsions with smaller droplet size were found to be more effective in controlling pests compared with larger droplet sizes [1, 20, 41]. In consequence, our study highlights that there are other factors such as cosurfactant concentration that need to be considered.

The mortality rate of NCTD nanoemulsions was obviously dependent on surfactant concentration and exposure time. The higher mortality of NCTD nanoemulsions was observed at lower SOR, coinciding with the finding of a previous study that the release of drug was relatively slow due to the protection of droplets structure by sufficient surfactant [25]. However, the highest mortality was obtained from the nanoemulsions with SOR 4:6. This phenomenon may be related to the appearance of micelles, attributed to the excess surfactant, which hinders the NCTD release [42, 43].

Compared the insecticidal activity of optimized NCTD nanoemulsion with the NCTD, the NCTD nanoemulsion was found to be more effective than its acetone solution. This finding was in agreement with many previous showed that nanoparticles can improve the bioactivity or bioaccessibility of active materials [1, 20, 21]. A pesticide with a lower LC$_{50}$ indicates it takes less amount of the pesticide to kill half of the test insects; therefore nanoemulsions may be a good choice as a potent and selective delivery system for pesticides.

Conclusion

In this investigation, nanoemulsion containing poorly water-soluble NCTD was optimized and prepared using spontaneous emulsification method to prevent the *P. xylostella* larvae. The compositions for the NCTD-nanoemulsion were selected by a solubility study, emulsification ability analysis and ternary phase diagrams construction. The surfactant and cosurfactant concentration significantly impacted the insecticidal activity of NCTD nanoemulsions. Surfactant concentration notably affects the oil-water interface structure and hinders the drug release from nano-droplets. However, nanoemulsions containing smaller droplets had better oral toxicity was not found in our study. In consequence, evaluation of the relationship between exposure time and drug release mechanism of nanoemulsions is a promising orientation for our further study. The composition of NCTD-nanoemulsion was optimized as 60% triacetin, 20% Cremophor EL and 20% butanol, which showed a higher efficacy as an insecticidal agent against *P. xylostella* when compared to NCTD acetone solution. It can be speculated that nanoemulsions may be a good option to other pesticides for the agriculture applications.

Methods

Materials

Norcantharidin (NCTD) was obtained from Alfa Aesar (> 98%, Ward Hill, MA). Soybean oil, olive oil, ethyl oleate, isopropyl myristate, triacetin, tributyrin, Tween 80, Tween 20, ethanol, ethylene glycol, 1-propanol, isopropanol, propylene glycol, 1-butanol, glycerol and PEG 400 were purchased from Aladdin (> 98%, Shanghai, China). Medium chain triglyceride (MCT), Cremophor EL and Cremophor RH 40 were obtained from BASF (Ludwigshafen, Germany). All other chemicals and solvent were of analytical grade and used without further purification.

Deionized water was generated from the Milli-Q gradient system of Millipore (Synergy, Millipore SAS, Molsheim, France).

Selection of nanoemulsions composition

The oil was determined by the saturation solubility of NCTD using shake flask method [44]. Excess amount of NCTD was added into each vial with various oils under vortex mixing for 30s. Then, mixtures were incubated in a shaking table at constant temperature (37 ± 3 °C, 50 strokes/min) to reach equilibrium states. After 3 days, mixtures were centrifuged (5000 rpm, 5 min) and the supernatants were filtered through a Millipore membrane filter (0.45 μL). The quantify analysis was implemented by the Agilent Gas Chromatograph [10, 42].

Surfactants and co-surfactants were screened for their emulsification ability according to the published research [44] with slight modification. Briefly, oil (500 mg) was homogenized with surfactant (500 mg) followed by warming at 50 °C for 30 s. Various isotropic mixtures (200 mg) were diluted to 50 mL deionized water to yield

fine emulsions. Emulsification time was employed to evaluate the emulsification ability of different surfactants. Similarly, screening of co-surfactants was implemented using the same method as described above. The selected surfactant (300 mg) was mixed with different co-surfactant (200 mg) and then blended with optimal oil (500 mg).

Construction of ternary phase diagrams

The nanoemulsions region was determined using ternary phase diagrams. Each corner corresponded to 100% of water, surfactant and oil. Various compositions were prepared by altering the water concentration in the order of 5% while the amount of surfactant and oil remained fixed. The homogeneous mixtures were then diluted with distilled water and evaluated transmittance, droplet size and polydispersity index (PDI). The mixtures formed emulsions with droplet size < 100 nm and PDI < 0.2 were considered to be in the nanoemulsions region.

Preparation of nanoemulsions

Nanoemulsions were prepared by spontaneous emulsification method [22] with slight modification. Spontaneous emulsification was performed by adding the oil phase into and aqueous phase under mildly stirring (500 rpm, 15 min) at room temperature (25 ± 2 °C). The oil phase was composed of oil, surfactant and co-surfactant.

Droplet size and size distribution

The droplet size and PDI were determined by dynamic light scattering (Nano ZS, Malvern Instrument Ltd., Worcestershire, UK) at a fixed scattering angle of 173° and with the laser wavelength of 633 nm. Measurements were conducted at 25 °C and each measurement was made with three readings per sample. In addition, measurements were carried out for all samples after equilibrating for 2 h.

Thermodynamic stability study

The thermodynamic stability of formulated nanoemulsions were subjected to the following stresses: (1) Centrifugation: the formulated nanoemulsions were centrifuged at 5000 rpm for 30 min and observed for phase separation, creaming or cracking. (2) Heating-cooling cycle: formulated nanoemulsions, after centrifugation tests, were stored at refrigerated temperature of 4 °C and 40 °C for 48 h respectively. This cycle was repeated six times. (3) Freeze-thaw cycle: formulated nanoemulsions were kept at − 20 °C and 25 °C for 48 h at each temperature. The cycle was repeated three times.

Insecticidal activity

Plutella xylostella (Lepidotera: Plutellidae) was chosen as the test insect. It is a worldwide pest of cruciferous vegetables and an important focus of research. The susceptible strain was obtained from the Key Lab of Plant

Protection Resources & Pest Management of Ministry of Education, Northwest A&F University (Yangling, China). The larvae were fed with pakchoi (*Brassica chinensis* L.) (25 ± 2 °C, RH 50 ± 5%, L: D = 16:8).

Third-instar larvae of *P. xylostella* were randomly chosen and separated in three treatment groups. One group was topically applied with nanoemulsions contain NCTD, while negative control groups were treated with blank nanoemulsions and NCTD solution (NCTD was dissolved in mixed solvent (water: acetone: dimethyl sulfoxide = 20: 19: 1, *V*/V) with 0.5% Tween 80 to different concentration) respectively. The bioactivities of all groups were tested using seedlings of pakchoi (without roots). For all test groups, the seedlings were dipped for 5 s, then placed on a blotting paper to drain superfluous fluid and left to dry in the room temperature.

Ten treated seedlings were put in the Petri plate (9 cm diametr) with wetting filter paper to keep them fresh. Larvae were released on the plate after fasting for 3 h and then kept in growth cabinet (25 ± 2 °C, RH 50 ± 5%, L: D = 16:8). All experiments were repeated at least five replicates with samples from 10 larvae. Percentage mortalities were determined 24 and 48 h post-treatment. Test insects were considered dead when, prodded with fine small brush (maximum three times), they showed no appendage movement.

Statistical analysis

All statistical test data were expressed as the mean ± standard deviation (SD). The LC_{50} value and the associated 95% confidence interval ($P < 0.05$) was determined using the Probit method [12]. The observed mortalities and effluent concentrations were transformed into a probit transformation and log_{10}, respectively. This involves estimation of linear regression parameters by an interactive approach. The LC 50 and associated confidence interval were calculated from the estimated linear regression parameters.

Additional file

Additional file 1: Table S1. The performance of nanoemulsions with different cosurfactants. **Table S2.** Regression equation of NCTD against the 3rd-instar larvae of *P. xylostella*. **Table S3.** Oral toxicity of different surfactant-to-cosurfactant mass ratio (Smix) against the 3rd- instar larvae of *P. xylostella* (SOR =4:6). **Table S4.** Oral toxicity of different surfactant-to-cosurfactant mass ratio (Smix) against the 3rd- instar larvae of *P. xylostella* (SOR = 6:4). **Table S5.** Oral toxicity of different surfactant-to-cosurfactant mass ratio (Smix) against the 3rd- instar larvae of *P. xylostella* (SOR = 3:7). Figures.xlsx Sheet1 Fig. 1 Ternary phase diagrams of water/nonionic surfactant/triacetin systems at 25 °C (NE: nanoemulsion region). Figures.xlsx Sheet2 Fig. 2 Ternary phase diagrams of water/Cremophor EL-butanol (Smix)/triacetin systems at 25 °C (NE: nanoemulsion region). Figures.xlsx Sheet3 Fig. 3 Droplet size, size distribution and mortality rate (48 h) of triacetin/Cremophor EL/butanolt/water systems at 25 °C (triacetin: Smix = 5:5, water = 96 wt.%). Figures.xlsx Sheet4 Fig. 4 The effect of different surfactant-to-oil mass ratio (SOR) on the oral toxicity of NCTD-nanoemulsion (Smix = 1:1, NCTD 200 mg/L) against the third-instar of *P. xylostella*. (ZIP 31 kb)

Abbreviations

NCTD: Norcantharidin; NE: Nanoemulsion region; PPs: Serine/threonine protein phosphatases; Smix: Surfactant-to-cosurfactant mass ratio; SOR: Surfactant-to-oil mass ratio

Acknowledgements

The authors are grateful to the School of Life Sciences (Northwestern Polytechnical University, China) for supporting the use of the Malvern Zetasizer photo correlation spectroscopy (Nano ZS, Malvern Instrument, UK).

Funding

This research is supported by The Chinese Natural Science Research Council (31601696, performed the design of the study), The Hunan Natural Science Research Council (2017NK2360, 2018JJ3198, performed the collection, analysis and interpretation of data) and Fundamental Research 410 Funds for the Introduce Talents (111011803003, performed the design of the study and the writing of the manuscript).

Authors' contributions

LZ and XL conceived the project, design experiments, interpreted data and wrote the manuscript. LZ performed the nanoemulsions preparation, optimization and characterization; YL performed bioactivity assay of NCTD nanoemulsion; JP advised and supported antimicrobial activity against *P. xylostella*. All authors read and approved the final manuscript.

Ethics approval and consent to participate
Not applicable.

Consent for publication
Not applicable.

Competing interests
The authors declare that they have no competing interests.

References

1. Pant M, Dubey S, Patanjali PK, Naik SN, Sharma S. Insecticidal activity of eucalyptus oil nanoemulsion with karanja and jatropha aqueous filtrates. Int Biodeter Biodeg. 2014;91:119–27.
2. Li YM, Casida JE. Cantharidin-binding protein: identification as protein phosphatase 2A. P Natl Acad Sci USA. 1992;89:11867–70.
3. Moed L, Shwayder TA, Chang M. Cantharidin revisited: a blistering defense of an ancient medicine. Jama Dermat. 2001;137:1357–60.
4. Shen M, Wu MY, Chen LP, Zhi Q, Gong FR, Chen K, Li DM, Wu Y, Tao M, Li W. Cantharidin represses invasion of pancreatic cancer cells through accelerated degradation of MMP2 mRNA. Sci Rep. 2015;5:11836.
5. Huang Z, Zhang Y. Chronic sublethal effects of cantharidin on the diamondback moth *plutella xylostella* (Lepidoptera: Plutellidae). Toxins. 2015; 7:1962.
6. Wu ZW, Yang XQ, Zhang YL. The toxicology and biochemical characterization of cantharidin on cydia pomonella. J Econ Entomol. 2015; 108:237–44.
7. Rashid M, Khan RA, Zhang Y. Physiological and population responses of armyworm *mythimna separata* (Lepidoptera: Noctuidae) to a sublethal dose of cantharidin-AC. J Econ Entomol. 2013;106:2177.
8. Hong CY, Huang SC, Lin SK, Lee JJ, Chueh LL, Lee CHK, Lin JH, Hsiao M. Norcantharidin-induced post-G2/M apoptosis is dependent on wild-type p53 gene. Biochem Bioph Res Co. 2000;276:278–85.
9. Huang Y, Liu Q, Liu K, Yagasaki K, Zhang G. Suppression of growth of highly-metastatic human breast cancer cells by norcantharidin and its mechanisms of action. Cytotechnology. 2009;59:201.
10. Zhao L, Yang G, Bai H, Zhang M, Mou D. NCTD promotes Birinapant-mediated anticancer activity in breast cancer cells by downregulation of c-FLIP. Oncotarget. 2017;8:26886–95.

11. Bertini I, Calderone V, Fragai M, Luchinat C, Talluri E. Structural basis of serine/threonine phosphatase inhibition by the archetypal small molecules cantharidin and norcantharidin. J Med Chem. 2009;52:4838–43.

12. Sun W, Liu Z, Zhang Y. Cantharidin and its anhydride-modified derivatives: relation of structure to insecticidal activity. Int J Mol Sci. 2013;14:1.

13. Guo Y. Toxicity analysis of emamectin benzoate and chlorpyrifos as well as their mixtures on *Plutella xylostella*. Entomol J East Cha. 2005;14:371–74.

14. Wang Y, Sun W, Zha S, Wang H, Zhang Y. Synthesis and biological evaluation of norcanthridin derivatives possessing an aromatic amine moiety as antifungal agents. Molecules. 2015;20:19782.

15. Tandon S, Sand NK. Rapid estimation of pyroquilon from its formulations by RP-HPLC. Pestology. 2015;39:32–4.

16. Singh A, Tandon S, Sand NK. Active ingredient estimation of clopyralid formulation by reversed phase HPLC. J Chromatogr Sep Tech. 2014;6:257–9.

17. Mehrotra N, Tandon S, Kanaujia S, Sand NK. Estimation of chlorpyrifos from paraben 20 EC by reverse phase HPLC. Indian J Plant Prot. 2009;37:11–3.

18. Mehrotra N, Tandon S, Sand NK. Rapid analytical method for determination of chlorpyrifos in soil by reverse phase HPLC. Pestology. 2007;31:69–70.

19. Anjali CH, Sudheer Khan S, Margulis-Goshen K, Magdassi S, Mukherjee A, Chandrasekaran N. Formulation of water-dispersible nanopermethrin for larvicidal applications. Ecotox Environ Safe. 2010;73:1932–6.

20. Jerobin J, Sureshkumar RS, Anjali CH, Mukherjee A, Chandrasekaran N. Biodegradable polymer based encapsulation of neem oil nanoemulsion for controlled release of Aza-a. Carbohyd Polym. 2012;90:1750–6.

21. Ha TVA, Kim S, Choi Y, Kwak HS, Lee SJ, Wen J, Oey I, Ko S. Antioxidant activity and bioaccessibility of size-different nanoemulsions for lycopene-enriched tomato extract. Food Chem. 2015;178:115–21.

22. Bouchemal K, Briançon S, Perrier E, Fessi H. Nano-emulsion formulation using spontaneous emulsification: solvent, oil and surfactant optimisation. Int J Pharm. 2004;280:241–51.

23. Saberi AH, Fang Y, McClements DJ. Fabrication of vitamin E-enriched nanoemulsions: factors affecting particle size using spontaneous emulsification. J Colloid Interf Sci. 2013;391:95–102.

24. Guttoff M, Saberi AH, McClements DJ. Formation of vitamin D nanoemulsion-based delivery systems by spontaneous emulsification: factors affecting particle size and stability. Food Chem. 2015;171:117–22.

25. Shafiq S, Shakeel F, Talegaonkar S, Ahmad FJ, Khar RK, Ali M. Development and bioavailability assessment of ramipril nanoemulsion formulation. Eur J Pharm Biopharm. 2007;66:227–43.

26. Davidov-Pardo G, McClements DJ. Nutraceutical delivery systems: resveratrol encapsulation in grape seed oil nanoemulsions formed by spontaneous emulsification. Food Chem. 2015;167:205–12.

27. Solans C, Izquierdo P, Nolla J, Azemar N, Garcia-Celma MJ. Nano-emulsions. Curr Opin Colloid In. 2005;10:102–10.

28. Singh Y, Meher JG, Raval K, Khan FA, Chaurasia M, Jain NK, Chourasia MK. Nanoemulsion: concepts, development and applications in drug delivery. J Control Release. 2017;252:28–49.

29. Lixin W, Haibing H, Xing T, Ruiying S, Dawei C. A less irritant norcantharidin lipid microspheres: formulation and drug distribution. Int J Pharm. 2006;323:161–7.

30. Burakova Y, Shi J, Schlup JR. Impact of oil composition on formation and stability of emulsions produced by spontaneous emulsification. J Disper Sci Technol. 2017;38:1749–54.

31. Komaiko J, McClements DJ. Low-energy formation of edible nanoemulsions by spontaneous emulsification: factors influencing particle size. J Food Eng. 2015;146:122–8.

32. Liu Z, Sun W, Zhang Y. Structure-activity relationship of cantharidin-related compounds on stomach poisoning of *Plutella xylostella*. J NWSUAF (Nat. Sci. Ed.). 2013;41:85–90.

33. Zeng L, Zhang Y. Impact of short-chain alcohols on the formation and stability of nano-emulsions prepared by the spontaneous emulsification method. Colloid Surface A. 2016;509:591–600.

34. Wang Z, Zhang Y, Huang M. Application of 5% norcantharidin microemulsion against *Plutella xylostella*. J NWSUAF (Nat Sci Ed). 2014;42:181–5.

35. Anton N, Vandamme TF. The universality of low-energy nano-emulsification. Int J Pharm. 2009;377:142–7.

36. Saberi AH, Fang Y, McClements DJ. Stabilization of vitamin E-enriched nanoemulsions: influence of post-homogenization cosurfactant addition. J Agr Food Chem. 2014;62:1625–33.

37. Hu Q, Gerhard H, Upadhyaya I, Venkitanarayanan K, Luo Y. Antimicrobial eugenol nanoemulsion prepared by gum arabic and lecithin and evaluation of drying technologies. Int J Biol Macromol. 2016;87:130–40.

38. Zeng L, Xin X, Zhang Y. Development and characterization of promising Cremophor EL-stabilized o/w nanoemulsions containing short-chain alcohols as a cosurfactant. RSC Adv. 2017;7:19815–27.

39. Zeeb B, Herz E, McClements DJ, Weiss J. Impact of alcohols on the formation and stability of protein-stabilized nanoemulsions. J Colloid Interf Sci. 2014;433:196–203.

40. AS O, DD G. Effects of pH and ehanol on the kinetics of destabilisation of oil-in-water emulsions containing milk proteins. J Sci Food Agr. 1996;72:448–54.

41. Sugumar S, Clarke SK, Nirmala MJ, Tyagi BK, Mukherjee A, Chandrasekaran N. Nanoemulsion of eucalyptus oil and its larvicidal activity against Culex quinquefasciatus. B Entomol Res. 2014;104:393–402.

42. Zeng L, Zhang Y. Development, optimization and in vitro evaluation of norcantharidin loadedself-nanoemulsifying drug delivery systems (NCTD-SNEDDS). Pharm Dev Technol. 2017;22:399–408.

43. Mohan P, Rapoport N. Doxorubicin as a molecular nanotheranostic agent: effect of doxorubicin encapsulation in micelles or nanoemulsions on the ultrasound-mediated intracellular delivery and nuclear trafficking. Mol Pharm. 2010;7:1959–73.

44. Zhou L, Yang L, Tilton S, Wang J. Development of a high throughput equilibrium solubility assay using miniaturized shake-flask method in early drug discovery. J Pharm Sci-US. 2007;96:3052–71.

15

Improved CRISPR/Cas9 gene editing by fluorescence activated cell sorting of green fluorescence protein tagged protoplasts

Bent Larsen Petersen[1]*[iD], Svenning Rune Möller[1,2], Jozef Mravec[1], Bodil Jørgensen[1], Mikkel Christensen[1,3], Ying Liu[1], Hans H. Wandall[4], Eric Paul Bennett[4] and Zhang Yang[4]

Abstract

Background: CRISPR/Cas9 is widely used for precise genetic editing in various organisms. CRISPR/Cas9 editing may in many plants be hampered by the presence of complex and high ploidy genomes and inefficient or poorly controlled delivery of the CRISPR/Cas9 components to gamete cells or cells with regenerative potential. Optimized strategies and methods to overcome these challenges are therefore in demand.

Results: In this study we investigated the feasibility of improving CRISPR/Cas9 editing efficiency by Fluorescence Activated Cell Sorting (FACS) of protoplasts. We used *Agrobacterium* infiltration in leaves of *Nicotiana benthamiana* for delivery of viral replicons for high level expression of gRNAs designed to target two loci in the genome, *NbPDS* and *NbRRA*, together with the Cas9 nuclease in fusion with the 2A self-splicing sequence and GFP (Cas9-2A-GFP). Protoplasts isolated from the infiltrated leaves were then subjected to FACS for selection of GFP enriched protoplast populations. This procedure resulted in a 3–5 fold (from 20 to 30% in unsorted to more than 80% in sorted) increase in mutation frequencies as evidenced by restriction enzyme analysis and the Indel Detection by Amplicon Analysis, which allows for high throughput profiling and quantification of the generated mutations.

Conclusions: FACS of protoplasts expressing GFP tagged CRISPR/Cas9, delivered through *A. tumefaciens* leaf infiltration, facilitated clear CRISPR/Cas9 mediated mutation enrichment in selected protoplast populations.

Keywords: Precise genetic editing, Genome engineering, CRISPR/Cas9, Protoplasting, Fluorescence activated cell sorting, Mutation enrichment, *Nicotiana benthamiana*

Background

CRISPR/Cas has emerged as a powerful tool for precise genetic editing (PGE) in a wide range of organisms [1], including plants [2]. CRISPR/Cas relies on the Cas DNA nuclease being guided by the small guide RNA (gRNA), to make a double stranded break (DSB) at the desired place in the genome (reviewed in [3]) leading to activation of inherent repair mechanisms (Non-Homologous End Joining (NHEJ) or Homologous Recombination (HR) if a DNA molecule with identical flanking sequences is co-delivered. CRISPR/Cas mediated PGE in plants may be complicated by the presence of complex

and high ploidy genomes or by inefficient or poorly controlled delivery of PGE components to gamete cells or cells with regenerative potential. Moreover, subsequent regeneration and tissue culturing post PGE is often lengthy, labor-intensive and prone to produce random somatic mutations and targeted insertion mediated mutagenesis through homologous recombination is still a main challenge within PGE [2]. There is therefore a demand for optimizing PGE in plants towards efficient generation and propagation of stable heritable editing at the organism level.

Nucleic acids may be introduced into plant cells/tissues by biolistic particle bombardment [4], which, however, often results in insertion of multiple copies at multiple sites in the genome [5]. Other strategies include transformation of protoplast by chemical means using

* Correspondence: blp@plen.ku.dk
[1]Department of Plant and Environmental Sciences, University of Copenhagen, DK-1871 Frederiksberg C, Denmark

polyethylene glycol (PEG) in combination with calcium ions or by electroporation (reviewed in [5]), where the latter requires elaborate tissue culturing for regeneration to fertile plants and may introduce genetic instability and resulting somaclonal variation. PEG-mediated transformation, in particular, has been used to deliver constructs encoding the PGE components, incl. Zinc Finger-Nucleases (ZFNs) [6], Transcription activator-like effector nucleases (TALENs) [7, 8] and CRISPR/Cas9 [8, 9] and lately also for delivery of the Cas9 enzyme and associated gRNA into plant cell protoplasts in vitro [10]. Excess DNA is regularly used for PEG mediated transformation of protoplasts (typically in molar ratios of 1: $1-2 \times 10^7$ (protoplast: plasmid DNA) [11]) and has been reported to confer unintended random integrations in the recipient genomes [12]. *Agrobacterium*-mediated transformation on the other hand is generally perceived to be an efficient and a more controlled way of delivering transgenes [13] and the use of strains, with putatively downregulated integration capacity [14], in combination with down-regulation of host factor integration genes may facilitate alternative ways of non-integrative delivery of PGE components. Also, *Agrobacterium* may in some cases be the only viable option for delivering of transgenes. In recent years, *Agrobacterium*-mediated delivered viral constructs has attracted increasing interest because of their high copy number and resulting expression capabilities [15, 16]. Deconstructed viral vectors (replicons) have proved extremely effective for rapid, high-yield production of a number of pharmaceutical proteins, of which some are currently undergoing clinical evaluation [16]. As efficient gene editing relies on PGE component expression, virus replicons have likewise attracted attention as delivery vehicles [17]. Deconstructed geminivirus type replicons (as delivery vehicles) have been shown to generate mutations in the solanaceous species *Nicotiana benthamiana* [17] and *Solanum lycopersicum* (tomato) [18] and recently in *Triticum aestivum* (wheat) [19]. *N. benthamiana* can be grown in high density and still produce large amounts of biomass in a matter of weeks [16], and has a track record for production of therapeutic glycoproteins in mg scale ([20–24]) through the use of leaf or leaf disc infiltration [25]. In addition, *N. benthamiana* may readily be subjected to protoplast transformation [26] and explant/protoplast regeneration [27, 28]. Several approaches have been reported to confer enrichment of PGE mutations in cells. Fluorescence Activated Cell Sorting (FACS) of edited cells, for example, is regularly used as means of PGE mutation enrichment in mammalian cell systems [29], and the present study addresses the feasibility of applying this strategy to plant cells.

So far reports on FACS and post FACS cultivation of plant protoplasts are relatively scarce [30], due to the removal of the rigid and structure providing cell wall,

which otherwise stabilize the cell integrity [31–33]. The present study explores the combined use of *Agrobacterium*-mediated delivery of viral replicons for expression of GFP tagged gRNA/Cas9 in leaves of *N. benthamiana* with FACS in order to obtain protoplast populations with significantly increased gene editing.

Results

The overall strategy for *Agrobacterium*-mediated delivery of deconstructed replicons expressing gRNA/Cas9-2A-GFP in leaves of *N. benthamiana* combined with FACS of GFP expressing protoplasts is outlined in Fig. 1.

gRNA and replicon construct design

In the present study we targeted the *Nicotiana benthamiana PHYTOENE DESATURASE* (*NbPDS*) and *REDUCED RESIDUAL ARABINOSE* arabinosyl transferase (*NbRRA*) loci, orthologous to the *Arabidopsis thaliana* arabinosyltransferase encoding genes involved arabinosylation of plant cell wall extensins (*AtRRA1–3*) [41, 42], which have a proven [43] and an untested CRISPR/Cas9 editing record, respectively (Fig. 2a). gRNA target sequences were confined to early exons and identified on the basis of in silico prediction analysis (http://portals.broadinstitute.org/gpp/public/analysis-tools/sgrna-design, [45]), and the presence of a Restriction Enzyme (RE) recognition sequence spanning the predicted cut site of *Sp*Cas9–3 bp upstream of the protospacer adjacent motif (PAM) [34] for RE-mediated mutation screening.

A deconstructed immobilized mild strain of the bean yellow dwarf virus (*BeYDV*), allowing for a high replicon copy number in the nucleus, has recently been used to construct an *Agrobacterium* T-DNA that integrates into the host cell chromosome and delivers a geminivirus replicon (GVR) [17, 46]. The minimal immobilized replicons are delivered by *Agrobacterium* infiltration (here to *N. benthamiana* leaves) along with co-infiltrated constructs for expression of replicon trans-acting replication initiation proteins (Rep or RepA) [47] (Fig. 1a). While the replicons are non-integrative and transiently expressed the initial *Agrobacterium* T-DNA (LB-RB) delivery of the replicon is integrative [17]. Lately, GVRs were constructed and used to propagate and express PGE components, such as ZFNs and TALENs and CRISPR/Cas9 [17]. In the present study, we inserted the *Streptococcus pyogenes* Cas9 enzyme (*Sp*Cas9) [48] in translational fusion with the 2A self-splicing sequence of the foot-and-mouth disease virus [37, 38] and GFP [49] (*Sp*Cas9-2A-GFP) under control of the CMV 35S promoter and the gRNAs under control of the *AtU6* promoter [35, 36] in the *BeYDV* GVR replicon [17] as depicted in Fig. 1a and detailed in the Methods section.

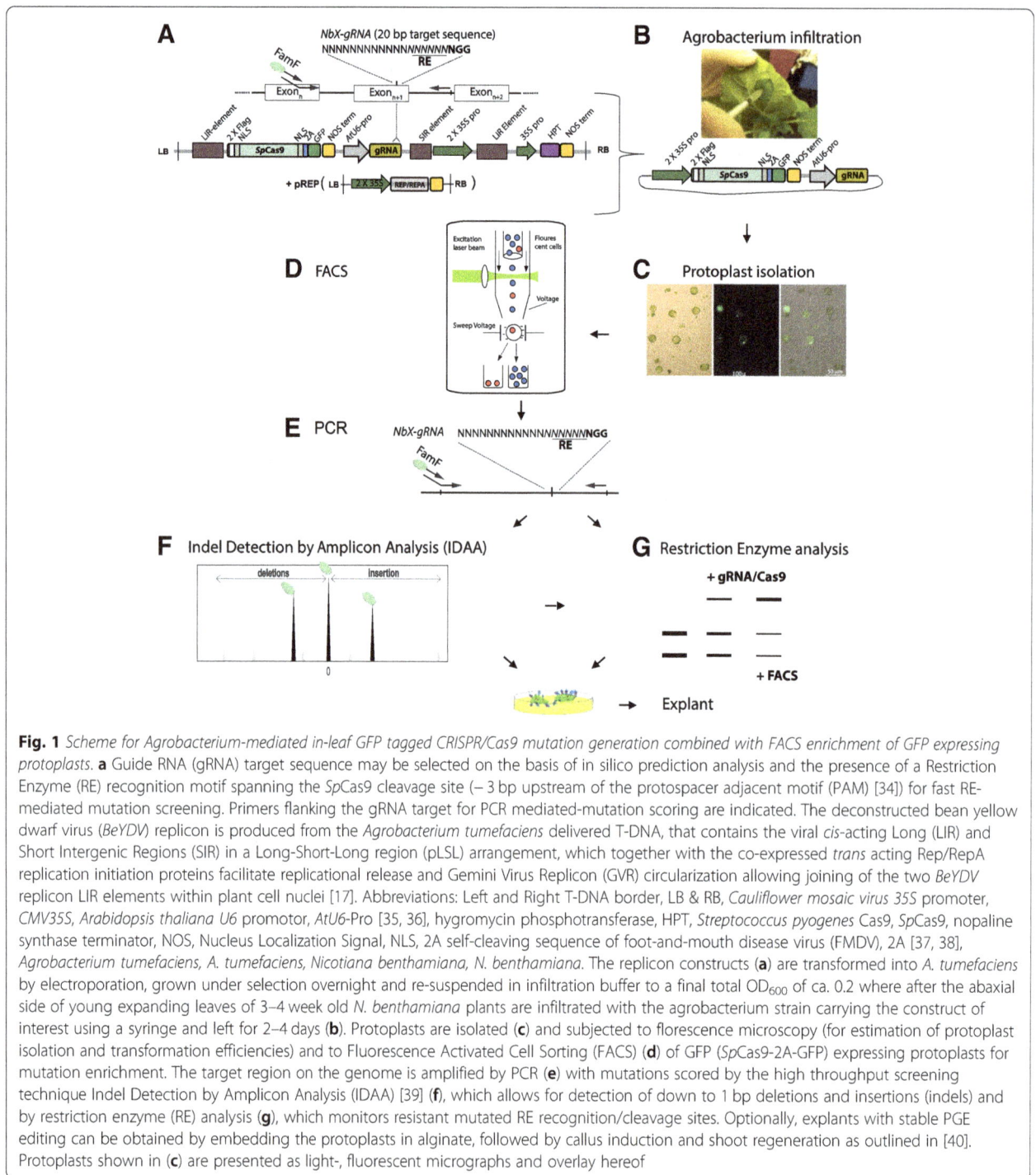

Fig. 1 *Scheme for Agrobacterium-mediated in-leaf GFP tagged CRISPR/Cas9 mutation generation combined with FACS enrichment of GFP expressing protoplasts.* **a** Guide RNA (gRNA) target sequence may be selected on the basis of in silico prediction analysis and the presence of a Restriction Enzyme (RE) recognition motif spanning the *Sp*Cas9 cleavage site (− 3 bp upstream of the protospacer adjacent motif (PAM) [34]) for fast RE-mediated mutation screening. Primers flanking the gRNA target for PCR mediated-mutation scoring are indicated. The deconstructed bean yellow dwarf virus (*BeYDV*) replicon is produced from the *Agrobacterium tumefaciens* delivered T-DNA, that contains the viral *cis*-acting Long (LIR) and Short Intergenic Regions (SIR) in a Long-Short-Long region (pLSL) arrangement, which together with the co-expressed *trans* acting Rep/RepA replication initiation proteins facilitate replicational release and Gemini Virus Replicon (GVR) circularization allowing joining of the two *BeYDV* replicon LIR elements within plant cell nuclei [17]. Abbreviations: Left and Right T-DNA border, LB & RB, *Cauliflower mosaic virus 35S* promoter, *CMV35S, Arabidopsis thaliana U6* promotor, *AtU6*-Pro [35, 36], hygromycin phosphotransferase, HPT, *Streptococcus pyogenes* Cas9, *Sp*Cas9, nopaline synthase terminator, NOS, Nucleus Localization Signal, NLS, 2A self-cleaving sequence of foot-and-mouth disease virus (FMDV), 2A [37, 38], *Agrobacterium tumefaciens, A. tumefaciens, Nicotiana benthamiana, N. benthamiana*. The replicon constructs (**a**) are transformed into *A. tumefaciens* by electroporation, grown under selection overnight and re-suspended in infiltration buffer to a final total OD$_{600}$ of ca. 0.2 where after the abaxial side of young expanding leaves of 3–4 week old *N. benthamiana* plants are infiltrated with the agrobacterium strain carrying the construct of interest using a syringe and left for 2–4 days (**b**). Protoplasts are isolated (**c**) and subjected to florescence microscopy (for estimation of protoplast isolation and transformation efficiencies) and to Fluorescence Activated Cell Sorting (FACS) (**d**) of GFP (*Sp*Cas9-2A-GFP) expressing protoplasts for mutation enrichment. The target region on the genome is amplified by PCR (**e**) with mutations scored by the high throughput screening technique Indel Detection by Amplicon Analysis (IDAA) [39] (**f**), which allows for detection of down to 1 bp deletions and insertions (indels) and by restriction enzyme (RE) analysis (**g**), which monitors resistant mutated RE recognition/cleavage sites. Optionally, explants with stable PGE editing can be obtained by embedding the protoplasts in alginate, followed by callus induction and shoot regeneration as outlined in [40]. Protoplasts shown in (**c**) are presented as light-, fluorescent micrographs and overlay hereof

In-leaf gRNA/Cas9 generated mutations

The *Sp*Cas9-2A-GFP/gRNA expressing GVR replicons (Fig. 1a) targeting the *NbPDS* and *NbRRA* loci (Fig. 2a) were electroporated into *Agrobacterium tumefaciens* and grown under selection overnight and re-suspended in infiltration buffer to a final total OD$_{600}$ of 0.2 where after the abaxial sides of young expanding leaves of *N. benthamiana* were subjected to *Agrobacterium* infiltration. The

infiltrated plants were left for 2–4 days allowing for gRNA/Cas9 expression and mutation generation within the intact leaves. Western blot analysis of total proteinacious extracts, using anti-Flag and anti-GFP mAbs as primary antibodies against the Flag- and GFP-tagged *Sp*Cas9 revealed the presence of a distinct band at the expected MW (154 kDa) of mature *Sp*Cas9 with a faint band corresponding to the uncleaved fusion protein (*Sp*Cas9-2A-GFP, ca 180 kDa) in

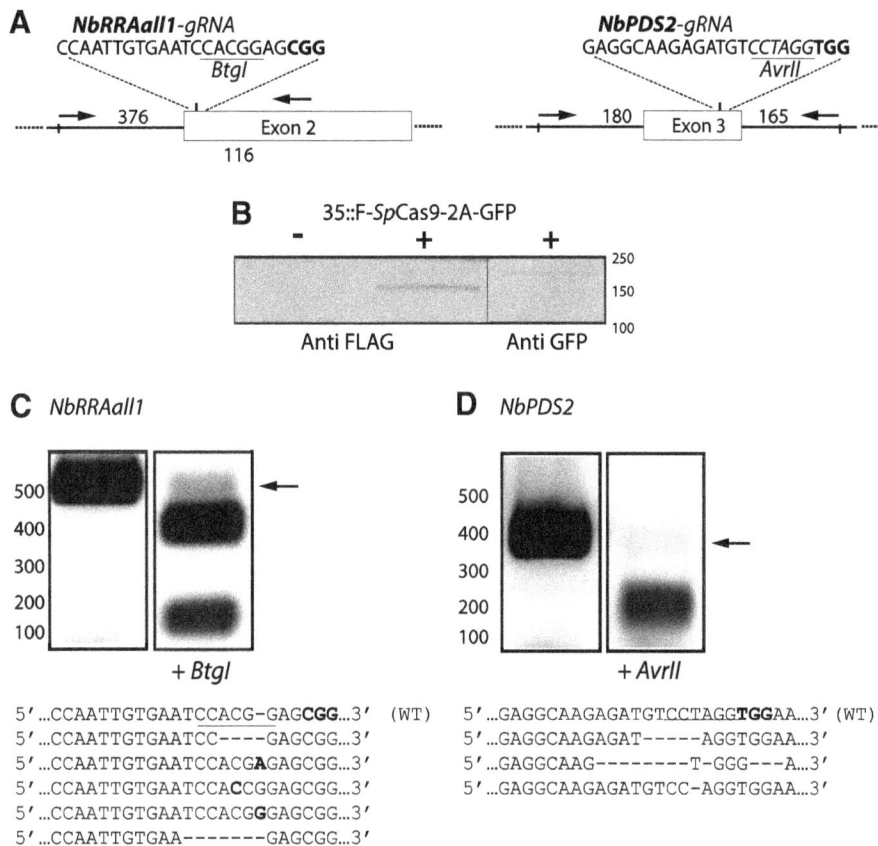

Fig. 2 *NbRRAall1/NbPDS2-gRNA generated indels.* **a** gRNA targets of the *N. benthamiana* loci, *REDUCED RESIDUAL ARABINOSE* arabinosyl transferase (*NbRRA*) and *PHYTOENE DESATURASE* (*NbPDS*), were a *Btg*I and a *Avr*II site is situated 2 and 0 bp upstream of the protospacer adjacent motif (PAM), respectively. Given the predicted cut site of *Sp*Cas9, 3 bp upstream of the PAM sequence [44], all of the *NbPDS2*-gRNA derived mutation combinations will destroy the *Avr*II site in the *NbPDS* target site and only insertions starting with 'G' at the cut site, i.e. less than one fourth the insertions possible, will restore the *Btg*I site in the *NbRRAall1* target site. Primers, flanking the gRNA targets, are indicated by arrows. **b** Western blot analysis of day 4 post infiltration leaves using anti Flag and anti GFP mAbs, cross-reacting to *Sp*Cas9 (154 kDa) and to a faint protein band corresponding to the un-cleaved fusion protein (*Sp*Cas9-2A-GFP, ca 180 kDa), respectively. **c, d** DNA from 4 days post infiltration leaf samples of *NbRRAall1-* and *NbPDS2*-gRNA/Cas9 infiltrations were isolated, PCR amplified and subjected to restriction enzyme digestions using *Btg*I (*NbRRAall1*) and *Avr*II (*NbPDS2*), respectively, with the resistant bands (indicated by arrow) isolated, cloned into pJet and 12 clones of each target sequenced revealing the resulting indels depicted

infiltrated leaves demonstrating expression and efficient 2A mediated auto cleavage of *Sp*Cas9-2A-GFP (Fig. 2b).

RE-mediated mutation analysis of PCR fragments, using primers flanking the gRNA target sites, revealed the presence of non-digestible bands indicative of mutated RE recognition/cleavage sequence for the two target sites (Fig. 2c and d). The RE resistant band of each locus was isolated, sub-cloned and sequenced with the presence of insertions or deletions (indels) demonstrated (Fig. 2c and d).

Protoplast isolation and FACS-mediated mutation enrichment

Protoplasts of WT and infiltrated leaves were essentially obtained using the protocol devised by Dovzhenko et al. 1998 [27] with minor modifications as outlined in the Methods section. Protoplast quality and yield varied

significantly apparently influenced by growth conditions pre- and post-infiltration. Here a temperature of 22–24 °C, a 16 h/8 h (light/dark) regime of moderate sunlight (See 'Growth conditions', Methods section) generally conferred a high amount of intact protoplasts. Protoplast integrity and transformation were assessed by comparative bright field and fluorescent microscopy frequently with varying estimated transformation rates of 20 - > 80% (Additional file 1: Figure S1). GFP fluorescence accumulated in particular in the cytoplasmic strands and the contours of the cell (Additional file 1: Figure S1), which is in agreement with a cytoplasmic 2A-mediated release of GFP. This was corroborated by the western blot analysis (Fig. 2b) showing presence of the mature *Sp*Cas9 with only traces un-cleaved product. Also, in agreement with soluble non-tagged GFP being able to pass into and accumulate in nuclei [50], some accumulation in nuclei structures was observed (Additional file 1: Figure S1).

FACS of fluorescent protoplasts were done using a FAC-SAria III (BD Biosciences) apparatus with settings to accommodate for the approximate size of *N. benthamiana* protoplasts [51] as described in the Methods section. Two fluorescent enriched populations, protoplasts with medium GFP intensity (P4) and with high intensity (P5), were selected for sorting corresponding to 17% & 10 and 14% & 5% of the total population for the *NbRRA*all1-gRNA/*Sp*Cas9-2A-GFP and *NbPDS2*-gRNA/*Sp*Cas9-2A-GFP infiltrations, respectively (Fig. 3b). RE analysis of PCR amplicons suggested an estimated indel frequency of unsorted, P4 and P5 sorted populations of 20–30, 50% and

70–80% for the *NbRRA*all1-gRNA and 40, 50 and > 80% for the *NbPDS2*-gRNA (Fig. 3c). This was corroborated by Indel Detection by Amplicon Analysis (IDAA) (Fig. 3d) and sequence analysis of the cloned PCR fragments of the two P5 populations (10 clones of each), which showed indel to WT ratio of 60 and 70%, respectively. The indel distributions obtained for the *NbRRA*all1- and *NbPDS2*-gRNA infiltrations – 3(1), – 1(2) & + 1(4) and – 1(3) & + 1(3) (Fig. 3e), respectively, are in agreement with earlier findings for *Sp*Cas9 mediated mutations in plants [52].

Bright field microscopy suggested that 10–20% of a WT protoplast population was intact after FACS when using PBS buffer as sheath fluid and MMM550 as recipient buffer (Additional file 2: Figure S2). Viability of post FACS GFP

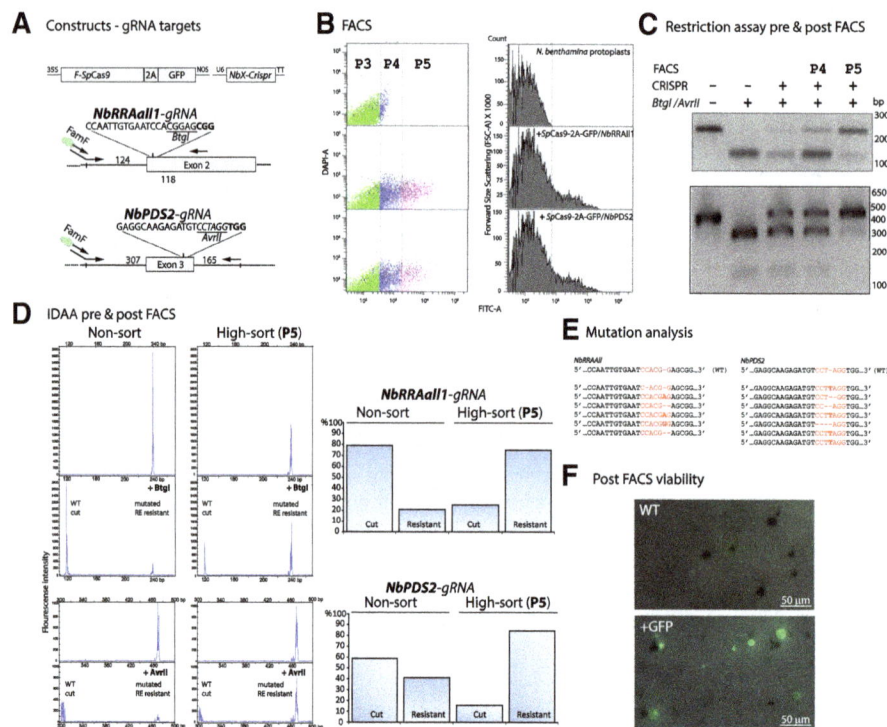

Fig. 3 *FACS mediated enrichment of gRNA/SpCas9 expressing protoplast cells and resulting mutations.* 3–5 *N. benthamiana* leaves were infiltrated with *Agrobacterium*-delivered replicons expressing *Sp*Cas9-2A-GFP together with *NbRRA*all1-gRNA or *NbPDS2*-gRNA (**a**), respectively, and left for 2–4 days. **b** WT protoplasts and protoplasts expressing *Sp*Cas9-2A-GFP and *NbRRA*all1- or *NbPDS2*-gRNA were subjected to GFP mediated FACS. The DAPI and FITC intensities for protoplasts were recorded and three populations, P3, P4 & P5, with the P3 population corresponding to non-transformed cells and the P4 and P5 populations representing intermediary and high stringently sorted cell populations, respectively, were selected from the Dot Scattering Chromatograms. Transfected protoplasts were defined as FITC-positive events and gates were set to separate WT and GFP enriched protoplast populations, using the WT sample to define non-transfected wild type populations (P3) in the transfected samples. P4 and P5 (GFP enriched populations) were gated with medium and high FITC signal intensity. **c** RE analysis of PCR amplified target regions using *Btg*I and *Avr*II for *NbRRA*all1- and *NbPDS2* gRNAs, respectively, demonstrating indel formation in unsorted and indel enrichment in FACS sorted (P4 and P5) populations. Indel enrichment in P5 populations were cooperated by the IDAA technique (**d**) where the additional restriction enzyme digest allows for visualization of the mutated population without the presence of non-mutated PCR amplicons ('mutated / RE resistant' designates the RE site were mutated rendering it resistant for digestion while 'WT/cut' designates WT sites that were cut and moved downstream in the chromatogram). **e** Sequence analysis of RE-resistant PCR fragments of the two P5 populations. **f** Post viability of protoplasts was assessed in WT protoplasts (dark circular objects) without detectable GFP signal and GFP fluorescence in Cas9-2A-GFP sorted protoplasts (presented as an overlay of light and fluorescent micrographs). FACS was carried out using a FACSAria III (BD Biosciences) apparatus with procedure and parameters as outlined in the Methods section and IDDA as described in [39]. For the viability test shown in **f** crude protoplasts were prepared and sorted on a Sony Cell sorter SH800S with sorting gating parameters similar to those used on the BD FACSAria III sorter and with the W5 buffer as recipient buffer

positive protoplasts was assessed by bright field and fluorescence microscopy (Fig. 3f) and confirmed by propidium iodide exclusion assays (Additional file 2: Figure S2). Sorting into PBS as recipient buffer resulted in instant lysis as evidenced by bright field microscopy (data not shown). In agreement with ribonucleoprotein, i.e. in vitro transcribed gRNA and heterologous expressed Cas9, conferring nuclease activity in-vitro [10, 53], we tested (post FACS) for PBS lysis mediated editing activity and found an 2–3 fold increased editing when the PBS lysed protoplasts were left in PBS for 2 h at room temperature (Additional file 3: Figure S3). All post FACS protoplast samples were immediately incubated on ice accordingly. Potential continued editing in the timespan from FACS to further processing may on the other hand likewise increase the 'in cell' editing.

Embedment of GFP transformed protoplasts in alginate with initial callus formation (Additional file 4: Figure S4) demonstrated the feasibility of obtaining gene edited lines using an explant shoot regeneration systems as described in [40].

Discussion

The use of PGE in plants may be complicated by the presence of complex genomes and inefficient or poorly controlled delivery of PGE components to gamete cells or recipient pluripotent cells. DNA encoding PGE components may be delivered to the plant cell either directly, i.e. by biolistic transformation or protoplast transformation (reviewed by [5]), or indirectly, mainly via bacteria, usually *Agrobacterium tumefaciens* or (less commonly) *Agrobacterium rhizogenes* [54], which is generally perceived to be a controlled way of delivering transgenes [55]. Virus replicons provide high copy number of expression units and thus a means of significantly boosting PGE component expression levels [46, 56] and methods for increasing identification/selection of PGE-edited cells have been introduced and applied successfully e.g. for mammal cells [29].

In the present study we combined *Agrobacterium*-mediated delivery of a viral replicon expressing GFP labeled gRNA/*Sp*Cas9 for generation of in-leaf mutations with the use of FACS of GFP-fluorescent protoplasts for enrichment of mutated protoplast populations. *BeYDV* GVR replicons, expressing gRNAs targeting the *NbPDS* and the *NbRRA* locus in *N. benthamiana*, respectively, together with the *Sp*Cas9 nuclease, fused to the 2A self-splicing sequence and GFP (SpCas9-2A-GFP), were introduced into leaves of *N. benthamiana* by *Agrobacterium*-mediated infiltration and left for expression and mutation generation within the intact leaf. In leaf expression of the GFP- and Flag-tagged *Sp*Cas9 enzyme was readily verified by western blot analysis and generated mutations as evidenced by the presence of restriction enzyme (RE) resistant bands of PCR amplicons

comprising the mutated recognition site were cooperated by cloning and sequence analysis of the RE resistant bands. Indel distribution was found to be in accordance with earlier studies for *Sp*Cas9 mediated genome editing in plants [52]. With the aim of selecting and concentrating edited cells GFP-expressing protoplasts were isolated and subjected to FACS. The two fluorescence-enriched populations were selected for FACS with the most stringently sorted population yielding a 3–5 fold enrichment in mutations as evidenced by RE-mediated mutation analysis of PCR amplicons and sequence analysis.

The IDAA method allows for fast and direct assessment of indel prevalence and distribution [39]. In the current study, IDAA was combined with RE analysis for visualization of the isolated mutation population, where potential single nucleotide substitutions within the RE recognition site will otherwise co-migrate with the WT peak. While the observed > 50% reduction of FAM-fluorescence signal in IDAA analyses of overnight RE digestions may complicate absolute peak quantification between samples, quantification of the WT peak and indel peak(s) within single samples, provides a means of estimating relative mutation efficacies between samples. The combined use of the RE analysis and the IDAA technique adds an extra analytical layer to the versatile IDAA technique. 10–20% of the protoplasts appeared to be intact in post FACS populations, when PBS and the MMM550 buffer were used as sheath fluid and recipient buffer, respectively. This ratio may, however, be increased by replacing the sheath fluid PBS buffer with a more osmotically favorable buffer and, if feasible on the FACS apparatus used, decrease shearing forces by lowering the psi. Here FACS on a Sony SH3800S cell sorter yielded ample intact protoplasts post FACS probably due to the available 130 μm sorting chip with accordingly lower psi. Isolation of non-ruptured protoplasts through the use of a sucrose gradient significantly aids the identification of protoplast populations with and without GFP-expression. Once an initial delineation of the protoplast populations on the cell sorter has been established, this step may potentially be omitted.

Extracellular gRNA/Cas9 activity from lysed protoplasts, e.g. mediated by FACS sorting, was significant and this residual activity, which may lead to an over estimated indel frequency, was abolished by incubation on ice or FACS sorting into an RNAse containing or protein denaturing buffer.

Also, in this study protoplast yields were generally highly variable. A recent study on *Agrobacterium* infiltration-mediated expression of a reporter in leaves of *N. benthamiana* recommended infiltration of more plants but less leaves and sample more positions on the leaf as opposed to run a high number of technical replicates [57]. In addition, it is conceivable that *Agrobacterium* infection/pathogenesis may affect intact protoplasts yield.

Recently, an unexpected high level of integrations in the recipient genomes associated with PEG-mediated plasmid transformation of protoplasts was reported [12]. Further optimization of the here devised PGE approach may include exploring the use of integration deficient *Agrobacterium* strains [58] or Virus Induced Gene Silencing (VIGS) mediated down-regulation of host plant factors [59] also important for T-DNA integration, as means of non-integrative delivery of the PGE components [60]. The obtained mutation enrichment may facilitate mutation detection e.g. in situations where the activity of a particular gRNA is weak and reduce laborious explant generation and screening steps. Alternatively, the protoplast based PGE system may be used e.g. in promoter-reporter editing test-screens.

Conclusions

The present study outlines a strategy for enrichment of CRISPR/Cas9 editing in leaf protoplasts. GFP tagged gRNA/Cas9 (gRNA/Cas9-2A-GFP) was delivered by *Agrobacterium*-infiltration to leaves of *N. benthamiana* and protoplasts isolated. Subsequent FACS of GFP expressing protoplasts resulted in several fold mutation enrichment in the selected fluorescence enriched protoplast populations.

Methods

Growth conditions

Seeds of wild-type *Nicotiana benthamiana* were sown and grown in soil (Pindstrup substrate number 2) for 4 weeks in greenhouse with a 16/8 h light/dark cycle, app. 70% relative humidity and a day/night temperature cycle of 24 and 17 °C.

2 days prior to infiltration plants were subjected to regular sunlight at a photosynthetic flux of 20–40 μmol photons $m^{-2}s^{-1}$, Photosynthetic Active Radiation (PAR): 20.5 $μE.m^{-2}s^{-1}$, Red – Far Red ratio (R:FR): 1,69), 22–24 °C temperature, an app. 16 h/8 h (light/dark) diurnal rhythm and 70% relative humidity, which were also imposed in the post-infiltration period.

Vectors and construct designs

Descriptive naming of vectors, constructs, primers and primer sequences are provided in Additional file 5: Table S1. The vector pLSLGFP-R (V82), described in [17], containing GFP insert in front of the CMV35S promoter and Gateway destination site in front of CMV 35S promoter-LIR, respectively, was kindly provided by Nicholas Baltes, Michigan University, US. The gRNA Gateway entry vector V26 (pUC57_attL1-*AtU6:BbsI-BbsI*-tracr-TT_AttL2) was synthesized by Genscript. To obtain insertion of

*NbRRA*all1- or *NbPDS2* gRNAs V26 was linearized with *Bbs*I and gRNA targets *NbRRA*all1 and *NbPDS2* were inserted by ligation of the annealed oligonucleotides P042 & P043 and P149 & P150, respectively, yielding V207 (attL1-*AtU6*: *NbPDS2*-tracr-TT_AttL2) and V208 (attL1-*AtU6*: *NbRRA*all1-tracr-TT_AttL2). V207 and V208 were linearized using *Eco*RI and cloned together with the *Streptococcus pyogenes* Cas9 (*Sp*Cas9) fragment [11], which was PCR amplified from HBT-Cas9 (Gift from Jen Sheen, Harvard Medical School) using the primer-set P077 & P212, the GFP-Nos fragment amplified from pLSLGFP. R using the primerset L1 & L2, all together using the In-fusion cloning kit (Clontech), yielding V197 (pUC57_AttL1-*Sp*Cas9-2A-GFP-Nos; *AtU6*-*NbRRA*all1-gRNA-TT.AttL2) and V198 (pUC57_AttL1-*Sp*Cas9-2A-GFP-Nos; *AtU6*-*NbPDS2*-gRNA-TT-AttL2). V197 & V198 was gateway cloned using pLSL_v2 as destination vector yielding V199 (pLSL_V2_ LIR-AttB1-*Sp*Cas9-2A-GFP-Nos; *AtU6*-*NbRRA*all1-gRNA-TT-AttB2 SIR-35S-LIR) and V200 (pLSL_V2_ LIR-AttB1-*Sp*Cas9-2A-GFP-Nos; *AtU6*-*NbPDS2*-gRNA-TT-AttB2 SIR-35S-LIR), respectively. V199 & V200 will, when co-expressed with pREP, express *Sp*Cas9 in fusion with the 2A self-splicing sequence of the foot-and-mouth disease virus (FMDV) [37, 38] and GFP [49] (*Sp*Cas9-2A-GFP) under control of the CMV35S promoter.

For GFP expression only V82 (pLSLGFP-R_v2) was used.

PDS (NbPDS) and RRA (NbRRA) target loci in the N. benthamiana chromosome

N. benthamiana genes were obtained from https://solgenomics.net/tools/blast/?db_id=266 [61] based on homology with the *Arabidopsis thaliana* genes. As *N. benthamiana* is allotetraploid both chromosome variations of a gene in the given locus are obtained. In contrast to e.g. the presence of 1 and 3 isogenes of *AtPDS* [62] and *AtRRA* [41, 42] in diploid Arabidopsis, respectively, *NbPDS* and *NbRRA* appear to be single gene loci in allotetraploid *N. benthamiana*.

The *NbRRA* gene SolGenomics: Niben101Scf18348 with exons (33526..33687, 35895..36708 & 36767..37113) and Niben101Scf09172 with exons (260530..260692, 261438..262553) with the *NbRRA*all1-gRNA situated in exon 2 (35905..35924, 261512..261531).

The *NbPDS* gene SolGenomics: Niben101Scf14708 with exons (13814..14036, 14118..14251, 15346..15435, 16328.. 16386, 16604..16760, 17017.. 17166, 17412..17532, 17695.. 17909 & 18003..18104) and Niben101Scf01283 with exons (198006..198228, 198317..198449, 199413..199501, 200074.. 200127, 200369..200501, 200792..200940, 201104..201223, 201388..201601 201694..201796, 202066..202113 & 202983.. 203028) and with the *NbPDS2*-gRNA situated in Exon 3 (15409..15428, 199476..199495).

Agrobacterium mediated leaf infiltration and expression in Nicotiana benthamiana

Agrobacterium tumefaciens pGV3850, harboring constructs (pREP, p19 and (pLSL_V2_ LIR-AttB1-*Sp*Cas9-2A-GFP-Nos; *AtU6-NbRRA*all1-gRNA-TT- AttB2 SIR-35S-LIR (V199) or pLSL_V2_LIR-AttB1-*Sp*Cas9-2A-GFP-Nos; *AtU6-NbPDS2*-gRNA-TT-AttB2 SIR-35S-LIR (V200) and empty vector control were inoculated in 5 mL YEP media with kanamycin (50 mg/L) and rifampicillin (50 mg/L) and incubated at 28 °C, 250 rpm for 24 h. Cells were harvested by centrifugation for 20 min at 4000×g and re-suspended in infiltration buffer (10 mM MES (Sigma-Aldrich), 10 mM $MgCl_2$ and 10 µM acetosyringone (3′,5′-Dimethoxy-4′-Hydroxyacetophenone, Sigma-Aldrich) to a final OD_{600} of ~ 0.2 and incubated for 3 h at room temperature.

The abaxial side of 3–5 young expanding leaves (4–6 × 6–8 cm (Width, Length)) of *N. benthamiana* was infiltrated with *A. tumesfaciens* pGV3850 containing the various constructs and co-infiltrated with the p19 construct [63] (Final OD_{600} = 0.2) essentially as described by Sainsbury and Lomonossoff (2008) [64], and left for 2–4 days depending on the experimental setting.

Protoplast isolation

Protoplasts were obtained using the protocol devised by Dovzhenko et al. 1998 [27]. Inoculated *N. benthamiana* leaves for subsequent protoplast-alginate embedment were sterilized by dipping in 96% ethanol and floating in 1.5% hypochlorite solution for 15 min. 3–5 leaves were cut into 0.5–1 mm strips with a scalpel and submerged in 10 ml enzyme solution (400 mM mannitol, 20 mM MES-KOH, pH 5.7, 20 mM KCl, supplemented with 1% Cellulase R10 (w/v) (Duchefa Biochemie, C8001), 0.25% Macerozyme (Duchefa Biochemie, C8002), heated to 55 °C, 10 min, then supplemented with 10 mM $CaCl_2$ and 0.1% BSA) and incubated 2–5 h at 26 °C, 100 rpm, then filtered through a 100 µm filter into a 50 ml Falcon tube, centrifuged for 5 min at 100×g, where after the supernatant was poured off and the protoplast-containing pellet was re-suspended in 3 ml of 10 mM $MgSO_4$, 10 mM $MgCl_2$, 10 mM MES-KOH, pH 5.8, buffer, 0.5 M mannitol (MMM550) which was carefully layered on top on 8 ml 0.6 M sucrose cushion and spun down at 100×g, 2 min, at room temperature. Intact protoplasts at the interface were collected and spun down at 100×g for 2 mins then re-suspended in MMM550 –– if used for alginate imbedding this step was repeated three times.

For viability test the protoplast-containing pellet was re-suspended in 5 ml 2.5 mM MES-KOH, pH 5.7, 125 mM $CaCl_2$, 154 mM NaCl, 5 mM KCl, 0.5 mM glucose (W5), centrifuged for 5 min at 100×g, the supernatant poured off, and the pellet re suspended in 0.5 ml W5 and

placed on ice until FACS, which was initiated immediately after the wash step.

Embedment of GFP-fluorescent protoplast in alginate

Protoplast embedment in alginate was essentially done as described in [27] except the thin alginate layer was formed using the 'droplet on Ca-agar' method as described in [65]. Briefly, protoplasts re-suspended in 200 µl MMM550 were mixed with 200 µl alginate solution (MMM550 + 2.8% alginate (low viscosity)). A 300 µl droplet was left on a Ca-Agar plate (0.4 M mannitol, 50 mM $CaCl_2$, 1% plant agar (Duchefa 1001.5000)) which was tilted to spread out the droplet, and after 30 min a floating solution (0.4 M mannitol, 50 mM $CaCl_2$) was added to the plates to allow for movement of the layer. The layer was taken up by a spatula and moved to small Petri dishes containing F-PCN (described in [8]) .

gDNA extraction

A single fully infiltrated leaf was thoroughly ground in liquid nitrogen and DNA was extracted using DNeasy Plant Mini Kit (Qiagen).

PCR of genome target NbRRA and NbPDS loci

PCR-amplicons containing the *NbRRA* & *Nb*PDS targets were amplified using nested PCR: First 5 µl of protoplast suspension (obtained as described in 'Protoplast isolation') was used in a 50 µl PCR reaction using Phire Plant Direct PCR Master Mix (ThermoFisher F160S) with the cycle parameters: 5 min at 98 °C followed by 40 cycles of 10 s at 98 °C, 10 s at (65 °C for *RRA* and 62 °C for *PDS*) and 40 s at 72 °C followed by 7 min at 72 °C using the primers P348 & P232 and P346 & P342 for *NbPDS2* and *NbRRA*all1, respectively. Nested *NbRRA* PCR was performed in a 50 µl reaction using X7 polymerase [66] with 1:100 diluted 1′th PCR reaction as template and the cycle parameters: 5 min at 94 °C followed by 25 cycles of 30 s at 94 °C, 30 s at 58 °C and 30 s at 72 °C followed by 7 min at 72 °C and the primers P319 and P320.

Nested PCR of *NbPDS2* was done in a 25 µl reaction using ClonAMP HiFi master mix 2x (Takara 639,298) with the cycle parameters: 5 min at 98 °C followed by 20 cycles of 30 s at 98 °C, 30 s at 65 °C and temperature dropping 0.5 °C per cycle and 30 s at 72 °C followed by 20 cycles of 30 s at 98 °C, 30 s at 58 °C and 30 s at 72 °C followed by 3 min at 72 °C and the primers P321 and P322.

Primers for scoring in leaf mutations were P321 & P322 (*NbRRA*all1) and P232 & P233 (*NbPDS2*).

Cloning in pJet and sequencing

10 µl of PCR product was digested ON in a 50 µl reaction with *Btg*I (*NbRRA*all1 amplicon) and *Avr*II (*Nb*PDS amplicon). Enzyme Resistant bands were isolated from agarose gels using NucleoSpin® Gel and Monarch® DNA

Gel Extraction Kit (New England Biolabs) and cloned into pJet1.2 using CloneJET PCR Cloning Kit #K1232. Sequences were aligned using CLC Workbench.

Indel detection by amplicon analysis (IDAA) and semi quantification of IDAA peaks

Indel Detection by Amplicon Analysis (IDAA) was done essentially as described in and outlined in [39] and in the Method section '*PCR of genome target NbRRA and NbPDS loci*'. Briefly A tri-primer PCR setup which relies of the incorporation of a florescent universal 6-FAM 5′-labelled primer (FamF), with the corresponding non-labeled primer in a 1:10 diluted concentration, was used for FAM labeling of PCR amplicons. PCR amplification of the *NbRRA*all1 and *NbPDS2* regions were done using the ClonAMP HiFi master mix 2x (Takara 639,298) in a 25 µl reaction with the cycle parameters: 5 min at 95 °C followed by 30 cycles of 30 s at 95 °C, 30 s at 58 °C and 30 s at 72 °C followed by 3 min at 72 °C. Primers were P230 & P231 (*NbRRA*all1) and P232 & P233 (*NbPDS2*), where bold designates FAM primer overhang (Additional file 5: Table S1).

Mutation frequencies, as identified by quantification of peak area in IDAA chromatograms, were estimated using the Open Source Software program ImageJ (https://imagej.nih.gov/ij/) from and with areas identified as described (http://www.openwetware.org/wiki/Protein_Quantification_Using_ImageJ).

Fluorescence microscopy

Fluorescence imaging (presence GFP) was carried out with an epifluorescence microscope Olympus BX41 equipped with a CCD camera (FITC filter for GFP fluorescence and DAPI filter for FDA staining) or a laser scanning confocal microscope Leica SP5 equipped with an Argon (448 nm) and a Argon laser (448 nm).

Western blot analysis

App. 50 µl of seedling powder, crushed in liquid N_2, was boiled in 50 µl 2 × SDS-PAGE loading buffer (280 mM SDS, 400 mM Tris, 40% glycerol, 1.4 M mercaptoethanol, 0.6 mM Bromophenol Blue) for 15 min and separated (200 V, 50 min) on 12% Criterion XT Bis-Tris gels (Bio-rad). Proteins were electrotransferred onto polyvinylidene difluoride (PVDF) membranes (Bio-rad) using a Trans-Blot® Turbo™ Blotting instrument (Bio-rad). The membrane was blocked in blocking solution (PBS pH 7.5, 5% non-fat dry milk) overnight at 4 °C under mild shaking. The membrane was probed with anti-GFP mouse IgG (Roche) and Anti-Flag M2 mouse IgG (Sigma) at 1 : 1000 dilution in blocking solution overnight at 4 °C, followed by 3 × 5′ wash in PBS buffer (PBS pH 7.5). The membrane was then incubated with goat anti-mouse IgG conjugated to Alkaline Phosphatase

(AP) (Sigma) (1 : 1000 dilution in blocking solution) for 1 h at room temperature, and rinsed 3 × 5′ with PBST. Pre-mixed NBT/BCIP AP solution (UCPH, DK) was added to the blot and incubated for color development.

Post FACS residual gRNA activity of lysed protoplasts

20 µl of protoplasts expressing gRNA-*NbPDS2*/*Sp*Cas9 were added to 80 µl PBS, briefly vortexed and left at room temperature for 2 h; 20 µl of protoplasts were added to PBS buffer with 5 µl RNAseA/T1 (Thermo fisher #EN0551), briefly vortexed and left at room temperature for 2 h; and 20 µl of protoplasts was flash frozen with immediately addition of 80 µl PBS which were then heated for 3 min at 95 °C. Flanking primers used were P233 and P232.

Flow cytometry and fluorescence activated cell sorting (FACS) of N. benthamiana protoplasts

The protoplast solution was first passed through 50-µm filcons (BD Biosciences) to achieve a single-cell suspension. Protoplast suspensions were cytometrically analyzed and sorted with a FACSAria III (BD Biosciences) fitted with a 100-µm nozzle and using phosphate-buffered saline (PBS) as a sheath fluid. The procedure and setting used were as described in [29] with a large nozzle size (100 µm) to provide optimal survival for most cell types and sorting based on ~ 10,000 events. Briefly, the sheath pressure was set at 20 psi, and the defection plate voltage was set at 5000 V (default "low" setting). A 488 nm Coherent Sapphire Solid State laser was used for excitation, and emission was measured at 530 nm for GFP. The photomultiplier tube voltage was set at 183 V for forward scatter, 286 V for side scatter, 308 V for GFP, and 518 V for Allophycocyanin. The threshold value for event detection was set at 8835 on forward scattering. The drop drive frequency was set to approximately 30 kHz, and the amplitude was set to approximately 45 V; the drop delay value was approximately 26 (these settings will vary slightly with day-to-day operation of the FACSAria III). Identification of viable, single protoplasts through the use of forward scatter (FSC) and side scatter (SSC) as a first gating strategy, which is routinely used for gating mammalian cells, was not attempted due to the high variability of protoplast size. Instead, the FITC and DAPI intensities were recorded as represented in dot plots. 10,000 events are displayed in each plot. Gates were set to separate and thus enable enrichment of WT and GFP transfected protoplasts, using the WT sample to define non-transfected wild type populations in the transfected samples. Transfected protoplasts were defined as FITC-positive events. Data were processed using the FACSDiva 8.0.1 software (BD Biosciences).

Viability test was done on a Sony SH800S Cell sorter, with automated setup for 130 μm microfluidics sorting chips, psi 9. PBS was used as sheath fluid, with samples sorted into flat-bottomed 96 well microtiter-plates containing 200 μl W5 buffer. For visualization purposes protoplasts were layered at the bottom of the microtiter plate by a brief centrifugation step, 100×g, 5 min.

Gating strategy on Sony SH800S cell sorter was similar to those used on the BD FACSAria III sorter.

Additional files

Additional file 1: Figure S1. *Localized GFP-fluorescence of SpCas9-2A-GFP/NbPDS2-gRNA.* GFP-fluorescence of *Sp*Cas9-2A-GFP/*NbPDS2*-gRNA from *Agrobacterium* infiltrated in leaves of *N. benthamiana* 3 days post infiltration was evident in the contours epidermis cells of intact leaves (**A**). Overlay of bright field and fluorescence (FITC filter) microscopy of isolated *Sp*Cas9-2A-GFP/*NbPDS2*-gRNA transformed protoplasts regularly showed 60 - > 80% transformation efficiency (**B**). GFP fluorescence was seen in cytoplasmic strands with some nuclei accumulation (**A** and **C**), which both are in accordance with a primarily cytoplasmic localization of the GFP. (PDF 9243 kb)

Additional file 2: Figure S2. *WT N. benthamiana protoplasts pre and post FACS.* Protoplasts were isolated as described in the Methods section, stored in buffer MMM550 on ice and immediately FACS sorted (total population sorted) into the MMM550 buffer and stored on ice. An estimated survival rate of ca 10–20% (concentric intact protoplast) was observed as evidenced by bright field (**A**) microscopy. (**B, C**) GFP expression analysis using confocal microscopy and viability test using propidium iodide (PI). Left panels are scan of the protoplasts expressing *Sp*Cas9-2A-GFP construct. Right panels non-transformed control. (**B**) Distinguishable GFP signal can be observed in transformed protoplasts (arrowhead). (**C**) Viability analysis using propidium iodide (PI). Non PI stained protoplast expressing GFP were observed. (PDF 7907 kb)

Additional file 3: Figure S3. *Post FACS residual gRNA/Cas9 activity of lysed protoplasts.* Ribonucleoprotein, i.e. in vitro transcribed gRNA mixed with heterologous expressed *Sp*Cas9 enzyme, delivered by PEG transformation, have been shown to confer efficient nuclease activity in *Arabidopsis thaliana*, tobacco, lettuce and rice protoplasts [10, 53]. We tested whether PBS mediated protoplast lysis could mediate additional extra-cellular derived indel formation resulting in an over-estimated gRNA/*Sp*Cas9 activity. Incubation 2 h at room temperature in PBS buffer resulted in a 2–3 fold increased indel formation, compared to immediate activity abolishment through flash freezing/boiling or RNAse addition, as judged by resistant RE band intensities. Lanes: Pos Ctrl (*NbPDS2*-gRNA/*Sp*Cas9 positive from leaves), Neg ctrl (WT without *NbPDS2*-gRNA/*Sp*Cas9), RT 2 h (PBS mediated lysis followed by 2 h incubation at room temperature), flash freezing (flash freezing in liquid N₂ followed by boiling), +RNAase (RNAase addition). For experimental setup see Method section (PDF 191 kb)

Additional file 4: Figure S4. *GFP-fluorescent protoplasts embedded in alginate.* Single fluorescent protoplasts are visible as evidenced by fluorescent (FITC filter) microscopy before (**A**) and after alginate embedment (**B**). Calli formation (**C**) of a single protoplast as evidenced by bright field microscopy. Protoplast embedment in alginate is described in the Methods section. (PDF 4381 kb)

Additional file 5: Table S1. *Vector construct and primer list (DOCX 18 kb)*

Abbreviations
A. tumefaciens: Agrobacterium tumefaciens; *AtU6-Pro*: Arabidopsis thaliana U6 promoter; *BeYDV*: Bean yellow dwarf virus; CMV35S: Cauliflower mosaic virus 35S promoter; CRISPR-Cas: Clustered Regularly Interspaced Short Palindromic Repeats (CRISPR)/CRISPR-associated systems (Cas)); FACS: Fluorescence Activated Cell Sorting; FMDV, 2A: Self-cleaving sequence of foot-and-mouth disease virus, 2A; gRNA: guide RNA; GVR: Gemini Virus Replicon;

HPT: Hygromycin phosphotransferase; HR: Homologous Recombination; IDAA: Indel Detection by Amplicon Analysis; indels: Deletions and insertions; LB & RB: Left and Right T-DNA border; *N. benthamiana*: Nicotiana benthamiana; *NbPDS*: Phytoene desaturase; *NbRRA*: REDUCED RESIDUAL ARABINOSE arabinosyl transferase; NHEJ : Non-Homologous End Joining; NLS: Nucleus Localisation Signal; NOS: Nopaline synthase terminator; PAM: Protospacer adjacent motif; PGE : Precise genome editing; RE: Restriction Enzyme; *Sp*Cas9: Streptococcus pyogenes Cas9; TALENs: Transcription activator-like effector nucleases; ZFNs: Zinc Finger-Nucleases

Acknowledgements
Nicholas Baltes, Minnesota University, U.S.A, is acknowledged for providing the Gemini Virus Replicon vector constructs. Jen Sheen, Harvard Medical School, U.S.A, is acknowledged for providing the HBT-Cas9 construct.

Authors' contributions
All authors have seen and approved the manuscript. RSM, BLP, JM, MC, YL and ZY conducted the experiments and BLP, RSM, BJ, HHW, EPB and ZY designed the experiments. BLP wrote the paper. All authors have read and approved the manuscript.

Funding
This work was supported by The Danish Councils for Strategic and Independent Research (12–125709, 12–131859) (PGE for controlled yeast and plant and glycosylation), The Danish National Research Foundation (DNRF107), The Copenhagen University Excellence Program for Interdisciplinary Research (CDO2016) PGE for controlled plant glycosylation, Villum Foundation (00017489) (Confocal imaging and microscopy).

Ethics approval and consent to participate
Nothing to declare.

Consent for publication
Not applicable.

Competing interests
The authors declare that they have no competing interests.

Author details
¹Department of Plant and Environmental Sciences, University of Copenhagen, DK-1871 Frederiksberg C, Denmark. ²Present Address: Centre for Novel Agricultural Products, University of York, Woodsmill Quay, Skeldergate, York YO1 6DX, UK. ³Present Address: UIT - Department of Chemistry, The Arctic University of Norway, Forskningsparken. 3, 9019 Tromsø, Norway. ⁴Copenhagen Center for Glycomics, Department of Molecular and Cellular Medicine and School of Dentistry, Faculty of Health Sciences, University of Copenhagen, DK-2200 Copenhagen N, Denmark.

References
1. Tsai SQ, Joung JK. Defining and improving the genome-wide specificities of CRISPR-Cas9 nucleases. Nat Rev Genet. 2016;17(5):300–12.
2. Gao C. The future of CRISPR technologies in agriculture. Nat Rev Mol Cell Biol. 2018;19(5):275–6.
3. Ding Y, Li H, Chen LL, Xie K. Recent advances in genome editing using CRISPR/Cas9. Front Plant Sci. 2016;7:703.
4. Sanford JC, Smith FD, Russell JA. Optimizing the biolistic process for different biological applications. Methods Enzymol. 1993;217:483–509.
5. Rivera AL, Gomez-Lim M, Fernandez F, Loske AM. Physical methods for genetic plant transformation. Phys Life Rev. 2012;9(3):308–45.
6. Townsend JA, Wright DA, Winfrey RJ, Fu F, Maeder ML, Joung JK, Voytas DF. High-frequency modification of plant genes using engineered zinc-finger nucleases. Nature. 2009;459(7245):442–5.
7. Zhang Y, Zhang F, Li X, Baller JA, Qi Y, Starker CG, Bogdanove AJ, Voytas DF. Transcription activator-like effector nucleases enable efficient plant genome engineering. Plant Physiol. 2013;161(1):20–7.

8. Mao Y, Zhang H, Xu N, Zhang B, Gou F, Zhu JK. Application of the CRISPR-Cas system for efficient genome engineering in plants. Mol Plant. 2013;6(6): 2008–11.

9. Li JF, Norville JE, Aach J, McCormack M, Zhang DD, Bush J, Church GM, Sheen J. Multiplex and homologous recombination-mediated genome editing in Arabidopsis and Nicotiana benthamiana using guide RNA and Cas9. Nat Biotechnol. 2013;31(8):688–91.

10. Woo JW, Kim J, Il Kwon S, Corvalan C, Cho SW, Kim H, Kim SG, Kim ST, Choe S, Kim JS. DNA-free genome editing in plants with preassembled CRISPR-Cas9 ribonucleoproteins. Nat Biotechnol. 2015;33(11):1162–U1156.

11. Li JF, Norville JE, Aach J, McCormack M, Zhang D, Bush J, Church GM, Sheen J. Multiplex and homologous recombination-mediated genome editing in Arabidopsis and Nicotiana benthamiana using guide RNA and Cas9. Nat Biotechnol. 2013;31(8):688–91.

12. Clasen BM, Stoddard TJ, Luo S, Demorest ZL, Li J, Cedrone F, Tibebu R, Davison S, Ray EE, Daulhac A, et al. Improving cold storage and processing traits in potato through targeted gene knockout. Plant Biotechnol J. 2016; 14(1):169–76.

13. Gelvin SB. Agrobacterium-mediated plant transformation: the biology behind the "gene-jockeying" tool. Microbiol Mol Biol Rev. 2003;67(1):16–37 table of contents.

14. Nester EW. Agrobacterium: nature's genetic engineer. Front Plant Sci. 2014; 5:730.

15. Gleba YY, Tuse D, Giritch A. Plant viral vectors for delivery by agrobacterium. Curr Top Microbiol Immunol. 2014;375:155–92.

16. Peyret H, Lomonossoff GP. When plant virology met agrobacterium: the rise of the deconstructed clones. Plant Biotechnol J. 2015;13(8):1121–35.

17. Baltes NJ, Gil-Humanes J, Cermak T, Atkins PA, Voytas DF. DNA replicons for plant genome engineering. Plant Cell. 2014;26(1):151–63.

18. Cermak T, Baltes NJ, Cegan R, Zhang Y, Voytas DF. High-frequency, precise modification of the tomato genome. Genome Biol. 2015;16:232.

19. Gil-Humanes J, Wang Y, Liang Z, Shan Q, Ozuna CV, Sanchez-Leon S, Baltes NJ, Starker C, Barro F, Gao C, et al. High-efficiency gene targeting in hexaploid wheat using DNA replicons and CRISPR/Cas9. Plant J. 2017;89(6): 1251–62.

20. Schneider JD, Marillonnet S, Castilho A, Gruber C, Werner S, Mach L, Klimyuk V, Mor TS, Steinkellner H. Oligomerization status influences subcellular deposition and glycosylation of recombinant butyrylcholinesterase in Nicotiana benthamiana. Plant Biotechnol J. 2014;12(7):832–9.

21. Alkanaimsh S, Karuppanan K, Guerrero A, Tu AM, Hashimoto B, Hwang MS, Phu ML, Arzola L, Lebrilla CB, Dandekar AM, et al. Transient expression of tetrameric recombinant human Butyrylcholinesterase in Nicotiana benthamiana. Front Plant Sci. 2016;7:743.

22. Dirnberger D, Steinkellner H, Abdennebi L, Remy JJ, van de Wiel D. Secretion of biologically active glycoforms of bovine follicle stimulating hormone in plants. Eur J Biochem. 2001;268(16):4570–9.

23. Le Mauff F, Mercier G, Chan P, Burel C, Vaudry D, Bardor M, Vezina LP, Couture M, Lerouge P, Landry N. Biochemical composition of haemagglutinin-based influenza virus-like particle vaccine produced by transient expression in tobacco plants. Plant Biotechnol J. 2015;13(5): 717–25.

24. Dicker M, Tschofen M, Maresch D, Konig J, Juarez P, Orzaez D, Altmann F, Steinkellner H, Strasser R. Transient Glyco-engineering to produce recombinant IgA1 with defined N- and O-Glycans in plants. Front Plant Sci. 2016;7:18.

25. Yang Y, Li R, Qi M. In vivo analysis of plant promoters and transcription factors by agroinfiltration of tobacco leaves. Plant J. 2000;22(6):543–51.

26. Shen J, Fu J, Ma J, Wang X, Gao C, Zhuang C, Wan J, Jiang L. Isolation, culture, and transient transformation of plant protoplasts. Curr Protoc Cell Biol. 2014;63:2.8.1–17.

27. Dovzhenko A, Bergen U, Koop HU. Thin-alginate-layer technique for protoplast culture of tobacco leaf protoplasts: shoot formation in less than two weeks. Protoplasma. 1998;204(1–2):114–8.

28. Shepard JF, Totten RE. Isolation and regeneration of tobacco Mesophyll cell protoplasts under low osmotic conditions. Plant Physiol. 1975;55(4):689–94.

29. Lonowski LA, Narimatsu Y, Riaz A, Delay CE, Yang Z, Niola F, Duda K, Ober EA, Clausen H, Wandall HH, et al. Genome editing using FACS enrichment of nuclease-expressing cells and indel detection by amplicon analysis. Nat Protoc. 2017;12(3):581–603.

30. Bargmann BO, Birnbaum KD. Fluorescence activated cell sorting of plant protoplasts. J Vis Exp. 2010;(36):1673.

31. Birnbaum K, Jung JW, Wang JY, Lambert GM, Hirst JA, Galbraith DW, Benfey PN. Cell type-specific expression profiting in plants via cell sorting of protoplasts from fluorescent reporter lines. Nat Methods. 2005;2(8):615–9.

32. Kirchhoff J, Raven N, Boes A, Roberts JL, Russell S, Treffenfeldt W, Fischer R, Schinkel H, Schiermeyer A, Schillberg S. Monoclonal tobacco cell lines with enhanced recombinant protein yields can be generated from heterogeneous cell suspension cultures by flow sorting. Plant Biotechnol J. 2012;10(8):936–44.

33. Carqueijeiro I, Guimaraes AL, Bettencourt S, Martinez-Cortes T, Guedes JG, Gardner R, Lopes T, Andrade C, Bispo C, Martins NP, et al. Isolation of cells specialized in anticancer alkaloid metabolism by fluorescence-activated cell sorting. Plant Physiol. 2016;171(4):2371–8.

34. Nishimasu H, Ran FA, Hsu PD, Konermann S, Shehata SI, Dohmae N, Ishitani R, Zhang F, Nureki O. Crystal structure of Cas9 in complex with guide RNA and target DNA. Cell. 2014;156(5):935–49.

35. Waibel F, Filipowicz W. U6 snRNA genes of Arabidopsis are transcribed by RNA polymerase III but contain the same two upstream promoter elements as RNA polymerase II-transcribed U-snRNA genes. Nucleic Acids Res. 1990; 18(12):3451–8.

36. Li X, Jiang D-H, Yong K, Zhang D-B. Varied transcriptional efficiencies of multiple Arabidopsis U6 small nuclear RNA genes. J Integr Plant Biol. 2007; 49(2):222–9.

37. El Amrani A, Barakate A, Askari BM, Li X, Roberts AG, Ryan MD, Halpin C. Coordinate expression and independent subcellular targeting of multiple proteins from a single transgene. Plant Physiol. 2004;135(1):16–24.

38. Szymczak AL, Workman CJ, Wang Y, Vignali KM, Dilioglou S, Vanin EF, Vignali DA. Correction of multi-gene deficiency in vivo using a single 'self-cleaving' 2A peptide-based retroviral vector. Nat Biotechnol. 2004; 22(5):589–94.

39. Yang Z, Steentoft C, Hauge C, Hansen L, Thomsen AL, Niola F, Vester-Christensen MB, Frodin M, Clausen H, Wandall HH, et al. Fast and sensitive detection of indels induced by precise gene targeting. Nucleic Acids Res. 2015;43(9):e59.

40. Vasil V, Vasil IK. Regeneration of tobacco and petunia plants from protoplasts and culture of corn protoplasts. In Vitro. 1974;10:83–96.

41. Egelund J, Obel N, Ulvskov P, Geshi N, Pauly M, Bacic A, Petersen BL. Molecular characterization of two Arabidopsis thaliana glycosyltransferase mutants, rra1 and rra2, which have a reduced residual arabinose content in a polymer tightly associated with the cellulosic wall residue. Plant Mol Biol. 2007;64(4):439–51.

42. Velasquez SM, Ricardi MM, Dorosz JG, Fernandez PV, Nadra AD, Pol-Fachin L, Egelund J, Gille S, Harholt J, Ciancia M, et al. O-glycosylated cell wall proteins are essential in root hair growth. Science. 2011;332(6036):1401–3.

43. Doench JG, Hartenian E, Graham DB, Tothova Z, Hegde M, Smith I, Sullender M, Ebert BL, Xavier RJ, Root DE. Rational design of highly active sgRNAs for CRISPR-Cas9-mediated gene inactivation. Nat Biotechnol. 2014; 32(12):1262–7.

44. Garneau JE, Dupuis ME, Villion M, Romero DA, Barrangou R, Boyaval P, Fremaux C, Horvath P, Magadan AH, Moineau S. The CRISPR/Cas bacterial immune system cleaves bacteriophage and plasmid DNA. Nature. 2010; 468(7320):67–71.

45. Gao Y, Zhang Y, Zhang D, Dai X, Estelle M, Zhao Y. Auxin binding protein 1 (ABP1) is not required for either auxin signaling or Arabidopsis development. Proc Natl Acad Sci. 2015;112(7):2275–80.

46. Regnard GL, Halley-Stott RP, Tanzer FL, Hitzeroth II, Rybicki EP. High level protein expression in plants through the use of a novel autonomously replicating geminivirus shuttle vector. Plant Biotechnol J. 2010;8(1):38–46.

47. Hefferon KL, Dugdale B. Independent expression of rep and RepA and their roles in regulating bean yellow dwarf virus replication. J Gen Virol. 2003; 84(12):3465–72.

48. Jinek M, Chylinski K, Fonfara I, Hauer M, Doudna JA, Charpentier E. A programmable dual-RNA-guided DNA endonuclease in adaptive bacterial immunity. Science. 2012;337(6096):816–21.

49. Siemering KR, Golbik R, Sever R, Haseloff J. Mutations that suppress the thermosensitivity of green fluorescent protein. Curr Biol. 1996;6(12):1653–63.

50. Chiu W, Niwa Y, Zeng W, Hirano T, Kobayashi H, Sheen J. Engineered GFP as a vital reporter in plants. Curr Biol. 1996;6(3):325–30.

51. Takebe I, Otsuki Y, Aoki S. Isolation of tobacco mesophyll cells in intact and active state. Plant Cell Physiol. 1968;9(1):115–24.

52. Feng Z, Mao Y, Xu N, Zhang B, Wei P, Yang DL, Wang Z, Zhang Z, Zheng R, Yang L, et al. Multigeneration analysis reveals the inheritance, specificity,

and patterns of CRISPR/Cas-induced gene modifications in Arabidopsis. Proc Natl Acad Sci U S A. 2014;111(12):4632–7.

53. Liu Y, Tao W, Wen S, Li Z, Yang A, Deng Z, Sun Y. In vitro CRISPR/Cas9 system for efficient targeted DNA editing. mBio. 2015;6(6):e01714–5.

54. Tzfira T, Yarnitzky O, Vainstein A, Altman A. Agrobacterium rhizogenes-mediated DNA transfer inPinus halepensis mill. Plant Cell Rep. 1996;16(1–2): 26–31.

55. Ziemienowicz A, Shim YS, Matsuoka A, Eudes F, Kovalchuk I. A novel method of transgene delivery into triticale plants using the agrobacterium transferred DNA-derived nano-complex. Plant Physiol. 2012;158(4):1503–13.

56. Yin K, Han T, Liu G, Chen T, Wang Y, Yu AYL, Liu Y. A geminivirus-based guide RNA delivery system for CRISPR/Cas9 mediated plant genome editing. Sci Rep-Uk. 2015;5:14926.

57. Bashandy H, Jalkanen S, Teeri TH. Within leaf variation is the largest source of variation in agroinfiltration of Nicotiana benthamiana. Plant Methods. 2015;11:47.

58. Mysore KS, Bassuner B, Deng XB, Darbinian NS, Motchoulski A, Ream W, Gelvin SB. Role of the agrobacterium tumefaciens VirD2 protein in T-DNA transfer and integration. Mol Plant-Microbe Interact. 1998;11(7):668–83.

59. Anand A, Krichevsky A, Schornack S, Lahaye T, Tzfira T, Tang Y, Citovsky V, Mysore KS. Arabidopsis VIRE2 INTERACTING PROTEIN2 is required for agrobacterium T-DNA integration in plants. Plant Cell. 2007;19(5):1695–708.

60. Waltz E. Tiptoeing around transgenics. Nat Biotechnol. 2012;30(3):215–7.

61. Bombarely A, Rosli HG, Vrebalov J, Moffett P, Mueller LA, Martin GB. A draft genome sequence of Nicotiana benthamiana to enhance molecular plant-microbe biology research. Mol Plant-Microbe Interact. 2012;25(12):1523–30.

62. Qin GJ, Gu HY, Ma LG, Peng YB, Deng XW, Chen ZL, Qu LJ. Disruption of phytoene desaturase gene results in albino and dwarf phenotypes in Arabidopsis by impairing chlorophyll, carotenoid, and gibberellin biosynthesis. Cell Res. 2007;17(5):471–82.

63. Voinnet O, Rivas S, Mestre P, Baulcombe D. An enhanced transient expression system in plants based on suppression of gene silencing by the p19 protein of tomato bushy stunt virus. Plant J. 2003;33(5):949–56.

64. Sainsbury F, Lomonossoff GP. Extremely high-level and rapid transient protein production in plants without the use of viral replication. Plant Physiol. 2008;148(3):1212–8.

65. Maćkowska K, Jarosz A, Grzebelus E. Plant regeneration from leaf-derived protoplasts within the Daucus genus: effect of different conditions in alginate embedding and phytosulfokine application. Plant Cell, Tissue Organ Cult (PCTOC). 2014;117(2):241–52.

66. Nørholm MH. A mutant Pfu DNA polymerase designed for advanced uracil-excision DNA engineering. BMC Biotechnol. 2010;10(1):21.

16

Efficient gene transfer into T lymphocytes by fiber-modified human adenovirus 5

Yun Lv[1,2,3], Feng-Jun Xiao[2], Yi Wang[3], Xiao-Hui Zou[3], Hua Wang[2], Hai-Yan Wang[4], Li-Sheng Wang[2,4*] and Zhuo-Zhuang Lu[3*]

Abstract

Background: The gene transduction efficiency of adenovirus to hematopoietic cells, especially T lymphocytes, is needed to be improved. The purpose of this study is to improve the transduction efficiency of T lymphocytes by using fiber-modified human adenovirus 5 (HAdV-5) vectors.

Results: Four fiber-modified human adenovirus 5 (HAdV-5) vectors were investigated to transduce hematopoietic cells. F35-EG or F11p-EG were HAdV-35 or HAdV-11p fiber pseudotyped HAdV-5, and HR-EG or CR-EG vectors were generated by incorporating RGD motif to the HI loop or to the C-terminus of F11p-EG fiber. All vectors could transduce more than 90% of K562 or Jurkat cells at an multiplicity of infection (MOI) of 500 viral particle per cell (vp/cell). All vectors except HR-EG could transduce nearly 90% cord blood CD34+ cells or 80% primary human T cells at the MOI of 1000, and F11p-EG showed slight superiority to F35-EG and CR-EG. Adenoviral vectors transduced CD4+ T cells a little more efficiently than they did to CD8+ T cells. These vectors showed no cytotoxicity at an MOI as high as 1000 vp/cell because the infected and uninfected T cells retained the same CD4/CD8 ratio and cell growth rate.

Conclusions: HAdV-11p fiber pseudotyped HAdV-5 could effectively transduce human T cells when human EF1a promoter was used to control the expression of transgene, suggesting its possible application in T cell immunocellular therapy.

Keywords: Gene therapy, Adenovirus 5 vector, Human hematopoietic cells

Background

T lymphocytes play an important role in adaptive immunity. It is highly desirable to introduce exogenous wild type or mutant genes into primary T cells for studying their growth, differentiation, death, and interaction with other immunocytes [1]. T cells are also important targets for gene therapy of numerous human diseases, including cancer, diabetes, arthritis and AIDS [2–5]. Gene delivery to T cells has been achieved by retroviruses including gamma-retrovirus, lentivirus and alpha-retrovirus [2, 6]. Retrovirus is able to integrate into the host's genome, and the transgene stably expresses. However, retrovirus based vector has drawbacks: it can hardly be produced on a large scale; purification operation cannot substantially improve its performance and is often skipped; transfection efficiency is relatively low; and the integration property can possibly cause genetic toxicity such as transformation of host cells.

Adenoviral vector is widely used in gene therapy and vaccine development [7–9]. Adenovirus is non-integrating vector. It can be amplified to very high titer and be conveniently purified. The cloning capacity is higher and the expression of exogenous gene is relatively more efficient. When combined with other gene transfer technique such as transposon, EBV nuclear antigen 1 (EBNA-1) or scaffold/matrix attachment region (S/MAR) element, adenovectors can be used to stably express transgene [10–12]. There are needs to transiently transduce T cells, such as genome editing of disease associated gene [13, 14]. However, adenovirus is seldom used in transducing T cells due to the low gene transfer efficiency.

* Correspondence: luzz@ivdc.chinacdc.cn; lishengwang@ymail.com
[2]Department of Experimental Hematology, Beijing Institute of Radiation Medicine, 27 Taiping Road, Beijing, China
[3]State Key Laboratory of Infectious Disease Prevention and Control, National Institute for Viral Disease Control and Prevention, Chinese Center for Disease Control and Prevention, 100 Ying Xin Jie, Beijing, China

Adenoviruses, containing a genome of double-stranded DNA, are nonenveloped icosahedral particles with fibers projecting from the vertices. Human adenoviruses (HAdV) belong to the genus of mastadenovirus and are divided into A-G species [15, 16]. The commonly-used adenovectors are constructed based on HAdV-5, a type of HAdV-C. HAdV-5 attaches host cell through the interaction of viral fiber with primary cellular receptor CAR. Binding of the RGD motif of penton base to coreceptor integrin (avb3 or avb5) further facilitates virus entry to cells by triggering endocytosis [17].

Hematopoietic cells including T cells can be poorly transduced by HAdV-5 vectors due to paucity of CAR on the cellular surface [17–19]. HAdV-35, a type of HAdV-B, uses CD46 as the cellular receptor, which is extensively expressed in hematopoietic cells. Therefore, HAdV-35 is able to infect human hematopoietic cells. Adenoviral fiber consists of three domains of knob, shaft and tail. Knob recognizes and binds to the cellular receptor, tail interacts with the penton base and makes the fiber implant in virion, and the shaft links knob and fiber. Adenoviral vector system of HAdV-5 has been extensively studied and is the most robust system. In order to construct adenovirus vector that can infect hematopoietic cells, the fiber knob and shaft domains of HAdV-5 was replaced with that of HAdV-35 and the resulted adenovirus was called HAdV-5F35, which could be rescued and amplified conveniently with the HAdV-5 vector system [20, 21]. HAdV-11p, another type of HAdV-B, is different from HAdV-35 in that HAdV-11p binds to both CD46 and DSG-2 on the surface of host cells [22, 23]. It was reported that HAdV-11p was more efficient than HAdV-35 when infecting CD34+ hematopoietic cells [24]. We constructed HAdV-5F11p vectors [25]. However, HAdV-5F35 and HAdV-5F11p have not be systematically compared in the ability of gene delivery to human hematopoietic cells, especially T cells. Insertion of RGD motif to the HI loop of HAdV-5 fiber knob domain could improve the ability of HAdV-5 to infect T cells [26–28]. In this study, we attempted to evaluate the gene transfer efficiency of HAdV-5F35 and HAdV-5F11p and to access whether insertion of RGD motif could further improve the capability of HAdV-5F11p.

Results
Construction of adenoviral vectors
We constructed 5 first-generation adenovectors, in which the E1/E3 regions were deleted and GFP expression cassette including the human EF1a promoter, GFP coding sequence and SV40 polyA signal were inserted into the E1 region. The five vectors were the same except the fiber protein: HAdV5-EG contained the original fiber of HAdV-5 and served as a control vector; HAdV5F35-EG contained a chimeric fiber of HAdV-5 tail and HAdV-35 shaft and knob; HAdV5F11p-EG contained a chimeric

fiber of HAdV-5 tail and HAdV-11p shaft and knob; HAdV5F11pHR-EG was different from HAdV5F11p-EG in that RGD4C peptide was inserted into the HI loop of the knob domain; and HAdV5F11pCR-EG was different from HAdV5F11p-EG in that RGD4C peptide was fused to the C-terminal of the fiber with a [GGGGS]3 linker (Fig. 1). The five vectors were abbreviated as Ad5-EG, F35-EG, F11p-EG, HR-EG and CR-EG, respectively. The genomic region that encoded the modified fiber was confirmed by DNA sequencing. All five vectors were rescued, amplified, purified and titrated. The particle-to-infectious unit ratio was 11 for Ad5-EG, 20 for F35-EG and F11p-EG, 10 for HR-EG, and 400 for CR-EG, implying various gene delivery ability on 293 cells.

Transduction of hematopoietic cell lines
Four cell lines of U937, K562, Jurkat and HL-60 were chosen for evaluating the gene transfer ability of these adenoviral vectors. Cells were infected and the GFP fluorescence was analyzed 2 days later. The abilities of Ad5-EG and F11p-EG to transduce hematopoietic cell lines were compared firstly (Fig. 2). Both vectors could hardly transduce HL-60 cells. For U937 and K562 cells, F11p-EG was strikingly superior to Ad5-EG. For Jurkat cells, the percentages of GFP-positive cells were very close between Ad5-EG and F11p-EG groups. However, if we looked more closely, we would see the difference. When Jurkat cells were infected with F11p-EG at an MOI of 100 vp/cell, more than 94% cells were GFP-positive; and the percentage of GFP+ cells was close to 94% while Ad5-EG was used at an MOI of 500 vp/cell. When the data of the geometric mean of fluorescence intensity of GFP-positive cells were checked, the result was the same, suggesting that F11p-EG was five times more efficient than Ad5-EG on Jurkat cells. The cells were infected without virus removal, or viruses were removed after 6 h' incubation (Fig. 2). For F11p-EG, prolonging incubation time had little influence on the transduction. In agreement with previous publications, HAdV-5 with original fiber was a poor gene transfer vector for hematopoietic cells. In order to simplify the procedure, only four vectors of F35-EG, F11p-EG, HR-EG and CR-EG were compared to find a better one for T cells transduction in following experiments.

For the four fiber-modified HAdV-5, HL-60 is the most insensitive cell line. When the MOI was increased to 500 vp/cell, only 44% of HL-60 cells could be infected by HR-EG. For the other 3 viruses, the infection efficiency was less than 5%. Although HR-EG was the most effective vector for HL-60, it was the weakest one for U937. When infected by HR-EG at an MOI of 100 vp/cell, 35% of U937 cells was GFP-positive. However, more than 98% cells were GFP-positive when U937 was transduced by the other 3 viruses. The gene transfer ability was comparable for all 4 vectors when infecting K562 or

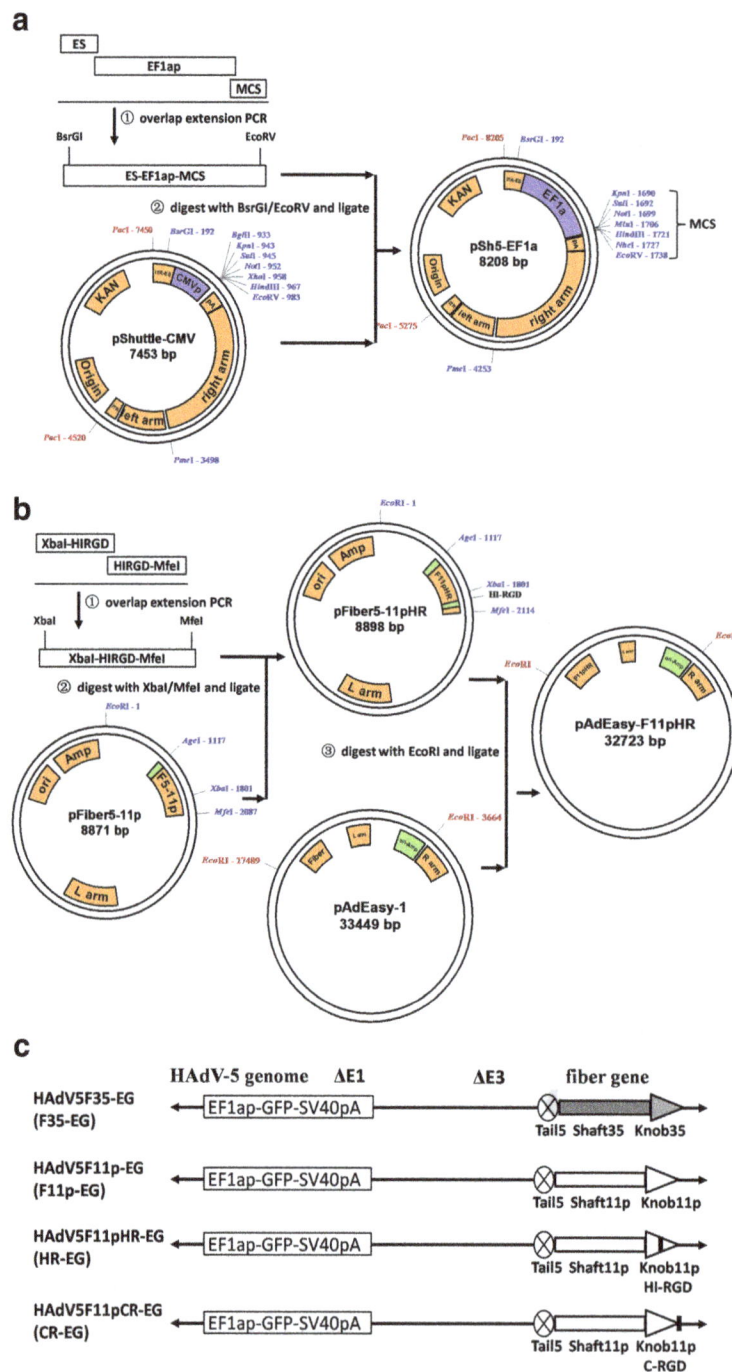

Fig. 1 Schematic diagram of the construction of fiber-modified human adenovirus 5 (HAdV-5) vectors. **a** Construction of the shuttle plasmid carrying human EF1a promoter. **b** Construction of the backbone plasmid carrying modified fiber gene. **c** Fiber-modified HAdV-5 vectors.All vectors contained the same human EF1a promoter controlled GFP expression cassette inserted into the E1 region and differently modified fiber genes. CMVp, CMV promoter; C-RGD, RGD4C fused to the C-terminus of HAdV-11p fiber; EF1ap, human EF1a promoter; ES, encapsidation signal; HI-RGD, RGD4C inserted into the HI loop of HAdV-11p fiber Knob; ITR, inverted terminal repeat; Knob11p, knob of HAdV-11p fiber; Knob35, knob of HAdV-35 fiber; MCS, multiple cloning site; pA, SV40 polyA signal; Shaft11p, shaft of HAdV-11p fiber; Shaft35, shaft of HAdV-35 fiber; Tail5, tail of HAdV-5 fiber

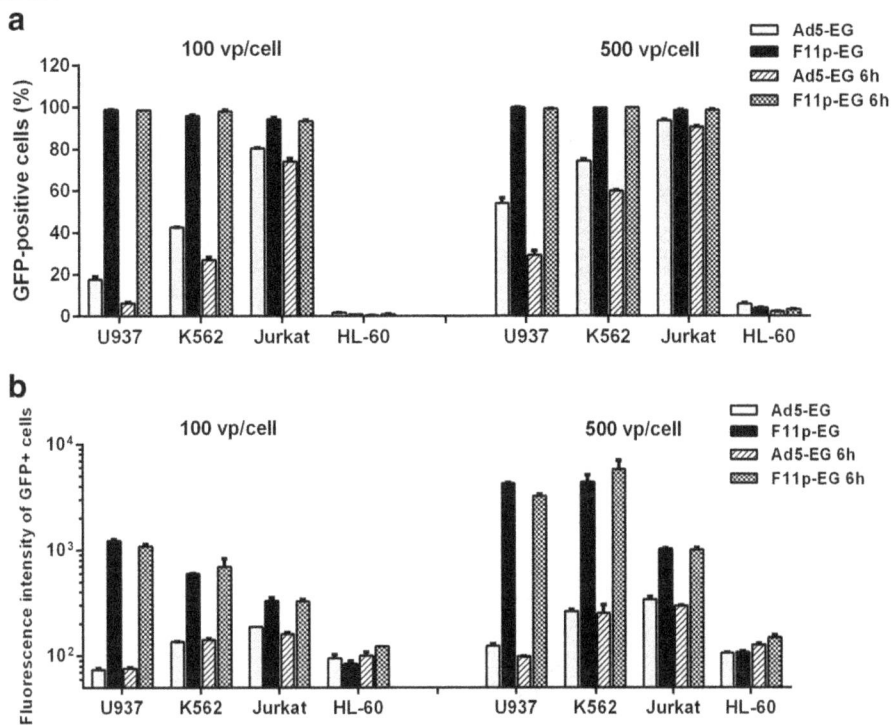

Fig. 2 Transduction of hematopoietic cell lines by Ad5-EG and F11p-EG. Cells were infected with adenoviral vectors at MOIs of 100 or 500 vp/cell without virus removal (Ad5-EG and F11p-EG), or viruses were discarded by centrifugation and washing after 6 h' incubation (Ad5-EG 6 h and F11p-EG 6 h). GFP expression was analyzed with flow cytometry assay 2 days post infecton. The percentage of GFP+ cells (**a**) and the mean fluorescence intensity of GFP+ cells (**b**) were compared among different vectors and cell lines

Jurkat cells (more than 90% GFP+ cells at the MOI of 500 vp/cell), while F11p-EG had slight superiority at the MOI of 100 (Fig. 3a). To further evaluate the difference among cell lines, mean fluorescence intensity of GFP+ cells was compared. When the percentage of GFP+ cells was close to 100%, GFP had the strongest expression (highest fluorescence intensity) in U937 cells (Fig. 3b).

Transduction of human CD34+ cord blood cells

CD34+ cells were isolated from cord blood mononuclear cells and infected with the 4 vectors at an MOI of 1000 vp/cell. As shown in Fig. 4, the purity of the cells was high (CD34+ cells were more than 95%), and the gene transfer efficiency was acceptable. HR-EG had the lowest efficiency of 55%, F35-EG had a gene transfer efficiency of 88%, while F11p-EG and CR-EG could transduce as much as 93% of CD34+ cells.

Transduction of primary human T cells

HR-EG could hardly transduce primary T cells considering that less than 2% cells were GFP-positive when infected with an MOI of 500 vp/cell. F11p-EG was superior to F35-EG or CR-EG. F11p-EG could tranduce 80% T cells while F35-EG or CR-EG could infect less than 70% when an MOI of 500 vp/cell was used (Fig. 5a).

Furthermore, F11p-EG-infected cells had the strongest GFP expression (Fig. 5b).

Transduction of T cell subgroups

T cells were infected by 4 vectors, respectively. Two days post infection, cells were labelled with fluorescein-conjugated anti-CD3, anti-CD4 and anti-CD8 antibodies, and analyzed with flow cytometry. As shown in Fig. 6a, lymphocytes were gated from the cellular debris on the FSC-H vs. SSC-H dot plot and analyzed for the expression of CD3. Nearly all cells were CD3-positive, indicating a high purity of cultured T cells. Gated CD3+ cells were further divided into CD4 + CD8- and CD4-CD8+ subsets, and GFP expression was separately analyzed. As shown in Fig. 6b, HR-EG could hardly transduce either CD4+ or CD8+ cells. The other 3 viruses could transduce considerable amount of T cells while F11p-EG was superior to F35-EG or CR-EG (Fig. 6b). When comparing the gene transfer efficiency between T cell subgroups, it was true for all vectors that they could transduce more CD4+ cells than CD8+ cells.

Viral infection might influence the host cell growth. Although the 4 vectors transduced the T cells with different efficiency, the percentages of CD4+ and CD8+ cells did not significantly vary among different virus-infected groups (Fig. 6c), suggestion that virus infection did not alter the inherent growth pattern of T

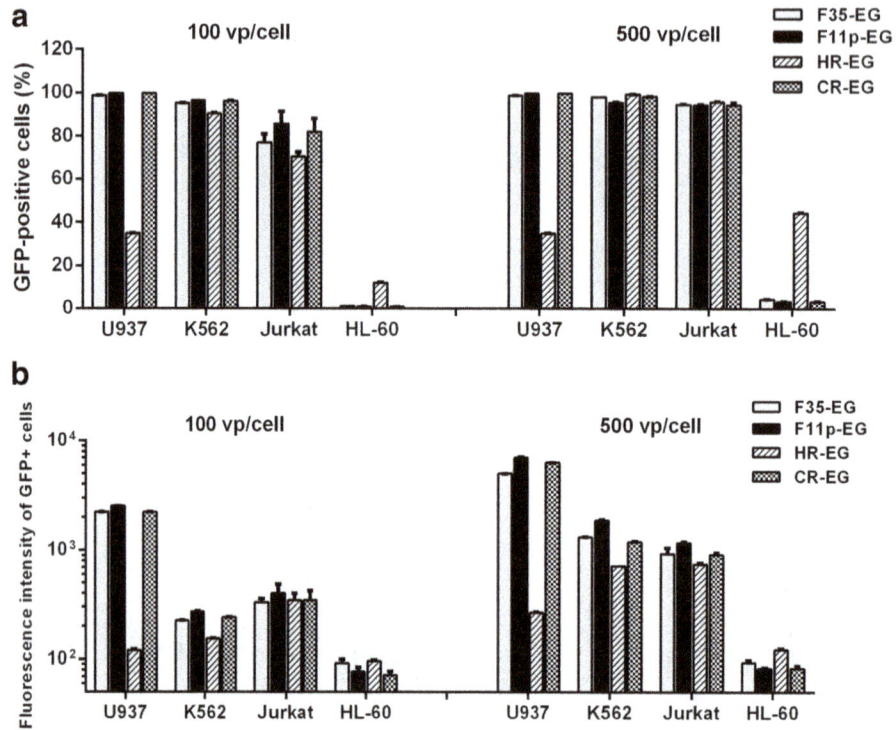

Fig. 3 Transduction of hematopoietic cell lines by fiber-modified HAdV-5 vectors. Cells were infected with adenoviral vectors at MOIs of 100 or 500 vp/cell, and GFP expression was analyzed with flow cytometry 2 days post infecton. The percentage of GFP+ cells (**a**) and the mean fluorescence intensity of GFP+ cells (**b**) were compared among different vectors and cell lines

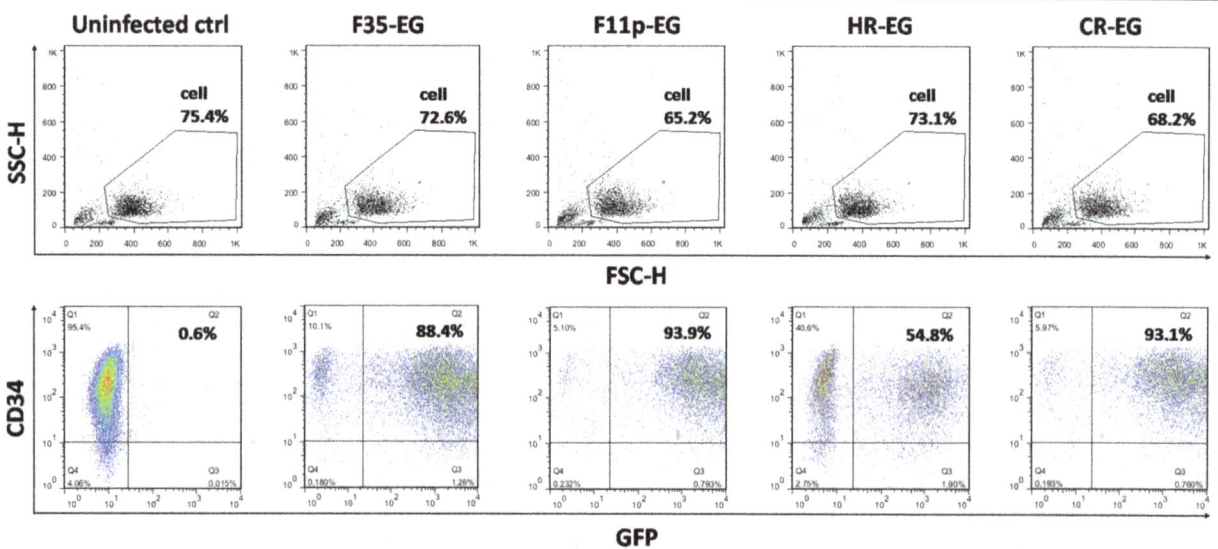

Fig. 4 Transduction of cord blood CD34+ cells by fiber-modified HAdV-5 vectors. Isolated cord blood CD34+ cells were infected by adenoviral vectors at an MOI of 1000 vp/cell. Two days post infection, cells were labelled with APC-conjugated anti-CD34 antibody, and the GFP and APC fluorescences were analyzed with flow cytometry

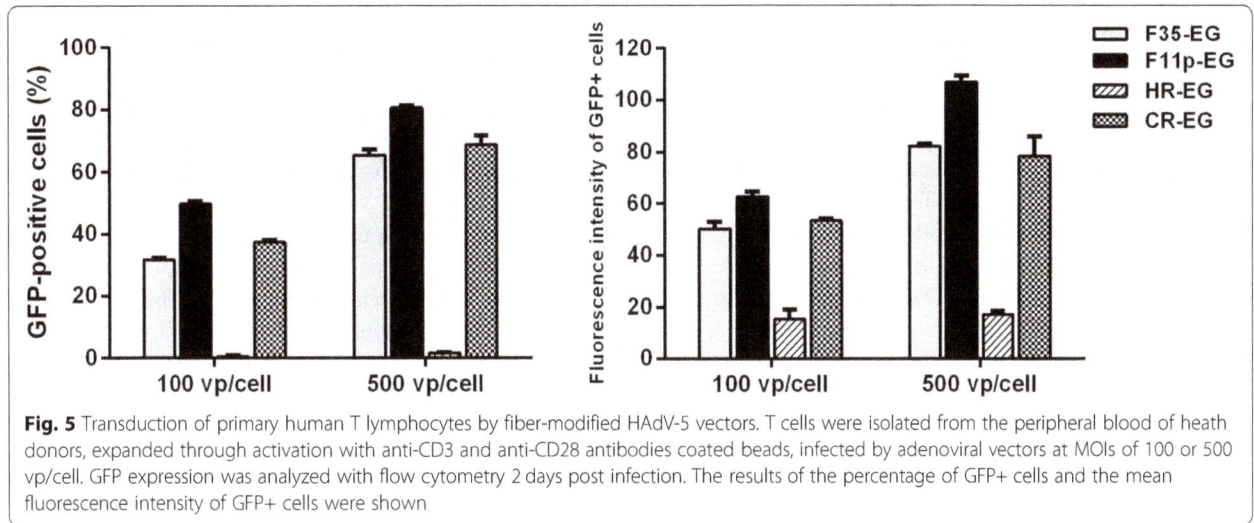

Fig. 5 Transduction of primary human T lymphocytes by fiber-modified HAdV-5 vectors. T cells were isolated from the peripheral blood of heath donors, expanded through activation with anti-CD3 and anti-CD28 antibodies coated beads, infected by adenoviral vectors at MOIs of 100 or 500 vp/cell. GFP expression was analyzed with flow cytometry 2 days post infection. The results of the percentage of GFP+ cells and the mean fluorescence intensity of GFP+ cells were shown

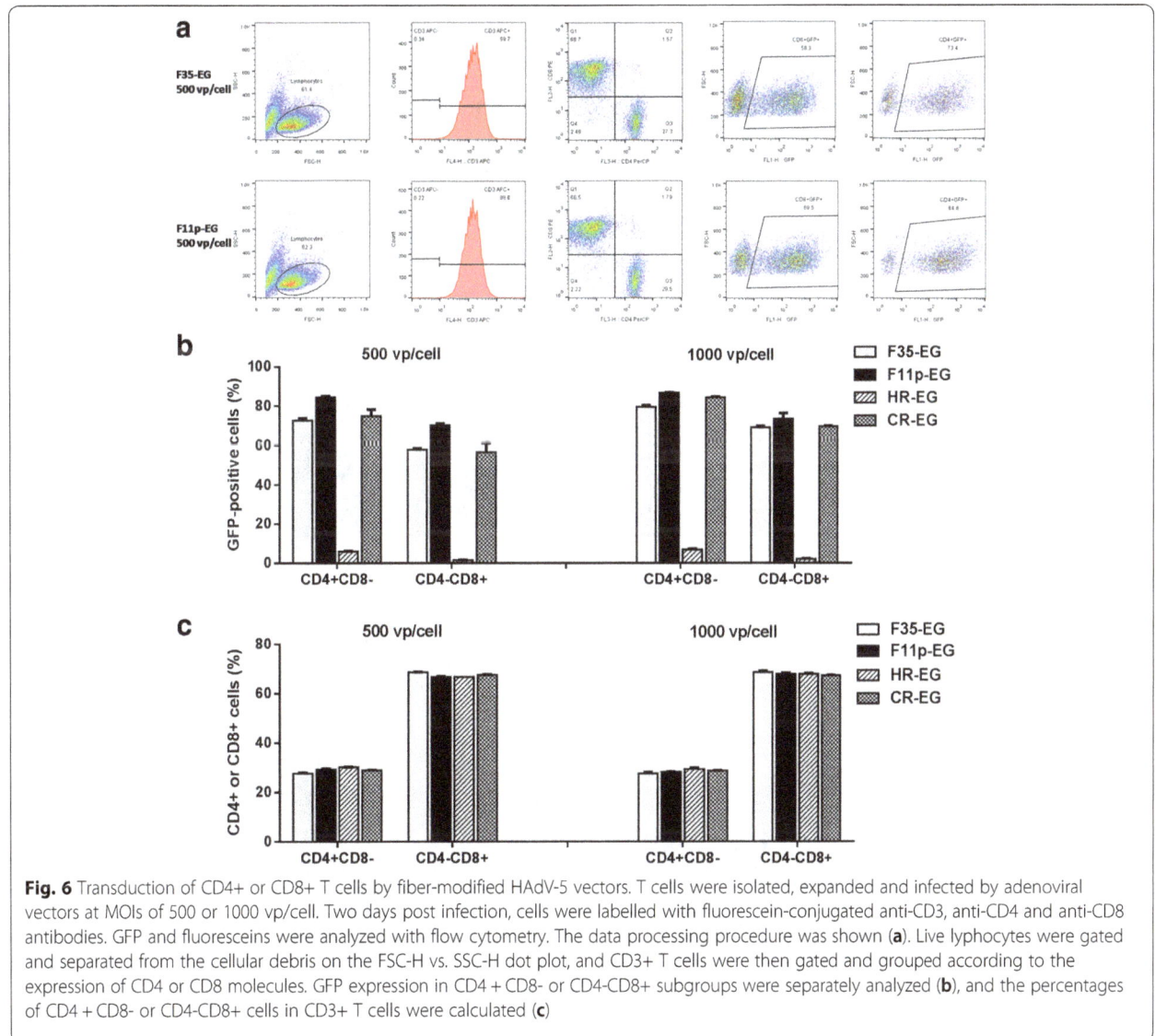

Fig. 6 Transduction of CD4+ or CD8+ T cells by fiber-modified HAdV-5 vectors. T cells were isolated, expanded and infected by adenoviral vectors at MOIs of 500 or 1000 vp/cell. Two days post infection, cells were labelled with fluorescein-conjugated anti-CD3, anti-CD4 and anti-CD8 antibodies. GFP and fluoresceins were analyzed with flow cytometry. The data processing procedure was shown (**a**). Live lyphocytes were gated and separated from the cellular debris on the FSC-H vs. SSC-H dot plot, and CD3+ T cells were then gated and grouped according to the expression of CD4 or CD8 molecules. GFP expression in CD4 + CD8- or CD4-CD8+ subgroups were separately analyzed (**b**), and the percentages of CD4 + CD8- or CD4-CD8+ cells in CD3+ T cells were calculated (**c**)

cell subgroups. This topic was further investigated in the next paragraph.

Effect of viral infection on the ratio of CD4/CD8 T cells and dynamic expression of transgene

The viruses transduced T cell subgroups with different efficiency. They might also have different effects on the growth of T cell subsets. We observed the dynamics of CD4+ and CD8+ percentage in a period of 72 h after infection. As shown in Fig. 7a and b, the CD4+ percentage had a trend of decrease while the CD8+ cell proportion had a trend of increase as the culture time extended. Because cells in the uninfected group experienced exactly the same trends, it could be concluded that the change of CD4/CD8 ratio resulted from the inherent property of these cells or from the culture system but not from the virus infection.

Adenovirus is a vector for transient expression, we observed the dynamic expression of GFP in this experiment. As shown in Fig. 7c, the GFP+ percentage reached the peak 48 h post infection (hpi) and started to slightly decrease 72 hpi, which was in agreement with the feature of adenoviral transduction of rapidly growing cells.

Effect of viral infection on cell growth

Infection of adenovirus at very high MOI would cause cytotoxicity to host cells. Effect of viral infection on T cell growth was investigated with the method of proliferation analysis. Proliferation index (PI) was the average number of cells that one original cell became. Compared with the uninfected T cells, cells which were infected with adenoviruses at an MOI of 1000 vp/cell showed similar PI values in the first 2 days (Fig. 8a). As the

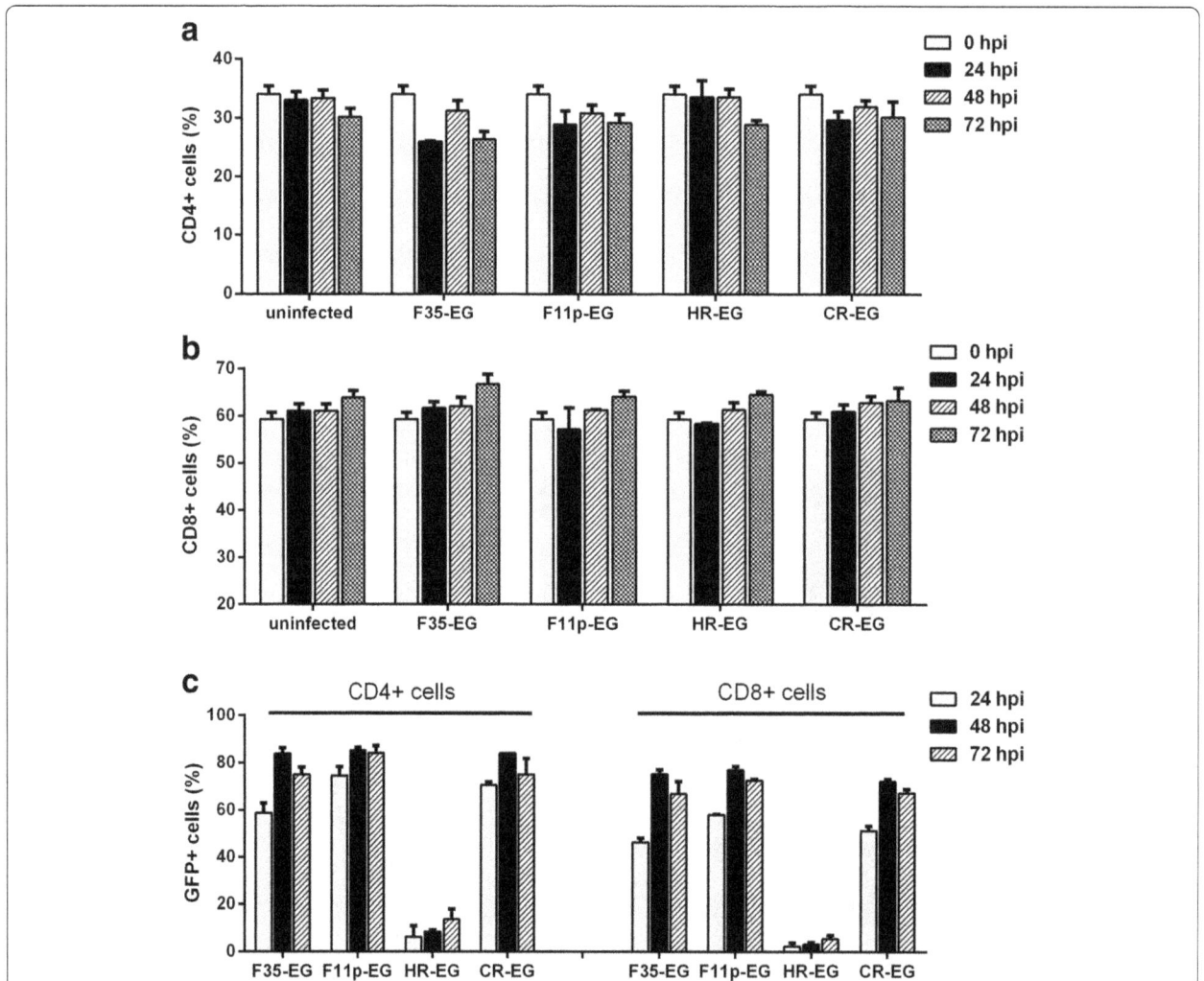

Fig. 7 Effect of viral infection on the ratio of CD4/CD8 T cells. T cells were isolated, expanded and infected by adenoviral vectors at an MOI of 1000 vp/cell. At 0, 24, 48 and 72 h post infection (hpi), cells were harvested and labelled with fluorescein-conjugated anti-CD4 and anti-CD8 antibodies. GFP and fluoresceins were analyzed with flow cytometry. The percentages of CD4 + CD8- (a) or CD4-CD8+ (b) T cells were calculated and sequentially displayed according to the order of culture time. The data of the uninfected group served as a control. The percentage of GFP+ cells in CD4+ or CD8+ T cell subsets was displayed to show the dynamic expression of GFP (c)

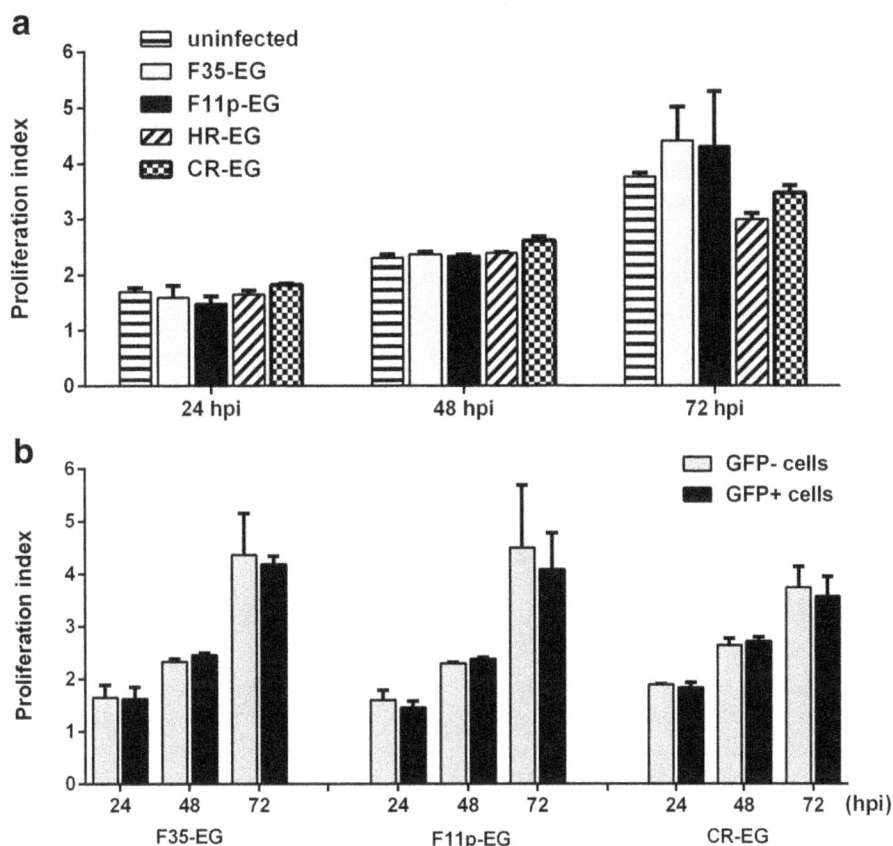

Fig. 8 Effect of viral infection on the growth of T cells. T cells were isolated, expanded, stained with Dye eFluor 670, and infected by adenoviral vectors at an MOI of 1000 vp/cell. At 0, 24, 48 and 72 h post infection (hpi), cells were harvested and fixed in 1.5% paraformaldehyde in PBS. Fluorescence of GFP and eFluor 670 were analyzed with flow cytometry at the end of the experiment. Proliferation index (PI) of the total cells were calculated (**a**), or PI values of GFP- or GFP+ cells were separately calculated (**b**)

culture time extended to 3 days, the PI values displayed some volatility among uninfected and infected groups, which might result from experimental deviation. We chose to analyze the data in another way. There were GFP+ and GFP- cells in individual culture wells. Because GFP- cells could be treated as uninfected and serve as the internal control, we gated GFP- and GFP+ cells through flow cytometry analysis and separately calculated the PI value. As shown in Fig. 8b, the results indicated that the viral infection had little influence on T cell growth.

Discussion

Hematopoietic cells are an important class of target cells for gene therapy. It is well-known that HAdV-B fiber pseudotyped HAdV-5 vectors could transduce hematopoietic cell lines, CD34+ blood cells and primary leukemia cells. However, the gene transfer ability of these vectors has not been thoroughly investigated in human T cells, the valuable target cells for immunocellular therapy in recent years. We compared HAdV-35 and HAdV-11p fiber pseudotyped HAdV-5 (F35-EG and F11p-EG),

and found that both of them could transduce hematopoietic cells efficiently while F11p-EG showed slim advantage in most cell lines, CD34+ cord blood cells, and primary T cells. The interaction between RGD motif in HAdV penton and coreceptor integrin leads to the viral entry to host cells. Combination of fiber substitution and RGD-modification could have synergistic effect on viral gene transduction. Therefore, we added the RGD4C peptide to the HI loop or the C-terminus of HAdV-11p fiber knob to generate HR-EG or CR-EG viruses. RGD-modification of the HI loop of HAdV-11p fiber did increase the viral infection of HL-60 and 293 cells. However, HR-EG virus showed a significantly decreased ability to transduce U937, CD34+ and T cells. It is possible that incorporation of RGD4C peptide in HI loop impaired the interaction of the fiber with its original receptor although the modified fiber obtained the feature of binding new molecules. Fusing RGD4C to the C-terminus of fiber had negligible influence on gene transduction of hematopoietic cells. Because fiber C-terminus was a free end of the tertiary structure and located at the border, such modification retained integrity of the knob domain and applied no damage to the fiber-receptor inter-action. However, C-terminal fusion of RGD4C brought no benefit for gene transfer, implying steric obstacles restricted RGD4C from binding integrins. Based on previous publications and our results, it could be concluded that the effect of RGD4C incorporation was dependent on the serotype of fiber knob, the length of fiber shaft and the position of insertion [26–29].

CMV promoter is widely applied to control transgene expression. However, it was reported that CMV promoter inclined to be silenced when used in hematopoietic cells. Transgene could be expressed at a very low level or could not be detected [28]. Activity of many other promoters has been tested in hematopoietic cells [30–34]. Human EF1a promoter contributes to an effective house-keeping expression. Replacing CMV promoter with that of EF1a dramatically enhanced the expression of target gene in cord blood CD34+ cells if we compared results here with that of our previous study [25]. HAdV5f11p-GFP with the CMV promoter could hardly transduced primary lymphoid leukemic cells while F11p-EG transduced 80% primary T cells at an MOI as low as 500 vp/cell (Fig. 5a), suggesting that vectors based on HAdV5F11p with the EF1a promoter could satisfy the reqirement of transient transduction of primary T cells.

Gene transduction of T cell by fiber-modified HAdV-5 was studied in detail in this study. We found that adenovirus could transduce more CD4+ cell than CD8+ cells (Fig. 6b), and viral infection did not change the ratio of CD4+/CD8+ (Fig. 7a and b). Furthermore, F11p-EG had no cytotoxicity when being used at an MOI as high as 1000 vp/cell where approximately 80% T cells could be transduced (Fig. 8).

Conclusions

HAdV-11p fiber pseudotyped HAdV-5 could transduce human primary T cells efficiently when the human EF1a promoter was used to control the expression of transgene. Such fiber-modified HAdV-5 has the potential to be used in T cell-based immunocellular therapy. Modification of the knob domain of HAdV-11p fiber with RGD motif did not further improve the gene transfer ability.

Methods

Donors

Six human peripheral blood samples and 3 human cord blood samples were obtained from anonymous adult donors after informed consent in accordance with the local ethics committee (Medical ethics committee of affiliated hospital of Qingdao university).

Construction of adenoviral plasmids

The information of PCR primers and templates was summarized in Table 1. Adenoviral plasmids were constructed by modifying the AdEasy system (Fig. 1a and b). Firstly, the CMV promoter in pShuttle-CMV was replaced by human EF1a promoter to generate the shuttle plasmid pSh5EF1a. Fragments of Encapsidation signal (ES), EF1a promoter (EF1ap) and multiple cloning site (MCS) were amplified by PCR or obtained by self-annealing of two DNA oligos. The 3 fragments were combined and fused to form one DNA fragment (ES-EF1ap-MCS) by overlap extension PCR. ES-EF1ap-MCS was digested with BsrGI/EcoRV and inserted into the corresponding sites of pShuttle-CMV to generate pSh5EF1a. The coding sequence (CDS) of GFP gene was amplified by PCR and was inserted into the KpnI/HindIII site of pSh5EF1a to generate shuttle plasmid pSh5EF1a-GFP (shortened as pSh5EG). Secondly, the fiber gene in the backbone plasmid pAdEasy-1 was modified to generate new backbone plasmids of HAdV-5 vector. The plasmids pFiber5-11p and pAdEasy-F11p, which carrying chimeric fiber gene of HAdV-5 and HAdV-11p, were constructed previously [25]. As illustrated in Fig. 1b, DNA sequence encoding RGD4C peptide (CDCRGDCFC) was integrated into the XbaI-HIRGD-MfeI fragment by PCR and overlap extension PCR using primers and templates described in Table 1 [35, 36]. XbaI/MfeI digested PCR product was inserted into the corresponding sites of pFiber5-11p to generate pFiber5-11pHR. pFiber5-11pHR was digested with EcoRI, dephosphorized, and used to substitute the corresponding part of pAdEasy-1 to generate pAdEasy-F11pHR backbone plasmid. In pAdEasy-F11pHR, RGD4C was added to the HI loop of chimeric fiber of HAdV-5 and HAdV-11p. To add RGD4C peptide to the C-terminal of the chimeric fiber, similar operation was performed except that different PCR primers was designed and used (Table 1), and the generated backbone plasmid was named pAdEasy-F11pCR. In pAdEasy-F11pCR, [GGGGS]3 linker was used to connect RGD4C peptide with the C-terminus of the fiber [29, 37]. To construct backbone plasmid carrying chimeric fiber of HAdV-5 and HAdV-35, chimeric fiber gene was synthesized according to genbank AC_000019, digested with AgeI/MfeI and used to substitute the corresponding part of pFiber5-11p to generate pFiber5–35 plasmid [25]. The backbone plasmid pAdEasy-F35 was similarly constructed (Fig. 1b). Finally, adenoviral plasmids were generated with the method of homologous recombination by electroporating E. coli BJ5183 strain with backbone plasmid and linearized pSh5EG [38].

Cell culture

The cell line 293 (ATCC no. CRL-1573) was cultured in Dulbecco's modified Eagle's medium (DMEM) plus 8% fetal bovine serum (FBS; HyClone, Logan, UT, USA). Human leukemic cell lines U937 (promonocytic leukemia), K562 (chronic myelogenous leukemia), Jurkat (T-cell

Table 1 summary of PCR information

Fragment	Primers code	Primers sequence	Template	Length of PCR product (bp)	restriction enzyme
ES	1411Sh5EF1aF1	ccggtgtaca caggaagtga caat	pShuttle	181	BsrGI
	1411Sh5EF1aR1	cttttgtatg aattactcga cgtcagtatt acgcgctatg agtaacacaa			AatII
EF1ap	1411Sh5EF1aF2	cgcgtaatac tgacgtcgag taatt catac aaaaggactc gc	pLVX-EF1a-Tet3G	1360	AatII
	1411Sh5EF1aR2	acggtacctc acgacacctg aaatg gaaga a			KpnI
MCS	1411Sh5EF1aF3	ttccatttca ggtgtcgtga ggtaccg tcg acgcggccgc acgcgttcta	self-anneal	80	KpnI
	1411Sh5EF1aR3	ggccgatatc ttagctagca agctta ggtc tagaacgcgt gcggccgcgt			EcoRV
ES-EF1ap-MCS		overlap extension PCR		1558	
GFP	1703GFP-kf	ggccggtacc atggtgagca aggg cgagga g	pLEGFP-C1	748	KpnI
	1703GFP-hr	ggccaagctt tagagtccgg acttg tacag ctcgt			HindIII
XbaI-HIRGD	1702F11pRGD1	ccagcacgac tgcctatcct tt	pFiber5-11p	164	XbaI
	1702F11pHIRGD2	gaaacagtct ccgcggcagt cacaat ttat tgctcttcgg ttaagcatg			
HIRGD-MfeI	1702F11pHIRGD3	tgtgactgcc gcggagactg tttctgc gac gagacatcat attgtattcg tataac	pFiber5-11p	240	
	1702F11pRGD4	ctgaatgaaa aatgacttga aattttct			MfeI
XbaI-HIRGD-MfeI		overlap extension PCR		380	
XbaI-CRGD	1702F11pRGD1	ccagcacgac tgcctatcct tt	pFiber5-11p	284	XbaI
	1702F11pCRGD2	tgaaccgcca ccacctgagt cgtcttctct gatgtagtaa aaggta			
CRGD	1702F11pCRGD3	gaagacgact caggtggtgg cggttcag gc ggaggtggct ctggcggtgg cggat	self-anneal	90	
	1702F11pCRGD4	ggctcagcag aaacagtctc cgcggcag tc acacgatccg ccaccgccag agcca			
CRGD-MfeI	1702F11pCRGD5	cgcggagact gtttctgctg agcccaagaa taaagaatcg	pFiber5-11p	105	
	1702F11pRGD4	ctgaatgaaa aatgacttga aattttct			MfeI
XbaI-CRGD-MfeI		overlap extension PCR		428	

leukemia), and HL-60 (acute myelogenous leukemia) were cultured with RPMI 1640 medium plus 10% FBS. All cells were maintained at 37 °C with 5% CO_2 in a humidified incubator and regularly split every 3 to 4 days.

Cord blood CD34+ cell isolation

Mononuclear cells (MNCs) were harvested from fresh buffy coats by Ficoll-Paque density gradient separation from pooled human cord blood samples of healthy donors. Medical ethics committee of affiliated hospital of Qingdao university approved all of the experiments. CD34 + cells were isolated from MNCs by using a CD34+ progenitor cell positive isolation kit (CD34 MicroBead Kit, CAT# 130–046-703; Miltenyi Biotech). Purity was routinely > 95% as assessed by flow cytometric analysis. CD34 + cells were maintained in serum-free medium (StemSpan SFEM, CAT#09650; Stemcell Technologies) supplemented with cytokine cocktail (50 ng/ml interleukin-3; 100 ng/ml interleukin-6; 100 ng/ml Flt-3 ligand; 50 ng/ml stem cell factor and 100 ng/ml thrombopoietin). Two days after isolation, cells were infected with adenoviral vectors.

Human T cell isolation

MNCs were collected from fresh buffy coats by Ficoll-Paque density gradient separation from peripheral blood samples of healthy donors. Medical ethics committee of affiliated hospital of Qingdao university approved all of the experiments.T cells were isolated from MNCs by using a T cell negative isolation kit (Dynabeads Untouched Human T Cells Kit, CAT#11344D; Life Technologies). Isolated T cells were cultured in X-VIVO 15 medium (CAT#04-418Q; Lonza) supplemented with

10% FBS (CAT#ASM-5007; Applied StemCell) and 400 IU/ml rIL-2 (Beijing SL Pharmaceutical) and expanded by incubating with Dynabeads Human T-Activator CD3/CD28 according to the manufacturer's instructions (CAT#11131D; Life Technologies). Expanded T cells were maintained in X-VIVO 15 medium plus 10% FBS and 2000 IU/ml rIL-2, and used for viral infection 8 to 14 days after isolation.

Preparation of adenoviral vectors

Adenoviral plasmids were digested with PacI, recovered by ethanol precipitation and used to transfect 293 cells with Lipofectamine 3000 according to the manufacturer's instructions (Life technologies). Plaques occurred within 1 week post transfection. Rescued viruses was released by three rounds of freeze-and-thaw and amplified in 293 cells. Amplified virus was purified with the traditional method of CsCl ultracentrifugation. Particle titer was determined by quantifying the genomic DNA of purified virus, and the infectious titer was determined by limiting dilution assay on 293 cells [39].

Transduction of hematopoietic cells

Exponentially proliferating cells were counted with hemacytometer, diluted and seeded in 24-well (for cell lines) or 96-well plates (for CD34+ or T cells) with a density of 3×10^5 cell/well. Purified viruses were diluted with culture medium and added to each well in a volume to achieve indicated multiple MOI. The infection volume was adjusted to the half amount of routine culture, which was 0.25 ml for each well in 24-well plate and 0.1 ml for 96-well plate, respectively. The plates were transferred to cell culture incubator, and fresh medium in the half volume of routine culture was supplemented to each well without removal of virus 24 hpi unless otherwise indicated.

Flow cytometry analysis

Following antibodies were used in flow cytometry assay: APC Mouse Anti-Human CD3 (CAT#555335; BD Pharmingen), PerCP-Cyanine5.5 CD4 Monoclonal Antibody (OKT4, CAT#45−0048-42; Thermo Fisher Scientific), PE CD8a Monoclonal Antibody (OKT8, CAT#12−0086-42; Thermo Fisher Scientific), and APC CD34 Monoclonal Antibody (4H11,CAT# 17−0349-42; Thermo Fisher Scientific). Cells were transferred to 1.5-ml Eppendorf tube, washed once with PBS containing 1% FBS, labelled with fluorescein-conjugated antibodies for 20 min at room template, washed with and then suspended in PBS containing 1% FBS, and analyzed by flow cytometry.

Cell proliferation analysis

Dye eFluor 670 (CAT#65−0840; Thermo Fisher Scientific) was used to label T cells for proliferation analysis

according to the manufacturer's instructions. Labelled cells were mixed with viruses and aliquotted to wells in 96-well plate at a density of 2×10^5 cell/well. Fresh medium of 100 µl was added to each well 24 h later. At indicated time points, cells were washed with and suspended in PBS containing 1% FBS, dispersed into singles cells, supplemented with 4% paraformaldehyde in PBS to a final concentration of 1.5%, and reserved at 4 °C. After going through all the time points, fixed cells were analyzed with flow cytometry. PI, which was defined as the number of modeled cells (the number of cells at the time point of harvest) divided by the cells in the original culture (the number of cells at the time of seeding), was calculated with the software of ModFit LT (Verity Software House).

Statistical analysis

The Data were represented as the mean ± standard deviation (SD) of representative experiments. The statistical analysis was performed using the Analysis of Variance (ANOVA) test. A p-value less than 0.05 was considered to be significant.

Abbreviations

ANOVA: Analysis of Variance; CDS: Coding sequence; DMEM: Dulbecco's modified Eagle's medium; EBNA-1: EBV nuclear antigen 1; EF1ap: EF1a promoter; ES: Encapsidation signal; HAdV: Human adenoviruses; HAdV-5: Human adenovirus 5; Hpi: Hours post infection; MCS: Multiple cloning site; MNCs: Mononuclear cells; MOI: Multiplicity of infection; PI: Proliferation index; S/MAR: Scaffold/matrix attachment region; SD: Standard deviation; vp/cell: Viral particle per cell

Acknowledgements

Not applicable.

Funding

This work was supported by the National Key R&D Program of China (No.2017YFC1200503) and National Natural Science Foundation of China (No.81470320). These funds were used for the design of the study, collection of materials, analysis data, the interpretation and the writing/publication of the manuscript.

Authors' contributions

ZL and LW conceived and designed the experiments. YL, FX, YW, XZ, H-YW and HW performed the experiments. YL, FX and ZL analyzed the data. ZL and LW wrote the paper. All authors read and approved the final manuscript.

Ethics approval and consent to participate

Because the ways to get samples have no effect on the health of donors and the samples are not for clinical use, the local ethics committee (Medical ethics committee of affiliated hospital of Qingdao university) agreed to the protocols (Hematopoietic Cell Transduction 2018-03) of this project in which oral inform consent is needed. All donors are oral informed in conversation room under video record.

Consent for publication

Not applicable.

Competing interests

The authors declare that they have no competing interests.

Author details

[1]Graduate School of Anhui Medical University, 81 Meishan Road, Shu Shan Qu, Hefei, Anhui, People's Republic of China. [2]Department of Experimental Hematology, Beijing Institute of Radiation Medicine, 27 Taiping Road, Beijing, China. [3]State Key Laboratory of Infectious Disease Prevention and Control, National Institute for Viral Disease Control and Prevention, Chinese Center for Disease Control and Prevention, 100 Ying Xin Jie, Beijing, China. [4]Affiliated Hospital of Qingdao University, 16 JiangSu Road, Qingdao, People's Republic of China.

References

1. Ciucci T, Vacchio MS, Bosselut R. Genetic tools to study T cell development. Methods Mol Biol. 2016;1323:35–45. https://doi.org/10.1007/978-1-4939-2809-5_3.
2. Field AC, Qasim W. Engineered T cell therapies. Expert Rev Mol Med. 2015; 17:e19. https://doi.org/10.1017/erm.2015.14.
3. Milone MC, Bhoj VG. The pharmacology of T cell therapies. Mol Ther Methods Clin Dev. 2018;8:210–21. https://doi.org/10.1016/j.omtm.2018.01.010.
4. Wijesundara DK, Ranasinghe C, Grubor-Bauk B, Gowans EJ. Emerging Targets for Developing T Cell-Mediated Vaccines for Human Immunodeficiency Virus (HIV)-1. Front Microbiol. 2017;8:2091. https://doi.org/10.3389/fmicb.2017.02091.
5. Johnson MC, Wang B, Tisch R. Genetic vaccination for re-establishing T-cell tolerance in type 1 diabetes. Hum Vaccin. 2011;7:27–36.
6. Morgan RA, Boyerinas B. Genetic modification of T cells. Biomedicines. 2016; 4. https://doi.org/10.3390/biomedicines4020009.
7. Fougeroux C, Holst PJ. Future prospects for the development of cost-effective adenovirus vaccines. Int J Mol Sci. 2017;18. https://doi.org/10.3390/ijms18040686.
8. Yamamoto Y, Nagasato M, Yoshida T, Aoki K. Recent advances in genetic modification of adenovirus vectors for cancer treatment. Cancer Sci. 2017; 108:831–7. https://doi.org/10.1111/cas.13228.
9. Crystal RG. Adenovirus: the first effective in vivo gene delivery vector. Hum Gene Ther. 2014;25:3–11. https://doi.org/10.1089/hum.2013.2527.
10. Richter M, Saydaminova K, Yumul R, Krishnan R, Liu J, Nagy EE, et al. In vivo transduction of primitive mobilized hematopoietic stem cells after intravenous injection of integrating adenovirus vectors. Blood. 2016;128: 2206–17. https://doi.org/10.1182/blood-2016-04-711580.
11. Voigtlander R, Haase R, Muck-Hausl M, Zhang W, Boehme P, Lipps HJ, et al. A Novel Adenoviral Hybrid-vector System Carrying a Plasmid Replicon for Safe and Efficient Cell and Gene Therapeutic Applications. Mol Ther Nucleic Acids. 2013;2:e83. https://doi.org/10.1038/mtna.2013.11.
12. Gil JS, Gallaher SD, Berk AJ. Delivery of an EBV episome by a self-circularizing helper-dependent adenovirus: long-term transgene expression in immunocompetent mice. Gene Ther. 2010;17:1288–93. https://doi.org/10.1038/gt.2010.75.
13. Tebas P, Stein D, Tang WW, Frank I, Wang SQ, Lee G, et al. Gene editing of CCR5 in autologous CD4 T cells of persons infected with HIV. N Engl J Med. 2014;370:901–10. https://doi.org/10.1056/NEJMoa1300662.
14. Li C, Guan X, Du T, Jin W, Wu B, Liu Y, et al. Inhibition of HIV-1 infection of primary CD4+ T-cells by gene editing of CCR5 using adenovirus-delivered CRISPR/Cas9. J Gen Virol. 2015;96:2381–93. https://doi.org/10.1099/vir.0.000139.
15. Khare R, Chen CY, Weaver EA, Barry MA. Advances and future challenges in adenoviral vector pharmacology and targeting. Curr Gene Ther. 2011;11:241–58.
16. Campos SK, Barry MA. Current advances and future challenges in Adenoviral vector biology and targeting. Curr Gene Ther. 2007;7:189–204.
17. Kremer EJ, Nemerow GR. Adenovirus tales: from the cell surface to the nuclear pore complex. PLoSPathog. 2015;11:e1004821. https://doi.org/10.1371/journal.ppat.1004821.
18. Arnberg N. Adenovirus receptors: implications for targeting of viral vectors. Trends Pharmacol Sci. 2012;33:442–8. https://doi.org/10.1016/j.tips.2012.04.005.
19. Yotnda P, Zompeta C, Heslop HE, Andreeff M, Brenner MK, Marini F. Comparison of the efficiency of transduction of leukemic cells by fiber-modified adenoviruses. Hum Gene Ther. 2004;15:1229–42. https://doi.org/10.1089/hum.2004.15.1229.
20. Matsui H, Sakurai F, Katayama K, Kurachi S, Tashiro K, Sugio K, et al. Enhanced transduction efficiency of fiber-substituted adenovirus vectors by the incorporation of RGD peptides in two distinct regions of the adenovirus serotype 35 fiber knob. Virus Res. 2011;155:48–54. https://doi.org/10.1016/j.virusres.2010.08.021.
21. Nilsson M, Ljungberg J, Richter J, Kiefer T, Magnusson M, Lieber A, et al. Development of an adenoviral vector system with adenovirus serotype 35

tropism; efficient transient gene transfer into primary malignant hematopoietic cells. J Gene Med. 2004;6:631–41. https://doi.org/10.1002/jgm.543.
22. Wang H, Li ZY, Liu Y, Persson J, Beyer I, Moller T, et al. Desmoglein 2 is a receptor for adenovirus serotypes 3, 7, 11 and 14. Nat Med. 2011;17:96–104. https://doi.org/10.1038/nm.2270.
23. Tuve S, Wang H, Ware C, Liu Y, Gaggar A, Bernt K, et al. A new group B adenovirus receptor is expressed at high levels on human stem and tumor cells. J Virol. 2006;80:12109–20. https://doi.org/10.1128/JVI.01370-06.
24. Mei YF, Segerman A, Lindman K, Hornsten P, Wahlin A, Wadell G. Human hematopoietic (CD34+) stem cells possess high-affinity receptors for adenovirus type 11p. Virology. 2004;328:198–207. https://doi.org/10.1016/j.virol.2004.07.018.
25. Lu ZZ, Ni F, Hu ZB, Wang L, Wang H, Zhang QW, et al. Efficient gene transfer into hematopoietic cells by a retargeting adenoviral vector system with a chimeric fiber of adenovirus serotype 5 and 11p. Exp Hematol. 2006; 34:1171–82. https://doi.org/10.1016/j.exphem.2006.05.005.
26. Borovjagin AV, Krendelchtchikov A, Ramesh N, Yu DC, Douglas JT, Curiel DT. Complex mosaicism is a novel approach to infectivity enhancement of adenovirus type 5-based vectors. Cancer Gene Ther. 2005;12:475–86. https://doi.org/10.1038/sj.cgt.7700806.
27. Zhang WF, Wu FL, Shao HW, Wang T, Huang XT, Li WL, et al. Chimeric adenoviral vector Ad5F35L containing the Ad5 natural long-shaft exhibits efficient gene transfer into human T lymphocytes. J Virol Methods. 2013; 194:52–9. https://doi.org/10.1016/j.jviromet.2013.07.052.
28. Ye Z, Shi M, Chan T, Sas S, Xu S, Xiang J. Engineered CD8+ cytotoxic T cells with fiber-modified adenovirus-mediated TNF-alpha gene transfection counteract immunosuppressive interleukin-10-secreting lung metastasis and solid tumors. Cancer Gene Ther. 2007;14:661–75. https://doi.org/10.1038/sj.cgt.7701039.
29. Tyler MA, Ulasov IV, Borovjagin A, Sonabend AM, Khramtsov A, Han Y, et al. Enhanced transduction of malignant glioma with a double targeted Ad5/3-RGD fiber-modified adenovirus. Mol Cancer Ther. 2006;5:2408–16. https://doi.org/10.1158/1535-7163.MCT-06-0187.
30. Weber EL, Cannon PM. Promoter choice for retroviral vectors: transcriptional strength versus trans-activation potential. Hum Gene Ther. 2007;18:849–60. https://doi.org/10.1089/hum.2007.067.
31. Sakurai F, Kawabata K, Yamaguchi T, Hayakawa T, Mizuguchi H. Optimization of adenovirus serotype 35 vectors for efficient transduction in human hematopoietic progenitors: comparison of promoter activities. Gene Ther. 2005;12:1424–33. https://doi.org/10.1038/sj.gt.3302562.
32. Dupuy FP, Mouly E, Mesel-Lemoine M, Morel C, Abriol J, Cherai M, et al. Lentiviral transduction of human hematopoietic cells by HIV-1- and SIV-based vectors containing a bicistronic cassette driven by various internal promoters. J Gene Med. 2005;7:1158–71. https://doi.org/10.1002/jgm.769.
33. Serafini M, Bonamino M, Golay J, Introna M. Elongation factor 1 (EF1alpha) promoter in a lentiviral backbone improves expression of the CD20 suicide gene in primary T lymphocytes allowing efficient rituximab-mediated lysis. Haematologica. 2004;89:86–95.
34. Salmon P, Kindler V, Ducrey O, Chapuis B, Zubler RH, Trono D. High-level transgene expression in human hematopoietic progenitors and differentiated blood lineages after transduction with improved lentiviral vectors. Blood. 2000;96:3392–8.
35. Dmitriev I, Krasnykh V, Miller CR, Wang M, Kashentseva E, Mikheeva G, et al. An adenovirus vector with genetically modified fibers demonstrates expanded tropism via utilization of a coxsackievirus and adenovirus receptor-independent cell entry mechanism. J Virol. 1998;72:9706–13.
36. Shen YH, Yang F, Wang H, Cai ZJ, Xu YP, Zhao A, et al. Arg-Gly-Asp (RGD)-Modified E1A/E1B Double Mutant Adenovirus Enhances Antitumor Activity in Prostate Cancer Cells In Vitro and in Mice. PLoS One. 2016;11:e0147173. https://doi.org/10.1371/journal.pone.0147173.
37. Trinh R, Gurbaxani B, Morrison SL, Seyfzadeh M. Optimization of codon pair use within the (GGGGS)3 linker sequence results in enhanced protein expression. Mol Immunol. 2004;40:717–22.
38. He TC, Zhou S, da Costa LT, Yu J, Kinzler KW, Vogelstein B. A simplified system for generating recombinant adenoviruses. Proc Natl Acad Sci U S A. 1998;95:2509–14.
39. Chen DL, Dong LX, Li M, Guo XJ, Wang M, Liu XF, et al. Construction of an infectious clone of human adenovirus type 41. Arch Virol. 2012;157:1313–21. https://doi.org/10.1007/s00705-012-1293-z.

17

Purification of the recombinant green fluorescent protein from tobacco plants using alcohol/salt aqueous two-phase system and hydrophobic interaction chromatography

Jie Dong[1,2,3], Xiangzhen Ding[1,2,3] and Sheng Wang[1,2,3]*

Abstract

Background: The green fluorescent protein (GFP) has been regarded as a valuable tool and widely applied as a biomarker in medical applications and diagnostics. A cost-efficient upstream expression system and an inexpensive downstream purification process will meet the demands of the GFP protein with high-purity.

Results: The recombinant GFP was transiently expressed in an active form in agoinoculated *Nicotiana benthamiana* leaves by using *Tobacco mosaic virus* (TMV) RNA-based overexpression vector (TRBO). The yield of recombinant GFP was up to ~ 60% of total soluble proteins (TSP). Purification of recombinant GFP from the clarified lysate of *N. benthaniana* leaves was achieved by using an alcohol/salt aqueous two-phase system (ATPS) and following with a further hydrophobic interaction chromatography (HIC). The purification process takes only ~ 4 h and can recover 34.1% of the protein. The purity of purified GFP was more than 95% and there were no changes in its spectroscopic characteristics.

Conclusions: The strategy described here combines the advantages of both the economy and efficiency of plant virus-based expression platform and the simplicity and rapidity of environmentally friendly alcohol/salt ATPS. It has a considerable potential for the development of a cost-efficient alternative for production of recombinant GFP.

Keywords: Green fluorescent protein, Plant virus, Transient gene expression, Aqueous two-phase system, Hydrophobic interaction chromatography

Background

Green fluorescent protein (GFP) was originally derived from jellyfish *Aequorea victoria* species, which exhibit an intensely natural fluorescence [1]. GFP has been regarded as a valuable tool in the field of biology and biotechnology [2]. Due to its widespread application as a molecular biomarker [3, 4], there is an increase in the demand for GFP with high-purity.

Through the application of DNA recombinant technology, GFP has successfully been produced by a variety of hosts [5]. Currently, the commercially available GFP produced by *Escherichia coli* costs approximately US$ 2000.00 per mg [6]. A cost-efficient upstream expression system and an inexpensive downstream purification process will be able to reduce the production costs and thereby meet the demands of the GFP with high-purity. Plants have been regarded as excellent biofactories for producing recombinant proteins of interest for research, pharma and industry [7]. It was estimated that proteins can be produced in plants at a cost of 10–50 fold less than in *Escherichia coli* [8]. Virus-based expression system can express the target proteins in plants at an extremely high level because of viral

* Correspondence: wang_s@nxu.edu.cn
[1]Key Laboratory of Ministry of Education for Protection and Utilization of Special Biological Resources in the Western China, Yinchuan 750021, People's Republic of China
[2]Key Laboratory of Modern Molecular Breeding for Dominant and Special Crops in Ningxia, Yinchuan 750021, People's Republic of China

amplification [9]. In addition, plant platform offers an eco-friendly way to produce recombinant proteins largely due to low energy requirements and CO_2 emission [10].

In order to achieve a high level of purity, diverse chromatographic techniques have been used to purify the recombinant GFP. In general, these chromatographic methods involve multistep, time-consuming and complicated operations, resulting in a higher purification cost [5]. Thus, an inexpensive method for GFP purification is highly needed. Aqueous two-phase system (ATPS) has been widely regarded as an alternative way for the separation and purification of proteins and other biomolecules [11]. Significant efforts have been made to develop different type of ATPSs and their applications in purification of various biomaterials [12]. Alcohol/salt ATPS is one of the promising members of the ATPS family [13]. The advantages of alcohol/salt ATPS include low cost, fast phase separation, simple operational procedures and easy scale-up [14]. Furthermore, this type of ATPS has an environmental friendliness aspect as ethanol and salt can be recycled via conventional processes [15].

Considering the excellent capabilities of plant viral expression vector and alcohol/salt ATPS, this work aimed to develop a cost-effective alternative for production of recombinant GFP. Plant viral amplicon-based gene expression system [16] was employed to transiently express recombinant GFP in *Nicotiana benthamiana* leaves by agroinfiltration. Subsequently, purification of GFP was achieved by combining an alcohol/salt ATPS stage with a further hydrophobic interaction chromatography (HIC) step. The GFP extraction efficiencies of each step were determined, and their purification aptitudes were evaluated. The fluorescence characterization of purified GFP was measured by using both gel-based imaging and the spectrofluorometric method.

Results

Transient expression of recombinant GFP in *N. benthamiana* leaves

The pJL TRBO-G vector (Fig. 1) was agoinoculated into *N. benthamiana* leaves in the presence of the suppressor of silencing P19. At 4–8 days after inoculation, high intensity of green fluorescence in the inoculated leaves was observed after illumination with long wave UV light (Fig. 2a). The cells exhibiting strong GFP signal could be seen in almost all cells in the agroinfected leaf area when examined under a fluorescence microscope (Fig. 2b). A protein corresponding to the expected molecular weight (27 kDa) was detected in the total soluble proteins extracted from the inoculated leaf tissues by both Coomassie stained polyacrylamide gel (Fig. 2c) and Western blot analysis (Fig. 2d). No signals were detected in samples from non-inoculated leaves (Fig. 2d). The GFP yield was up to ~ 60% of total soluble proteins (Table 1). All results together indicated that the recombinant GFP was successfully and efficiently expressed in the *N. benthamiana* leaves. The overexpression of GFP by plant amplicon-based vector provided a good foundation for the downstream purification of recombinant GFP.

Purification of GFP from *N. benthamiana* leaves

A procedure using alcohol/salt ATPS and HIC was applied for the isolation and purification of GFP from the inoculated *N. benthamiana* leaves.

The alcohol/salt ATPS was performed by a two-step procedure. In the first step, the GFP was exclusively extracted into ethanol phase. The GFP fluorescence in the upper ethanol phase was clearly observed upon UV illumination after phase separation (Fig. 3a2). A thin layer of host cellular proteins was observed at the interphase (Fig. 3a2). In addition, analysis by Coomassie-stained polyacrylamide gel electrophoresis

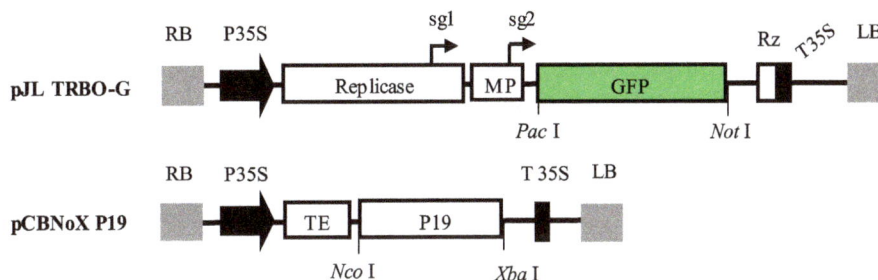

Fig. 1 Structures of the expression vectors used in this study. pJL TRBO-G, TMV-based vectors used to express green fluorescent protein (GFP) in *N. benthamiana* leaves; RB, the right T-DNA border; P35S, duplicated *Cauliflower mosaic virus* (CaMV) 35S promoter; Replicase, RNA-dependent RNA polymerase of TMV; sg1 and sg2, subgenomic mRNA1 and mRNA2 promoter of the TMV; MP, movement protein of TMV; GFP, green fluorescent protein; Rz, ribozyme; T35S, CaMV polyA signal sequence/terminator; LB, the left T-DNA border; pCBNoX P19, RNA silencing suppressor expression vector used to co-infiltrate with pJL TRBO-G; TE, translational enhancer of *Tobacco etch virus* (TEV); P19, 19-kDa RNA silencing suppressor from TBSV; *Pac* I, *Not* I, *Nco* I and *Xba* I, restriction enzyme recognition sites

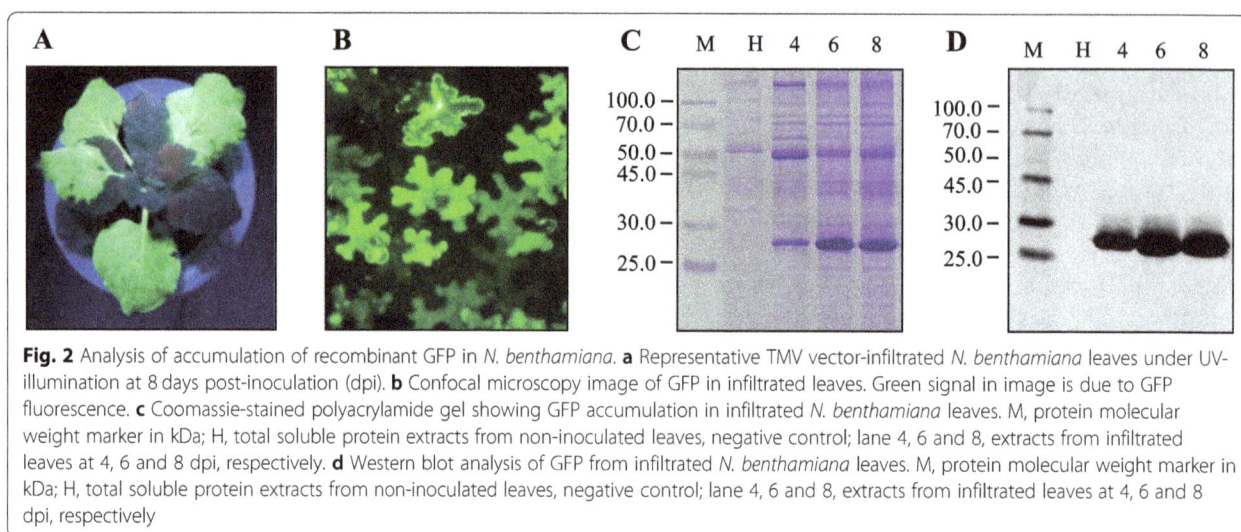

Fig. 2 Analysis of accumulation of recombinant GFP in *N. benthamiana*. **a** Representative TMV vector-infiltrated *N. benthamiana* leaves under UV-illumination at 8 days post-inoculation (dpi). **b** Confocal microscopy image of GFP in infiltrated leaves. Green signal in image is due to GFP fluorescence. **c** Coomassie-stained polyacrylamide gel showing GFP accumulation in infiltrated *N. benthamiana* leaves. M, protein molecular weight marker in kDa; H, total soluble protein extracts from non-inoculated leaves, negative control; lane 4, 6 and 8, extracts from infiltrated leaves at 4, 6 and 8 dpi, respectively. **d** Western blot analysis of GFP from infiltrated *N. benthamiana* leaves. M, protein molecular weight marker in kDa; H, total soluble protein extracts from non-inoculated leaves, negative control; lane 4, 6 and 8, extracts from infiltrated leaves at 4, 6 and 8 dpi, respectively

showed a reduction in plant proteins after ethanol extraction (Fig. 3b), indicating that GFP was partly purified by the step. The 27-kDa band was verified to be GFP by Western blot analysis (Fig. 3c). The purity of GFP was increased from 59.9% to about 87.3%, with a yield of 58.5% (Table 1). In the second stage of this proceedure, the GFP was recovered into the aqueous phase by addition of *n*-butanol, which is more hydrophobic than ethanol. After configuration, the two phases were separated, and GFP effectively partitioned into the lower aqueous phase (Fig. 3a3). The volume of aqueous phase decreased, indicating that the GFP solution was concentrated simultaneously, as verified by both the Coomassie-stained polyacrylamide gel (Fig. 3b) and Western blot (Fig. 3c). This step also provided modest additional purification, because the purity of GFP was raised from 87.3% to about 89.4%, with a yield of 51.2% (Table 1).

Because the aqueous phase may contain the residuals of organic solvents and salts, HIC chromatography was employed for their removal. A single peak was eluted (Fig. 3d) and silver-stained polyacrylamide gel analysis of purified recombinant GFP showed unique band of 27 kDa, even when high levels of GFP were examined (Fig. 3e). Overall purification resulted

in GFP with purity above 95% and a yield of 34.1% (Table 1).

The fluorescence characterization of purified GFP

Fluorescence characterization of purified GFP was carried out using both gel-based imaging [17] and conventional spectrofluorometer-based method.

Because some kinds of alterations in GFP structure do not affect its chromophore fluorescence [18] and are not detectable using the spectrofluorometric method [17], gel-based imaging method, which is able to differentiate the alterations in GFP structure by relating the observed changes in the position of fluorescent bands, was employed to assess the structural changes of the purified GFP, which might take place during the purification process. GFP dilution samples, without heat treatment, were separated in a native discontinuous polyacrylamide gel. After electrophoresis, fluorescent image of GFP on the gel was captured using the Gel Doc XR (Bio-Rad) under UV illumination. The GFP bands were clearly visible in the gel under UV illumination, and no additional bands were observed (Fig. 4a), suggesting that the purification process did not cause detectable changes in GFP structure.

Table 1 Quantitative specifications of the principal stages of GFP purification

Processing step	TSP[a] (mg)	GFP[b] (mg)	Purity[c] (%)	GFP Yield [d] (%)
Extract	13.7	8.2	59.9	100.0
Ethanol extraction	5.5	4.8	87.3	58.5
n-butanol extraction	4.7	4.2	89.4	51.2
HiScreen Capto Butyl	2.9	2.8	96.6	34.1

[a]Amount of the TSP was determined according to BCA
[b]Amount of the GFP was measured by the ELISA
[c]Purity is defined as the amount of GFP divided by the amount of TSP in the same sample
[d]GFP Yield is defined as the amount of GFP recovered divided by initial amount of GFP in the crude extract of inoculated leaves

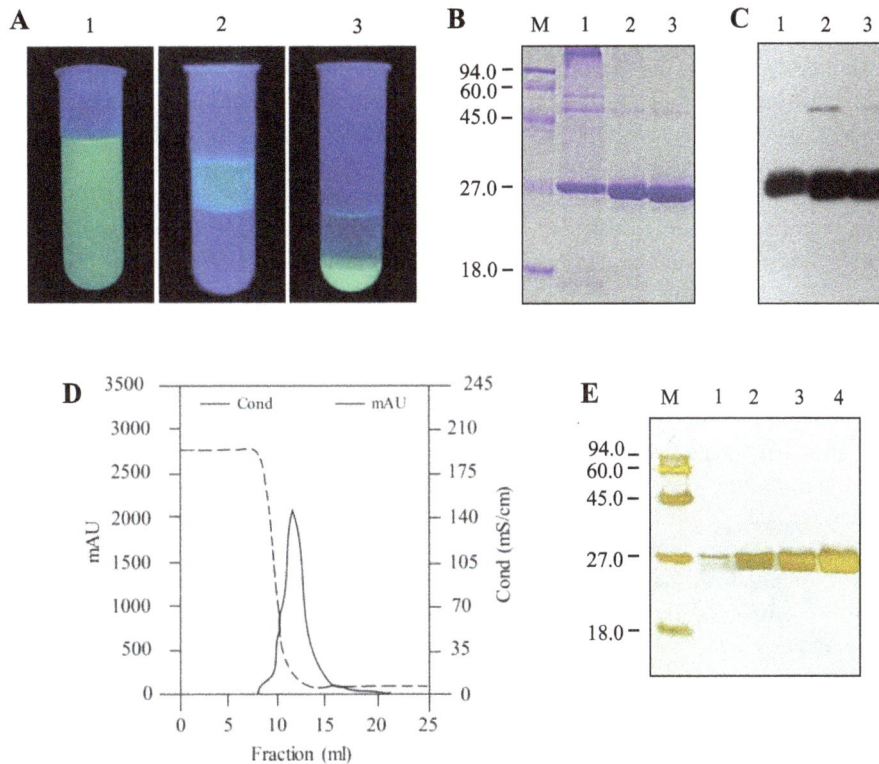

Fig. 3 Purification of GFP by alcohol/salt ATPS and HIC. **a** Successive fractions in the course of alcohol/salt ATPS procedures under UV-illumination. 1, the supernatant of homogenate of infiltrated *N. benthamiana* leaves after centrifugation; 2, the phase separation showing GFP in the upper ethanol phase after ethanol extraction; 3, the phase separation showing GFP in the lower water phase after addition of *n*-butanol to the ethanol extract. **b** Analysis of the various GFP fractions from ATPS procedure by Coomassie-stained polyacrylamide gel. M, protein molecular weight marker in kDa; lanes 1, 2 and 3, the samples corresponding to the fractions from ATPS, as shown in a. **c** Immunoblot of the various GFP fractions in the course of ATPS. M, protein molecular weight marker in kDa; lanes 1, 2 and 3, the samples corresponding to the fractions in the course of ATPS, as shown in b. **d** Elution curves of the GFP purified by HIC. **e** Silver-stained polyacrylamide gel showing the HIC purified GFP. M, Protein molecular weight marker in kDa; lane1, 2, 3 and 4, twofold serial dilution of purified GFP

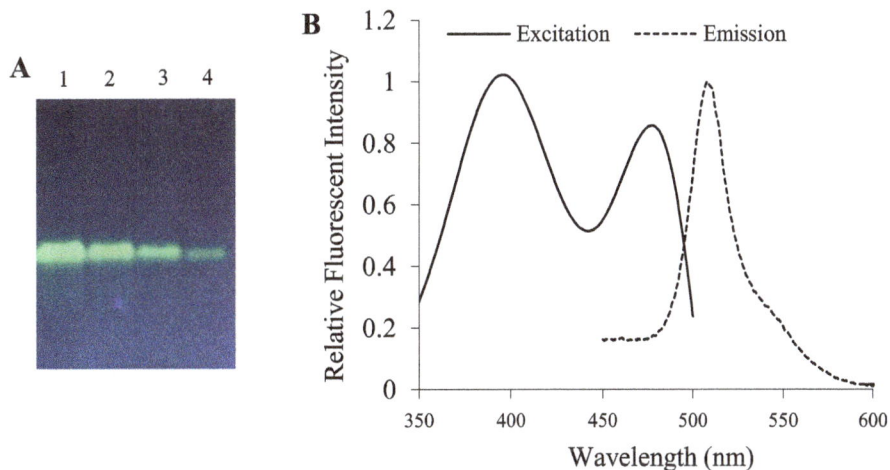

Fig. 4 The fluorescence characterization of GFP purified from *N. benthamiana* plants. **a** Polyacrylamide gel was viewed under UV light. Purified GFP was mixed with SDS loading buffer (without reducing agent) and fractionated by electrophoresis through a polyacrylamide gel without prior heat denaturation, then photographed under UV light. Lane1, 2, 3 and 4, twofold serial dilution of purified GFP. **b** The fluorescence spectrum characterization of purified GFP. The maximum fluorescence excitation peak is at 396 nm; fluorescence emission peak is at 508 nm. Spectra were obtained at a concentration of 0.1 mg/ml and normalized to chromospheres absorption maxima

For spectrofluorometric analysis, samples were measured at the optimal excitation and emission wavelengths for GFP. Figure 4b shows the excitation and emission spectra of purified GFP from infected *N. benthamiana*. The emission spectrum shows a maximum at 505 nm and the excitation spectrum shows peaks at 395 nm and 470 nm (Fig. 4b). These spectroscopic properties are similar to those observed when GFP is purified by other methods.

Discussion

Plants are an attractive alternative platform for the production of recombinant protein [19]. It offers numerous potential advantages; including low capital equipment, low energy requirements, easy scale-up, reduced risk of carrying pathogen contamination, and ability to post-translational modifications, etc. [9, 19]. Plant viral vectors are widely used as powerful tools for expressing heterologous proteins in plants with inexpensive production costs [20]. Here, we employed the viral expression vector to transiently express recombinant GFP in *N. benthamiana*, which is commonly used for producing target proteins by plant viral vectors [21]. Using TMV expression vector (pJL TRBO-G) with the help of RNA silencing suppressor, recombinant GFP in soluble form was expressed at an extremely high level (up to ~ 60% of TSP) in less than 1 week in the *N. benthamiana* leaves. In term of the yield, the plant-produced GFP is comparable with that obtained from *E. coli* (generally ranges from ~ 10% to ~ 50% of total protein). Moreover, it has been estimated that purification and downstream processing of recombinant proteins represents 80–90% of the cost of producing pharmaceuticals [22]. GFP could be produced in both soluble form and insoluble inclusion bodies from *E. coli* over-expression system, depending on culturing conditions (such as low growth temperatures, co-overexpression of molecular chaperones, etc.) [23, 24]. However, ~ 10% to ~ 20% of the recombinant GFP was found in the insoluble cell fraction even at optimal conditions [23]. Theoretically, the soluble form of plant-produced GFP may possess a good benefit for the cost reduction of the final GFP product. In addition, plant expression platform is more eco-friendly than the most of non-plant expression systems (such as bacteria, yeast, insect, and mammalian cell cultures, etc.). Altogether, we propose that the viral amplicon-based transient expression system is more suitable for the production of recombinant GFP than any other previously published method.

Currently, the various strategies, including chromatographic and non-chromatographic techniques, were developed to purify the recombinant GFP or its variants. Nevertheless, it is believed that the purification of GFP with the chromatographic methods generally involves multiple steps, time consuming, low throughput and of high cost [2]. In contrast, ATPS have been viewed as a potential alternative for protein purification because of its cost effectiveness and the simplicity of operation [5]. In addition, it is notable that alcohol/salt ATPS method is eco-friendly because ethanol and salt can be recycled. Although the yield of purified GFP could be considered modest in our study, alcohol/salt ATPS offers a considerably easier and faster way for purification of recombinant GFP. Otherwise, certain issues need to be considered while using alcohol/salt ATPS method in order to obtain effective GFP separation. Firstly, the soluble protein should be properly diluted before use, since GFP concentration exceeding 1.5 mg/ml can result in GFP co-precipitation with host proteins during phase separation. Secondly, operations should be performed at room temperature, because cooling can cause crystallization of ammonium sulfate. Lastly, exposure to ethanol can lead to a degree of protein denaturation. Thus, the step involving removing ethanol from the water-ethanol mixtures should be performed immediately.

The purpose of using HIC chromatography is initially to concentrate the sample and to remove the residuals of organic solvents and salts which may remain in the aqueous phase. However, we found that this step can further increase the GFP purity (from 89.4 to 96.6%). We speculate that some compounds, which may not be stained with coomassie dye, were removed by the HIC process. Moreover, we also observed that a little amount of fluorescent proteins were deposited on the top of the HIC column and they cannot be eluted even by low-salt buffer. Comparing with ATPS purified samples (Fig. 3b and c), a doubtful dimer GFP band was missed in the HIC purified sample (Fig. 3e), indicating that the HIC process probably also remove the oligomer version of recombinant GFP (soluble aggregates) [25]. One may argue that a single step of HIC can be used to purify GFP because it has advantage of handling large volume of samples and yielding a good result in terms of purity and yield [26]. Unfortunately, plant extracts and homogenates contain unique compounds, such as pigments, phenolics, and etc. [27]. Those compounds, especially phenolics, present obstacles in the downstream processes since they can interact with the target proteins during purification [27]. Moreover, the hydrophobic plant pigments can strongly bind to HIC media, thereby seriously reducing the useful life-time of the media to a few cycles of operation [28]. Otherwise, re-generation of the HIC media with high-frequency is costly since it requires the use of organic solvents [28]. Therefore, pretreatment is highly needed and a single step of HIC might be unattractive in practice for recovery and purification of recombinant proteins from plant tissues.

In this study, a strategy for producing GFP, which combines advantages of both plant virus-based expression platform and alcohol/salt ATPS, was developed. The high level of recombinant GFP was achieved by plant virus-based overexpression vector and the high yield adds to the economy of the downstream process. Moreover, the complete purification process requires only a few steps, takes only ~ 4 h (Fig. 5) and recovers 34.1% of the protein, equivalent to a yield of ~ 1.7 g of GFP per kilogram of *N. benthaniana* leaves. The purity of final GFP product was determined to be more than 95%, and there were no changes in its spectroscopic characteristics. Although we did not experimentally compare our strategy with *E. coli*-based commercial strategy, two green technologies are involved in our strategy, thus making it more eco-friendly than what of the previously established methodologies. Accordingly, the developed expression and purification process in this study offer a cost-effective and concise alternative for the production of GFP.

Conclusions

In summary, our study proposed a strategy for producing recombinant GFP. The procedures include overexpressing recombinant GFP in tobacco leaves via viral amplicon-based vectors and downstream purification by alcohol/salt ATPS and HIC. The strategy possesses cost-efficient and eco-friendly aspects and thereby has considerable potential for the development of an efficient process for large-scale production of recombinant GFP.

Methods

Plant materials and growth conditions

N. benthamiana seeds were kindly provided by Dr. Yongjiang Zhang of the Chinese Academy of Inspection and Quarantine (CAIQ). *N. benthamiana* seedlings were grown at 22–25 °C with 16 h dark period/8 h light period and 65% humidity in small 5 in. diameter plastic pots.

All seedlings were watered on daily basis and supplemented with Hoagland solution once a week when required.

Plasmids and *Agrobacterium* strains

The plasmids pJL TRBO-G, pCBNoX P19 and *Agrobacterium tumefaciens* GV3101 were kindly provided by Dr. John A. Lindbo of the Ohio State University. pJL TRBO-G, a *Tobacco mosaic virus* (TMV) RNA-based overexpression vector [16], was used to express GFP in *N. benthamiana* leaves using transient agroinfiltration. pCBNoX P19, expressing the 19-kDa silencing suppressor from *Tomato bushy stunt virus* (TBSV), was co-infiltrated with the pJL TRBO-G to prevent the RNAi-mediated gene silencing in plants. The plasmids were transformed into *A. tumefaciens* GV3101 using the freeze-thaw method.

Agroinfiltration of *N. benthamiana* plants

A. tumefaciens GV3101 carrying either pJL TRBO-G or pCBNoX P19, were grown in 4 ml LB medium for 24 h at 28 °C and shaking at 250 rpm. The cultures were then transferred into 100 ml LB medium having 200 μM of acetosyringone (Sigma-Aldrich) grown overnight at 28 °C and shaking at 250 rpm. Cells were harvested by centrifugation at 3000 g for 10 min and re-suspended in infiltration buffer (pH 5.6, 10 mM MES, 10 mM MgCl$_2$ and 200 μM acetosyringone) to achieve an OD$_{600}$ of 0.4. The pJL TRBO-G expression vector was mixed in a 1:1 volume ratio with the gene-silencing suppressor (pCBNoX P19). The mixed *Agrobacterium* suspensions were incubated in the dark at room temperature for 2–3 h before infiltration. The incubated *Agrobacterium* suspensions were infiltrated into the abaxial surface of leaves using a 1-ml syringe without needle. The agroinfiltrated plants were incubated in the growth chamber for 4–8 days after which the leaves were harvested.

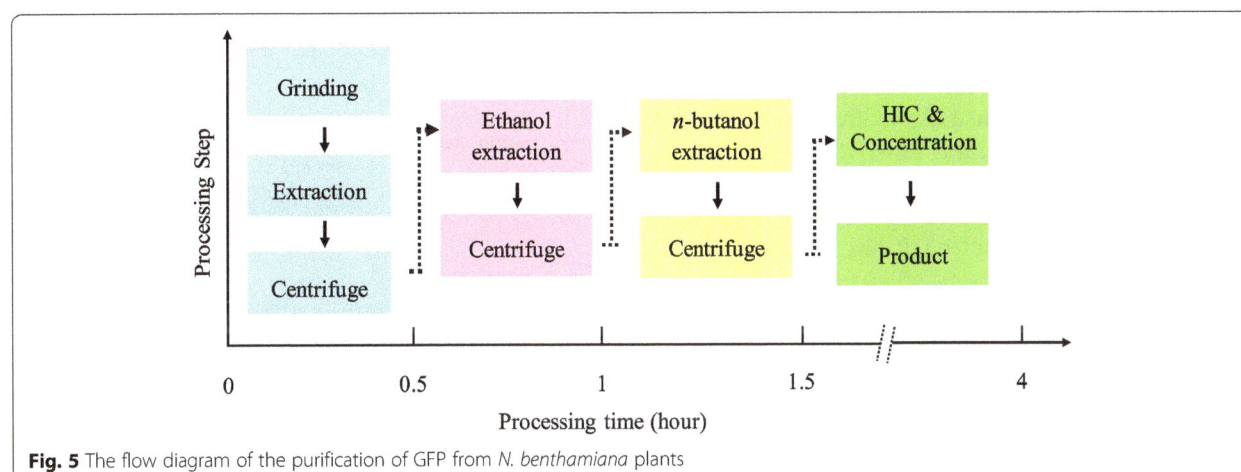

Fig. 5 The flow diagram of the purification of GFP from *N. benthamiana* plants

GFP imaging

The GFP fluorescence was monitored by illumination with a hand-held long-wave UV source (UVP Blak-Ray 100AP) and was photographed with a Cannon G6 digital camera. For microscopic analysis, GFP-positive leaf cells were visualized using confocal laser scanning microscopy (Leica TCS STED Microscopy).

Total soluble proteins (TSP) extraction and protein quantification

Protein samples were prepared by freezing agroinfiltrated leaves in liquid nitrogen and grinding to a fine powder with a mortar and pestle. The Extraction Buffer (50 mM Tris-HCl, pH 8.0, 150 mM NaCl, 10 mM EDTA) was added into the leaf powder at a ratio of 1: 4 (g/ml). Extracts were clarified by centrifugation at 12,000 g for 15 min at 4 °C. The supernatant containing TSP was recovered. Total protein content was determined by BCA protein assay kit (Pierce). The concentration of GFP was measured by GFP ELISA Kit (Abcam) according to the manufacturer's specifications.

Purification of GFP by alcohol/salt ATPS

0.3 ml of 5 M NaCl and 2.33 ml of saturated ammonium sulfate were added in turn to a 1-ml aliquot of TSP. Subsequently, the anhydrous ethanol was immediately added to the entire solution at a ratio of 1: 3 (volume-to-volume, v/v) and vigorously shaken for 30 s. The phases were separated by centrifugation at 3000 g for 5 min and the upper ethanol phase containing GFP was carefully collected. Afterward, n-butanol was added to the ethanol extracts at a 1: 4 volume-to-volume ratio. After shaking and centrifugation as described above, the lower aqueous phase containing GFP was recovered.

Purification of GFP by HIC

AKTA Purifier system (GE Healthcare) equipped with an HIC column was used for further purification of GFP. Ammonium sulfate was added to aqueous phase obtained from the previous step up to the final concentration of 1.7 M. The solution was filtered through a 0.22 μm filter (Millipore) and loaded onto 4.7-ml HiScreen Capto Butyl column (GE Healthcare) equilibrated with Binding Buffer (10 mM Tris-HCl, pH 8.0, 10 mM EDTA, and 1.7 M ammonium sulfate). After extensive washing with the Binding Buffer, the GFP was eluted at 1 ml/min with Elution Buffer (10 mM Tris-HCl, pH 8.0, 10 mM EDTA).

Sodium dodecyl sulfate-polyacrylamide gel electrophoresis (SDS-PAGE) and Western blot

Protein samples were subjected to 12% SDS-PAGE gels. After electrophoresis, gels were either stained with Coomassie brilliant blue or silver. For Western blot analysis, the proteins were transferred to a 0.45 μm nitrocellulose membrane (Sigma-Aldrich) using semi-dry electrophoresis transfer (Bio-Rad Trans-Blot SD system). The blot was developed with rabbit anti-GFP antiserum (Abcam) diluted 1:5000, followed by secondary goat anti-rabbit antibody conjugated with alkaline phosphatase (Abcam). Specific immunoreactive proteins were detected using a Western Blot ECL Plus kit (GE Healthcare).

Gel-based imaging

Serial 2-fold dilutions of the purified GFP were prepared with double-distilled water. The diluted samples were mixed with SDS-PAGE gel-loading buffer (without reducing agent). A native and discontinuous polyacrylamide gel (PAGE) electrophoresis system was employed to fractionate the samples, with no prior heat treatment. Polyacrylamide gel with a 4% (w/v) of stacking gel and a 15% (w/v) of resolving gel were used in this study. After electrophoresis, the gel was captured using the Gel Doc XR (Bio-Rad).

Fluorescence spectroscopy

Samples were diluted in 10 mM Tris-HCl, 10 mM EDTA, pH 8.0 buffers and fluorescence spectra were recorded on a Hitachi F-4500 spectrophotometer at room temperature. Excitation spectra were measured between 350 and 500 nm with the emission wavelength fixed at 508 nm. Emission spectra were measured between 450 and 600 nm with the excitation wavelength fixed at 396 nm.

Abbreviations

A. tumefaciens: *Agrobacterium tumefaciens*; ATPS: Aqueous two-phase system; CaMV: *Cauliflower mosaic virus*; GFP: Green fluorescent protein; HIC: Hydrophobic interaction chromatography; kDa: kilo-Dalton; *N. benthamiana*: *Nicotiana benthamiana*; rpm: rotation per minute; SDS-PAGE: Sodium dodecyl sulfate-polyacrylamide gel electrophoresis; TBSV: *Tomato bushy stunt virus*; TEV: *Tobacco etch virus*; TMV: *Tobacco mosaic virus*; TSP: Total soluble proteins; UV: Ultra-violet

Acknowledgements

We would like to thank Dr. John A. Lindbo at The Ohio State University for generously providing the plasmid pJL TRBO-G, pCBNoX P19 and *A. tumefaciens* GV3101. We also thank Dr. K. Andrew White from York University, CA for editing the manuscript.

Authors' contributions

Conceived and designed the experiments: SW. Performed the experiments: JD. Analysed the data: XD. All authors read and approved the final manuscript.

Funding

This work was supported by the National Natural Science Foundation of China (Project No. 31660037 and No. 31060023). No funding body had any role in the design of the study and collection, analysis, and interpretation of data and in writing the manuscript.

Ethics approval and consent to participate

Not applicable.

Consent for publication

Not applicable.

Competing interests

The authors declare that they have no competing interests.

Author details

[1]Key Laboratory of Ministry of Education for Protection and Utilization of Special Biological Resources in the Western China, Yinchuan 750021, People's Republic of China. [2]Key Laboratory of Modern Molecular Breeding for Dominant and Special Crops in Ningxia, Yinchuan 750021, People's Republic of China. [3]School of Life Science, Ningxia University, 539 W. Helanshan Road, Yinchuan, Ningxia 750021, People's Republic of China.

References

1. Shimomura O, Johnson FH, Saiga Y. Extraction, purification and properties of aequorin, a bioluminescent protein from the luminous hydromedusan, Aequorea. J Cell Comp Physiol. 1962;59:223–9.

2. Cher Pin S, Poh En L, Zora T, Schian Pei L, Chien WO. Purification of the recombinant green fluorescent protein using aqueous two-phase system composed of recyclable CO_2-based alkyl carbamate ionic liquid. Front Chem. 2018;6:529.

3. Wouters FS, Verveer PJ, Bastiaens PI. Imaging biochemistry inside cells. Trends Cell Biol. 2001;11:203–11.

4. Gerisch G, Albrecht R, Heizer C, Hodgkinson S, Maniak M. Chemoattractant-controlled accumulation of coronin at the leading edge of Dictyostelium cells monitored using a green fluorescent protein-coronin fusion protein. Curr Biol. 1995;5:1280–5.

5. Lo SC, Ramanan RN, Tey BT, Tan WS, Show PL, Ling TC, Ooi CW. Purification of the recombinant enhanced green fluorescent protein from, Escherichia coli, using alcohol+salt aqueous two-phase systems. Sep Purif Technol. 2018;192:130–9.

6. Dos Santos NV, Martins M, Santos-Ebinuma VC, Ventura SP, Coutinho JA, Valentini SR, Pereira JF. Aqueous biphasic systems composed of cholinium chloride and polymers as effective platforms for the purification of recombinant green fluorescent protein. ACS Sustain Chem Eng. 2018;6: 9383–93.

7. Moustafa K, Makhzoum A, Trémouillaux-Guiller J. Molecular farming on rescue of pharma industry for next generations. Crit Rev Biotechnol. 2016; 36:840–50.

8. Scheller J, Henggeler D, Viviani A, Conrad U. Purification of spider silk-elastin from transgenic plants and application for human chondrocyte proliferation. Transgenic Res. 2004;13:51–7.

9. Hefferon K. Plant virus expression vectors: a powerhouse for global health. Biomedicines. 2017;5:44.

10. Obembe OO, Popoola JO, Leelavathi S, Reddy SV. Advances in plant molecular farming. Biotechnol Adv. 2011;29:210–22.

11. Merchuk JC, Andrews BA, Asenjo JA. Aqueous two-phase systems for protein separation: studies on phase inversion. J Chromatogr B. 1998; 711:285–93.

12. Yau YK, Ooi CW, Ng EP, Lan JCW, Ling TC, Show PL. Current applications of different type of aqueous two-phase systems. Bioresour Bioprocess. 2015;2:49.

13. Amid M, Shuhaimi M, Sarker MZI, Manap MYA. Purification of serine protease from mango (Mangifera Indica cv. Chokanan) peel using an alcohol/salt aqueous two phase system. Food Chem. 2012;132:1382–6.

14. Fu H, Yang ST, Xiu Z. Phase separation in a salting-out extraction system of ethanol-ammonium sulfate. Sep Purif Technol. 2015;148:32–7.

15. Soares PA, Vaz AF, Correia MT, Pessoa A Jr, Carneiro-da-Cunha MG. Purification of bromelain from pineapple wastes by ethanol precipitation. Sep Purif Technol. 2012;98:389–95.

16. Lindbo JA. TRBO: a high-efficiency tobacco mosaic virus RNA-based overexpression vector. Plant Physiol. 2007;145:1232–40.

17. Chew FN, Tan WS, Tey BT. Fluorescent quantitation method for differentiating the nativity of green fluorescent protein. J Biosci Bioeng. 2011;111:246–8.

18. Melnik BS, Povarnitsyna TV, Melnik TN. Can the fluorescence of green fluorescent protein chromophore be related directly to the nativity of protein structure? Biochem Bioph Res Co. 2009;390:1167–70.

19. Lico C, Santi L, Twyman RM, Pezzotti M, Avesani L. The use of plants for the production of therapeutic human peptides. Plant Cell Rep. 2012;31:439–51.

20. Peyret H, Lomonossoff GP. When plant virology met Agrobacterium: the rise of the deconstructed clones. Plant Biotechnol J. 2015;13:1121–35.

21. Goodin MM, Zaitlin D, Naidu RA, Lommel SA. Nicotiana benthamiana: its history and future as a model for plant-pathogen interactions. Mol Plant-Microbe Interact. 2008;21:1015–26.

22. Sabalza M, Christou P, Capell T. Recombinant plant-derived pharmaceutical proteins: current technical and economic bottlenecks. Biotechnol Lett. 2014; 36:2367–79.

23. Vera A, González-Montalbán N, Arís A, Villaverde A. The conformational quality of insoluble recombinant proteins is enhanced at low growth temperatures. Biotechnol Bioeng. 2007;96:1101–6.

24. Martínez-Alonso M, Vera A, Villaverde A. Role of the chaperone DnaK in protein solubility and conformational quality in inclusion body-forming Escherichia coli cells. FEMS Microbiol Lett. 2007;273:187–95.

25. Gonzalez-Montalban N, Garcia-Fruitos E, Villaverde A. Recombinant protein solubility—does more mean better? Nat Biotechnol. 2007;25:718.

26. Noor SSM, Tey BT, Tan WS, Ling TC, Ramanan RN, Ooi CW. Purification of recombinant green fluorescent protein from Escherichia coli using hydrophobic interaction chromatography. J Liq Chromatogr R T. 2014;37: 1873–84.

27. Wilken LR, Nikolov ZL. Recovery and purification of plant-made recombinant proteins. Biotechnol Adv. 2012;30:419–33.

28. Armah GE, Achel DG, Acquaah RA, Belew M. Purification of miraculin glycoprotein using tandem hydrophobic interaction chromatography: U.S. Patent 5,886,155; 1999.

18

Fecal DNA isolation and degradation in clam *Cyclina sinensis*: noninvasive DNA isolation for conservation and genetic assessment

Min Zhang[1†], Min Wei[1,2†], Zhiguo Dong[1,2*], Haibao Duan[1], Shuang Mao[1], Senlei Feng[1], Wenqian Li[1], Zepeng Sun[1], Jiawei Li[1], Kanglu Yan[1], Hao Liu[1], Xueping Meng[1] and Hongxing Ge[1,2]

Abstract

Background: To avoid destructive sampling for conservation and genetic assessment, we isolated the DNA of clam *Cyclina sinensis* from their feces. DNA electrophoresis and PCR amplification were used to determine the quality of fecal DNA. And we analyzed the effects of different conditions on the degradation of feces and fecal DNA.

Results: The clear fecal DNA bands were detected by electrophoresis, and PCR amplification using clam fecal DNA as template was effective and reliable, suggesting that clam feces can be used as an ideal material for noninvasive DNA isolation. In addition, by analyzing the effects of different environmental temperatures and soaking times on the degradation of feces and fecal DNA, we found that the optimum temperature was 4 °C. In 15 days, the feces maintained good texture, and the quality of fecal DNA was good. At 28 °C, the feces degraded in 5 days, and the quality of fecal DNA was poor.

Conclusions: The clam feces can be used as an ideal material for noninvasive DNA isolation. Moreover, the quality of fecal DNA is negatively correlated with environmental temperature and soaking time.

Keywords: *Cyclina sinensis*, Feces; noninvasive DNA isolation, DNA degradation

Background

The clam *Cyclina sinensis* is an economically important marine bivalve that is abundant and widely distributed around the maritime coasts of Asia. *C. sinensis* is a kind of eurythermal and euryhaline filter-feeding clam, and its food source mainly includes planktonic microalgae (*Nannochloropsis oculata*, *Chaetoceros muelleri*, *Isochrysis galbana*,etc.) [1, 2] and the remains of organic debris by filtering water and sometimes opepods, facilitating the formation of fecal texture. *C. sinensis* has two hard and symmetrical shells on both sides, and it will quickly close the shells to protect itself from damage when it is stimulated by outside environment. Destructive and nondestructive sampling methods are often applied in scientific researches of clam [3, 4]. The former is conducted by taking parts of specific tissue after the experimental animals are dissected directly, whereas the latter is usually completed by means of a shell opener or a mini electric drill. Nevertheless, sampling using both methods will negatively influence the life of clams, even leading to their death.

Noninvasive sampling is a sampling method for genetic analysis by collecting exfoliated hair, feces, and urine without having to catch, handle, or even observe the animals [5]. It has been widely used in the field of conservation genetics because it is simple and does not harm experimental animals. At present, noninvasive sampling methods are being applied to fish and marine mammals by collecting body surface mucus [6], shedding scales [7], and feces [8, 9]. Among them, feces can be easily collected without disturbing or negatively affecting the

* Correspondence: dzg7712@163.com

†Min Zhang and Min Wei contributed equally to this work.

[1]Jiangsu Key Laboratory of Marine Biotechnology, Jiangsu Ocean University, Lianyungang 222005, Jiangsu, China

[2]Co-Innovation Center of Jiangsu Marine Bio-industry Technology, Jiangsu Ocean University, Lianyungang 222005, Jiangsu, China

normal life of experimental animals. Therefore, feces are potentially valuable research materials in noninvasive sampling. The main component of feces is undigested food residues, where intestinal epithelial cells adhere to when they pass through the intestine. Therefore, mitochondrial and nuclear genomic DNA can be isolated from the remaining epithelial cells in the feces [10]. Fecal molecular biotechnology provides a rapid and dependable way of sampling endangered animals [11–14]. In addition, with the development of molecular biology technology, fecal DNA is extensively used in genetic biology studies for species identification [15–17], individual identification [18–20], sex identification [21–25], population genetic structure [26–28], and genetic diversity evaluation [29]. However, fecal sampling has some problems, such as poor fecal DNA isolation quality and low success rate of PCR amplification [30]. Moreover, no study has performed fecal DNA extraction on invertebrates, especially shellfish. Studies on terrestrial animals have found that fecal DNA degradation occurs with the increase of exposure time [31] and is affected by many other factors, such as light, temperature, and humidity [32, 33]. Compared with those of terrestrial animals, the feces of aquatic animals are more vulnerable due to the external water environment, and their fecal DNA is easier to degrade. Therefore, to obtain good quality shellfish fecal DNA, an improved fecal DNA extraction method should be developed, and the optimal environmental conditions for fecal sampling should be investigated.

In this study, clam feces was used as an experimental material to isolate DNA noninvasively. Moreover, the effects of environmental temperature and soaking time on the degradation of feces and fecal DNA were analyzed. The results can be used as a basis for developing noninvasive DNA isolation technology of shellfish and provide a reference for optimal conditions of fecal sampling, providing technical support for further research on molecular biology and conservation genetics of shellfish.

Results

DNA isolation of fresh feces

To determine the quality of fecal DNA, electrophoresis was conducted, and the foot muscle DNA was chosen as the positive control. The results showed that all bands of the fecal DNA were clear but showed a slight tailing phenomenon (Fig. 1 and Additional file 1: Figure S1), which was proved by the results of A260/280 (Table 1). Moreover, the bands of fecal DNA in lanes 2, 4, 5, and 6 were very bright, similar to the foot DNA band (lane F).

PCR amplification

To determine the effectiveness of fecal DNA, PCR amplification was conducted using the specific primers designed on the basis of mitochondrial and nuclear genomic DNA of *C. sinensis*. The results revealed that the band size of fecal DNA was the same as that of foot DNA and consistent with the expected length of the target band (Fig. 2 and Additional file 2: Figure S2), which was also proved by the sequencing results.

Effects of soaking time and environmental temperature on fecal degradation

Fecal degradation was evaluated by observing changes in fecal texture using a stereoscope. The fecal texture changed over time and was influenced by the environmental temperature. The fresh fecal pellets (0 days) were yellowish green in color and cylindrical. They had a length of 700 μm and diameter of 450 μm (Fig. 3). In fecal samples stored at 28 °C (Fig. 3a), the surface texture became loose at 5 days, with filaments growing abundantly. The filaments grew in large numbers and gradually formed into microbial micelles. More bacteria attached to the microbial micelles, eventually forming bacterial micelles. At 10 and 15 days, the loose feces obviously broke apart, and the local fecal textures were decomposed. At 20 days, the breakage sites increased, and the fecal pellets became looser. At 25 days, the fecal pellets developed into bioflocs framed with filamentous fungi. In fecal

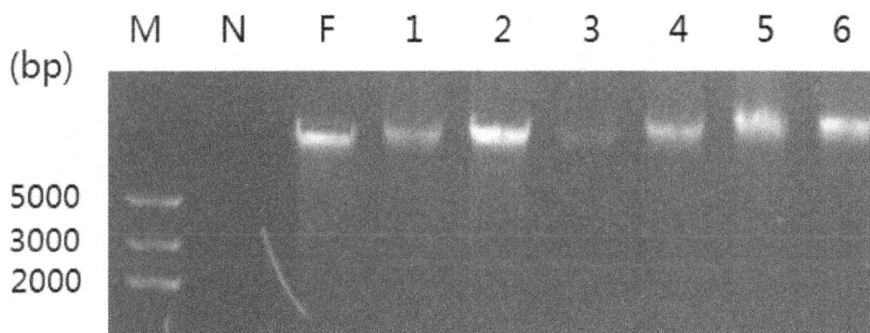

Fig. 1 Agarosegel electrophoresis of fecal DNA. Lane M, DNA marker; lane N, negative control; lane F, foot DNA; lanes 1–6, DNA of fresh feces. (N=6).

Table 1 DNA quality of fresh feces (N = 6)

Fecal DNA	A260/280	Concentration (ng/μL)
1	1.62	24.2
2	1.79	27.3
3	1.59	20.5
4	1.70	24.3
5	1.67	23.6
6	1.77	25.1

samples stored at 15 °C (Fig. 3b), some fecal textures were slightly decomposed on the 10th day. Fecal breakage sites gradually increased at 15–20 days, and large cracks were observed at 25 days. In fecal samples stored at 4 °C (Fig. 3c), the fecal texture was not loose until 10 days and became slightly loose at 15 days. At 20–25 days, some parts of the fecal pellets were slightly decomposed.

Effects of soaking time and environmental temperature on fecal DNA degradation

Under different soaking times and environmental temperatures, the degradation degree of fecal DNA was determined by agarose gel electrophoresis. At 28 °C, fecal DNA degradation occurred at 5 days after soaking the feces in seawater, but high-quality DNA could still be

isolated from few fecal samples (Fig. 4a and Additional file 3: Figure S3a). At 15 and 20 days after soaking, poor-quality DNA was obtained from fecal samples, and serious fecal DNA degradation was observed. At 15 °C, good-quality fecal DNA could still be extracted at 10 days after soaking (Fig. 4b and Additional file 3: Figure S3b); however, the sample degraded to varying degrees after 15 days. At 4 °C, high-quality DNA without tailing phenomenon could still be obtained from fecal samples at 15 days after soaking (Fig. 4c and Additional file 3: Figure S3c).

Discussion

Using the modified phenol/chloroform method for fecal DNA isolation and PCR verification

Feces is a very complex mixture of biotic and abiotic components. In this study, DNA was extracted from clam feces, and the quality of fecal DNA was identical to that of foot DNA. As shown in Fig. 2, four specific fragments of mitochondrial and nuclear genomic DNA from *C. sinensis* were amplified by PCR using the fecal DNA as template. The results suggest that the isolation of fecal DNA was successful and reliable, which were proved by sequencing results. Therefore, clam DNA can be nondestructively isolated from feces. However,

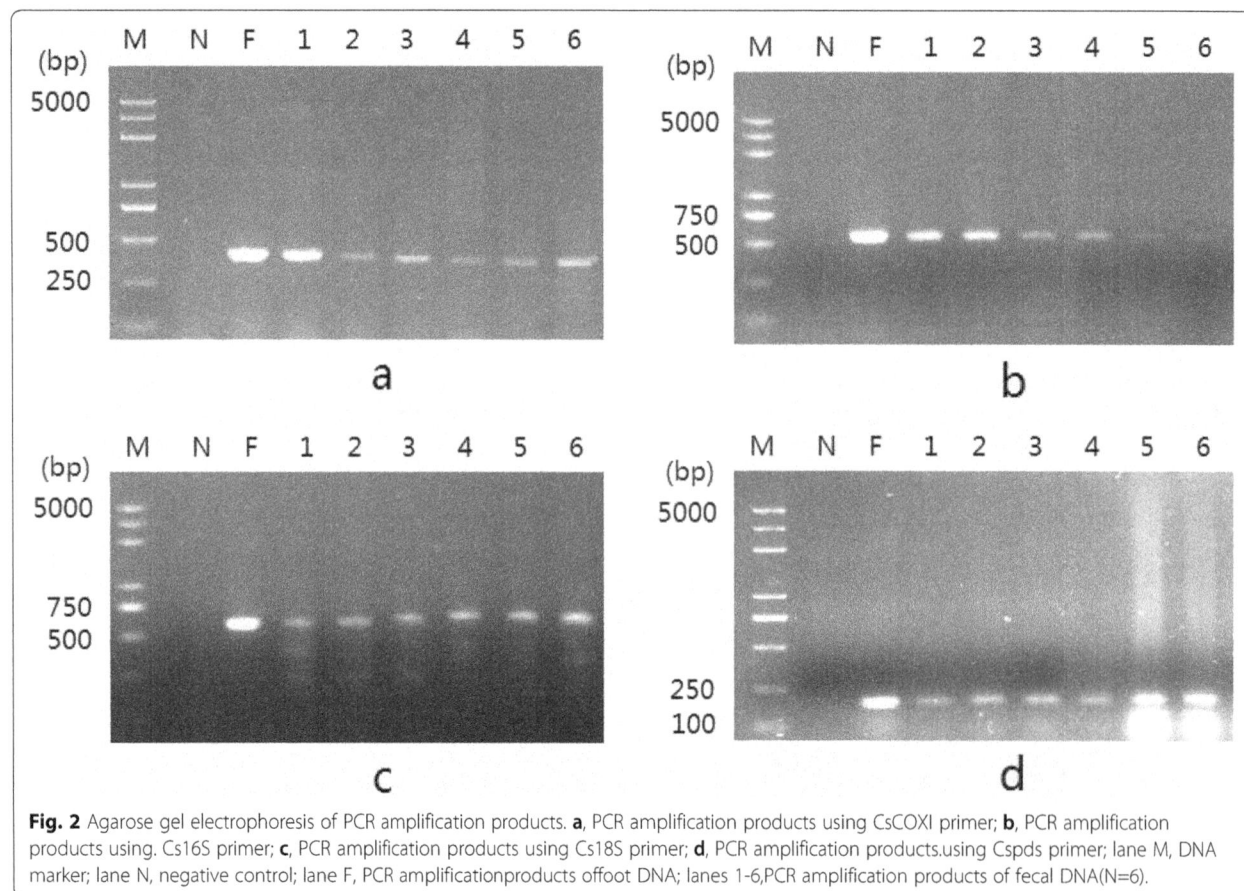

Fig. 2 Agarose gel electrophoresis of PCR amplification products. **a**, PCR amplification products using CsCOXI primer; **b**, PCR amplification products using. Cs16S primer; **c**, PCR amplification products using Cs18S primer; **d**, PCR amplification products.using Cspds primer; lane M, DNA marker; lane N, negative control; lane F, PCR amplificationproducts offoot DNA; lanes 1-6,PCR amplification products of fecal DNA(N=6).

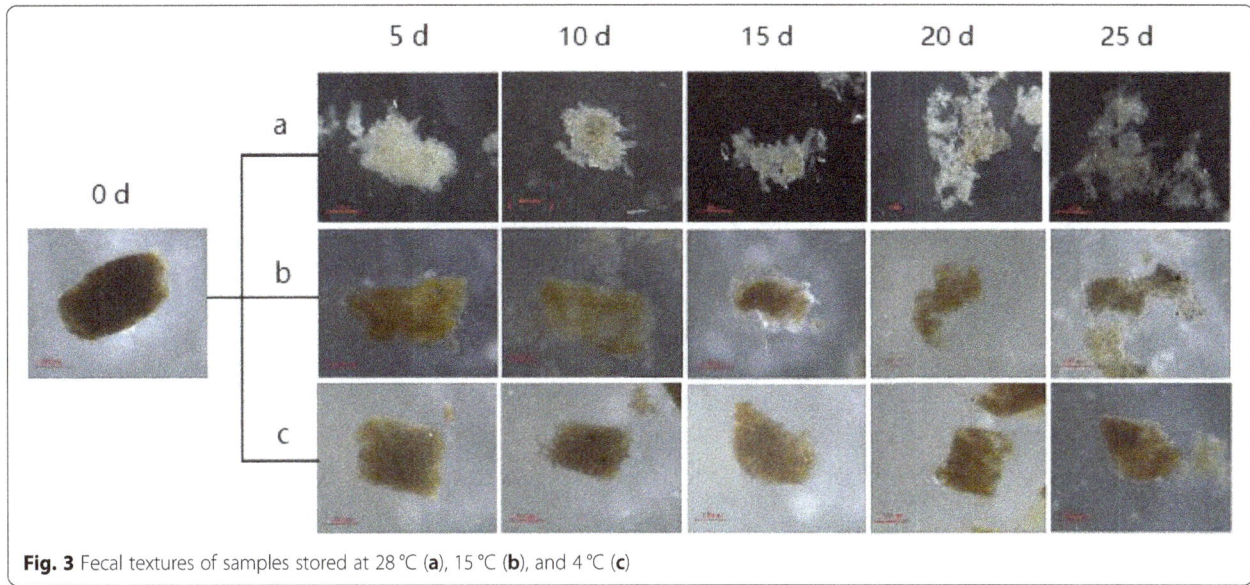

Fig. 3 Fecal textures of samples stored at 28 ℃ (**a**), 15 ℃ (**b**), and 4 ℃ (**c**)

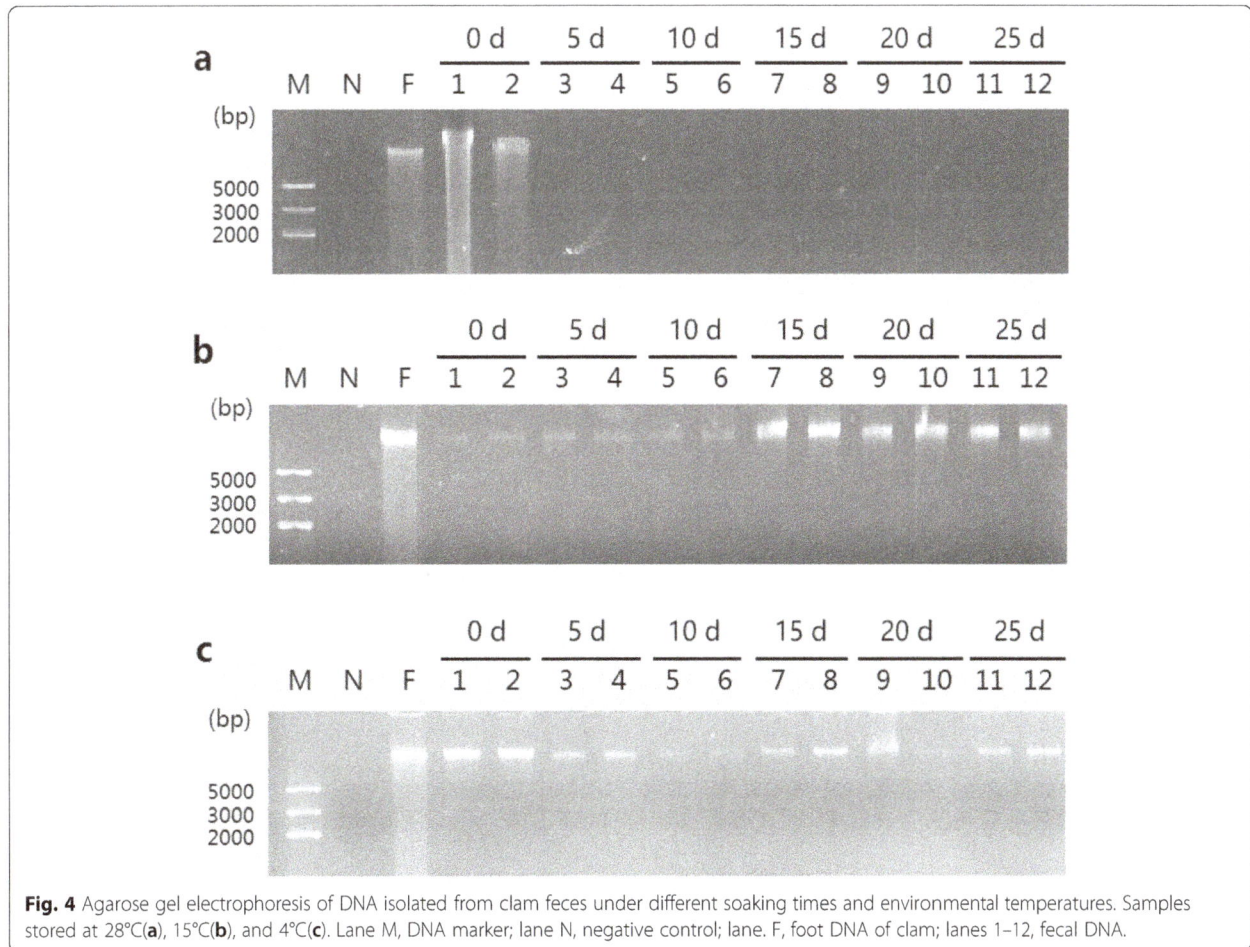

Fig. 4 Agarose gel electrophoresis of DNA isolated from clam feces under different soaking times and environmental temperatures. Samples stored at 28℃(**a**), 15℃(**b**), and 4℃(**c**). Lane M, DNA marker; lane N, negative control; lane. F, foot DNA of clam; lanes 1–12, fecal DNA.

differences in the quality of fecal DNA were still observed among different fecal samples, which may be due
to the quantity variance of intestinal cells adhered to
feces. Besides intestinal cells, feces contain undigested
food, digestive enzymes, mucus, and other blockers,
which affect Taq DNA polymerase activity [34–36]. Furthermore, by extracting fecal DNA from diet-restricted
brown bears, Murphy et al. found that the diet has a significant effect on the success rate of PCR amplification
using fecal DNA as template [22]. In this study, the success rate of PCR amplification using fecal DNA as template was 100%. This finding can be explained by two
factors: (i) *C. sinensis* have special feeding habits (mostly
microalgae), which may cause their feces to contain few
inhibitors to Taq DNA polymerase. (ii) The modified
phenol/chloroform method was effective in isolating
DNA from feces. Taken together, the findings indicate
that the modified phenol/chloroform method is effective
in isolating DNA from feces, and fecal DNA from herb-
ivorous animals can be used as a template for PCR amplification, which is also supported by the findings of
Zhang et al.'s study on fecal DNA of pandas [37].

Degradation of feces and fecal DNA under different soaking times and environmental temperatures

Bioflocs are composed of microorganisms, protozoa,
algae, filamentous bacteria, and organic matter in water
[38]. With the decomposition of feces into flocs, the microbial community structure was altered. In the process
of degradation, the structure of fecal pellets became
loose, which increased the contact area with seawater,
thereby increasing the attachments for protozoa and
leading to a looser structure [39]. As shown in Fig. 3,
differences were observed in the rate of fecal texture
changes at different temperatures. The degradation rate
at 28 °C was significantly higher than that at 15 °C and
4 °C after the same soaking time, which was probably
due to the high temperature (28 °C) appropriate for the
growth and reproduction of microorganisms in feces
and improvement of the activity of fecal degradation-
related enzymes [40]. Moreover, fecal degradation was
often accompanied by fecal DNA degradation (Fig. 4),
and the degree of fecal DNA degradation was also affected by the environmental temperature. Similar phe-
nomena were observed in feces of other animals. The
quality of DNA isolated from feces of *Canis lupus* in
winter remained significantly higher than that in summer [41]. Moreover, the quality of fecal DNA from ape
was negatively correlated with fecal environmental
temperature [42]. These findings suggest that fecal sampling should be conducted in seasons with low
temperature to obtain good-quality fecal DNA [43–45].

The degradation of feces and fecal DNA was also affected by soaking time in seawater. The longer the

soaking time, the more serious the degradation (Fig. 4).
Several studies suggest that fecal DNA degradation is affected by water. The rate of fecal DNA degradation is
significantly accelerated by rain wash [46], and removing
water from feces can essentially prevent the activation of
nuclease in feces [47]. Moreover, *Cyclina sinensis* is a
kind of marine shellfish, and its feces are soaked in seawater. Seawater is a very complicated multicomponent
aqueous solution containing various organic, inorganic,
dissolved, and suspended substances, which may be the
reason for the degradation of feces and fecal DNA from
clam. Therefore, the fresher the fecal samples collected,
the higher the quality of DNA [48].

Conclusions

In this study, clam feces were used as experimental material to isolate DNA noninvasively. The isolation of
fecal DNA was found to be successful and reliable by
PCR amplification. The effects of different environmental temperatures and soaking times on the degradation
of feces and fecal DNA were investigated. The results
suggest that fresh fecal samples stored at low environmental temperature (~ 4 °C) were beneficial to the isolation of fecal DNA with good quality. This study provides
technical support for further molecular biology research
and conservation genetics research of shellfish.

Methods
Sampling and processing

Healthy clams *C. sinensis* were collected from a clam farm
in Jiangsu, China. They were cultured in seawater for two
weeks at room temperature and fed with 0.005 g/mL of
Chlorella once a day. Natural seawater was filtered with a
double-layer 500-mesh sieve after precipitation, disinfection,
and aeration for culturing clams and replaced once a day.

Forty-eight healthy clams (body weight, 10.09 ± 2.81 g;
shell length, 3.01 ± 0.38 cm) were randomly selected and
divided into 12 parallel groups (labeled 1 to 12), each
groups containing four clams. Feeding was withheld for
2 days in continuously oxygenated seawater. Thereafter,
the clams were fed with 6×10^5 cells/L of *Chlorella* until
waste matter (feces) was completely expelled. During
this period, clam defecation was observed every 2 h. The
feces were collected from the bottom of the beaker using
siphon method.

DNA isolation

Total DNA was isolated from clam feces and foot tissue
(used as positive control). Fecal DNA isolation was performed using the phenol/chloroform method in accordance with a previous study of Sambrook [49] with some
modifications as follows:

1) Place the feces on a 200-mesh silk screen, and wash it slowly with double-distilled water (ddH$_2$O) to remove impurities on the fecal surface.

2) Transfer each 200 mg fecal sample into a new 1.5 mL Eppendorf tube.

3) Add 100 μL ddH$_2$O, blow the feces repeatedly with a straw to make it homogenate, and then vortex fully.

4) Centrifuge for 3 min at 800×g, and then transfer the supernatant into a new 1.5 mL tube. Add 400 μL of 10% SDS and 10 μL of proteinase K into each tube, and then vortex fully.

5) Incubate the tubes for 1 h in a 65 °C water bath with occasional shaking (~ 10 min).

6) Add 10 μL of 20 mg/mL RNase, and incubate the tubes in a 37 °C water bath for 10 min. Centrifuge for 3 min at 12,000×g, and then transfer the supernatant into a new 1.5 mL tube.

7) Add an equal volume of ice-cold Tris-saturated phenol (pH 7.9), and mix upside down and store at room temperature for 5 min.

8) Centrifuge at 12,000×g for 12 min, and then transfer the supernatant into a new 1.5 mL tube.

9) Add an equal volume of chloroform, and mix upside down. Centrifuge at 12,000×g for 10 min, and then transfer the supernatant into a new 1.5 mL tube.

10) Add an equal volume of isopropanol, mix upside down, and store at room temperature for 3 min. Centrifuge at 12,000×g for 12 min, and then remove the supernatant completely.

11) Wash the DNA pellet twice with 1 mL ice-cold 70% ethanol.

12) Air dry.

13) Resuspend the DNA pellet in 30 μL of TE buffer, and then store at − 40 °C before use.

Primer design and PCR amplification

To determine DNA quality, PCR amplification was conducted with primers designed on the basis of mitochondrial and nuclear genomic DNA sequences. The sequences of mitochondrial (COXI and 16S rRNA) and nuclear genomic DNA (18S rRNA and partial sequence of nuclear DNA) were retrieved and downloaded from NCBI (https://www.ncbi.nlm.nih.gov/). PCR primers were designed by Primer Premier 5.0 software and are shown in Table 2. PCR amplification was conducted in a 15 μL reaction volume, containing 1.0 μL of DNA template, 0.2 μL of Taq (Takara, Dalian, China), 0.8 μL of primers (including forward and reverse primers), 1.0 μL of dNTPs, 1.5 μL of 10× buffer, and 10.5 μL of ddH$_2$O. The PCR amplification procedure was conducted as follows: initial denaturation at 95 °C for 5 min, followed by 30 cycles of denaturation at 94 °C for 1 min, annealing for 30 s, extension at 72 °C for 30 s, and final extension at 72 °C for 10 min. The PCR amplification products were detected by 1.5% agarose gel electrophoresis and captured with a gel imaging system (Universal Hood II, Bio-Rad, USA). The purified PCR products were sequenced by Shanghai Map Biotech Co., Ltd. The sequencing results were checked by Chromas software and blasted by BLAST online software (https://blast.ncbi.nlm.nih.gov/Blast.cgi).

Effects of soaking time and environmental temperature on the degradation of feces and fecal DNA

To explore the effects of environmental temperature and soaking time on fecal texture changes and fecal DNA degradation, the fecal samples were collected immediately after the clams defecated. They were then soaked in clean seawater and stored at 28 °C, 15 °C, and 4 °C. To observe fecal texture changes, the fecal samples were placed on clean slides, observed, and photographed with a stereoscope (Nikon SME 1500, Nikon, Japan) at 0, 5, 10, 15, 20, and 25 days after soaking in seawater. Fecal DNA isolation was conducted using the modified phenol/chloroform method mentioned above, and the fecal DNA quality was determined by Ultramicro Nucleic Acid Analyser (Eppendorf BioPhotometer® D30, Eppendorf, Germany), electrophoresis, and PCR amplification.

Table 2 Primers and sequences

Primer	Sequence (5′–3′)	Source	Gene ID	Product size/bp
CsCOXI	F:TGGTGGTTTAACTGGTGTTGTT	Mitochondrial DNA	26,898,108	404
	R:AAAACACCAAACCACGCTGAG	from C. sinensis		
Cs16S	F:GATCGTACCTGCCCTGTGAT	Mitochondrial DNA	26,898,076	548
	R:ACCACTCTAGCTTACGCCGA	from C. sinensis		
Cs18S	F:TGCGTTCAAGGTGTCGATGT	Nuclear genomic	unregistered	581
	R:GGGGCCGACATGAAATGAAA	DNA from C. sinensis		
Cspds	F: ACTTCAGAATTCAGAATTCAG	Nuclear genomic	unregistered	187
	R: GTCACGCACAATGTAACG	DNA from C. sinensis		

Data analysis

The DNA purity was confirmed by Ultramicro Nucleic Acid Analyser (Eppendorf BioPhotometer® D30, Eppendorf, Germany). The DNA and PCR amplification products were detected by 1 and 1.5% agarose gel electrophoresis respectively, and the gel images were observed and captured with a gel imaging system (Universal Hood II, Bio-Rad, America).

Supplementary information

Additional file 1: Figure S1. Agarose gel electrophoresis of fecal DNA. Lane M, DNA marker; lane N, negative control; lane F, foot DNA; lanes 1–20, DNA of fresh feces ($N = 20$).

Additional file 2: Figure S2. Agarose gel electrophoresis of PCR amplification products Lane M, DNA marker; lane N, negative control; lane F, PCR amplification products of foot DNA; lanes 1–20, PCR amplification products of fecal DNA ($N = 20$).

Additional file 3: Figure S3. Agarose gel electrophoresis of PCR amplification products from clam feces under different soaking times and environmental temperatures. Samples stored at 28 ℃ (a), 15 ℃ (b), and 4 ℃ (c). Lane M, DNA marker; lane N, negative control; lane F, foot DNA of clam; lanes 1–12, fecal DNA.

Abbreviations
C. sinensis: *Cyclina sinensis*; PCR: Polymerase chain reaction; rDNA: Ribosomal DNA; rRNA: Ribosomal RNA

Acknowledgments
The authors would like to express our appreciation to Guangen Xu for trial sample preparation, and thanks to Essaystar for modification-polish.

Authors' contributions
MZ, MW and ZD conceived and designed the experiments. MZ performed the experiments. KY, HL, SM, HD, SF, ZS, WL, JL contributed materials. MZ and MW analyzed the data and wrote the manuscript. XM and HG guided the experiments and writing. All authors read, reviewed and approved the manuscript.

Funding
This study was supported by grants from China Agriculture Research System (CARS–49), the Priority Academic Program Development of Jiangsu, the Natural Science Foundation of the Jiangsu Higher Education Institutions of China (No. 18KJA240001), the Project of Jiangsu Fisheries Science and Technology (Y2018–27), the Project of Jiangsu Key Laboratory of Marine Biotechnology (No. HS2018002), the Postgraduate Research & Practice Innovation Program of Jiangsu Province (No. SJCX18_0932), and the Undergraduate Innovation and Entrepreneurship Training Program of Jiangsu Ocean University (Z201911641105018). The funding bodies did not play any role in the design of the study, the collection, analysis, and interpretation of data and in writing the manuscript.

Ethics approval and consent to participate
All applicable institution and/or national guidelines for the care and use of animals were followed.

Consent for publication
Not applicable.

Competing interests
The authors declare that they have no competing interests.

References

1. Nakamura Y, Kerciku F. Effects of filter-feeding bivalves on the distribution of water quality and nutrient cycling in a eutrophic coastal lagoon. J Mar Syst. 2000;26:209–21.
2. Lin TT, Zhou K, Liu X, Lai QF, Zhang D, Shi LY. Effects of clam size, food type, sediment characteristic, and seawater carbonate chemistry on grazing capacity of Venus clam *Cyclina sinensis* (Gmelin, 1791). Chin J Oceanol Limnol. 2016;35:1239–47.
3. Taberlet P, Waits LP, Luikart G. Noninvasive genetic sampling: look before you leap. Trends Ecol Evol. 2008;14:323–7.
4. Cheng HL, Xia DQ, Wu TT, Meng XP, Ji HJ, Dong ZG. Study on sequences of ribosomal DNA internal transcribed spacers of clams belonging to the Veneridae Family (Mollusca: Bivalvia). Acta Genet Sin. 2006;33(8):702–10.
5. Morin PA, Woodruff DS. Noninvasive genotyping for verteb rate conservation. Mol Genet Approaches Conserv. 1996. p. 298–313.
6. Fernández-Alacid L, Sanahuja I, Ordóñez-Grande B, Sánchez-Nuño S, Viscor G, Gisbert E, Herrera M, Ibarz A. Skin mucus metabolites in response to physiological challenges: a valuable non-invasive method to study teleost marine species. Sci Total Environ. 2018;644:1323–35.
7. Wasko AP, Martins C, Oliveira C, Foresti F. Non-destructive genetic sampling in fish. An improved method for DNA extraction from fish fins and scales. Hereditas. 2010;138(3):161–5.
8. Reed JZ, Tollit DJ, Thompson PM, Amos W. Molecular scatology: the use of molecular genetic analysis to assign species, sex and individual identity to seal faeces. Mol Ecol. 1997;6:225–34.
9. Parsons KM. Reliable microsatellite genotyping of dolphin DNA from faeces. Mol Ecol Notes. 2001;l:341–4.
10. Chancellor RL, Langergraber K, Ramirez S, Rundus AS, Vigilant L. Genetic sampling of unhabituated chimpanzees (*Pan troglodytes schweinfurthii*) in Gishwati forest reserve, an isolated forest fragment in Western Rwanda. Int J Primatol. 2012;33:479–88.
11. Lathuillière M, Ménard N, Gautier-Hion A, Crouau-Roy B. Testing the reliability of noninvasive genetic sampling by comparing analyses of blood and fecal samples in Barbary macaques (*Macaca sylvanus*). Am J Primatol. 2010;55:151–8.
12. King SRB, Schoenecker KA, Fike J, Oyler-McCance SJ. Long-term persistence of horse fecal DNA in the environment makes equids particularly good candidates for noninvasive sampling. Ecology Evolution. 2018;8:4053–64.
13. Zhu Y, Liu HY, Yang HQ, Li YD, Zhang HM. Factors affecting genotyping success in giant panda fecal samples. Peer J. 2017;5(5):e3358.
14. Zhang BW, Li M, Ma LC, Wei FW. A widely applicable protocol for DNA isolation from fecal samples. Biochem Genet. 2006;44(11–12):494–503.
15. Yamashiro A, Yamashiro T, Baba M, Endo A, Kanmada M. Species identification based on the faecal DNA samples of the Japanese serow (*Capricornis crispus*). Conserv Genet Resour. 2010;2:409–14.
16. Laguardia A, Wang J, Shi FL, Shi K, Riordan P. Species identification refined by molecular scatology in a community of sympatric carnivores in Xinjiang, China. Zool Res. 2015;36(2):72–8.
17. Modi S, Mondol S, Ghaskadbi P, Hussain Z, Nigam P, Habib B. Noninvasive DNA-based species and sex identification of Asiatic wild dog (*Cuon alpinus*). J Genet. 2018;97(5):1457–61.
18. Brinkman TJ, Person DK, Schwartz MK, Pilgrim KL, Colson KE, Hundertmark KJ. Individual identification of Sitka blacktailed deer (*Odocoileus hemionus sitkensis*) using DNA from fecal pellets. Conserv Genet Resour. 2010;2:115–8.
19. Ruiz-González A, Madeira MJ, Randi E, Urra F, Gómez-Moliner BJ. Non-invasive genetic sampling of sympatric marten species (*Martes martes* and *Martesfoina*): assessing species and individual identification success rates on faecal DNA genotyping. Eur J Wildl Res. 2013;59:371–86.
20. Yamashiro A, Kaneshiro Y, Kawaguchi Y, Yamashiro T. Species, sex, and individual identification of Japanese serow (*Capricorni scrispus*) and sika deer (*Cervus nippon*) in sympatric region based on the fecal DNA samples. Conserv Genet Resour. 2017;9:333–8.
21. Liu J, Bao YX, Wang YN, Zhang WC, Chen XN, He WP, Shi WW. Individual and sexual identification for the wild black muntjac (*Muntiacus crinifrons*) based on fecal DNA. Acta Ecol Sin. 2014;34:13–8.
22. Murphy MA, Waits LP, Kendall KC. The influence of diet on faecal DNA amplification and sex identification in brown bears (*Ursus arctos*). Mol Ecol. 2010;12:2261–5.
23. Yamauchi K, Hamasaki S, Miyazaki K, Kikusui T, Takeuchi Y, Mori Y. Sex determination based on fecal DNA analysis of the amelogenin gene in sika deer (*Cervus nippon*). J Vet Med Sci. 2000;62(6):669–71.

24. Liu X, Yang YY, Wang XM, Liu ZS, Wang ZH, Ding YZ. Sex identification based on AMEL gene PCR amplification from blue sheep (Pseudois nayaur) fecal DNA samples. Genet Mol Res. 2015;14(3):9045–52.

25. Yamazaki S, Motoi Y, Nagai K, Ishinazaka T, Asano M, Suzuki M. Sex determination of sika deer (Cervus nippon yesoensis) using nested PCR from feces collected in the field. J Vet Med Sc. 2011;73(12):1611–6.

26. Saito W, Amaike Y, Sako T, Kaneko Y, Masuda R. Population structure of the raccoon dog on the grounds of the imperial palace, Tokyo, revealed by microsatellite analysis of fecal DNA. Zool Sci. 2016;33(5):485–90.

27. Wultsch C, Caragiulo A, Dias-Freedman I, Quigley H, Rabinowitz S, Amato G. Genetic diversity and population structure of Mesoamerican jaguars (Panthera onca): implications for conservation and management. PLoS One. 2016;11(10):e0162377.

28. Mekonnen A, Rueness EK, Stenseth NC, Fashing PJ, Bekele A, Hernandez-Aguilar RA, Missbach R, Haus T, Zinner D, Roos C. Population genetic structure and evolutionary history of bale monkeys (Chlorocebus djamdjamensis) in the southern Ethiopian highlands. BMC Evol Biol. 2018;18(1):106.

29. Sugimoto T, Aramilev VV, Kerley LL, Nagata J, Miquelle DG, McCullough DR. Noninvasive genetic analyses for estimating population size and genetic diversity of the remaining far eastern leopard (Panthera pardus orientalis) population. Conserv Genet. 2014;15:521–32.

30. Espinosa MI, Bertin A, Squeo FA, Cortés A, Gouin N. Comparison of DNA extraction methods for polymerase chain reaction amplification of guanaco (Lama guanicoe) fecal DNA samples. Genet Mol Res. 2015;14(1):400–6.

31. Frantzen MA. SilkJB, Ferguson JW, Wayne RK, Kohn MH. Empirical evaluation of preservation methods for faecal DNA. Mol Ecol. 1998;7:1423–8.

32. Femando P, Vidya TN, Rajapakse C, Dangolla A, Melnick JD. Reliable noninvasive genotyping: fantasy of reality? J Hered. 2003;94:115–23.

33. Lonsinger RC, Gese EM, Dempsey SJ, Kluever BM, Johnson TR, Waits LP. Balancing sample accumulation and DNA degradation rates to optimize noninvasive genetic sampling of sympatric carnivores. Mol Ecol Resour. 2015;15(4):831–42.

34. Iacovacci G, Serafini M, Berti A, Lago G. STR typing from human faeces: a modified DNA extraction method. Int Congr Ser. 2003;1239:917–20.

35. Sidransky D, Tokino T, Kinzler SR, Hamilton KW, Levin B, Frost P, Vogelstein B. Identification of ras oncogene mutations in the stool of patients with curable colorectal tumors. Science. 1992;256:102–5.

36. Carss DN, Parkinson SG. Errors associated with otter (Lutra lutra) fecal analysis. I. Assessing general diet from spraints. J Zool. 1996;238:301–17.

37. Zhang BW, Wei FW, Li M, Lv XP. A simple protocol for DNA extraction from faeces of the giant panda and lesser panda. Acta Zool Sin. 2004;50:452–8.

38. Xia Y, Yu EM, Xie J, Yu DG, Wang GJ, Li ZF, Wang HY, Gong WB. Analysis of bacterial community structure of bio-Floc by PCR-DGGE. J Fish China. 2012; 36:1563–71.

39. Poulsen L, Iversen M. Degradation of copepod fecal pellets: key role of protozooplankton. Mar Ecol Prog. 2008;367:1–13.

40. Saborowski R, Friedrich M, Dietrich U, Gutow L. The degradation of organic material from invertebrate fecal pellets by endogenous digestive enzymes-effects of pH and temperature, 2nd annual meeting: biological impacts of ocean acidification. Germany: Bremen; 2011.

41. Lucchini V, Fabbri E, Marucco F, Ricci S, Boitani L, Randi E. Noninvasive molecular tracking of colonizing wolf (Canis lupus) packs in the western Italian Alps. Mol Ecol. 2002;11:857–68.

42. Nsubuga AM, Robbins MM, Roeder AD, Morin PA, Boesch C, Vigilant L. Factors affecting the amount of genomic DNA extracted from ape faeces and the identification of an improved sample storage method. Mol Ecol. 2004;13:2089–94.

43. Maudet C, Luikart G, Dubray D, Von Hardenberg A, Taberlet P. Low genotyping error rates in wild ungulate faeces sampled in winter. Mol Ecol Notes. 2004;4:772–5.

44. DeMay SM, Becker PA, Eidson CA, Rachlow JL, Johnson TR, Waits LP. Evaluating DNA degradation rates in faecal pellets of the endangered pygmy rabbi. Mol Ecol Resour. 2013;13:654–62.

45. Hu YB, Nie YG, Wei W, Ma TX, Horn VR, Zheng XG, Swaisgood RR, Zhou ZX, Zhou WL, Yan L, Zhang ZJ, Wei FW. Inbreeding and inbreeding avoidance in wild giant pandas. Mol Ecol. 2017;26:5793–806.

46. Agetsuma YY, Inoue E, Agetsuma N. Effects of time and environmental conditions on the quality of DNA extracted from fecal samples for genotyping of wild deer in a warm temperate broad-leaved forest. Mammal Res. 2017;62:1–7.

47. Beja-Pereira A. OliveiraR, Alves PC, Schwartz MK, Luikart G. advancing ecological understandings through technological transformations in noninvasive genetics. Mol Ecol Resour. 2009;9:1279–301.

48. Reddy PA, Bhavanishankar M, Bhagavatula J, Harika K, Mahla RS, Shivaji S. Improved methods of carnivore faecal sample preservation, DNA extraction and quantification for accurate genotyping of wild tigers. PLoS One. 2012;7: e46732.

49. Sambrook J, Fritsch EF, Maniatis T. Molecular cloning: a laboratory manual. 2nd ed. Cold Spring Harbor: Cold Spring Harbor Laboratory Press; 1989.

19

Simultaneous hydrolysis with lipase and fermentation of rapeseed cake for iturin A production by *Bacillus amyloliquefaciens* CX-20

Wenchao Chen[1,2,4,5], Xuan Li[1], Xuli Ma[1], Shouwen Chen[3], Yanping Kang[1], Minmin Yang[1], Fenghong Huang[1,2,4,5] and Xia Wan[1,2,4,5]*

Abstract

Background: Rapeseed cake (RSC), as the intermediate by-product of oil extraction from the seeds of *Brassica napus*, can be converted into rapeseed meal (RSM) by solvent extraction to remove oil. However, compared with RSM, RSC has been rarely used as a raw material for microbial fermentation, although both RSC and RSM are mainly composed of proteins, carbohydrates and minerals. In this study, we investigated the feasibility of using untreated low-cost RSC as nitrogen source to produce the valuable cyclic lipopeptide antibiotic iturin A using *Bacillus amyloliquefaciens* CX-20 in submerged fermentation. Especially, the effect of oil in RSC on iturin A production and the possibility of using lipases to improve the iturin A production were analyzed in batch fermentation.

Results: The maximum production of iturin A was 0.82 g/L at the optimal initial RSC and glucose concentrations of 90 and 60 g/L, respectively. When RSC was substituted with RSM as nitrogen source based on equal protein content, the final concentration of iturin A was improved to 0.95 g/L. The production of iturin A was further increased by the addition of different lipase concentrations from 0.1 to 5 U/mL into the RSC medium for simultaneous hydrolysis and fermentation. At the optimal lipase concentration of 0.5 U/mL, the maximal production of iturin A reached 1.14 g/L, which was 38.15% higher than that without any lipase supplement. Although rapeseed oil and lipase were firstly shown to have negative effects on iturin A production, and the effect would be greater if the concentration of either was increased, their respective negative effects were reduced when used together.

Conclusions: Appropriate relative concentrations of lipase and rapeseed oil were demonstrated to support optimal iturin A production. And simultaneous hydrolysis with lipase and fermentation was an effective way to produce iturin A from RSC using *B. amyloliquefaciens* CX-20.

Keywords: Rapeseed cake, Rapeseed oil, Lipase, Iturin A, *Bacillus amyloliquefaciens*

Background

In conventional agricultural cultivation, easy-to-use chemical pesticides and fungicides are commonly used to control plant diseases, which are a major threat to food security worldwide. However, the overuse of these chemicals has raised great concerns due to their side effects such as environmental contamination and harm to human health by the presence of chemical residues in

food [1]. Biological control offers an eco-friendly and sustainable alternative and is become increasingly prevalent in modern agriculture. Among a variety of microorganisms, the genus *Bacillus* has been widely used as biological control agent due to its production of antimicrobial substances such as the cyclic lipopeptides iturin, fengycin and surfactin [2]. Iturin A, a prominent member of the iturin group, was found to suppress many plant diseases via a combination of its broad-spectrum antifungal activity and the activation of plant defense systems [3]. This makes iturin A an ideal potential biological control agent for reducing the use of artificial chemical pesticides in agriculture. In recent years,

* Correspondence: wanxia@oilcrops.cn
[1]Oil Crops Research Institute of the Chinese Academy of Agricultural Sciences, Wuhan 430062, People's Republic of China
[2]Key Laboratory of Biology and Genetic Improvement of Oil Crops, Ministry of Agriculture, Wuhan 430062, People's Republic of China

many attempts have been made to use agricultural by-products, such as rapeseed meal (RSM) [4, 5], soybean curd residue [6] and fish meal [7], to realize low-cost and large-scale production of iturin A. For example, Jin et al. [4, 5] used untreated RSM as the nitrogen source for the production of iturin A using *B. subtilis* and enhanced the production two-fold by using a two-stage glucose feeding strategy in liquid fermentation. Yao et al. [8] used *B. subtilis* to co-produce iturin A and poly-γ-glutamic acid from RSM in solid-state fermentation.

Rapeseed cake (RSC) and RSM are the major by-products in the production of rapeseed oil and are mainly composed of proteins, carbohydrates and minerals. Both are good protein resources for animal feed, with a favorable balance of essential and sulfur-containing amino acids [9]. The production process of RSM from RSC involves solvent extraction to remove oil and heating to remove the organic solvent. Although this procedure increases the protein content of RSM compared with RSC, and RSC only contains 30–40% crude protein in dry matter [10], RSC is considered more suitable for animal diets than solvent-extracted RSM by many scholars. The reasons are as follows: (1) the production technology of RSC is more environmentally friendly [11], (2) RSC has a higher metabolizable energy value due to its higher residual oil content [12], and (3) organic animal production precludes the use of oilseed meal due to the exposure to chemical solvents during the extraction [13]. However, compared with RSM [4, 5, 14–18], there are fewer reports on the use of RSC as a nutrient for microbial fermentations. In addition to obtaining more oil and profits by chemical solvent-extraction, one of the key factors is the high content of oil in RSC, which can influence the titer and productivity of some secondary metabolites [19]. Therefore, if the problem of high oil content in RSC can be solved, it is likely that RSC can be used for microbial production similarly to RSM.

Lipases, which not only hydrolyze triacylglycerides to form glycerol and fatty acids, but can also catalyze the synthesis of esters under certain conditions [20], constitute one of the most important families of industrial enzymes. Following carbohydrases and proteases, lipases are considered the third largest group of enzymes based on total sales volume [21], and are widely applied in the manufacture of foods, fine chemicals, cosmetics, pharmaceuticals, detergents, wastewater treatment, leather processing and biomedical assays [22]. However, unlike carbohydrases and proteases [16], lipases have not been widely used for

the pretreatment of rapeseed by-products to improve their value for microbial production. It has been reported that RSC could be used as a valuable raw material for producing lipases and proteases due to its high content of lipids and proteins [23]. Therefore, it seems feasible to use lipases to solve the problem of high oil content in RSC.

RSM has been demonstrated as a more effective nitrogen source than two different commercial nitrogen sources for iturin A production by *Bacillus* [4]. However, to our best knowledge, there are no reports on the production of iturin A by the fermentation of RSC. In this study, we investigated the feasibility of using untreated low-cost RSC as nitrogen source to produce the valuable cyclic lipopeptide antibiotic iturin A using *B. amyloliquefaciens* CX-20 in submerged fermentation. Especially, the effect of oil in RSC on iturin A production and the possibility of using lipases to improve the iturin A production were analyzed in batch fermentation.

Methods
Raw material and enzyme
RSC and RSM used in the experiments were kindly supplied by the Oil Crops Research Institute of the Chinese Academy of Agricultural Sciences (Wuhan, China) and were milled in a commercial plant. The composition of RSC and RSM flour was determined and was shown in Table 1.

The lipase used in this study was kindly supplied by Wuhan Sunhy Biology Co., Ltd. (China). The product is a lipase mixture used as a feed additive with a nominal activity of 10,000 U/g.

Microorganisms and media
B. amyloliquefaciens CX-20 (CCTCC No. M2018794), a iturin A high- production strain, was kindly provided by professor Shouwen Chen (College of Life Sciences, Hubei University, Wuhan, China). Luria-Bertani (LB) medium (10 g tryptone, 5 g yeast extract, 10 g NaCl, and 1 L H_2O) was used for seed cultures of *Bacillus*. The fermentation medium was composed of 60 g glucose, 1 g $K_2HPO_4 \cdot 3H_2O$, 0.5 g $MgSO_4 \cdot 7H_2O$, 0.005 g $MnSO_4 \cdot H_2O$, 90 g RSC or 76.52 g RSM, and 1 L H_2O. Flask experiments were performed in 250 mL flasks with 20 mL of medium. Due to the insolubility of RSM/RSC, 1.80 g RSC or 1.53 g RSM was weighed and placed into each 250 mL flask in advance, respectively. And then mix with distilled water containing above mentioned concentrations of inorganic salts and glucose to a final volume of 20 mL. The initial pH of the medium was

Table 1 The components of RSM and RSC

	Moisture (%)	Ash (%)	Crude Protein (%)	Crude Fat (%)	Crude fiber (%)	Neutral detergent fiber (%)	Acid detergent fiber (%)
RSM	8.3	7.3	39.4	1.6	8.8	26.6	12.8
RSC	4.7	7.0	33.5	14.4	7.1	16.6	11.7

adjusted to 7.0 and it was autoclaved at 121 °C for 30 min. The inoculum size was 5% (v/v). All fermentations were carried out at 28 °C under constant orbital shaking at 220 rpm.

Simultaneous hydrolysis and fermentation

Batch submerged fermentation experiments were performed in 250 mL flasks with an initial working volume of 20 mL. The medium for submerged fermentation contained (per liter): 60 g glucose, 1 g $K_2HPO_4 \cdot 3H_2O$, 0.5 g $MgSO_4 \cdot 7H_2O$, 0.005 g $MnSO_4 \cdot H_2O$, 90 g RSC or 76.52 g RSM. Due to the insolubility of RSM/RSC, 1.80 g RSC was weighed and placed into each 250 mL flask in advance, respectively. And then mix with distilled water containing above mentioned concentrations of inorganic salts and glucose to a final volume of 20 mL. Submerged fermentation was started by adding 5% (v/v) of exponentially growing cells and lipase. Submerged fermentation was conducted at 28 °C with a rotating speed of 220 rpm. The effects of lipase loading (0, 0.1, 0.5, 1, 5, 10 U/mL) and rapeseed oil concentrations (0, 6, 12 and 24 g/L) on submerged fermentation were investigated.

Extraction and quantitation of iturin A

Iturin A was extracted according to a reported method [4, 5, 24]. Briefly, 0.3 mL of the mixed fermentation broth was added into a 2 mL glass tube containing 1.2 mL of methanol, shaken well and incubated for 60 min. The mixture was centrifuged at 12,000 rpm for 20 min, and the supernatant was filtered through a 0.22 μm pore-size hydrophobic polytetrafluoroethylene (PTFE) type disposable syringe. The iturin A concentration was quantified using a Waters 2695 HPLC system equipped with an ACQUITY UPLC BEA C18 column (1.7 μm 2.1 × 100 mm, Waters, USA). A mixture of acetonitrile and 10 mM ammonium acetate (35:65, v/v) was used as the mobile phase at a flow rate of 0.3 mL/min, and the elution was monitored at 210 nm. The concentration of iturin A was analyzed and quantified using an authentic reference standard (Sigma Chemicals, St. Louis, MO, USA). The content of iturin A was measured using triplicate samples.

Determination of reducing sugar, free ammonium nitrogen concentrations and the viable cells

The concentrations of reducing sugar in the fermentation were determined by the DNS method using 3, 5-dinitrosalicylic acid reagent [25]. The concentration of free ammonium nitrogen (FAN) was determined using the ninhydrin colorimetric method [26]. The viable cell count during submerged fermentation was determined as follows: 0.5 mL of the sample was placed into a sterile 10 mL test tube, mixed thoroughly with 4.5 ml of sterile distilled water and shaken at 150 rpm on a vortex for 5 min at room temperature. The mixture was then serially

diluted and spread onto LB agar plates. After 24 h of incubation at 28 °C, the number of colonies was counted and expressed as colony forming units (CFU).

Determination of the iturin A stability to lipase

First, submerged fermentation were performed in 250 mL flasks with 20 mL medium, containing (per liter): 60 g glucose, 1 g $K_2HPO_4 \cdot 3H_2O$, 0.5 g $MgSO_4 \cdot 7H_2O$, 0.005 g $MnSO_4 \cdot H_2O$, and 76.52 g RSM. After 72 h of fermentation, the fermentation broth was divided into two groups. Therein, one group supplemented with 5 U/mL lipase while the other without lipase addition. Then submerged fermentation was continued to be conducted at 28 °C with a rotating speed of 220 rpm. And the concentrations of iturin A at different time (0, 24, 48 and 72 h) were measured.

Statistical analysis

All experiments were performed in triplicate and the data were processed using Origin v8.6 software (Origin Lab Corp., Northampton, MA, USA).

Results

Influence of initial RSC and glucose concentrations on the production of iturin A

In our previous study [4], the optimal initial RSM and glucose concentrations for iturin A production by *B. subtilis* 3–10 in submerged fermentation were found to be 90 and 20 g/L, respectively. Therefore, 20 g/L initial glucose was used to test the influence of initial RSC concentration on iturin A production by *B. amyloliquefaciens* CX-20. As shown in Fig. 1a, the maximum iturin A final concentration 0.39 g/L was obtained when the initial RSC concentration was 90 g/L. With the increase of RSC concentration from 30 to 90 g/L, the production of iturin A increased. However, the iturin A production started to decrease upon further increase of RSC concentration from 90 to 150 g/L. An interesting phenomenon was that the final reducing sugar concentrations increased almost linearly with the initial RSC concentrations. This might be positively related to the reducing sugar released from RSC during fermentation process. The initial reducing sugar concentration slightly decreased, which was speculated to be related to the Maillard reaction caused by the high-temperature sterilization due to the covalent bonds formed between a free reactive -NH2 group of an amino acid and the carbonyl group of a reducing sugar [27]. The concentration of final free ammonium nitrogen (FFAN) also increased linearly from 162.50 to 1266.95 mg/L with the RSC concentration increasing from 30 to 150 g/L. Compared with the FFAN concentration, the initial free ammonium nitrogen (IFAN) concentration increased slightly (Fig. 1b). This indicated that *B. amyloliquefaciens* CX-20 had a strong ability to hydrolyze the insoluble nitrogen source in RSC to produce soluble form, which was not

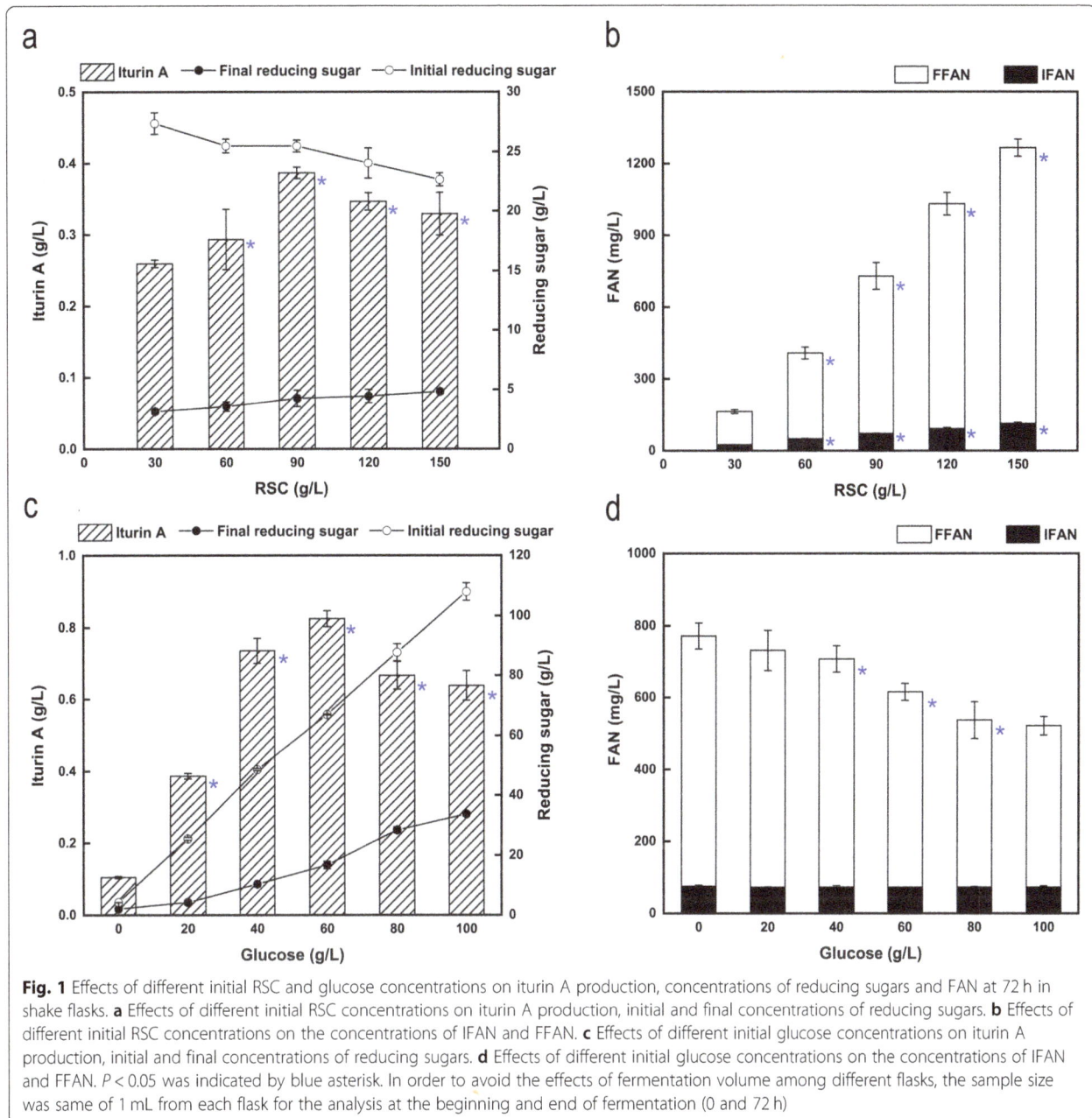

Fig. 1 Effects of different initial RSC and glucose concentrations on iturin A production, concentrations of reducing sugars and FAN at 72 h in shake flasks. **a** Effects of different initial RSC concentrations on iturin A production, initial and final concentrations of reducing sugars. **b** Effects of different initial RSC concentrations on the concentrations of IFAN and FFAN. **c** Effects of different initial glucose concentrations on iturin A production, initial and final concentrations of reducing sugars. **d** Effects of different initial glucose concentrations on the concentrations of IFAN and FFAN. $P < 0.05$ was indicated by blue asterisk. In order to avoid the effects of fermentation volume among different flasks, the sample size was same of 1 mL from each flask for the analysis at the beginning and end of fermentation (0 and 72 h)

surprisingly considering the strong intrinsic protease activity of many *Bacillus* [4]. Genome and transcriptome analysis of *B. amyloliquefaciens* CX-20 demonstrated that it could not only produce proteases that hydrolyze proteins into peptides and amino acids, but also phytase, xylanase, cellulase and lipase enzyme, which was similar to *Aspergillus oryzae* and could result in the release of phosphate and the production of simple sugars to be used as carbon source for the growth of the microorganisms [28]. Therefore, *Bacillus* has intrinsic advantages for direct bio-utilization of RSC for the production of microbial metabolites [4, 5].

According to above results, the initial RSC concentration of 90 g/L was fixed to determine the optimum concentration of glucose and further improve the production of iturin A. Subsequently, the effects of different glucose concentrations ranging from 0 to 100 g/L on iturin A production were explored. It was clearly demonstrated that the optimal initial glucose concentration for iturin A production was 60 g/L, and the corresponding maximum iturin A concentration reached 0.82 g/L. Nevertheless, 0.10 g/L of iturin A was still produced even without adding glucose (Fig. 1c). As a complex mixture, RSM can not only be used as a nitrogen source [4, 5, 15, 17, 29], but can also provide carbon [14, 16] for the growth and metabolism of microorganisms. According to our results, RSC could also be used as both a carbon source and a nitrogen source. As a carbon source, RSC seemed to be efficient, since the final concentration of reducing sugars was 2.05 g/L from initial 4.22 g/L after 72 h fermentation. This further proved that *B. amyloliquefaciens* CX-20 could produce many enzymes to hydrolyze the carbohydrates of RSC into reducing sugars. It has been reported that rapeseed oil could be used as a source of carbon to ferment microbial products such as lipase [23], erythromycin [30] and isocitric acid [31]. However, whether the residual oil in RSC could be used as a carbon source for iturin A production by *B. amyloliquefaciens* CX-20 was still unclear. But RSC was insufficient to support the fermentation and synthesis of iturin A by *B. amyloliquefaciens* CX-20 as the sole carbon source. Accordingly, with the increase of initial glucose concentration from 0 to 60 g/L, the production of iturin A continued rising, but decreased when the initial glucose concentration was raised above 60 g/L. Although the final concentration of reducing sugars increased with the increase of initial glucose concentration, the change trend of the FFAN concentration (decreased from 771 to 522 mg/L) was opposite (Fig. 1d). This indicated that higher glucose or RSC concentrations could mutually promote the corresponding substrate consumption, but might not be necessary for improving iturin A production.

Influence of lipase loading on the production of iturin A from RSC

Different concentrations of lipase (ranging from 0 to 10 U/mL) were added into the medium containing the optimal concentrations of 90 g/L RSC and 60 g/L glucose at the beginning of the process to enable simultaneous hydrolysis and fermentation. As shown in Fig. 2, with the increase of lipase concentration from 0 to 0.5 U/mL, the production of iturin A gradually increased. However, with the further increase of lipase concentration (from 0.5–10 U/mL), the production of iturin A began to decrease. When the concentration of lipase was 0.5 U/mL, the iturin A production reached a maximum of 1.14 g/L, which represented a 38.15% increase over the fermentation without any lipase addition. By contrast, when the concentration of lipase reached 10 U/mL, the final concentration of iturin A decreased to 0.59 g/L, which was even 27.94% lower than that without any lipase addition. The change trend of the final reducing sugar concentration was similar to that of iturin A production. Lipases are a group of enzymes that hydrolyze the ester bonds in triacylglycerides to form fatty acids and glycerol, or catalyze the synthesis of esters under certain conditions [32]. The lipase used in this study was a mixture used as a feed additive to aid animal digestion. Therefore, we speculated that the appropriate addition of lipase might help hydrolyze the residual oil in RSC into fatty acids and glycerol. Notably, glycerol has been proved to be a more suitable carbon source for iturin A production than glucose (data not shown). This helps explain why the FFAN concentrations with added lipase was lower than that without lipase, since the released glycerol could also promote the consumption of the nitrogen source (Fig. 2b). However, excess lipase had a negative effect on the production of iturin A.

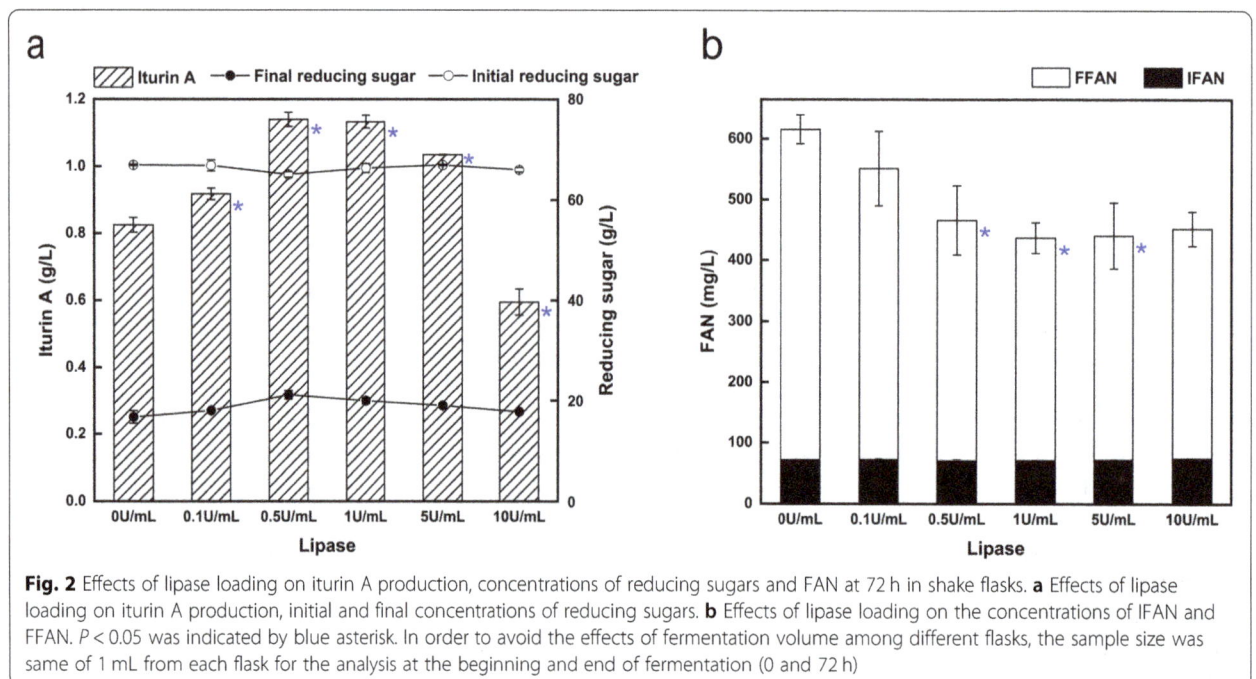

Fig. 2 Effects of lipase loading on iturin A production, concentrations of reducing sugars and FAN at 72 h in shake flasks. **a** Effects of lipase loading on iturin A production, initial and final concentrations of reducing sugars. **b** Effects of lipase loading on the concentrations of IFAN and FFAN. $P < 0.05$ was indicated by blue asterisk. In order to avoid the effects of fermentation volume among different flasks, the sample size was same of 1 mL from each flask for the analysis at the beginning and end of fermentation (0 and 72 h)

Influence of lipids and lipase on the fermentation of iturin A

As shown in Table 1, the content of protein in RSC was 33.5% while the protein content in RSM was 39.4%. Therefore, the protein content in 90 g/L RSC was equal to that of 76.52 g/L RSM, and the latter was added to substitute 90 g/L RSC to simulate the protein content in the original medium. Because the crude fat content of RSM was 1.6% while that of RSC was 14.4%, the crude fat content of the medium containing 90 g/L RSC was 12.96 g/L, while that of the medium containing 76.52 g/L RSM was 1.22 g/L. The difference of crude fat between the two media was 11.74 g/L. In order to explore the effect of crude fat on iturin A production, 0, 6, 12 or 24 g/

L of natural rapeseed oil was added into the medium containing 76.52 g/L RSM (Fig. 3a). After 72 h of fermentation, the final concentration of iturin A without added oil was 0.95 g/L, which was 15.56% higher than that of the 90 g/L RSC medium. With the increase of oil concentration, the trend of iturin A production was gradually decreasing. When 12 g/L rapeseed oil was added into the medium, the iturin A production decreased to 0.79 g/L. Although the production of iturin A decreased 17.46% compared to that produced without any oil addition, its value was very close to that produced with 90 g/L RSC (0.82 g/L). When 6 g/L rapeseed oil was added to the medium, the iturin A production decreased to 0.93 g/L, which was very close to the value obtained without adding

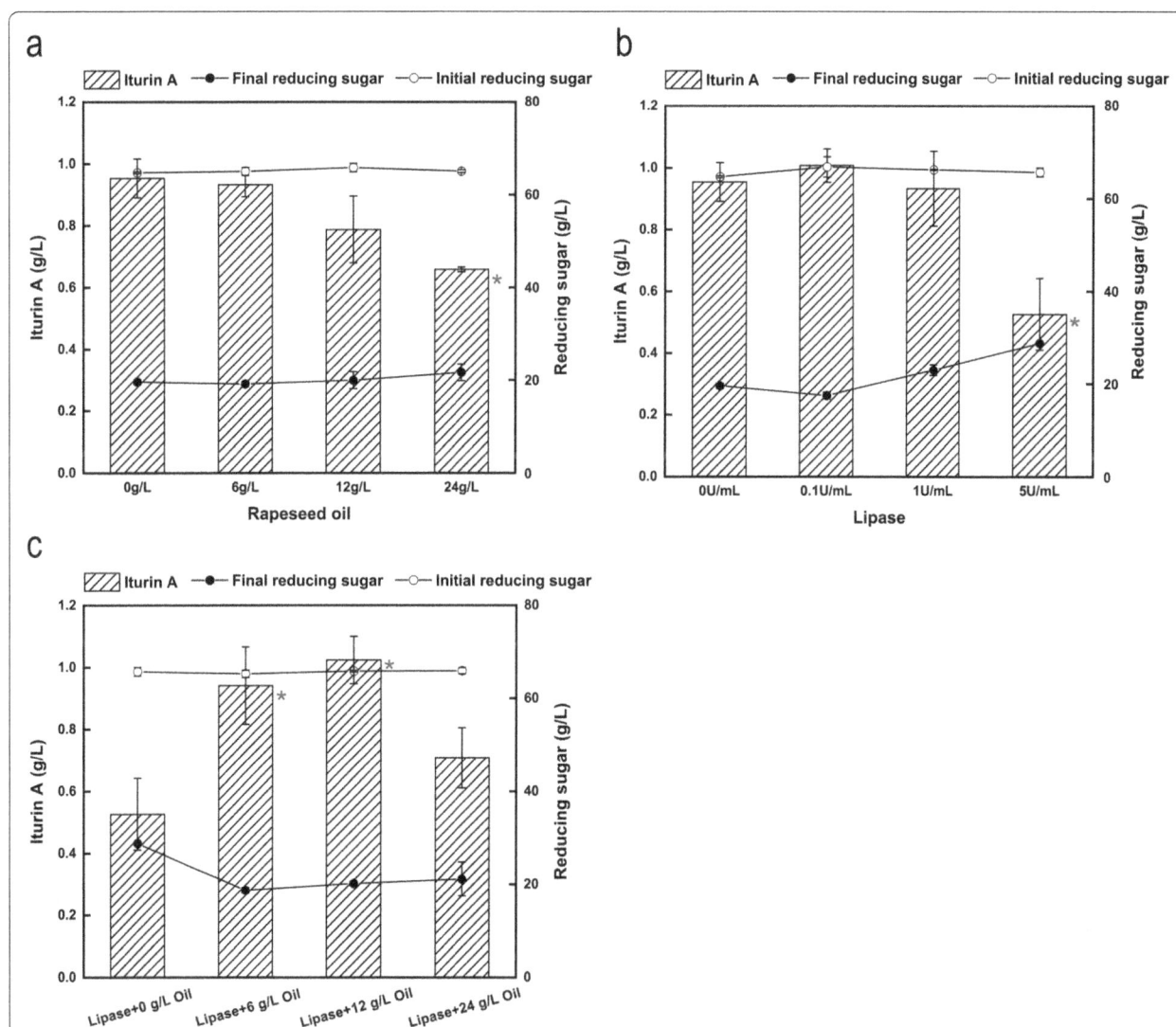

Fig. 3 Effects of different rapeseed oil and lipase concentrations. **a** Effects of different rapeseed oil concentrations on iturin A production, initial and final concentrations of reducing sugars. **b** Effects of different initial lipase concentrations on iturin A production, initial and final concentrations of reducing sugars. **c** Effects of different ratio of lipase and rapeseed oil on iturin A production, initial and final concentrations of reducing sugars. $P < 0.05$ was indicated by blue asterisk. In order to avoid the effects of fermentation volume among different flasks, the sample size was same of 1 mL from each flask for the analysis at the beginning and end of fermentation (0 and 72 h)

oil. Because the lipopeptide products possess surfactin activity and cause foam formation, it is difficult to control the fermentation, which also restricts the industrialization of lipopeptide production [2]. Rapeseed oil has been found to be an efficient antifoam compound [33]. Our results also demonstrated that when the concentration of added rapeseed oil was lower than 0.6%, there was almost no negative effect on iturin A production. Therefore, rapeseed oil seemed to also be a suitable antifoam compound for iturin A production. However, when 24 g/L rapeseed oil was added into the medium, the iturin A production decreased to 0.66 g/L.

From the results shown in Fig. 2, we found that an appropriate concentration of lipase could improve the production of iturin A. Conversely, its production would be reduced when the concentration of lipase was too high. The oil content of RSM was lower than that of RSC (Table 1). Therefore, in theory, the optimal concentration of lipase for RSM should decrease accordingly. As expected, when the concentration of lipase was 0.1 U/mL, the iturin A production had a slight increase, from 0.95 g/L to 1.01 g/L, but when the lipase concentration was increased to 1 U/mL, the final concentration of iturin A was slight lower than that without any lipase added (0.95 g/L vs. 0.93 g/L). When the lipase concentration was increased to 5 U/mL, the final concentration of iturin A drastically decreased to 0.53 g/L, which was only 55.14% of that produced without any added lipase (Fig. 3b). At the same time, the final reducing sugar concentration gradually increased with the increase of lipase concentration when the lipase concentration was more than 0.1 U/mL.

Although the production of iturin A decreased slightly when RSC was used as nitrogen source if the lipase concentration was increased to 5 U/mL, the production of iturin A apparently decreased when RSM containing an equal protein content was used as the nitrogen source. Therefore, proper proportions of rapeseed oil and lipase appeared to be crucial for optimal iturin A production. As shown in Fig. 3b, a lipase concentration of 5 U/mL had a significant influence on the production of iturin A from RSM, and was chosen to explore the appropriate ratio of rapeseed oil to lipase. As shown in Fig. 3c, with the increase of rapeseed oil concentration from 0 to 12 g/L, iturin A production continued rising and reached a maximum of 1.02 g/L, which was almost equal to the iturin A production (1.03 g/L) produced from 90 g/L RSC with 5 U/mL lipase, when the addition of rapeseed oil was 12 g/L. However, when the concentration of rapeseed oil was increased to 24 g/L, the final concentration of iturin A decreased to 0.61 g/L, which was only 59.42% of that obtained with 12 g/L rapeseed oil.

There are many possible explanations why lipase and rapeseed oil could reduce each other's negative effects on iturin A production. The most likely one is related to microbial growth. Therefore, we examined the growth curves of Bacillus under several representative conditions. As shown in Fig. 4, when 12 g/L rapeseed oil or 5 U/mL lipase was added separately into the medium containing 76.52 g/L RSM, both the specific growth rates (0.56 and 0.52 h^{-1} from 1.03 h^{-1}) and the ultimate maximum viable cell count (about 8×10^8 and 3×10^8 mL^{-1} from about 8×10^9 mL^{-1}) were significantly reduced compared with the medium without either rapeseed oil or lipase. However, when 12 g/L rapeseed oil and 5 U/mL lipase were added at the same time, both the specific growth rate (0.80 h^{-1}) and the ultimate maximum viable cell count (about 2×10^9 mL^{-1}) showed an obvious recovery.

Discussion

The feasibility of using RSM as nitrogen sources in microbial fermentation processes has been reported [4, 5, 14–18]. However, as the low-cost by-product of the processing of oil crops, RSC also contains a high content of oil in addition to its protein content. Whether the residual oil in RSC might affect the production of microbial metabolites including iturin A, and whether lipase could be useful for improving their productions, were still not clear.

In this study, the feasibility of using untreated RSC as nitrogen source to produce the valuable cyclic lipopeptide antibiotic iturin A using B. amyloliquefaciens CX-20 in submerged fermentation was first investigated. The maximum production of iturin A was 0.82 g/L at the optimal initial RSC and glucose concentrations of 90 and 60 g/L, respectively. However, when RSC was substituted with

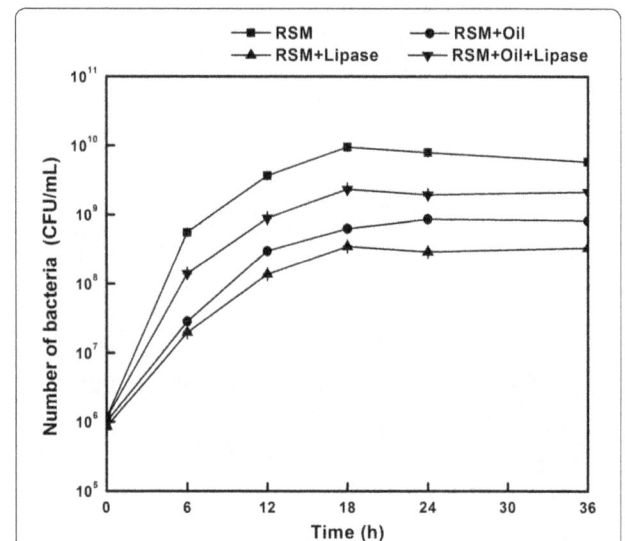

Fig. 4 Effects of 12 g/L rapeseed oil and 5 U/mL lipase on the growth of Bacillus amyloliquefaciens. In order to avoid the effects of fermentation volume among different flasks, the sample size was same of 0.5 mL from each flask for the analysis at 0, 6, 12, 18, 24 and 36 h, respectively

RSM as nitrogen source based on equal protein content, the final concentration of iturin A was improved to 0.95 g/L. Excess rapeseed oil over 6 g/L would suppress the production of iturin A in *B. amyloliquefaciens* CX-20. Kamzolova et al. [31] used rapeseed oil as a source of carbon and energy to produce isocitric acid in the unconventional yeast *Yarrowia lipolytica*. The first step of utilization of rapeseed oil by the yeast was its hydrolysis by extracellular lipases that produce glycerol and fatty acids. Moreover, the fatty acid profile of rapeseed oil (%, by mass) was found to be C16:0, 4.0; C18:0, 1.2; C18:1, 58.8; C18:2, 28.1; C18:3, 5.9 with a total unsaturated fatty acid mass fraction of 93.6% [34]. Therein, palmitic acid was proved to be a useful precursor whose addition could enhance the production of iturin A [19]. Oils are the essential components of industrial fermentation media and have been routinely added to media for the production of secondary metabolites. They have been used as antifoams, sole carbon sources, auxiliary carbon sources, to provide precursors for antibiotic synthesis and to remove the antibiotic from bacterial access and reduce its suppressive effect on antibiotic production [30]. However, it was considered possible that oil might form a thin film on the surface of the medium, decreasing the oxygen dissolution efficiency, which may influence the production of iturin A [35]. Wu et al. [19] verified the negative effects on iturin A production after the addition of either palm or soybean oil. Our results confirmed that rapeseed oil also had a negative effect on iturin A production, which might be a key factor influencing the utilization of RSC.

The nutrients in rapeseed oil by-products cannot be directly assimilated by the majority of industrial microorganisms without pretreatment due to its particular physical and chemical structure [27, 36]. Although many studies investigated ways of using enzymes such as proteases, cellulase and viscozyme to enhance the value of nitrogen from the rapeseed by-products for microbial fermentations [16, 29], we are not aware of any studies on utilizing lipase to improve the substrates' value for microbial production. To our best knowledge, this might be the first study on directly using lipase for the production of iturin A from rapeseed oil by-products. At the optimal lipase concentration of 0.5 U/mL, the maximal production of iturin A from RSC reached 1.14 g/L, which was 38.15% higher than that without any lipase supplement. The experiments of RSC substitution with RSM based on equal protein content further proved that excess concentration of rapeseed oil or lipase would have a negative effect on iturin A production. However, the proper ratio of these two elements for simultaneous fermentation by *B. amyloliquefaciens* CX-20 would mitigate the effects of each, and could even boost the amount of iturin A produced from RSC. These variation trends seemed similar to that of microbial growth.

Therefore, we concluded that the negative effect caused by rapeseed oil or lipase might be related to their corresponding effects on cell growth, although this was likely not the only reason. After all, when both were added at the same time, although the growth rate and the ultimate maximum viable count were still lower than that of the control, the production of iturin A was slightly higher. This was consistent with the study by Jin et al. [4] whose research demonstrated that ammonium nitrate was a good nitrogen source for *Bacillus* growth while it was not suitable for iturin A production. More work would be needed to further elucidate this phenomenon.

Iturin A consisted of a cyclic heptapeptide linked to a 14–17 carbons β-hydroxy fatty-acid chain. Therefore, it was speculated that lipase might cleave the fatty acid chain of iturin A so that reduced iturin A production. However, as shown in Fig. 5, when the deviations were taken into consideration, the addition of excess lipase (5 U/mL) after 72 h fermentation had almost no effect on the stability of iturin A from RSM, compared with that without any lipase supplement. Although lipopeptides have been reported to have excellent thermal and chemical stability [2], this might be the first study to investigate the good resistance of iturin A to lipase.

Conclusion

Compared with RSM, RSC has been rarely used as a raw material for microbial fermentation due to its high content of residual oil. This study not only demonstrated that the residual oil in RSC had a negative effect on iturin A production, but also provided an efficient

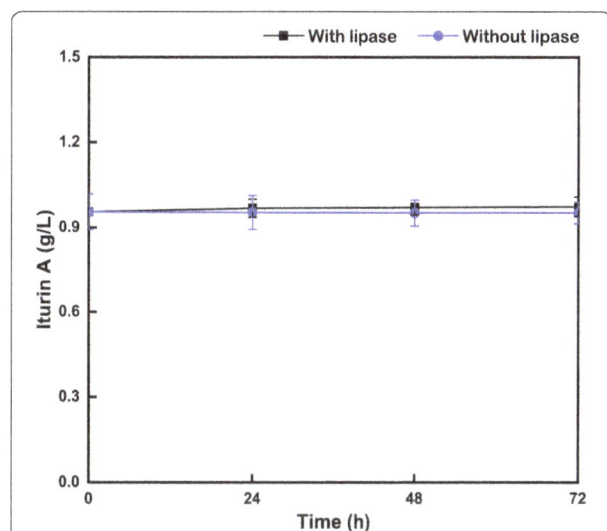

Fig. 5 Effects of excess lipase on the stability of iturin A. In order to avoid the effects of fermentation volume among different flasks, the sample size was same of 0.5 mL from each flask for the analysis at 0, 24, 48 and 72 h, respectively

means to solve this problem by adding a commercially-available feed-processing lipase for simultaneous hydrolysis and fermentation. When the optimal lipase concentration of 0.5 U/mL was added into the RSC medium, the final concentration of iturin A increased from 0.82 to 1.14 g/L, which was also higher than the 0.95 g/L produced in RSM medium containing an equal protein content. By using RSM and rapeseed oil to simulate RSC, excess rapeseed oil was proved to suppress the production of iturin A, which might be related to the oil forming a thin film on the surface of the medium, and thus decreasing the oxygen dissolution efficiency and cell growth. The proper supplementation ratio of the lipase for simultaneous hydrolysis and fermentation by *B. amyloliquefaciens* CX-20 was verified to mitigate the effect of rapeseed oil for the first time in this study. This research is important not only for improving the economic value of RSC and decreasing the production cost of iturin A, but also for establishing a new way to increase the value of rapeseed oil for microbial fermentation, either as antifoams, auxiliary carbon sources, or providing precursors for antibiotic synthesis.

Abbreviations
CFU: Colony forming units; FAN: Free ammonium nitrogen; FFAN: Final free ammonium nitrogen; IFAN: Initial free ammonium nitrogen; LB: Luria-Bertani; RSC: Rapeseed cake; RSM: Rapeseed meal

Acknowledgements
Not applicable

Authors' contributions
CWC, WX and HFH conceived and designed the experiments. CWC, LX, MXL, KYP, YMM performed experiments. CWC and LX analyzed the data and prepared the Figures and Tables and wrote the manuscript. WX, HFH and CSW corrected and proofread the manuscript. All authors read and approved the final manuscript.

Funding
The strain and enzyme acquisition, preliminary experiments and data analysis were funded by the Natural Science Foundation of Hubei Province (Grant Number 2019CFB378) and Ministry of Science and Technology of the People's Republic of China (Grant Number 2016YFD0501209). The experiment performance, data collection, analysis and interpretation of data were supported by the Chinese Academy of Agricultural Sciences (Grant Number CAAS-ASTIP-2016-OCRI).

Ethics approval and consent to participate
This article does not contain any studies with human participants or animals performed by any of the authors.

Consent for publication
Not applicable.

Competing interests
The authors declare that they have no competing interests.

Author details
[1]Oil Crops Research Institute of the Chinese Academy of Agricultural Sciences, Wuhan 430062, People's Republic of China. [2]Key Laboratory of Biology and Genetic Improvement of Oil Crops, Ministry of Agriculture, Wuhan 430062, People's Republic of China. [3]Hubei Collaborative Innovation Center for Green Transformation of Bio-Resources, Environmental Microbial Technology Center of Hubei Province, College of Life Sciences, Hubei University, Wuhan 430062, People's Republic of China. [4]Oil Crops and Lipids Process Technology National & Local Joint Engineering Laboratory, Wuhan 430062, People's Republic of China. [5]Hubei Key Laboratory of Lipid Chemistry and Nutrition, Wuhan 430062, People's Republic of China.

References
1. Khan N, Maymon M, Hirsch AM. Combating *Fusarium* infection using *Bacillus*-based antimicrobials. Microorganisms. 2017;5(4):E75.
2. Zhao H, Shao D, Jiang C, Shi J, Li Q, Huang Q, Rajoka MSR, Yang H, Jin M. Biological activity of lipopeptides from *Bacillus*. Appl Microbiol Biotechnol. 2017;101(15):5951–60.
3. Kawagoe Y, Shiraishi S, Kondo H, Yamamoto S, Aoki Y, Suzuki S. Cyclic lipopeptide iturin A structure-dependently induces defense response in *Arabidopsis* plants by activating SA and JA signaling pathways. Biochem Biophys Res Commun. 2015;460(4):1015–20.
4. Jin H, Zhang X, Li K, Niu Y, Guo M, Hu C, Wan X, Gong Y, Huang F. Direct bio-utilization of untreated rapeseed meal for effective iturin A production by *Bacillus subtilis* in submerged fermentation. PLoS One. 2014;9(10): e111171.
5. Jin H, Li K, Niu Y, Guo M, Hu C, Chen S, Huang F. Continuous enhancement of iturin A production by *Bacillus subtilis* with a stepwise two-stage glucose feeding strategy. BMC Biotechnol. 2015;15:53.
6. Mizumoto S, Hirai M, Shoda M. Enhanced iturin A production by *Bacillus subtilis* and its effect on suppression of the plant pathogen *Rhizoctonia solani*. Appl Microbiol Biotechnol. 2007;75(6):1267–74.
7. Zohora US, Rahman MS, Khan AW, Okanami M, Ano T. Improvement of production of lipopeptide antibiotic iturin A using fish protein. J Environ Sci (China). 2013;25(Suppl 1):S2–7.
8. Yao D, Ji Z, Wang C, Qi G, Zhang L, Ma X, Chen S. Co-producing iturin A and poly-γ-glutamic acid from rapeseed meal under solid state fermentation by the newly isolated *Bacillus subtilis* strain 3-10. World J Microbiol Biotechnol. 2012;28(3):985–91.
9. Shi C, He J, Yu J, Yu B, Huang Z, Mao X, Zheng P, Chen D. Solid state fermentation of rapeseed cake with *Aspergillus niger* for degrading glucosinolates and upgrading nutritional value. J Anim Sci Biotechnol. 2015;6(1):13.
10. Drazbo A, Ognik K, Zaworska A, Ferenc K, Jankowski J. The effect of raw and fermented rapeseed cake on the metabolic parameters, immune status, and intestinal morphology of turkeys. Poult Sci. 2018;97(11):3910–20.
11. Fang ZF, Peng J, Tang TJ, Liu ZL, Dai JJ, Jin LZ. Xylanase supplementation improved digestibility and performance of growing pigs fed Chinese double-low rapeseed meal inclusion diets: in vitro and in vivo studies. Asian-Australas J Anim Sci. 2007;11:1721–8.
12. Smulikowska S, Czerwiński J, Mieczkowska A. Effect of an organic acid blend and phytase added to a rapeseed cake-containing diet on performance, intestinal morphology, caecal microflora activity and thyroid status of broiler chickens. J Anim Physiol Anim Nutr (Berl). 2010;94(1):15–23.
13. Kaewtapee C, Mosenthin R, Nenning S, Wiltafsky M, Schäffler M, Eklund M, Rosenfelder-Kuon P. Standardized ileal digestibility of amino acids in European soya bean and rapeseed products fed to growing pigs. J Anim Physiol Anim Nutr (Berl). 2018;102(2):e695–705.
14. Almeida JM, Lima VA, Giloni-Lima PC, Knob A. Canola meal as a novel substrate for β-glucosidase production by *Trichoderma viride*: application of the crude extract to biomass saccharification. Bioprocess Biosyst Eng. 2015; 38(10):1889–902.
15. Chatzifragkou A, Papanikolaou S, Kopsahelis N, Kachrimanidou V, Dorado MP, Koutina AA. Biorefinery development through utilization of biodiesel industry by-products as sole fermentation feedstock for 1, 3-propanediol production. Bioresour Technol. 2014;159:167–75.
16. Chen KQ, Zhang H, Miao YL, Wei P, Chen JY. Simultaneous saccharification and fermentation of acid-pretreated rapeseed meal for succinic acid production using *Actinobacillus succinogenes*. Enzym Microb Technol. 2011; 48:339–44.
17. García IL, López JA, Dorado MP, Kopsahelis N, Alexandri M, Papanikolaou S, Villar MA, Koutinas AA. Evaluation of by-products from the biodiesel industry as fermentation feedstock for poly (3-hydroxybutyrateco-3-

hydroxyvalerate) production by *Cupriavidus necator*. Bioresour Technol. 2013;130:16–22.

18. Uckun Kiran E, Trzcinski A, Webb C. Microbial oil produced from biodiesel by-products could enhance overall production. Bioresour Technol. 2013;129:650–4.

19. Wu JY, Liao JH, Shieh CJ, Hsieh FC, Liu YC. Kinetic analysis on precursors for iturin A production from *Bacillus amyloliquefaciens* BPD1. J Biosci Bioeng. 2018;126(5):630–5.

20. Gupta R, Gupta N, Rathi P. Bacterial lipases: an overview of production, purification and biochemical properties. Appl Microbiol Biotechnol. 2004; 64(6):763–81.

21. Hasan F, Shah AA, Hameed A. Industrial applications of microbial lipases. Enzym Microb Technol. 2006;39:235–51.

22. Salihu A, Alam MZ, AbdulKarim MI, Salleh HM. Lipase production: an insight in the utilization of renewable agricultural residues. Resour Conserv Recycl. 2012;58:36–44.

23. Boratyński F, Szczepańska E, Grudniewska A, Gniłka R, Olejniczak T. Improving of hydrolases biosythesis by solid-state fermentation of *Penicillium camemberti* on rapeseed cake. Sci Rep. 2018;8(1):10157.

24. Chen W, Ma X, Wang X, Chen S, Rogiewicz A, Slominski B, Wan X, Huang F. Establishment of a rapeseed meal fermentation model for iturin A production by *Bacillus amyloliquefaciens* CX-20. Microb Biotechnol. 2019; 12(6):1417–29.

25. Miller GL. Use of dinitrosalicylic acid reagent for determination of reducing sugar. Anal Chem. 1959;31:426–8.

26. Lie S. The EBC-ninhydrin method for determination of free alpha amino nitrogen. J Inst Brew. 1973;79:37–41.

27. Salazar-Villanea S, Butré CI, Wierenga PA, Bruininx EMAM, Gruppen H, Hendriks WH, van der Poel AFB. Apparent ileal digestibility of Maillard reaction products in growing pigs. PLoS One. 2018;13(7):e0199499.

28. Wang RH, Shaarani SM, Godoy LC, Melikoglu M, Vergara CS, Koutinas A, Webb C. Bioconversion of rapeseed meal for the production of a generic feedstock. Enzym Microb Technol. 2010;47:77–83.

29. Uckun Kiran E, Salakkam A, Trzcinski AP, Bakir U, Webb C. Enhancing the value of nitrogen from rapeseed meal for microbial oil production. Enzym Microb Technol. 2012;50:337–42.

30. Hamedi J, Malekzadeh F, Saghafi-nia AE. Enhancing of erythromycin production by *Saccharopolyspora erythraea* with common anduncommon oils. J Ind Microbiol Biotechnol. 2004;31(10):447–56.

31. Kamzolova SV, Dedyukhina EG, Samoilenko VA, Lunina JN, Puntus IF, Allayarov RL, Chiglintseva MN, Mironov AA, Morgunov IG. Isocitric acid production from rapeseed oil by Yarrowia lipolytica yeast. Appl Microbiol Biotechnol. 2013;97(20):9133–44.

32. Ma RJ, Wang YH, Liu L, Bai LL, Ban R. Production enhancement of the extracellular lipase LipA in *Bacillus subtilis*: effects of expression system and sec pathway components. Protein Expr Purif. 2018;142:81–7.

33. Kougias PG, Tsapekos P, Boe K, Angelidaki I. Antifoaming effect of chemical compounds in manure biogas reactors. Water Res. 2013;47(16):6280–8.

34. Kamzolova SV, Allayarov RK, Lunina JN, Morgunov IG. The effect of oxalic and itaconic acids on threo-ds-isocitric acid production from rapeseed oil by *Yarrowia lipolytica*. Bioresour Technol. 2016;206:128–33.

35. Shih IL, Lin CY, Wu JY, Hsieh C. Production of antifungal lipopeptide from *Bacillus subtilis* in submerged fermentation using shake flask and fermentor. Korean J Chem Eng. 2009;26:1652–61.

36. Ramachandran S, Singh SK, Larroche C, Soccol CR, Pandey A. Oil cakes and their biotechnological applications--a review. Bioresour Technol. 2007;98:2000–9.

20

Optimized production of a biologically active *Clostridium perfringens* glycosyl hydrolase phage endolysin PlyCP41 in plants using virus-based systemic expression

Rosemarie W. Hammond[1]* ⓘ, Steven M. Swift[2], Juli A. Foster-Frey[2], Natalia Y. Kovalskaya[1,3] and David M. Donovan[2]

Abstract

Background: *Clostridium perfringens*, a gram-positive, anaerobic, rod-shaped bacterium, is the third leading cause of human foodborne bacterial disease and a cause of necrotic enteritis in poultry. It is controlled using antibiotics, widespread use of which may lead to development of drug-resistant bacteria. Bacteriophage-encoded endolysins that degrade peptidoglycans in the bacterial cell wall are potential replacements for antibiotics. Phage endolysins have been identified that exhibit antibacterial activities against several Clostridium strains.

Results: An *Escherichia coli* codon-optimized gene encoding the glycosyl hydrolase endolysin (*PlyCP41*) containing a polyhistidine tag was expressed in *E. coli*. In addition, The *E. coli* optimized endolysin gene was engineered for expression in plants (*PlyCP41p*) and a plant codon-optimized gene (*PlyCP41pc*), both containing a polyhistidine tag, were expressed in *Nicotiana benthamiana* plants using a potato virus X (PVX)-based transient expression vector. PlyCP41p accumulated to ~ 1% total soluble protein (100μg/gm f. wt. leaf tissue) without any obvious toxic effects on plant cells, and both the purified protein and plant sap containing the protein lysed *C. perfringens* strain Cp39 in a plate lysis assay. Optimal systemic expression of PlyCP41p was achieved at 2 weeks-post-infection. PlyCP41pc did not accumulate to higher levels than PlyCP41p in infected tissue.

Conclusion: We demonstrated that functionally active bacteriophage PlyCP41 endolysin can be produced in systemically infected plant tissue with potential for use of crude plant sap as an effective antimicrobial agent against *C. perfringens*.

Keywords: Alternative antimicrobial, Bacteriophage, Endolysin, *Nicotiana benthamiana*, Plant production of recombinant proteins, Plant virus-based gene expression, *Clostridium perfringens*, Potato virus X

Background

Clostridium perfringens is a Gram-positive, rod-shaped, spore-forming, anaerobic bacterium that is commonly found in the environment and is present in the intestines of animals and humans. The bacterium produces four major toxins and is the third leading cause of human foodborne illnesses; outbreaks are frequently associated with exposure to raw meat or poultry which has not been maintained properly [1, 2]. *C. perfringens* also causes gas gangrene in humans that have been subjected to severe injuries. In wild and domestic animals, it causes enteric diseases. In poultry, *C. perfringens* causes necrotic enteritis, characterized by necrotic lesions on the intestinal mucosa, which can be very costly to the poultry industry [3]. Control of clostridia in commercial poultry has commonly been by the feeding of sub-therapeutic amounts of antibiotics added to animal feed [4, 5], however concern that antibiotic resistance may

* Correspondence: rose.hammond@usda.gov
[1]USDA ARS NEA BARC Molecular Plant Pathology Laboratory, Beltsville, MD 20705, USA

develop from the continual use of antibiotics has led to reduced or banned use of antibiotics in some countries, resulting in increased cases of necrotic enteritis in poultry [6, 7]. Therefore, there is increasing interest in the development of alternative and specific antimicrobials to control *C. perfringens* and other bacterial animal pathogens.

Bacteriophage lysins are highly evolved, phage-encoded enzymes that hydrolyze peptidoglycans, the major structural component of bacterial cell walls. Bacteriophage and their derived lysins have been explored as tools to control bacterial infections [8–15]. Several bacteriophages of *C. perfringens* have been characterized [16] and putative phage endolysins have been identified to control *C. perfringens* [17]. Two recombinant, native lysins produced in and purified from *Escherichia coli*, Ply26F and Ply39O, lysed their parental *C. perfringens* host strains in addition to other strains of Clostridium, but did not lyse other bacterial species [18]. The demonstrated modular nature of endolysins [11, 19] led Swift et al. to design a thermally stable endolysin, a chimeric protein composed of the catalytic domain derived from an endolysin of the thermophilic bacteriophage, GVE2, fused to a cell wall binding domain derived from an endolysin of *C. perfringens* bacteriophage CP26F. The resulting protein, PlyG-VE2CpCWB, was active over a range of pH and salt conditions and was more resistant to elevated temperatures, demonstrating the ability to impart new properties to these catalytic enzymes [20].

To identify new lysins against *C. perfringens*, the genomes of 43 *C. perfringens* strains were searched for prophage regions predicted to encode endolysins. Sequence analysis and annotation resulted in the identification of a glycosyl hydrolase endolysin from the source strain Cp41, with the resulting endolysin designated PlyCP41 [21]. Bacterially-produced recombinant PlyCP41 lysed 75 strains of *C. perfringens*, which included isolates from chickens, pigs, and cows [21].

Plant production of antimicrobials is advantageous because of lower production costs, smaller risks of pathogen contamination, the ability to produce a large amount of protein, and their simplicity to produce and deliver in feed. Bacteriophage endolysins have been synthesized in stably transformed tobacco chloroplasts [22, 23], however this method is laborious and requires time and selection to identify transgene inheritance. Alternatively, plant virus-based transient expression, in which the inserted mRNA encoding a recombinant protein is replicated by the plant virus, can produce high levels of protein within a short period. For that reason, virus-based expression is an attractive alternative to transformation and has been used to produce many recombinant proteins, including antimicrobials and bacteriophage endolysins in *Nicotiana benthamiana* [24–28].

The purpose of this study was to produce PlyCP41 in plants (PlyCP41p) and to examine the activity of the purified protein and lysin-containing crude plant sap against *C. perfringens*. In our study, we compared the production of PlyCP41 in bacteria and plants and found that PlyCP41 was present in soluble bacterial fractions, eliminating the need for laborious re-solubilization and refolding steps required when recombinant proteins form inclusion bodies, leading to poor recoveries of active protein. PlyCp41p and the plant codon-optimized PlyCP41pc were expressed in plant sap at 1% total soluble protein in *N. benthamiana* leaves using a potato virus X (PVX)-based vector. PlyCP41 expressed both in bacteria and in plant tissues lysed *C. perfringens* in a plate lysis assay. In the future, phage lysins produced in plants could be added as lysates or dried plant tissue to animal feeds for reducing the bacterial colonization of the poultry gut to improve animal health and food safety.

Results

Expression and purification of recombinant PlyCP41 in bacteria

An expression construct encoding an *E. coli*-codon optimized gene was used to produce a histidine-tagged PlyCP41 in *E. coli* strain BL21(DE3). Analysis of protein fractions by SDS-PAGE following IPTG-induction revealed that PlyCP41 (335 amino acids, 38.5 kDa) was predominantly localized in the soluble fraction compared to a Lysin D (361 amino acids, 40.8 kDa) that was localized predominantly in the insoluble inclusion bodies (Fig. 1a). PlyCP41 was easily purified to almost complete homogeneity from the soluble fraction using a Ni-NTA resin under native conditions (Fig. 1b). The concentration of PlyCP41 in the second elution fraction was ~ 2 mg/mL. PlyCP41 also eluted in the wash buffers, resulting in a concentration of 100–200 μg/mL, and may have resulted from a wash stringency that was too high.

Expression of recombinant PlyCP41 in plant tissues

For plant expression, the *PlyCP41* gene was engineered into a PVX-plant virus expression vector, pGDPVXMCS. In the resulting construct, pGDPVXMCS: PlyCP41p, plant gene expression is under control of the *Cauliflower mosaic virus* 35S transcriptional promoter. Co-agroinfiltration of *A. tumefaciens* strain EHA105 containing pGDPVXMCS: PlyCP41p and pGDp19 into *N. benthamiana* leaves led to systemic PVX virus infection and synthesis of PlyCP41p protein in upper, systemically infected, symptomatic leaves at 9 days' post-infiltration (Fig. 2; CP41p#1), while there was no detectable PlyCP41p protein in an asymptomatic leaf of the same plant (Fig. 2; CP41p#2). pGDp19 encodes a plant virus-derived silencing suppressor protein that facilitates high

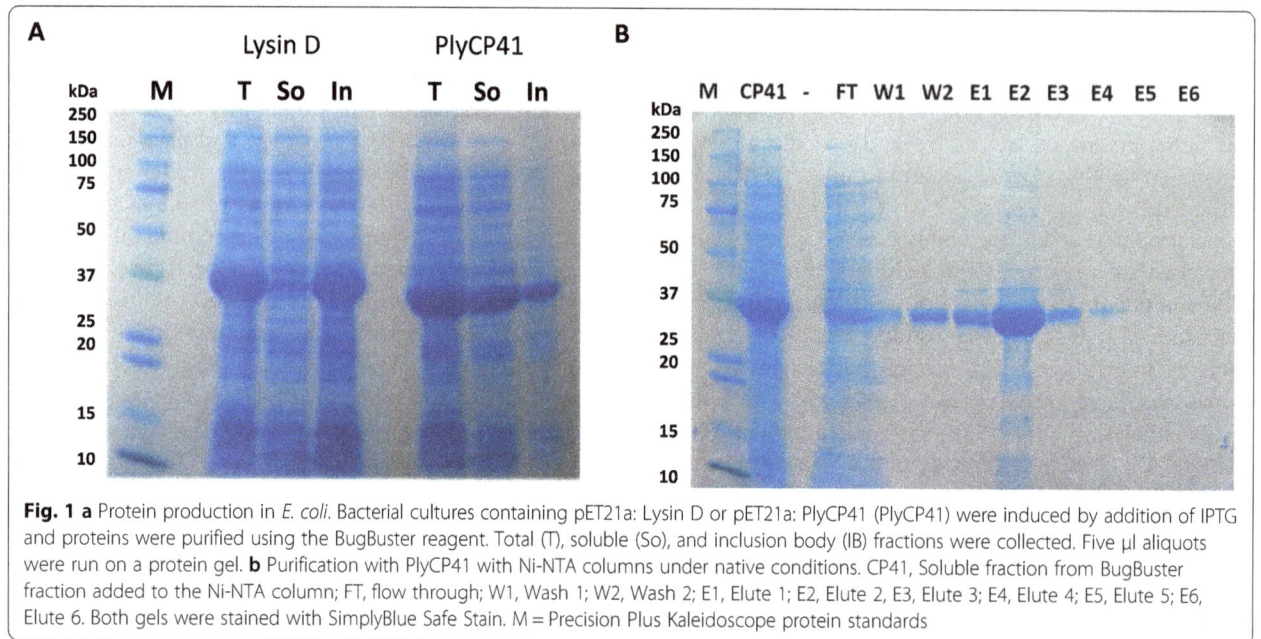

Fig. 1 a Protein production in *E. coli*. Bacterial cultures containing pET21a: Lysin D or pET21a: PlyCP41 (PlyCP41) were induced by addition of IPTG and proteins were purified using the BugBuster reagent. Total (T), soluble (So), and inclusion body (IB) fractions were collected. Five μl aliquots were run on a protein gel. **b** Purification with PlyCP41 with Ni-NTA columns under native conditions. CP41, Soluble fraction from BugBuster fraction added to the Ni-NTA column; FT, flow through; W1, Wash 1; W2, Wash 2; E1, Elute 1; E2, Elute 2, E3, Elute 3; E4, Elute 4; E5, Elute 5; E6, Elute 6. Both gels were stained with SimplyBlue Safe Stain. M = Precision Plus Kaleidoscope protein standards

levels of gene expression from the PVX construct [29]. There was no obvious phenotypic difference between plants infected with PVX and those infected with PVX containing the lysin insert (not shown).

Lytic activity of bacterial and plant-produced lysins

To ensure that the lysins were active against *C. perfringens* bacteria, a plate lysis assay was performed (Fig. 3). PlyCP41 fractions purified from bacteria (shown in Fig. 1b, E1–6) were all efficient in lysing the bacteria when compared to control, purified lysin (Fig. 3, F1–3). Crude sap extracts obtained from plants expressing PlyCP41p were also active against the bacteria (Fig. 3, B3–4, C3–6, D5–6, indicated by asterisks on the figure) while extracts from healthy plants and PVX-infected plants were not (Fig. 3, B1–2, C1–2). This assay revealed that clearing occurred within 30 min at room temperature after addition of the samples. The plant

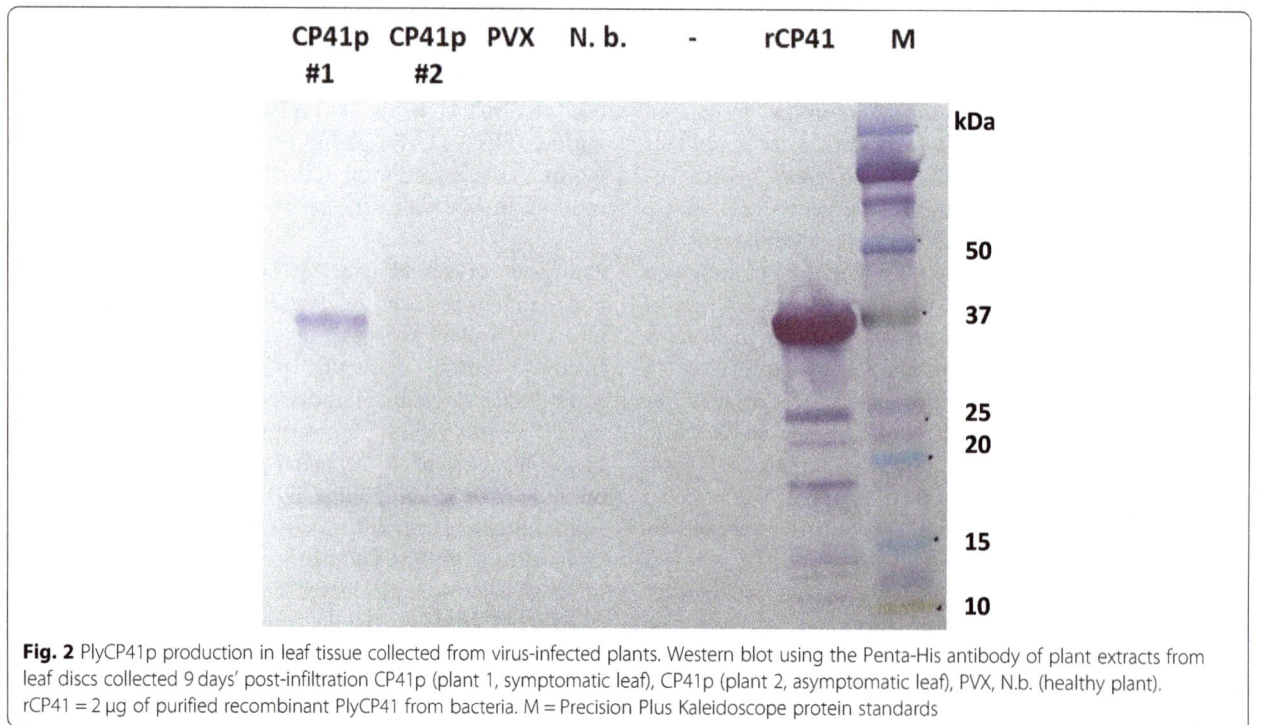

Fig. 2 PlyCP41p production in leaf tissue collected from virus-infected plants. Western blot using the Penta-His antibody of plant extracts from leaf discs collected 9 days' post-infiltration CP41 (plant 1, symptomatic leaf), CP41p (plant 2, asymptomatic leaf), PVX, N.b. (healthy plant). rCP41 = 2 μg of purified recombinant PlyCP41 from bacteria. M = Precision Plus Kaleidoscope protein standards

Optimized production of a biologically active Clostridium perfringens glycosyl hydrolase phage...

215

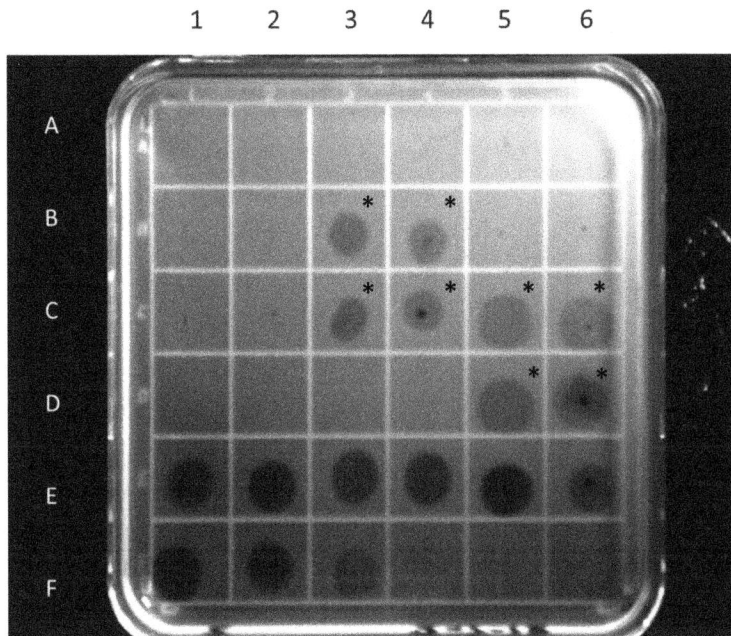

Fig. 3 Plate lysis assay of protein samples on *Clostridium perfringens* Cp39 confluent plates. Spot assays were conducted as described in the Materials and Methods and a photographic image of the plate was taken 30 min after application of the samples. The contents of the wells are as follows: A1-A6, Negative control, His-purified fractions from *E. coli*: BL21 pET21a; B1- B4- plant virus samples in PBS buffer; B1 & B2-empty PVX virus; B3 & B4-PVX virus with PlyCP41p; B5-PBS buffer; B6-elution buffer control; C1-C4- plant virus samples in "10:90" buffer (50 mM NaH$_2$PO$_4$ pH 7.0, 30 mM NaCl, 25 mM imidazole, 3% glycerol). C1&C2-empty PVX; C3&C4-PVX with PlyCP41p; C5–10 μL PVX with PlyCP41p in PBS buffer; C6–10 μL PVX with PlyCP41p in PBS buffer; D1-D4- empty; D5–10 μL PVX with PlyCP41p in "10:90 buffer"; D6–10 μL PVX with PlyCP41p in "10:90 buffer" E1-E6- *E. coli* PlyCP41 fractions purified on Ni-NTA column (FT, W1, W2, E1, E2, E3); F1- F4–10 μg, 1 μg, 0.1 μg, 0.001 μg of PlyCP41; F5 10:90 buffer. Asterisks indicate the plant extracts containing the PlyCP41p protein

extracts (spot equivalent to 2 mg of fresh weight (f. wt.) tissue and 200 ng of purified PlyCP41p) had similar clearing compared to 0.1 μg of purified, *E. coli*-produced PlyCP41 (Fig. 3, compare B-3 to F-3).

Solubility and optimization of the production capacity in PlyCP41p-expressing plants

For purification of PlyCP41p to homogeneity from crude plant sap using the Ni-NTA resin, the same native buffer conditions used to purify the bacterially-expressed lysin, described in Materials and Methods, were applied. PlyCP41p was easily recovered in the second elution fraction under native conditions (Fig. 4a), identical to what was observed for PlyCP41 from bacteria (Fig. 1b), suggesting that PlyCP41p is also soluble in plant tissues, thus facilitating its ease of purification for future studies. To quantify the amounts of PlyCP41p that could be recovered plant extracts, we determined that from 100 mg of leaf protein, we could purify 100 μg of PlyCP41p in elution fraction 2 at 200 ng/ul, or 1 g/1 kg f.wt. tissue (Fig. 4b, lanes E2).

Expression of PlyCP41p was substantially greater at 2 weeks' post-inoculation of the leaf compared to 3.5 weeks, and expression was further reduced at later time points (Fig. 5). This suggests that the optimal time to harvest infected leaf tissue to achieve the highest

recovery of PlyCP41p was within 2 weeks' post-inoculation (Fig. 5). Although the leaf tissue was positive for the PVX virus as assayed using the immunostrips, and PVX titer appeared to be equivalent to leaves in which the protein was expressed, protein expression in older plants was reduced. When older, infected leaf tissue was used to mechanically inoculate healthy *N. benthamiana* plants (Fig. 6, lane 7) indicating that the virus retained the lysin insert and can express the lysin protein when passaged.

To achieve the highest production of PlyCP41 protein in plants using the PVX-expression vector, we engineered a plant codon-optimized lysin gene (PlyCP41pc) into the same PVX vector and compared its accumulation to that of *E. coli* codon-optimized gene (PlyCP41p) in plants 13 days' post-infiltration. (Fig. 6). PlyCP41pc (lanes 1–4) did not accumulate to higher levels than PlyCP41p (lanes 5, 6), however we did find that using undiluted Agrobacterium cultures (neat, lanes 2, 4, 6) led to higher accumulation of the recombinant proteins in systemically infected leaf tissue than a 1:10 dilution (Fig. 6, lanes 1, 3, 5).

Discussion

In this study, we report the first successful expression of an active PlyCP41, a previously characterized phage-

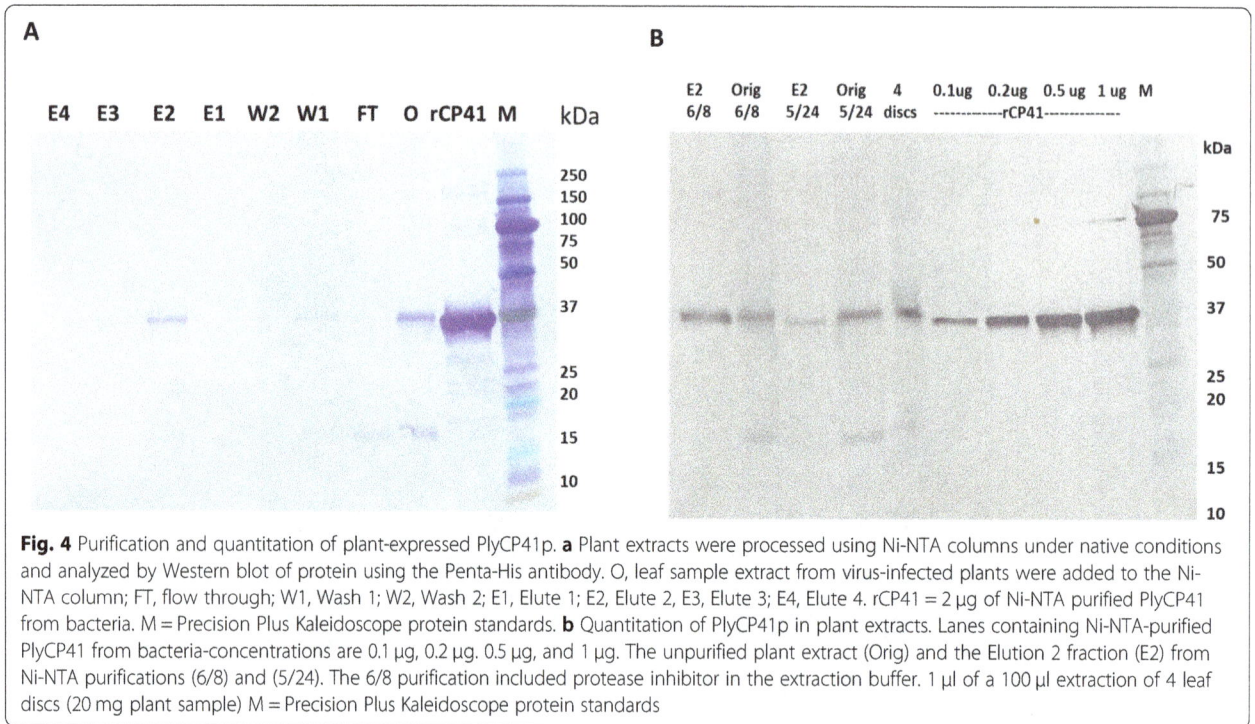

Fig. 4 Purification and quantitation of plant-expressed PlyCP41p. **a** Plant extracts were processed using Ni-NTA columns under native conditions and analyzed by Western blot of protein using the Penta-His antibody. O, leaf sample extract from virus-infected plants were added to the Ni-NTA column; FT, flow through; W1, Wash 1; W2, Wash 2; E1, Elute 1; E2, Elute 2, E3, Elute 3; E4, Elute 4. rCP41 = 2 µg of Ni-NTA purified PlyCP41 from bacteria. M = Precision Plus Kaleidoscope protein standards. **b** Quantitation of PlyCP41p in plant extracts. Lanes containing Ni-NTA-purified PlyCP41 from bacteria-concentrations are 0.1 µg, 0.2 µg. 0.5 µg, and 1 µg. The unpurified plant extract (Orig) and the Elution 2 fraction (E2) from Ni-NTA purifications (6/8) and (5/24). The 6/8 purification included protease inhibitor in the extraction buffer. 1 µl of a 100 µl extraction of 4 leaf discs (20 mg plant sample) M = Precision Plus Kaleidoscope protein standards

derived endolysin that targets *C. perfringens* [21] in systemically infected *N. benthamiana* plants using a PVX-based expression vector (Fig. 2). Cytoplasmic expression of the PlyCP41p gene in *N. benthamiana* led to expression levels of 1% TSP/ (0.1 mg/gm fresh weight plant tissue). In addition, we constructed and expressed a plant codon-optimized PlyCP41pc gene in plants and equivalent amounts of PlyCP41pc protein were produced. We demonstrated that for optimal PlyCP41p protein production, plant material needed to be harvested within 2 weeks' post-inoculation (Fig. 5) whether the initial inoculation occurred via agroinfiltration or inoculation of sap from virus-infected plants to healthy plants (Fig. 5).

Notably, there were no obvious additional local or systemic symptoms in plants infected with PVX expressing the lysin compared with those infected with PVX alone (data not shown), allowing recovery of the recombinant protein 2–3 weeks' post-inoculation. Although plants were infected with PVX for several weeks, the recombinant protein levels decreased even though the lysin gene was stable in the virus, as evidenced by the ability of sap extracts from these plants to generate new infections from which high levels of lysin were produced (Fig. 5). We have observed this phenomenon over several experiments and have found that we can reliably use sap from initially infected plants to scale up protein production in

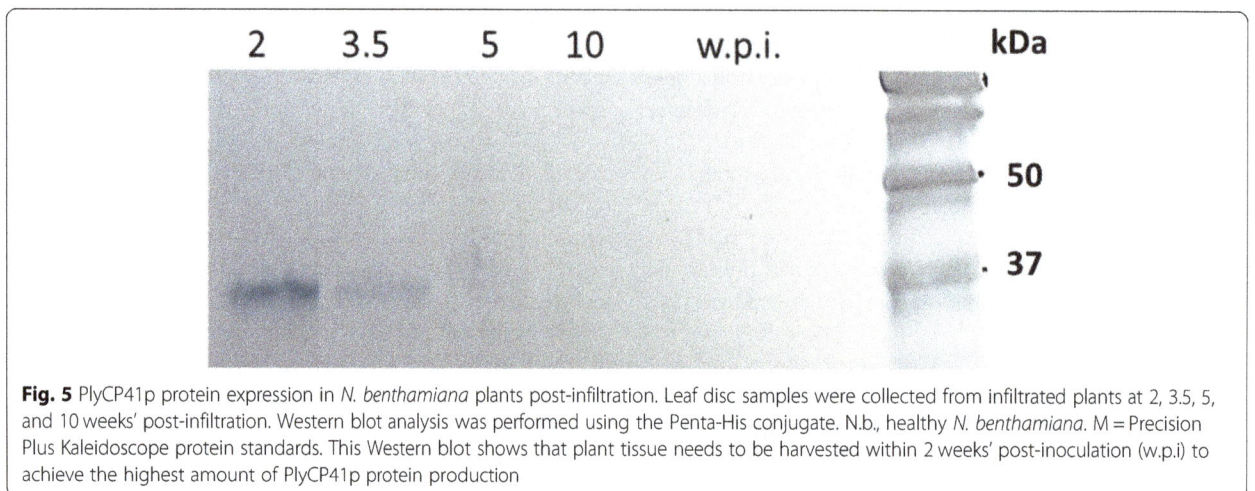

Fig. 5 PlyCP41p protein expression in *N. benthamiana* plants post-infiltration. Leaf disc samples were collected from infiltrated plants at 2, 3.5, 5, and 10 weeks' post-infiltration. Western blot analysis was performed using the Penta-His conjugate. N.b., healthy *N. benthamiana*. M = Precision Plus Kaleidoscope protein standards. This Western blot shows that plant tissue needs to be harvested within 2 weeks' post-inoculation (w.p.i) to achieve the highest amount of PlyCP41p protein production

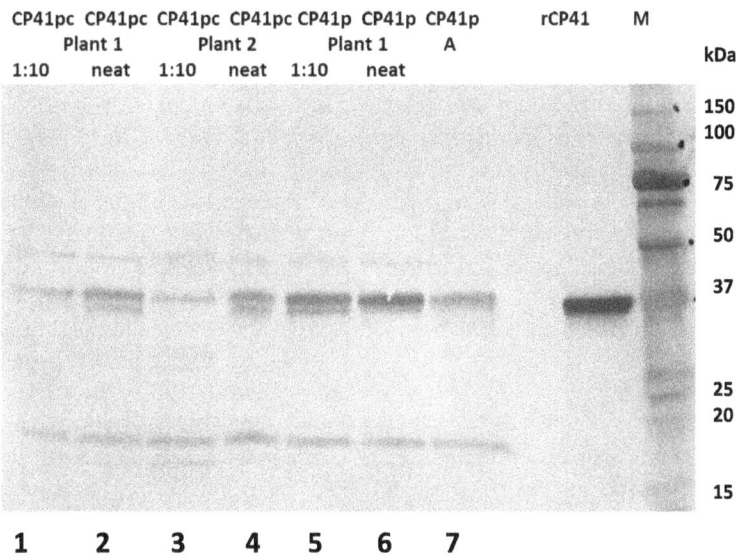

Fig. 6 Comparison of the expression of PlyCp41p and PlyCp41pc in plants. Western blot of plant tissue collected 13 days post-infiltration. Neat (O.D. 600 nm = 2.7) and 1:10 (O. D at 600 nm = .27) represent dilutions of Agrobacterium cultures used to infiltrate plants with pGDPVXMCS: PlyCP41pc (CP41pc) or pGDPVXMCS:PlyCP41p mixed in a 1:10 dilution with Agrobacterium containing pGDp19. CP41p (A) designates a plant that was mechanically inoculated from a plant 22 days' post-infiltration. This sample represents 7 days' post-infection. rCP41 = 2 μg. M = Precision Plus Kaleidoscope protein standards

subsequently inoculated plants. This observation might be explained by the plants RNA silencing response in virus-infected plants [29, 30], however we do not have experimental evidence to support that theory.

In the prokaryotic expression system that we utilized to produce the positive control, PlyCP41 expressed from the *E. coli*-optimized gene was predominantly located in the soluble fraction of lysed bacterial cells (Fig. 1b), in contrast to a similarly expressed endolysin (Lysin D) which was predominantly localized in the inclusion fraction. Solubility facilitated ease of purification using native conditions and nickel resin from the bacteria and may have facilitated ease of purification using native conditions from plant sap (Fig. 1b and Fig. 4a). There are numerous computational tools that predict protein solubility and aggregation as increasing solubility for industrial and therapeutic applications is of great value [31, 32]. Future examination of the protein sequence of the soluble PlyCP41 and other endolysins that were insoluble using these tools may aid in the design of future endolysins with improved solubility without impact on activity.

Substitution of codons encoding the same amino acid can affect the expression of proteins when attempting to express proteins in different hosts. Codon optimization of PlyCP41 for expression in plants, PlyCP41pc, did not result in increased levels of recombinant protein production (Fig. 6) in contrast to our earlier studies where poor expression in plants of an *E. coli*-optimized triple fusion protein, composed of phage lysin cassettes, was improved by plant codon optimization and allowed

accumulation of the protein to 0.12 mg/gm f. wt. tissue [26]. Although the impact of codon optimization on heterologous gene expression is unpredictable, there are several cases in the literature where 2 to 3-fold increases have been reported [33].

The plate lysis assay indicated that both the bacterial- and plant-expressed lysins possessed lytic activity against *C. perfringens* strain Cp39 (Fig. 3). Incubation of clarified plant sap and purified protein produced clear zones within 30 min of application to the bacterially-embedded agar revealing that plant components in the sap did not inhibit the lytic activity of PlyCP41.

Kazanavičiūtė et al. [28] recently reported the production of six biologically-active phage lysins in plants, using a deconstructed, transient tobacco mosaic virus-based vector expression system, by infiltration or spraying of leaves with the Agrobacterium constructs. Tissue was collected 5–6 days' post infiltration from the 'inoculated' leaves, and the lysins were active against a panel of *C. perfringens* serotypes. The authors estimated lysin production at 30% total soluble protein (TSP) based on visual inspection of stained gels, and purification from 150 to 1150 μg/gm f. wt. depending upon the lysin. Although they could produce up to 10-fold higher amounts of lysin in the infiltrated leaves, their system does not result in systemic virus infection. The advantage of our expression system is that we have production of the recombinant protein systemically and can passage the virus to young plants with resulting increased production and scale-up.

Our results suggest that crude plant lysates, or unprocessed plant tissue, containing recombinant phage lysins could be effective additives to animal feeds to control bacterial infections to improve animal health and food safety. Limitations of oral delivery, such as stability in the poultry gastrointestinal tract and effective concentration in the gut, will be addressed in follow-up in vitro and in vivo studies. Although we expressed PlyCP41p in *N. benthamiana*, expression of PlyCP41p in alternative plant species is being explored and will be used to generate the lysins using additional plant virus-based vectors.

Conclusions

Using a plant virus-based systemic expression system we produced, within two-week post-inoculation of plants, a biologically active phage endolysin with demonstrated activity against *C. perfringens*. As the lysin was active in unpurified, crude plant lysates, purification of the protein is not required. The lysates or dried plant tissue could be added to animal feeds for reducing *C. perfringens* colonization of the poultry gut to improve animal health and food safety and reduce production costs for the industry. This technology could be applied for the expression of other bacteriophage-derived endolysins for use as alternative antimicrobials for control of animal diseases.

Methods

Plasmid constructions

The PlyCP41 gene (GenBank KX884995) was synthesized as an *E. coli* codon optimized construct and cloned into pET21a (Novagen®, Millipore Sigma, Billerica, MA) by GenScript (Piscataway, NJ) [21]. PlyCP41 contains a C-terminal 6xHis-tag to facilitate purification using an Ni-NTA resin. pET21a: PlyCP41 was transformed and maintained in *E. coli* TOP10 cells (Life Technologies, Carlsbad, CA). The plasmid was transformed into *E. coli* strain BL21 (DE3) (Stratagene, La Jolla, CA) for protein production. As a control for expression in bacteria, plasmid pET21a: Lysin D (unpublished), which encodes another bacteriophage endolysin of similar size, was also transformed in BL21 cells for protein production only in *E. coli*. A plant codon-optimized CP41 gene containing a C-terminal 6xHis-tag (PlyCP41pc) with a CAI index of 0.92 was synthesized by Genscript USA (Piscataway, NJ) and was cloned in the pJET1.2 vector [33] (Additional file 1: Figure S1).

For expression in plants, the PlyCP41 coding region was amplified from the pET21a vector using primer pair BKEYS08 (5′- CC*GGATCC*AACAATGCTGAAGGGTAT CGACGTTAGC-3′) and BKEYS09 (5′- CC*AAGCTT*T CAGTGGTGGTGGTGGTGGTGCTCGAG-3′) and Ampli Taq DNA polymerase (Applied Biosystems, Foster City, CA) and the amplicon was cloned in the pCR4 vector (Life

Technologies) for sequence analysis. The gene was then isolated by restriction digestion using *Bam*HI and *Hin*dIII and inserted into a similarly digested intermediate pSKAS vector that is based on pBluescript SK+ and containing nt 4945 to nt 6541 of the pP2C2S PVX-based vector [34] and an expanded multiple cloning site [35]. The intermediate pSKAS vector allows engineering of insertions into a smaller plasmid vector from which the insertion can be transferred into the full-length virus-based vector. The resulting pSKAS: CP41 plasmid was digested with *Apa*I and *Spe*I and the insert was isolated and cloned into the similarly digested pGDPVXMCS plasmid (containing the full-length, PVX genome [36]), creating pGDPVXMCS: CP41p. The *E. coli* codon-optimized gene was maintained in this construct.

The plant codon-optimized PlyCP41pc gene was amplified from pJET: CP41p using oligonucleotide primers BKEYS17 (5′- CCC*ATGG*AACAATGCTTAAGGGGAAT TGATGTTTCTGAAC-3′) and BKEYS18 (5- CC*GA ATTCC*TAATGATGATGATGATGATGAAGTTTC 3′). The resulting amplicon was cloned into pCR4 vector. The PlyCP41pc gene was isolated from pCR4:CP41p by digestion with *Nco*I and *Eco*RI and cloned into the *Nco*I/*Eco*RI sites of the pSKAS vector, creating pSKAS: PlyCP41pc. Digestion of this plasmid with *Apa*I/*Spe*I released a fragment that was ligated into *Apa*I/*Spe*I digested pGDPVXMCS, creating pGDPVXMCS:PlyCP41pc. For all cloning, PCR products and gene fragments were gel purified from 1% agarose/TBE gels using the QIAquick Gel Extraction Kit (Qiagen GmbH, Hilden, Germany), vectors and inserts were ligated using T4 DNA ligase (New England Biolabs), and transformed into competent Top 10 *E. coli* cells (Life Technologies). The plasmid constructs were maintained the *E. coli* TOP 10 cells using appropriate antibiotics and plasmid DNAs were purified using the QIAprep Miniprep kit (Qiagen GmbH). All plasmids were sequenced for verification (Genscript USA).

Bacterial protein overexpression and purification using the BugBuster reagent

The pET21A:PlyCP41 construct was transformed into *E. coli* BL21 (DE3) cells for protein induction. Briefly, 5 mL of LB broth was inoculated with a loop of bacterial cells harboring the construct. After overnight incubation in a shaking incubator at 37 °C, 100 µl of cells was inoculated into 2 mL of LB and the culture was grown for 2 h. An aliquot of the culture was removed as a non-induced control. For protein induction, isopropyl-β-D-1-thiogalactopyranoside (IPTG) was then added to the cell cultures at a final concentration of 2 mM and the cultures were incubated with shaking at 37 °C for a further 2 h, during which time aliquots were removed at after 1 and 2 h for analysis by SDS-PAGE. For large scale protein purification of induced proteins, 500 µl of an overnight culture was added to 50 mL of LB in a 125 ml

Erlenmeyer flask. Bacterial pellets were recovered from 50 mL of IPTG-induced bacterial cultures grown at 37 °C by centrifugation at 4000 x g for 20 min in a Jouan CR422 centrifuge (Saint-Herblain, France) as previously described [24]. The BugBuster Master Mix Protein Extraction Reagent (Novagen, Madison, WI) and protease inhibitor cocktail for plant cells) (Sigma Chemical Co.) (1 μL cocktail per 100 μl of BugBuster Reagent) was added to the bacterial pellet to prepare the bacterial lysates and extract total proteins. The extraction was carried out per manufacturer's instructions to obtain total, soluble, and inclusion body fractions. For determination of protein concentrations, the Bradford assay using the Quick Start™ Bradford 1xDye Reagent and Quick Start™-Bovine Serum Albumin (BSA) Standard Set (Bio-Rad Laboratories, Hercules, CA) were used per manufacturer's instructions.

Agroinfiltration of *N. benthamiana* leaves

Agrobacterium tumefaciens strain EHA105 was transformed with the pGDPVXMCS, pGDPVXMCS:PlyCP41p, and pGDPVXMCS:PlyCP41pc plasmids and the bacteria were plated on Luria Broth-glucose (LBg) agar containing rifampicin and kanamycin at 50 μg/mL each. Colonies which appeared after incubation of the plates at 28 °C were inoculated into 5 mL of liquid LBg broth and grown overnight at 28 °C and 250 rpm in a shaking incubator. The cultures were centrifuged for 10 min at 4000 x g at 25 °C in a Jouan CR422 centrifuge. The bacterial pellets were gently resuspended in 2 mL of infiltration medium (10 mM MES, 10 mM $MgCl_2$, pH 5.7) and 4 μL of 1 M acetosyringone (Sigma Chemical Co.) was added. After incubation at ambient room temperature for 4 h, the cultures were individually mixed with a culture of similarly prepared EHA105 containing the plasmid pGDp19 (encoding a plant viral-encoded suppressor protein [29]) at a ratio of 1:10 (pGDp19:pGDPVXMCS construct). Three to four young leaves *N. benthamiana* plants at the 5–6 leaf stage were infiltrated on the abaxial side of the leaf using a needleless syringe. Plants were grown in the laboratory at 27 °C and were observed for symptom production and monitored for virus infection using PVX AgriStrips following manufacturer's instructions (Eurofins BioDiagnostics, Inc., Longmont, CO).

CP41 extraction from *N. benthamiana* plants

Four leaf discs (~ 20 mg tissue) were collected and placed into an eppendorf tube to test for PlyCP41p and PlyCP41pc protein production. The leaf samples were ground in 100 μl of the CellLytic™ P Plant Cell Lysis/Extraction Reagent (Sigma Chemical Co., Saint Louis, MO) containing 1 μl of plant protease inhibitor cocktail (Sigma Chemical Co.) using a blue pestle. Cell debris

was removed by centrifugation at 4 °C, and the supernatant was combined with an equal volume of Laemmli buffer (BioRad Laboratories, Hercules, CA). After boiling for 10 mins, an aliquot of the sample was applied to a 10–20% Tris-glycine gel as described below.

Protein gel electrophoresis and Western blot analysis

Proteins were resolved by SDS-PAGE analysis on a Novex 10–20% Tris-glycine gradient mini gels (Life Technologies) under denaturing conditions using manufacturer's instructions. The proteins were visualized by staining with SimplyBlue Safe Stain (Life Technologies). Alternatively, the proteins were transferred to a 0.45 μM nitrocellulose membrane (Life Technologies). The membranes were subsequently incubated with a 1:1000 dilution of Anti-His HRP Conjugate solution (Penta His HRP Conjugate Kit (Qiagen) following manufacturer's instructions followed by development using the TMP Membrane Peroxidase Substrate System (Kirkegaard and Perry, Gaithersburg, MD) to visualize the proteins.

Protein purification using nickel resin (IMAC) under native conditions

To purify bacterial and plant expressed His-tagged proteins under native conditions, we used the Ni-NTA His-Bind Resins Kit and the Ni-NTA Buffer kit (Novagen) following manufacturer's instructions to purify PlyCP41 from the soluble fraction obtained previously from *E. coli* using the Bug Buster reagent (above). The plant PlyCP41p protein was also purified from plant sap using the Ni-NTA His-Bind Resin under native conditions. Plant tissue was ground in a chilled mortar and pestle using Binding Buffer (BB; 50 mM NaH_2PO_4, pH 8.0; 300 mM NaCl; 10 mM imidazole) from the His-Bind Resins Kit, to which the plant protease inhibitor cocktail (Sigma Chemical Co.) was added in a ratio of 1 μl cocktail to 100 μl BB. The Bradford assay described above was used to determine protein concentrations in the fractions.

Testing the lytic activity of expressed proteins against *C. perfringens*

The plate lysis (spot) assay was performed essentially as described previously [20]. *C. perfringens* strain Cp39 cultures were propagated to mid-log phase (OD600 = 0.4–0.6) in 50 mL BHIB, where upon the cells were centrifuged at 5000 g for 30 min. The cell pellet was washed with 50 mL lysin buffer (50 mM NH_4OAc, 10 mM $CaCl_2$, 1 mM DTT, pH 6.2) and pelleted again. The cells were suspended in 1.0 mL lysin buffer. Ten milliliters of 50 °C semisolid BYC ss agar (37 g/L brain heart infusion powder, 5 g/L yeast extract, 0.5 g/L cysteine, 7 g/L Bacto agar) was added to the cells and then the cells were poured into a sterile 6 × 6 grid square petri dish. The plates sat 20 min at room temperature to solidify the agar. Ten µL of the purified endolysin or plant sap was then spotted onto the plate and allowed to air dry 20 min. The purified lysins were in diluted in "10:90" buffer (50 mM NaH_2PO4, pH 7, 30 mM NaCl, 2 mM imidazole, 3% glycerol). Plant sap was prepared from uninfected tissue, leaf tissue from plants infiltrated with the empty pGDPVXMCS, PVX-based vector, and plants infiltrated with the pGDPVXMCS: PlyCP41p plasmid by grinding 4 leaf discs ($\sim$ 20 mg) with a pestle in 100 µL 1 x PBS buffer (Bio-Rad, Hercules, CA) in an Eppendorf tube. After one round of centrifugation at 16,000 x g for 5 min, the supernatant was removed from the pellet containing cellular debris and 10 µL of plant sap was applied to the plate as described. A positive lytic reaction was determined by visible clearing of the turbidity of the bacterial cells. The plate was observed for development of visible clearing and then incubated overnight in an anaerobic chamber at 37 °C.

Supplementary information

Additional file 1: Figure S1. Plant codon-optimized PlyCP41pc gene. A. Nucleotide sequence of the PlyCP41pc gene and explanatory notes and encoded protein. B. Alignment of *E. coli* optimized gene PlyCP41 (and identical PlyCP41p) nucleotide sequence (lower line in black) and plant codon-optimized gene PlyCP41pc (upper line in red). Yellow boxes indicate the modified sequences in PlyCP41pc.

Abbreviations
CAI: Codon adaptation index; His: Histidine; PVX: Potato virus X; TSP: Total soluble proteins

Acknowledgments
We thank Ms. Breannah Keys and Nancy Kreger, D.V.M., for providing technical expertise. Mention of trade names or commercial products in this publication is solely for providing specific information and does not imply recommendation by the U. S. Department of Agriculture.

Authors' contributions
RWH, SMS, JAF-F, NYK, and DMD designed the experiments. RWH, SMS, JAF-F,and NYK performed the experiments and collected the data. RWH, SMS, NYK, and DMD interpreted the data and wrote the article. All authors have read and approved the manuscript.

Funding
The research was funded by the U. S. Department of Agriculture, Agricultural Research Service, USA. The funding agency was not involved in the design, experimentation, interpretation of the data, or writing of the manuscript.

Ethics approval and consent to participate
Not applicable.

Consent for publication
Not applicable.

Competing interests
The authors declare that they have no competing interests.

Author details
[1]USDA ARS NEA BARC Molecular Plant Pathology Laboratory, Beltsville, MD 20705, USA. [2]USDA ARS NEA BARC Animal Biosciences and Biotechnology Laboratory, Beltsville, MD 20705, USA. [3]Oak Ridge Institute for Science and Education, ORISE, Beltsville, MD 20705, USA.

References
1. Olsen SJ, MacKinon LC, Goulding JS, Bean NH, Slutsker L. Surveillance for foodborne-disease outbreaks- United States, 1993-1997. Morb Mortal Wkly Rep. 2000;49:1–51.
2. Scallan E, Hoekstra RM, Angulo FJ, Tauxe RV, Widdowson MA, Roy SL, Jones JL, Griffin PM. Food-borne illness acquired in the United States—major pathogens. Emerg Infect Dis. 2011;17:7–15.
3. McDevitt RM, Brooker JD, Acamovic T, Sparks NHC. Necrotic enteritis; a continuing challenge for the poultry industry. Worlds Poult Sci J. 2006;62: 221–47.
4. Devriese LA, Daube G, Hommez J, Haesebrouck F. In vitro susceptibility of *Clostridium perfringens* isolated from farm animals to growth-enhancing antibiotics. J Appl Bacteriol. 1993;75:55–7.
5. Watkins KL, Shryock TR, Dearth RN, Saif YM. In vitro antimicrobial susceptibility of *Clostridium perfringens* from commercial Turkey and broiler chicken origin. Vet Microbiol. 1997;54:195–200.
6. Casewell M, Friis C, Marco E, McMullin P, Phillips I. The European ban on growth-promoting antibiotics and emerging consequences for human and animal health. J Antimicrob Chemother. 2013;52:159–61.
7. Van Immerseel F, De Buck J, Pasmans F, Huyghebaert G, Haesebrouck F, Ducatelle R. *Clostridium endolysins* in poultry: an emerging threat for animal and public health. Avian Pathol. 2004;33:537–49.
8. Fischetti VA. Exploiting what phage have evolved to control gram-positive pathogens. Bacteriophage. 2011;1:188–94.
9. Dong H, Zhu C, Chen J, Ye X, Huang Y-P. Antibacterial activity of *Stenotrophomonas maltophilia* endolysin P28 against both gram-positive and gram-negative bacteria. Front Microbiol. 2015;6:1299.
10. Nakonieczna A, Cooper C, Gryko R. Bacteriophages and bacteriophage-derived endolysins as potential therapeutics to combat gram-positive spore forming bacteria. J Appl Microbiol. 2015;119:620–31. https://doi.org/10.1111/jam.12881.
11. Schmelcher M, Donovan DM, Loessner MJ. Bacteriophage endolysins as novel antimicrobials. Future Microbiol. 2012;7:1147–71.
12. Schmelcher M, Loessner MJ. Bacteriophage endolysins: applications for food safety. Curr Opin Biotechnol. 2016;37:76–87.
13. Miller RW, Skinner J, Sulakvelidze A, Mathis GF, Hofacre CL. Bacteriophage therapy for control of necrotic enteritis of broiler chickens experimentally infected with *Clostridium perfringens*. Avian Dis Dig. 2010;54:33–40.
14. Becker SC, Roach DR, Chauhan VS, Shen Y, Foster-Frey J, Powell AM, et al. Triple-acting lytic enzyme treatment of drug-resistant and intracellular *Staphylococcus aureus*. Sci Rep. 2016;6:25063.
15. Gervasi T, Horn N, Wegmann U, Dugo G, Narbad A, Mayer MJ. Expression and delivery of an endolysin to combat *Clostridium perfringens*. Appl Microbiol Biotechnol. 2014;98:2495–505.
16. Seal BS, Volozhantsev NV, Oakley BB, Morales CA, Garrish JK, Simmons M, Svetoch EA, Siragusa GR. Bacteriophages of Clostridium perfringens. In: Kurtboke I, editor. Bacteriophages: IntechOpen; 2012. https://doi.org/10.5772/33106. ISBN: 978-953-51-0272-4.
17. Seal BS. Characterization of bacteriophages virulent for *Clostridium perfringens* and identification of phage lytic enzymes as alternatives to antibiotics for potential control of the bacterium. Poult Sci. 2013;92:526–33.
18. Simmons M, Donovan DM, Siragusa GR, Seal BS. Recombinant expression of two bacteriophage proteins that lyse *Clostridium perfringens* and share identical sequences in the C-terminal cell wall binding domain of the molecules but are dissimilar in their N-terminal active domains. J Agric Food Chem. 2010;58:10330–7.
19. Oliveira H, Melo LD, Santos SB, Nobrega FL, Ferreira EC, Cerca N, et al. Molecular aspects and comparative genomics of bacteriophage endolysins. J Virol. 2013;87:4558–70.
20. Swift S, Seal B, Garrish J, Oakley B, Hiett K, Yeh H, Donovan D. A Thermophilic phage endolysin fusion to a *Clostridium perfringens*-specific cell wall binding domain creates an anti-Clostridium antimicrobial with improved thermostability. Viruses. 2015;7:3019–34.
21. Swift S, Waters JJ, Rowley DT, Oakley BB, Donovan DM. Characterization of two glycosyl hydrolases, putative prophage endolysins, that target

Optimized production of a biologically active Clostridium perfringens glycosyl hydrolase phage...

221

Clostridium perfringens. FEMS Microbiol Lett. 2018;363:fny 179. https://doi.org/10.1093/femsle/fny179.

22. Oey M, Lohse M, Kreikemeyer B, Bock R. Exhaustion of the chloroplast protein synthesis capacity by massive expression of a highly stable protein antibiotic. Plant J. 2009;57:436–45.

23. Stoffels L, Taunt HN, Charalambous B, Purton S. Synthesis of bacteriophage lytic proteins against *Streptococcus pneumoniae* in the chloroplast of *Chlamydomonas reinhardtii*. Plant Biotechnol J. 2017;15:1130–40.

24. Kovalskaya N, Hammond RW. Expression and functional characterization of the plant antimicrobial snakin-1 and defensin recombinant proteins. Protein Expr Purif. 2009;63:12–7.

25. Kovalskaya N, Foster-Frey J, Donovan DM, Bauchan G, Hammond RW. Antimicrobial activity of bacteriophage endolysin produced in *Nicotiana benthamiana* plants. J Microbiol Biotechnol. 2016;26:160–70.

26. Kovalskaya NY, Herndon EE, Foster-Frey JA, Donovan DM, Hammond RW. Antimicrobial activity of bacteriophage derived triple fusion protein against *Staphylococcus aureus*. AIMS Microbiol. 2019;5:158–75.

27. Starkevič U, Bortesi L, Virgailis M, Ružauskas M, Giritch A, Ražanskienė A. High-yield production of a functional bacteriophage lysin with antipneumococcal activity using a plant virus-based expression system. J Biotechnol. 2015;200:10–6.

28. Kazanavičiūtė V, Misiūnas A, Gleba Y, Giritch A, Ražanskienė A. Plant-expressed bacteriophage lysins control pathogenic strains of *Clostridium perfringens*. Sci Rep. 2018;8:10589. https://doi.org/10.1038/s41598-018-28838-4.

29. Qiu W, Park JW, Scholthof HB. Tombusvirus P19-mediated suppression of virus-induced gene silencing is controlled by genetic and dosage features that influence pathogenicity. Mol Plant-Microbe Interact. 2002;15:269–80.

30. Ghoshal B, Sanfaçon H. Symptom recovery in virus-infected plants: revisiting the role of RNA silencing mechanisms. Virology. 2015;479-480:167–79.

31. Broom A, Jacobi Z, Trainor K, Meiering EM. Computational tools help improve protein stability but with a solubility tradeoff. J Biol Chem. 2017;292:14349–61.

32. Chan P, Curtis RA, Warwicker J. Soluble expression of proteins correlates with a lack of positively-charged surface. Sci Rep. 2013;3:3333.

33. Webster GR, Teh AY, Ma JK. Synthetic gene design-the rationale for codon optimization and implications for molecular pharming in plants. Biotechnol Bioeng. 2016;114:492–502. https://doi.org/10.1002/bit.26183.

34. Chapman S, Kavanagh TA, Baulcombe DC. Potato virus X as a vector for gene expression in plants. Plant J. 1992;2:549–57.

35. Kovalskaya N, Zhao Y, Hammond RW. Antibacterial and antifungal activity of a snakin-defensin hybrid protein expressed in tobacco and potato plants. Open Plant Sci J. 2011;5:29–42.

36. Lim H-S, Vaira AM, Domier LL, Le SC, Kim HG, Hammond J. Efficiency of VIGS and gene expression in a novel bipartite potexvirus vector delivery system as a function of strength of TGB1 silencing suppression. Virology. 2010;402:149–63.

21

Identification and characterization of an Endo-glucanase secreted from cellulolytic *Escherichia coli* ZH-4

Jian Pang[1,3], Junshu Wang[2*], Zhanying Liu[1,3*], Qiancheng Zhang[1] and Qingsheng Qi[2]

Abstract

Background: In the previous study, the cellulolytic *Escherichia coli* ZH-4 isolated from bovine rumen was found to show extracellular cellulase activity and could degrade cellulose in the culture. The goal of this work was to identify and characterize the secreted cellulase of *E. coli* ZH-4. It will be helpful to re-understand *E. coli* and extend its application in industry.

Results: A secreted cellulase was confirmed to be endo-glucanase BcsZ which was encoded by *bcsZ* gene and located in the cellulose synthase operon *bcsABZC* in cellulolytic *E. coli* ZH-4 by western blotting. Characterization of BcsZ indicated that a broad range of pH and temperature tolerance with optima at pH 6.0 and 50 °C, respectively. The apparent Michaelis–Menten constant (K_m) and maximal reaction rate (V_{max}) for BcsZ were 8.86 mg/mL and 0.3 μM/min·mg, respectively. Enzyme activity of BcsZ was enhanced by Mg^{2+} and inhibited by Zn^{2+}, Cu^{2+} and Fe^{3+}. BcsZ could hydrolyze carboxymethylcellulose (CMC) to produce cello-oligosaccharides, cellotriose, cellobiose and glucose.

Conclusions: It is confirmed that extracellular cellulolytic capability of *E. coli* ZH-4 was attributed to BcsZ, which explained why *E. coli* ZH-4 can grow on cellulose. The endo-glucanase BcsZ from *E. coli*-ZH4 has some new characteristics which will extend the understanding of endo-glucanase. Analysis of the secretion characteristics of BcsZ provided a great reference for applying *E. coli* in multiple industrial fields.

Keywords: Cellulolytic *Escherichia coli* ZH-4, Secretory endo-glucanase, BcsZ, Enzyme characterization

Background

Cellulose biomass is the most abundant carbohydrate on the earth. It can be hydrolyzed to reducing sugars for production of biofuels and chemicals, and thus has a great economic and commercial potential [1]. Cellulose as the main component of plant cell wall consists of linear long chains of β-1, 4 glucose units. Hydrolyzing cellulose by cellulase is ideal and promising for its utilization in environmentally friendly and high efficiency manner [2]. However, the cooperative action of three kinds of cellulolytic enzymes (endo-glucanase, exo-glucanase, and β-glucosidase) is essential in hydrolysis of cellulose to glucose [3]. Among three kinds of cellulolytic enzymes, endoglucanases plays an important role in

the process of cellulose hydrolysis because it hydrolyzes the glycosidic bond randomly and shorten the cellulose chains in the initial stage of cellulose breakdown [4]. Cellulase is produced by various cellulolytic bacteria and fungi which have been isolated from different environment [5]. Isolating cellulolytic microorganisms from various environment and characterizating their cellulase are crucial for understanding the evolution mechanism of cellulolytic microorganisms and the hydrolysis mechanism of cellulase, which will promote their application in industry. In recent years, cellulase from bacteria was focused again because the glycoside hydrolases of cellulolytic bacteria are very diverse [6].

The previous study revealed that a cellulase (Cel-CD) from *Bacillus sp.* can be secreted into culture medium when Cel-CD was overexpressed in *E.coli* with or without its signal peptide, which indicated that *E. coli* has the capacity of secreting cellulase [7]. In addition, cellulolytic *E. coli* will be got when expressed this cellulase in

---

* Correspondence: wangjunshusdu@163.com; zyliu1979@163.com
[2]State Key Laboratory of Microbial Technology, Shandong University, Qingdao 266237, China
[1]School of Chemical Engineering, Inner Mongolia University of Technology, Hohhot 010051, Inner Mongolia, China
Full list of author information is available at the end of the article

E. coli, and the *E. coli* has a potential application to produce enzymes and chemicals directly from lignocellulose biomass [8].

In our previous study, a cellulolytic *E. coli* ZH-4 was isolated from the rumen [9]. *E. coli* ZH-4 is capable of converting corn straw to ethanol and hydrogen anaerobically. Extracellular endo-glucanase and β-glucosidase activity were detected. The results indicated that such enzymes were expressed and secreted in cellulolytic *E. coli* ZH-4. Genome sequence analysis of *E. coli* ZH-4 revealed an endoglucanase gene (Genbank accession number KY965823) encoding a BcsZ homolog.

From another point of view, cellulose is a major structural component in bacteria, which provides cell-surface and cell–cell interaction in various of biofilm models, and protects cells against chlorine treatment [10–12]. The previous study showed that inactivation of BcsZ altered the cellulose-associated phenotypes in *Salmonella enterica* serovar Typhimurium, such as rdar biofilm morphotype, cell clumping, biofilm formation, pellicle formation and flagella-dependent motility [10]. The hydrolase activity of BcsZ is hypothesized to mediate alignment of each β-1, 4 -glucan for proper cellulose microfibril formation [13]. *BcsZ* is a conserved component of the cellulose synthase operon *bcsABCZ*, which encodes the cellulose synthase BcsAB and the outer membrane porin for cellulose translocation and secretion [10, 14]. BcsZ belongs to Glycoside Hydrolase family 8 with endo-1,4-D-glucanase activity. BcsZ hydrolyzes glycosidic bonds by a pair of acidic residues inverting the anomeric configuration at the new reducing end [15]. The crystal structure analysis of BcsZ from *E. coli* showed an $(\alpha/\alpha)_6$-barrel fold. BcsZ binds 4 glucan moieties of cellopentaose via highly conserved residues exclusively on the non-reducing side of its catalytic center [13]. However, whether BcsZ is responsible for cellulolytic ability of *E. coli* ZH-4 is uncertain. Little is known about the characteristic of BcsZ-ZH-4. Enzymatic hydrolysate of BcsZ-ZH-4 from cellulose is unknow.

In this study, the cellulolytic ability of *E. coli* ZH-4 was verified from BcsZ. The endoglucanase was assessed through transcription, expression and secretion. BcsZ from *E. coli* ZH-4 was functionally expressed in *E. coli* BL21 (DE3), and the recombinant protein was purified and characterized.

Results
Identification and verification of secreted cellulase in cellulolytic *E. coli* ZH-4

The mature protein in the culture medium was analyzed through western blotting to identify the extracellular endo-glucanase in ZH-4. The coding gene was located on the operon of *bcsABZC*. Meanwhile, the extracellular protein was verified by western blotting using BcsZ antiserum (Fig. 1). The transcription level of *bcsZ* in cellulolytic *E. coli* ZH-4

was also found to be 2.6 ± 0.25 and 6.0 ± 0.26 fold higher than that of *E. coli* MG1655 under aerobic and anaerobic condition respectively (Fig. 2).

A signal peptide sequence of 1–21 amino acid residues was found in *E. coli* ZH-4 by the analysis of the protein sequence. In compared to *E. coli* MG1655, there were two amino acids difference in BcsZ of *E. coli* ZH-4: Ser63 to Phe (Ser-Phe) and Ala71to Val (Ala-Val). The *bcsZ* located in the downstream of *bcsB* and was supposed under the control of the *bcsB* promoter. DNA-binding transcriptional dual regulator FNR (Fnr) regulates *bcsBZ* operon expression under anaerobiosis, and the putative FNR-binding site was identified in upstream of this operon. Genetic analysis of the operon indicated that there was no difference with that of MG1655 in the regulation and transcription region.

Expression and purification of endo-glucanase BcsZ

The *bcsZ* gene amplified from *E. coli* ZH-4 was cloned in pET-28a vector, and then overexpressed in *E. coli* BL21 (DE3). BcsZ was detected in culture medium (Fig. 3a, line 1) with the recombinant cell by Sodium dodecyl sulfate polyacrylamide gel

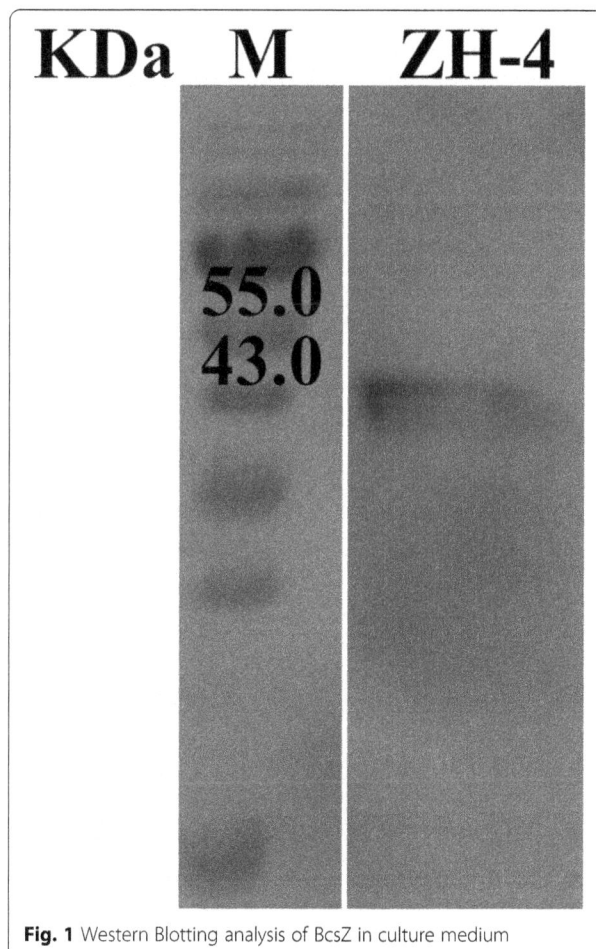

Fig. 1 Western Blotting analysis of BcsZ in culture medium

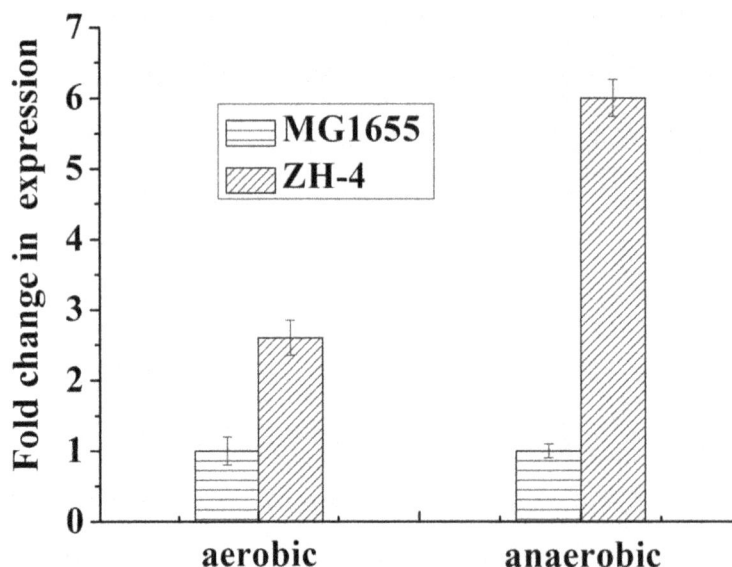

Fig. 2 The fold change in gene expression of *bcsZ* in *E. coli* ZH-4 and MG1655

electrophoresis (SDS-PAGE). The result indicated that endo-glucanase, BcsZ, can be secreted to the outside of cell. The crude protein of BcsZ was purified, and the purified protein appeared as a single protein band on SDS–PAGE gel with a molecular mass of 41.7 KD, which was consistent with prediction (Fig. 3a, line 3). The *E. coli* BL21 (DE3) carrying the pET-28a vector (empty) was used as control (Fig. 3b).

Characterization of the endo-glucanase BcsZ

The optimal temperature and pH of purified BcsZ were determined. As shown in Fig. 4a, the optimum temperature was 50 °C. With the rising of the temperature, the enzyme activity began to decrease, but retained over 70% at the temperature of till 65 °C. The enzyme activity sharply decreased when the temperature is over 65 °C, and was remained by 40% when the temperature was 80 °C. From Fig. 4b, the BcsZ displayed

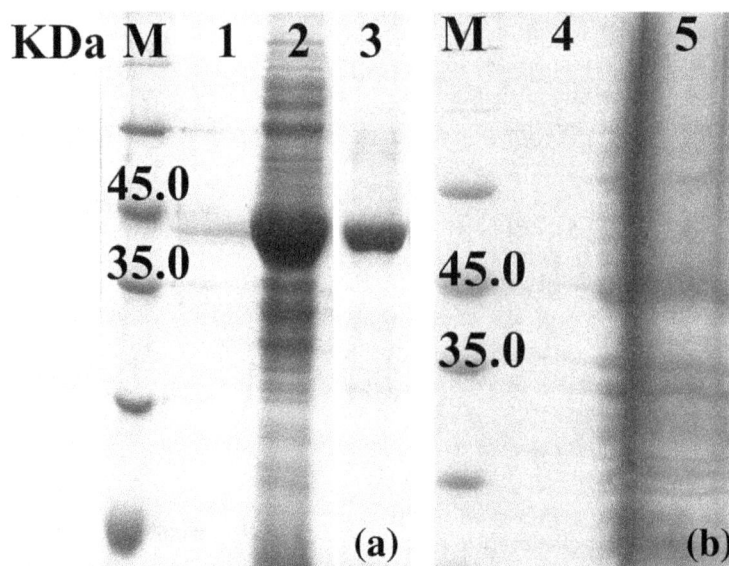

Fig. 3 SDS–PAGE analysis of recombinant BcsZ protein stained with coomassie blue (**a**). M: Protein molecular weight marker; Lane 1: BcsZ in culture medium; Lane 2: BcsZ in cells; Line 3: The purified BcsZ. The *E. coli* BL21 (DE3) carrying the empty plasmid was used as control (**b**). Line 4: culture medium; Line 5: cells

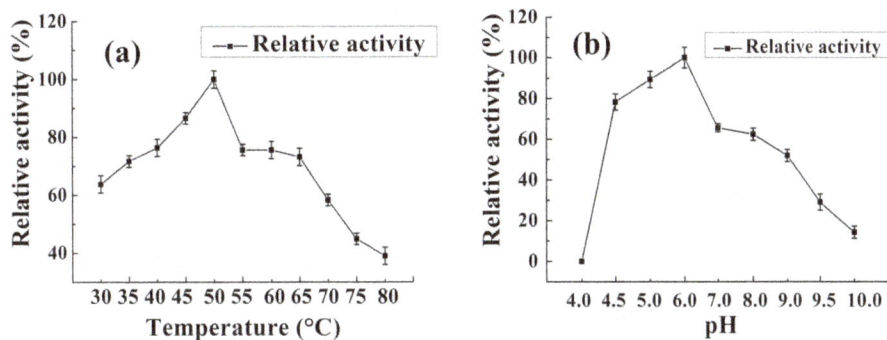

Fig. 4 The optimum temperature (**a**) and pH (**b**) of BcsZ

optimal activity at pH 6.0. It had relatively high enzyme activity at the broad pH range of 4.5–9.0. BcsZ was sensitive to the acidic condition. It remained about 80% activity when pH at 4.5, and enzyme activity was not detected when the pH was 4.0. The BcsZ protein has a tolerance to alkali solution because it retained over 15% of the maximum enzyme activity when the pH was 10.0.

The thermal stability result was shown in Fig. 5, and showed BcsZ displayed high thermal stability at 50 °C and 60 °C with about 90 and 70% of its enzyme activity remained, respectively. The enzyme activity wasn't detected at 70 °C after 0.5 h incubation.

The substrate specificity results were shown in Table 1. As shown in Table 1, BcsZ displayed enzyme activity with CMC, RAC and glucan (data not shown) as substrate. The enzyme showed a weak catalytic activity on Avicel, xylan, cellobiose, laminarin and chitin. The K_m value of BcsZ for CMC as substrate was

8.86 mg/mL, and the calculated V_{max} was 0.3 μM/min·mg.

Effect of metal ion on BcsZ activity

The effect of metal ions on the enzyme activity of BcsZ was examined (Fig. 6). The enzyme activity was improved 165% by Mg^{2+}. Fe^{3+} inhibited BcsZ, and 74% enzyme activity was lost with 10 mM Fe^{3+}. Zn^{2+} also inhibited the enzyme of BcsZ, and there was 66% of enzyme activity was left with 10 mM Zn^{2+}. The enzyme was strongly inhibited by Cu^{2+}, there is no enzyme activity detected when Cu^{2+} concentration reached 5 mM in the reaction system.

Analysis of hydrolysis products by TLC

The hydrolysis products of CMC and RAC were analyzed by TLC. As shown in Fig. 7, glucose was released when BcsZ hydrolyzed RAC (G9). The hydrolysis

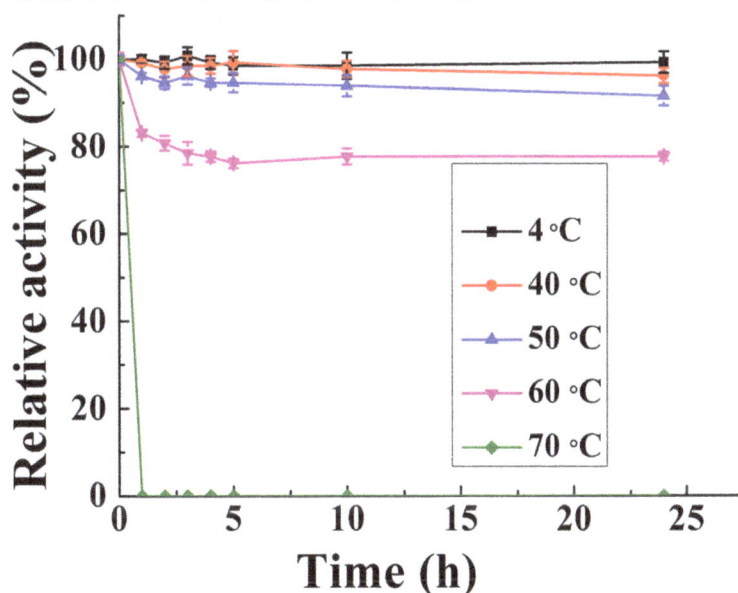

Fig. 5 The thermostability of BcsZ

Table 1 Substrate specificity of BcsZ

Substrate	Specific activity (IU/mg)
CMC	0.84 ± 0.05
RAC	0.52 ± 0.08
Avicel	< 0.01
xylan	< 0.01
cellobiose	< 0.01
laminarin	< 0.01
chitin	< 0.01

products were glucose, a small amount of cellobiose and cellotriose, and unknown cello oligosaccharides from CMC (G7) as substrate.

Discussion

In our previous study, a cellulolytic *E. coli* ZH-4 was isolated from bovine rumen, this strain could produce extracellular cellulases. In this study, secretion of BcsZ to culture medium was confirmed by western blotting (Fig. 1). These results explained the reason ZH-4 formed a clear zone in anaerobic Hungate roll tubes (containing cellulose and Congo red) [9]. While *E. coli* strain MG1655 and W3110 didn't generate the above phenotype at the same condition. Secretion of cellulase (BcsZ) and its contribution to the cellulolytic capacity of *E. coli* ZH-4 were confirmed, which were consistent with our previous supposition and the published [9]. The previous studies showed overexpression of BcsZ in *E. coli* leads to its secretion to extracellular space and formation of clear zone on congo red plates [16, 17]. Further analysis confirmed that the transcription level of *bcsZ* in ZH-4 was higher than that of MG1655, especially

under anaerobic condition. The increased transcription level of *bcsZ* is critical to the cellulolytic activity of *E. coli* ZH-4. Sequence analysis showed that the promoter, Fnr and FNR-binding site of *bcsZ* in *E. coli* ZH-4 is consistent with MG1655 and W3110. The genome of *E. coli* ZH-4 (5.3 Mb) is larger than *E. coli* MG1655 (4.6 Mb). It is supposed that some regulatory factors and elements regulate *bcsZ* transcription and expression in *E. coli* ZH-4. This may lead to the increased transcription level of *bcsZ* in *E. coli* ZH-4. Further elevated transcription level investigation of *bcsZ* may reveal the difference in BcsZ secretion between *E. coli* ZH-4 and MG1655.

The endo-glucanase BcsZ had the broad pH range and strong alkali tolerance. These characteristics of BcsZ might be affected by the rumen habitat and consistent with the previous reports. For instance, Gong et al. cloned and identified some novel hydrolase genes from a dairy cow rumen. The purified recombinant enzyme displayed optimal activity at pH 6.0 and 50 °C. It was stable over a broad pH range, from pH 4.0 to 10.0 [18]. Chang et al. reported an endoglucanase from yak rumen microorganisms and the optimal conditions for enzyme activity were 50 °C and pH 5.0 [19]. BcsZ is mesophilic enzyme which is in line with the other endoglucanse in some microorganisms such as *Bacillus sp.* HSH-810 [20], *Komagataeibacter xylinus* [21].

Metal ions significantly affected the enzymatic catalytic activity based on the above results. Endoglucanase of GH8 family has two conserved glutamate residues at the active site and the cellulase activity might be restrained when metal ions is bound to the radical of these residuals [22]. Mg^{2+} might enhance cellulase activity through altering the dimensional structure of BcsZ or stabilizing the enzyme structural conformation. Co^{2+}, Mn^{2+}, and

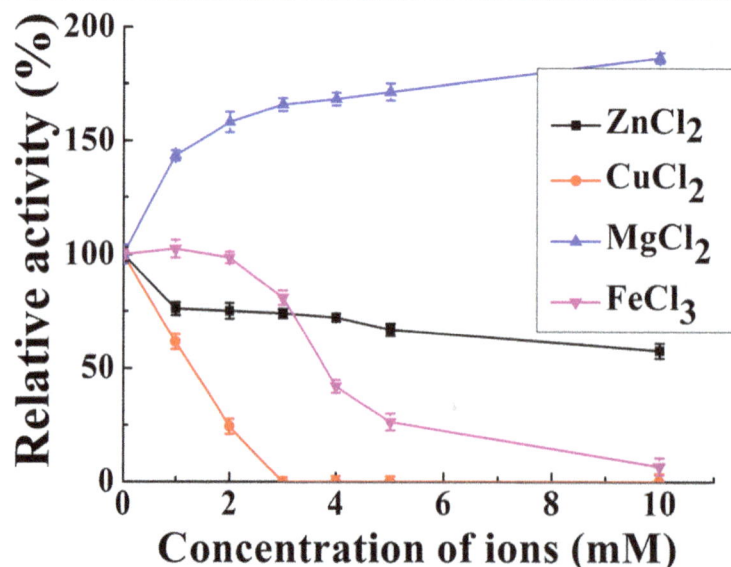

Fig. 6 The effect of metal ions on the activity of BcsZ

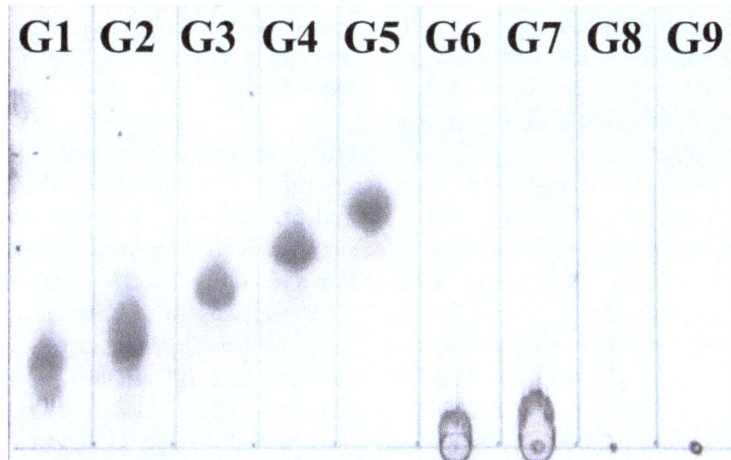

Fig. 7 TLC analysis of hydrolysates of CMC and RAC catalyzed by BcsZ. Standards of Cellopentose (G1), Cellotetraose (G2), Cellotriose (G3), Cellobiose (G4) and Glucose (G5); CMC incubated with inactivated BcsZ as the control (G6), CMC incubated with BcsZ (G7); RAC incubated with inactivated BcsZ as the control (G8), RAC incubated with BcsZ (G9)

Fe^{2+} enhanced the enzyme activity of GHF9 endogluca-nase from *Reticulitermes speratus*, but Pb^{2+} and Cu^{2+} inhibited its enzyme activity [23]. Meleiro et al cloned and overexpressed an endoglucanase (Egst) from *Scytalidium thermophilum* that showed high catalytic activity at harsh condition. NH^{4+}, Na^+, K^+, Ba^{2+}, Ca^{2+} and Mg^{2+} had little effect on Egst. Cu^{2+} presented a slight inhibitory effect and Hg^{2+} had a strong inhibitory effect on Egst [24].

BcsZ could hydrolyze CMC-Na, RAC and glucan. BcsZ had a higher affinity for CMC than Thcel9A (12.02 mg/mL) from *Thermobifida halotolerans* YIM 90462 T [25], the puri-fied endoglucanase (21.01 mg/mL) from *Aspergillus niger B03* [26] and C67–1 (37 mg/mL) from metagenomes of buf-falo rumens [27]. It had a weaker affinity for CMC than the endoglucanase from *Bacillus sp.* (0.8 mg/mL) [28]. The V_{max} value (0.3 µM/min·mg) of BcsZ was lower than Cell-1 (0.84 µmol/min·mg) [29] and an endoglucanse (1000 µM/min) [28]. The V_{max} value of BcsZ was low indicating this enzyme had a weak catalytic efficiency for CMC.

The BcsZ from *E. coli* ZH-4 could be classified as an endo-β-1, 4 glucanase according to its product pattern. Some other endo-β-1, 4 glucanases also have the above properties. An endoglucanase was expressed and characterized from *Serratia proteamaculans* CDBB-1961 by Cano-Ramírez, and then was applied to hydrolyze CMC to glucose and cello oli-gosaccharides [30]. The endo-β-1, 4 glucanase (E4–90) from *Termomonospora fusca* hydrolyzed CMC to produce cellobi-ose, cellotriose, cellotetraose and glucose [31]. As the previ-ous study, a purified β-1, 4 endoglucanase (LbGH5) could hydrolyze CMC and phosphoric acid swollen cellulose (PASC) with the enzymatic hydrolysate including cellobiose, cellotriose, cellotetraose and a small amount of cellopentaose and glucose [32]. In contrast, a thermophilic endo-1, 4-β-glucanase from *Sulfolobus shibatae* hydrolyzed cellotetraose

and cellopentose not cellobiose or cellotriose. The products of CMC hydrolysis were cellobiose, cellotriose, cellotetraose and cellopentose [33]. Cellobiose and cellotriose was pro-duced from PASC and Avicel respectively, and cellobiose, cellotriose and cellotetraose were produced from CMC with Endo-glucanase EG5C-1 [34]. The release of significant amount of glucose also explained why *E.coli* ZH-4 can grow on cellulose. Our experiments also showed that expression of BcsZ in *E.coli* BL21 resulted in the growth of *E.coli* BL21 on cellulose. These identification and characterization of BcsZ may improve understanding the adaptation of *E. coli* to different environment.

Conclusions

BcsZ was shown to be responsible for extracellular endo-glucanase activity and cellulolytic capability of *E. coli* ZH-4. Expression, purification and characterization of the endo-glucanase BcsZ showed that it was thermotolerant and pH tolerant. This enzyme could hydrolyze CMC to produce glucose, cellobiose, cellotriose and unknown cello oligosac-charides, which explained why *E. coli* ZH-4 can grow on cellulose.

Methods

Bacterial strains and culture conditions

The cellulolytic *E. coli* ZH-4 was isolated from the bovine rumen by our laboratory and it was preserved in China Gen-eral Microbiological Culture Collection Center (CGMCC) (Preservation No. 12427). *E. coli* DH5α was used for plasmid amplification, and *E. coli* BL21 (DE3) was used as the host for recombinant protein expression with pET28a as vector. *E. coli* strains were routinely cultured in Luria–Bertani (LB) medium at 37 °C with shaking at 220 rpm. Kanamycin was added when needed at a final concentration of 50 µg/mL.

Quantitative RT-PCR analysis

Total RNA was extracted using the RNAprep pure Cell/Bacteria Kit (TIANGEN) following manufacturer's instruction. The concentration and quality of RNA was measured using the Nano-Drop spectrophotometry (NanoDrop Technologies, Wilmington, DE, USA). The cDNA was synthesized according to the First Stand cDNA Synthesis Kit (TOYOBO) following the manufacturer's protocol. The primers were displayed in Table 2. The PCR reaction included 12.5 uL SYBR Green Realtime PCR Master Mix (TOYOBO), 2.5 uL diluted cDNA (500 mM) reaction mixture, 1uL each forward and reverse primer (10 umol) and 8.0 uL ddH$_2$O. The RT-PCR assays were performed on a 7900HT Fast Real-Time PCR System (Applied Biosystems, Carlsbad, California, USA). The relative gene expression calculated by the equation of $2^{(-\Delta\Delta Ct)}$ method (Applied Biosystems Research Bulletin No. 2 P/N 4303859).

Western blot analysis

Western blotting was performed as described by Sambrook [35]. Proteins separated by SDS-PAGE were transferred onto polyvinylidene difluoride (PVDF) membranes (Millipore, Bilerica, MA, USA) using the Bio-Rad semi-dry apparatus. The blots were incubated with primary anti-BcsZ serum in 1:2000 dilution in 2% (wt/vol) skimmed milk for 1 h with agitation at room temperature followed by washes with PBST buffer (PBS buffer with 1% Tween) for 3 times. The secondary antibody with horseradish peroxidase (HRP) conjugated goat anti-rabbit (1:2000 dilution) was performed as above. Blots were developed with the ECL Plus Kit (Thermo Scientific, US) following the manufacturer's directions. The in-gel identification of secreted protein by mass spectrometry analysis was performed by Shenzhen BGI gene co., LTD.

Sequence analysis

bcsZ gene sequences from *E. coli* ZH-4 and *E. coli* MG1655 was compared using the BLASTN program (https://blast.ncbi.nlm.nih.gov/Blast.cgi).

Plasmid construction

To overexpress BcsZ, *bscZ* gene was amplified from *E. coli* ZH-4 using primers BcsZ-F (5′- CGCGGATCCGGGTGTGAATTTGCGCATTCCT-3′) and BcsZ-R (5′-CATGCCATGGGCAATGTGTTGCGTAGTGGAAT-3′). The PCR products were then digested and cloned

Table 2 The primers of qRT-PCR

Primers	Sequences (5′-3′)
bcsZ-F	GAGAACAGTAAGTGGGAAGTGC
bcsZ-R	AACGCTGCTCTTTCCACAAACG
16S rRNA-F	GCTCAACCTGGGAACTGC
16S rRNA-R	CCACGCTTTCGCACCTGA

into *Bam*HI and *Nco*I sites of pET-28a plasmid. His-tag was located in C-terminus of *bcsZ*. The constructed expression vector pET-28a-BcsZ was verified by DNA sequencing.

Protein expression and purification

Overnight culture of *E. coli* BL21 (DE3) harboring plasmid pET-28a-BcsZ was inoculated to fresh LB medium supplemented with 50 µg/mL kanamycin in 1:100 dilution and cultured at 37 °C with shaking. The *E. coli* BL21 (DE3) carrying the empty plasmid was used as control. To induce *bscZ* gene expression, isopropyl-β- d-thiogalactoside (IPTG) was added to the culture at a final concentration of 0.25 mM when the optical density at 600 nm reached 0.4~0.6. Cells were harvested by centrifugation (8000×g, 10 min) after 24 h cultivation at 25 °C with shaking at 220 rpm. The resulting cells were suspended in 10 mM phosphate buffer saline (PBS) with PMSF and DNase I. The cells were lysed using JNBIO JN-3000 PLUS high-pressure cell press. Then, the crude cell lysate was prepared by centrifugation (12,000×g, 30 min) to remove the cell debris.

The BcsZ was further purified by affinity chromatography using HisPur Cobalt Resin (Thermo Fisher Scientific Inc) according to the manufacturers' instruction. The purity and homogeneity of the purified enzyme was detected by SDS–PAGE.

Enzyme assay and protein determination

The enzyme activity of BcsZ was determined by the standard DNS method [36]. The mixture of 1.5 mL 1% (w/v) CMC solubilized in 50 mM phosphate buffer (pH 6.0) and 0.5 mL purified BcsZ was incubated at 50 °C for 30 min. The the reaction was stopped by the addition of 3 mL of DNS. The release of reducing sugars was determined by the absorbance at 540 nm. One unit of activity is defined as the amount of enzyme that released 1 µmol of reducing sugars/min from CMC. Protein concentration was measured by Bradford method using bovine serum albumin as standard [37].

Basic biochemical character of BcsZ

The optimum pH was determined by measuring BcsZ activity in different buffers (50 mM) of pH ranging from 4.0 to 10.0: citric acid buffer for pH 4.0–6.0; phosphate buffer for pH 6.0–8.0; Glycine-NaOH buffer for pH 8.5–10.0. The optimum temperature was identified by incubating the enzyme in phosphate buffer (pH 6.0) at different temperatures (30–80 °C). For thermostability determination of BcsZ, it was incubated at different temperatures 4, 40, 50, 60 and 70 °C for 0.5, 1, 2, 3, 4, 5, 10 and 24 h. The residual endoglucanase activity was determined, respectively. The influence of metal ions on enzyme activity of BcsZ was measured in presence of tested metal ions at indicated concentration. All enzyme activity was measured as previously described. The highest enzymatic activity was used as benchmark of 100% activity when calculating relative activity.

Kinetic constants

The kinetic parameter values Michaelis-Menten constants (K_m) and maximum velocity (V_{max}) were determined to calculate the K_m value of BcsZ on CMC hydrolysis according to double-reciprocal Lineweaver–Burk plots (Eq. 1). The activity assay was performed with CMC at different concentrations at pH 6.0 and 50 °C for 10 min.

$$\frac{1}{V} = \frac{K_m}{V_{max}} \frac{1}{[S]} + \frac{1}{V_{max}} \tag{1}$$

Substrate specificity

The hydrolytic ability of BcsZ on 1% (w/v) of various substrates were determined by DNS method under optimal conditions. The substrate included Avicel, CMC-Na, regenerated amorphous cellulose (RAC), xylan, cellobiose, laminarin and chitin.

Thin layer chromatographic (TLC)

The products were analyzed by Thin Layer Chromatographic (TLC) as previously described when BcsZ hydrolyzed CMC and RAC [38].

Statistical analysis

All experiments were conducted in triplicate and the data were presented as mean values ± standard deviation.

Abbreviations

CGMCC: China General Microbiological Culture Collection Center; CMC: carboxymethylcellulose; DNS: dinitrosalicylic; GH: Glycoside hydrolase; HRP: horseradish peroxidase; IPTG: isopropyl-β- d-thiogalactoside; LB: Luria–Bertani; PBS: phosphate buffer saline; PVDF: polyvinylidene difluoride; RAC: Regenerated amorphous cellulose; SDS–PAGE: Sodium dodecyl sulfate–polyacrylamide gel electrophoresis; TLC: Thin layer chromatographic

Acknowledgments

Not applicable.

Authors' contributions

JP did the experimental work and wrote the draft manuscript. JSW revised and designed the draft manuscript. ZYL participated in the design of the study. QCZ took part in checking the results. QSQ participated in the design of the study and commented on the manuscript. All authors read and approved the final manuscript.

Funding

This work was supported by the National Natural Science Foundation of China [Grant number 31730003]; the National Natural Science Foundation of China (NSFC) [Grant number 61361016]; the Foundation of Talent Development of Inner Mongolia and the "Prairie talent" project of Inner Mongolia [Grant number CYYC20130034]; the National Natural Science Foundation of Inner Mongolia [Grant number 2018MS02019]; the Science and Technology Planning Project of Inner Mongolia; and the Graduate student education innovation planning project of Inner Mongolia [Grant number B2018111925]. These funding bodies had no role in the design of the study and collection, analysis, and interpretation of data and in writing the manuscript.

Ethics approval and consent to participate

Not applicable.

Consent for publication

Not applicable.

Competing interests

The authors declare that they have no competing interests.

Author details

[1]School of Chemical Engineering, Inner Mongolia University of Technology, Hohhot 010051, Inner Mongolia, China. [2]State Key Laboratory of Microbial Technology, Shandong University, Qingdao 266237, China. [3]Inner Mongolia Energy Conservation and Emission Reduction Engineering Research Center in Fermentation Industry, Hohhot 010051, China.

References

1. Ecem Oner B, Akyol C, Bozan M, Ince O, Aydin S, Ince B. Bioaugmentation with *Clostridium thermocellum* to enhance the anaerobic biodegradation of lignocellulosic agricultural residues. Bioresour Technol. 2018;249:620–5.
2. Zeng R, Hu Q, Yin XY, Huang H, Yan JB, Gong ZW, Yang ZH. Cloning a novel endo-1,4-β- d -glucanase gene from *Trichoderma virens* and heterologous expression in *E. coli*. AMB Express. 2016;6:108.
3. Seneesrisakul K, Guralp SA, Gulari E, Chavadej S. *Escherichia coli* expressing endoglucanase gene from Thai higher termite bacteria for enzymatic and microbial hydrolysis of cellulosic materials. Electron J Biotechnol. 2017;27:70–9.
4. Yennamalli RM, Rader AJ, Wolt JD, Sen TZ. Thermostability in endoglucanases is fold-specific. BMC Struct Biol. 2011;11:10.
5. Sadhu S, Maiti TK. Cellulase production by bacteria: a review. British Microbiol Res J. 2013;3:235.
6. Maki M, Leung KT, Qin W. The prospects of cellulase-producing bacteria for the bioconversion of lignocellulosic biomass. Int J Biol Sci. 2009;5:500.
7. Gao D, Wang S, Li H, Yu H, Qi Q. Identification of a heterologous cellulase and its N-terminus that can guide recombinant proteins out of *Escherichia coli*. Microb Cell Factories. 2015;14:49.
8. Gao D, Luan Y, Wang Q, Liang Q, Qi Q. Construction of cellulose-utilizing *Escherichia coli* based on a secretable cellulase. Microb Cell Factories. 2015;14:159.
9. Pang J, Liu ZY, Hao M, Zhang YF, Qi QS. An isolated cellulolytic *Escherichia coli* from bovine rumen produces ethanol and hydrogen from corn straw. Biotechnol Biofuels. 2017;10:165.
10. Ahmad I, Rouf SF, Sun L, Cimdins A, Shafeeq S, Le Guyon S, et al. BcsZ inhibits biofilm phenotypes and promotes virulence by blocking cellulose production in *Salmonella enterica serovar Typhimurium*. Microb Cell Factories. 2016;15:177.
11. Bokranz W, Wang X, Tschape H, Romling U. Expression of cellulose and curli fimbriae by *Escherichia coli* isolated from the gastrointestinal tract. J Med Microbiol. 2005;54:1171–82.
12. Grantcharova N, Peters V, Monteiro C, Zakikhany K, Romling U. Bistable expression of CsgD in biofilm development of *Salmonella enterica serovar typhimurium*. J Bacteriol. 2010;192:456–66.
13. Mazur O, Zimmer J. Apo- and cellopentaose-bound structures of the bacterial cellulose synthase subunit BcsZ. J Biol Chem. 2011;286:17601–6.
14. Standal R, Iversen TG, Coucheron DH, Fjaervik E, Blatny JM, Valla S. A new gene required for cellulose production and a gene encoding cellulolytic activity in *Acetobacter xylinum* are colocalized with the bcs operon. J Bacteriol. 1994;176:665–72.
15. Rabinovich ML, Melnick MS, Bolobova AV. The structure and mechanism of action of cellulolytic enzymes. Biochem Mosc. 2002;67:850–71.
16. Park Y, Yun H. Cloning of the *Escherichia coli* endo-1,4-D-glucanase gene and identification of its product. Mol Gen Genet. 1999;261:236–41.

17. Lin L, Fu C, Huang W. Improving the activity of the endoglucanase, Cel8M from *Escherichia coli* by error-prone PCR. Enzym Microb Technol. 2016;86:52–8.

18. Gong X, Gruninger RJ, Qi M, Paterson L, Forster RJ, Teather RM, et al. Cloning and identification of novel hydrolase genes from a dairy cow rumen metagenomic library and characterization of a cellulase gene. BMC Res Notes. 2012;5:566.

19. Chang L, Ding M, Bao L, Chen Y, Zhou J, Lu H. Characterization of a bifunctional xylanase/endoglucanase from yak rumen microorganisms. Appl Microbiol Biotechnol. 2011;90:1933–42.

20. Kim JY, Hur SH, Hong JH. Purification and characterization of an alkaline cellulase from a newly isolated alkalophilic *Bacillus sp.* HSH-810. Biotechnol Lett. 2005;27:313–6.

21. KOO HM. Expression and characterization of CMCax having β-1, 4-endoglucanase activity from *Acetobacter xylinum*. BMB Rep. 1998;31:53–7.

22. Demain AL, Newcomb M, Wu JH. Cellulase, clostridia, and ethanol. Microbiol Mol Biol Rev. 2005;69:124–54.

23. Zhang P, Yuan X, Du Y, Li JJ. Heterologous expression and biochemical characterization of a GHF9 endoglucanase from the termite *Reticulitermes speratus* in Pichia pastoris. BMC Biotechnol. 2018;18:35.

24. Meleiro LP, Carli S, Fonseca-Maldonado R, da Silva TM, Zimbardi A, Ward RJ, et al. Overexpression of a Cellobiose-glucose-halotolerant endoglucanase from *Scytalidium thermophilum*. Appl Biochem Biotechnol. 2018;185:316–33.

25. Zhang F, Chen JJ, Ren WZ, Nie GX, Ming H, Tang SK, Li WJ. Cloning, expression and characterization of an alkaline thermostable GH9 endoglucanase from *Thermobifida halotolerans* YIM 90462 T. Bioresour Technol. 2011;102:10143–6.

26. Dobrev GT, Zhekova BY. Biosynthesis, purification and characterization of endoglucanase from a xylanase producing strain *Aspergillus niger* B03. Braz J Microbiol. 2012;43:70–7.

27. Duan C, Xian L, Zhao GC, Feng Y, Pang PH, Bai XL, et al. Isolation and partial characterization of novel genes encoding acidic cellulases from metagenomes of buffalo rumens. J Appl Microbiol. 2009;107:246–56.

28. Sriariyanun M, Tantayotai P, Yasurin P, Pornwongthong P, Cheenkachorn K. Production, purification and characterization of an ionic liquid tolerant cellulase from *Bacillus sp.* isolated from rice paddy field soil. Electron J Biotechnol. 2016;19:23–8.

29. Zhou X, Kovaleva ES, Wu-Scharf D, Campbell JH, Buchman GW, Boucias DG, et al. Production and characterization of a recombinant beta-1,4-endoglucanase (glycohydrolase family 9) from the termite *Reticulitermes flavipes*. Arch Insect Biochem Physiol. 2010;74:147–62.

30. Cano-Ramirez C, Santiago-Hernandez A, Rivera-Orduna FN, Garcia-Huante Y, Zuniga G, Hidalgo-Lara ME. Expression, purification and characterization of an endoglucanase from *Serratia proteamaculans* CDBB-1961, isolated from the gut of Dendroctonus adjunctus (Coleoptera: Scolytinae). AMB Express. 2016;6:63.

31. Irwin D, Shin D-H, Zhang S, Barr BK, Sakon J, Karplus PA, et al. Roles of the catalytic domain and two cellulose binding domains of *Thermomonospora fusca* E4 in cellulose hydrolysis. J Bacteriol. 1998;180:1709–14.

32. Zhang F, Anasontzis GE, Labourel A, Champion C, Haon M, Kemppainen M, et al. The ectomycorrhizal basidiomycete Laccaria bicolor releases a secreted beta-1,4 endoglucanase that plays a key role in symbiosis development. New Phytol. 2018;220:1309–21.

33. Boyce A, Walsh G. Expression and characterisation of a thermophilic endo-1,4-beta-glucanase from *Sulfolobus shibatae* of potential industrial application. Mol Biol Rep. 2018;45:2201–11.

34. Wu B, Zheng S, Pedroso MM, Guddat LW, Chang S, He B, et al. Processivity and enzymatic mechanism of a multifunctional family 5 endoglucanase from *Bacillus subtilis* BS-5 with potential applications in the saccharification of cellulosic substrates. Biotechnol Biofuels. 2018;11:20.

35. Sambrook J, Fritsch EF, Maniatis T. Molecular cloning: a laboratory manual. CSH. 1989.

36. Miller GL. Use of dinitrosalicylic acid reagent for determination of reducing sugar. Anal Chem. 1959;31:426–8.

37. Bradford MM. A rapid and sensitive method for the quantitation of microgram quantities of protein utilizing the principle of protein-dye binding. Anal Biochem. 1976;72:248–54.

38. Zhang C, Wang Y, Li Z, Zhou X, Zhang W, Zhao Y, et al. Characterization of a multi-function processive endoglucanase CHU_2103 from *Cytophaga hutchinsonii*. Appl Microbiol Biotechnol. 2014;98:6679–87.

Permissions

List of Contributors

Melinda Moyon, Laetitia Gautreau-Rolland, Benjamin Navet, Jeanne Perroteau, Richard Breathnach and Xavier Saulquin
CRCINA, INSERM, CNRS, Université d'Angers, Université de Nantes, Nantes, France
LabEx IGO "Immunotherapy, Graft, Oncology", Nantes, France

Marie-Claude Gesnel and Marie-Claire Devilder
CRCINA, INSERM, CNRS, Université d'Angers, Université de Nantes, Nantes, France
LabEx IGO "Immunotherapy, Graft, Oncology", Nantes, France
Centre Hospitalier Universitaire Hôtel-Dieu, Nantes, France

Florent Delbos
HLA Laboratory, EFS Centre Pays de la Loire, Nantes, France

Yong Xu Mu
Interventional Department, the First Affiliated Hospital of Baotou Medical College, Inner Mongolia University of Science and Technology, Baotou, Inner Mongolia, China

Yu Xia Zhao, Hong Jing Bao, Hui Jiang, Xiao Lei Qi and Li Yun Bai
Department of Blood, the People's Hospital of Xing'an League, Xing'an League, Inner Mongolia, China

Bing Yao Li
Department of Medicine, Chifeng Cancer Hospital, Chifeng, Inner Mongolia, China

Yun Hong Wang and Xiao Yun Wu
Department of Technology, Stem Cell Medicine Engineering & Technology Research Center of Inner Mongolia, Huhhot, Inner Mongolia, China
Department of Research and Development, Beijing Jingmeng Stem Cell Technology CO., LTD, Beijing, China

Zhi Jie Ma
Department of Pharmacy, Beijing Friendship Hospital, Capital Medical University, Beijing, China

Steven Mayers and Julie Audet
Department of Chemical Engineering and Applied Chemistry, University of Toronto, Toronto, Canada

Institute of Biomaterials and Biomedical Engineering (IBBME), University of Toronto, Toronto, Canada

Pablo DiegoMoço, Talha Maqbool, Pamuditha N. Silva and Dawn M. Kilkenny
Institute of Biomaterials and Biomedical Engineering (IBBME), University of Toronto, Toronto, Canada

Buvani Murugesan, Janina Hoßbach, Stephanie K. Evans and Margaret C. M. Smith
Department of Biology, University of York, York, North Yorkshire YO10 5DD, UK

Hong Gao
Department of Biology, University of York, York, North Yorkshire YO10 5DD, UK
School of Science, Engineering & Design, Teesside University, Middlesbrough TS1 3BX, UK

W. Marshall Stark
Institute of Molecular, Cell and Systems Biology, University of Glasgow, Glasgow G12 8QQ, UK

Bruna S. Fernandes
Department of Civil and Environmental Engineering, Federal University of Pernambuco, Recife, PE, Brazil
Centre of Biological Engineering, Universidade do Minho, Braga, Portugal

Oscar Dias, Gisela Costa, Tiago F. C. Resende and Isabel Rocha
Centre of Biological Engineering, Universidade do Minho, Braga, Portugal

Antonio A. Kaupert Neto and Juliana V. C. Oliveira
Brazilian Bioethanol Science and Technology Laboratory (CTBE), Brazilian Centre of Research in Energy and Materials (CNPEM), Campinas, SP, Brazil

Diego M. Riaño-Pachón
Computational, Evolutionary and Systems Biology Laboratory, Center for Nuclear Energy in Agriculture, University of São Paulo, Piracicaba, São Paulo, Brazil

Marcelo Zaiat
Biological Processes Laboratory, Center for Research, Development and Innovation in Environmental Engineering, São Carlos School of Engineering (EESC), University of São Paulo, São Carlos, SP, Brazil

José G. C. Pradella
PRBiotec Ltd, São José dos Campos, SP, Brazil

François-Xavier Gillet, Marcelo Porto Bemquerer, Vanessa Olinto dos Santos and Maria Cristina Mattar Silva
Embrapa Genetic Resources and Biotechnology, Brasília, DF, Brazil

Guilherme Souza Prado, Joel Antônio Cordeiro de Abreu, Maria Fatima and Grossi-de-Sa
Embrapa Genetic Resources and Biotechnology, Brasília, DF, Brazil
Catholic University of Brasília, Brasília, DF, Brazil

Jean-Paul Brizard
IRD, CIRAD, Université Montpellier, Interactions Plantes Microorganismes et Environnement (IPME), Montpellier, France

Pingdwende Kader Aziz Bamogo, Martine Bangratz, Christophe Brugidou and Séverine Lacombe
IRD, CIRAD, Université Montpellier, Interactions Plantes Microorganismes et Environnement (IPME), Montpellier, France
INERA/LMI Patho-Bios, Institut de L'Environnement et de Recherches Agricoles (INERA), Laboratoire de Virologie et de Biotechnologies Végétales, Ouagadougou, Burkina Faso

Drissa Sérémé
INERA/LMI Patho-Bios, Institut de L'Environnement et de Recherches Agricoles (INERA), Laboratoire de Virologie et de Biotechnologies Végétales, Ouagadougou, Burkina Faso

Kristian Alsbjerg Skipper and Jacob Giehm Mikkelsen
Department of Biomedicine, HEALTH, Aarhus University, DK- 8000 Aarhus C, Denmark

Anne Kruse Hollensen
Department of Biomedicine, HEALTH, Aarhus University, DK- 8000 Aarhus C, Denmark
Department of Molecular Biology and Genetics, Science and Technology, Aarhus University, DK-8000 Aarhus C, Denmark

Michael N. Antoniou
Gene Expression and Therapy Group, King's College London, Faculty of Life Sciences & Medicine, Department of Medical and Molecular Genetics, 8th Floor Tower Wing, Guy's Hospital, London SE1 9RT, UK

Veronika Dill, Martin Beer and Michael Eschbaumer
Institute of Diagnostic Virology, Friedrich-Loeffler-Institut, Südufer 10, 17493 Greifswald, Insel Riems, Germany

Aline Zimmer
Merck KGaA, Merck Life Sciences, Upstream R&D, Frankfurter Straße, 250, 64293 Darmstadt, Germany

Min Zhang
Beijing Advanced Innovation Center for Food Nutrition and Human Health, Beijing Technology & Business University (BTBU), Beijing 100048, China

Xiaoyang Pang
Beijing Advanced Innovation Center for Food Nutrition and Human Health, Beijing Technology & Business University (BTBU), Beijing 100048, China
Institute of Food Science and Technology, Chinese Academy of Agricultural Science, Beijing 100193, China

Ziyang Jia, Jing Lu, Shuwen Zhang and Jiaping Lv
Institute of Food Science and Technology, Chinese Academy of Agricultural Science, Beijing 100193, China

Cai Zhang
Laboratory of Environment and Livestock Products, Henan University of Science and Technology, Luoyang 471023, China

Maria José Chiabai, Mariana Gabriela Dantas de Azevedo, Suelen Soares Fernandes and Isabel Garcia Sousa
Laboratório de Imunologia Molecular, Departamento de Biologia Molecular, Universidade de Brasília, Brasília, Distrito Federal, Brazil

Andrea Queiroz Maranhão and Marcelo Macedo Brigido
Laboratório de Imunologia Molecular, Departamento de Biologia Molecular, Universidade de Brasília, Brasília, Distrito Federal, Brazil
Instituto Nacional de Investigação em Imunologia, INCTii, Brasília, Distrito Federal, Brazil

Juliana Franco Almeida
Centro de Biotecnologia, Departamento de Biologia Celular e Molecular, Universidade Federal da Paraíba, João Pessoa, Paraíba, Brazil

Vanessa Bastos Pereira and Anderson Miyoshi
Laboratório de Tecnologia Genética, Departamento de Biologia Geral, Universidade Federal de Minas Gerais, Belo Horizonte, Minas Gerais, Brazil

Raffael Júnio Araújo de Castro, Márcio Sousa Jerônimo and Anamelia Lorenzetti Bocca
Laboratório de Imunologia Aplicada, Departamento de Biologia Celular, Universidade de Brasília, Brasília, Distrito Federal, Brazil

Leonora Maciel de Souza Vianna
Departmento de Patologia, Escola de Medicina, Universidade de Brasília, Brasília, Distrito Federal, Brazil

Da Sol Kim, Seon Woong Kim and Soon Young Kim
Department of Biological Sciences, Andong National University, Andong, South Korea

Jae Min Song
Department of Global Medical Science, Health & Wellness College, Sungshin University, Seoul, South Korea

Kwang-Chul Kwon
MicroSynbiotiX Ltd, 11011 N Torrey Pines Rd Ste. #135, La Jolla, CA 92037, USA

Luisa Hildebrand and Ernst A. Wimmer
Department of Developmental Biology, Johann-Friedrich-Blumenbach-Institute of Zoology and Anthropology, Göttingen Center for Molecular Biosciences, Georg-August-University Göttingen, 37077 Göttingen, Germany

M. Alejandro Carballo-Amador
Facultad de Ciencias, Universidad Autónoma de Baja California, Km. 103 Carretera Tijuana–Ensenada, Pedregal Playitas, 22860 Ensenada, Baja California, Mexico

Hassan M. M. Ahmed
Department of Developmental Biology, Johann-Friedrich-Blumenbach-Institute of Zoology and Anthropology, Göttingen Center for Molecular Biosciences, Georg-August-University Göttingen, 37077 Göttingen, Germany
Department of Crop Protection, Faculty of Agriculture-University of Khartoum, 13314 Khartoum, Khartoum North, Sudan

Edward A. McKenzie and Jim Warwicker
School of Chemistry, Manchester Institute of Biotechnology, University of Manchester, 131 Princess Street, Manchester M1 7DN, UK

Alan J. Dickson
Faculty of Science and Engineering, Manchester Institute of Biotechnology, University of Manchester, 131 Princess Street, Manchester M1 7DN, UK

Liya Zeng, Yongchang Liu, Jun Pan and Xiaowen Liu
Key Laboratory of Comprehensive Utilization of Advantage Plants Resources in Hunan South, College of Chemistry and Bioengineering, Hunan University of Science and Engineering, Yongzhou, Hunan, China

Bent Larsen Petersen, Jozef Mravec, Bodil Jørgensen and Ying Liu
Department of Plant and Environmental Sciences, University of Copenhagen, DK-1871 Frederiksberg C, Denmark

Svenning Rune Möller
Department of Plant and Environmental Sciences, University of Copenhagen, DK-1871 Frederiksberg C, Denmark
Centre for Novel Agricultural Products, University of York, Woodsmill Quay, Skeldergate, York YO1 6DX, UK

Mikkel Christensen
Department of Plant and Environmental Sciences, University of Copenhagen, DK-1871 Frederiksberg C, Denmark
UIT - Department of Chemistry, The Arctic University of Norway, Forskningsparken. 3, 9019 Tromsø, Norway

Hans H. Wandall, Eric Paul Bennett and Zhang Yang
Copenhagen Center for Glycomics, Department of Molecular and Cellular Medicine and School of Dentistry, Faculty of Health Sciences, University of Copenhagen, DK-2200 Copenhagen N, Denmark

Yun Lv
Graduate School of Anhui Medical University, 81 Meishan Road, Shu Shan Qu, Hefei, Anhui, People's Republic of China
Department of Experimental Hematology, Beijing Institute of Radiation Medicine, 27 Taiping Road, Beijing, China
State Key Laboratory of Infectious Disease Prevention and Control, National Institute for Viral Disease Control and Prevention, Chinese Center for Disease Control and Prevention, 100 Ying Xin Jie, Beijing, China

Feng-Jun Xiao and Hua Wang
Department of Experimental Hematology, Beijing Institute of Radiation Medicine, 27 Taiping Road, Beijing, China

Li-Sheng Wang
Department of Experimental Hematology, Beijing Institute of Radiation Medicine, 27 Taiping Road, Beijing, China
Affiliated Hospital of Qingdao University, 16 JiangSu Road, Qingdao, People's Republic of China

Yi Wang, Xiao-Hui Zou and Zhuo-Zhuang Lu
State Key Laboratory of Infectious Disease Prevention and Control, National Institute for Viral Disease Control and Prevention, Chinese Center for Disease Control and Prevention, 100 Ying Xin Jie, Beijing, China

Hai-Yan Wang
Affiliated Hospital of Qingdao University, 16 JiangSu Road, Qingdao, People's Republic of China

Jie Dong, Xiangzhen Ding and Sheng Wang
Key Laboratory of Ministry of Education for Protection and Utilization of Special Biological Resources in the Western China, Yinchuan 750021, People's Republic of China
Key Laboratory of Modern Molecular Breeding for Dominant and Special Crops in Ningxia, Yinchuan 750021, People's Republic of China
School of Life Science, Ningxia University, 539 W. Helanshan Road, Yinchuan, Ningxia 750021, People's Republic of China

Min Zhang, Senlei Feng, Wenqian Li, Zepeng Sun, Jiawei Li, Kanglu Yan, Hao Liu, Xueping Meng, Haibao Duan and Shuang Mao
Jiangsu Key Laboratory of Marine Biotechnology, Jiangsu Ocean University, Lianyungang 222005, Jiangsu, China

Min Wei, Zhiguo Dong and Hongxing Ge
Jiangsu Key Laboratory of Marine Biotechnology, Jiangsu Ocean University, Lianyungang 222005, Jiangsu, China
Co-Innovation Center of Jiangsu Marine Bio-industry Technology, Jiangsu Ocean University, Lianyungang 222005, Jiangsu, China

Xuan Li, Xuli Ma, Yanping Kang and Minmin Yang
Oil Crops Research Institute of the Chinese Academy of Agricultural Sciences, Wuhan 430062, People's Republic of China

Fenghong Huang, and Xia Wan and Wenchao Chen
Oil Crops Research Institute of the Chinese Academy of Agricultural Sciences, Wuhan 430062, People's Republic of China

Key Laboratory of Biology and Genetic Improvement of Oil Crops, Ministry of Agriculture, Wuhan 430062, People's Republic of China
Oil Crops and Lipids Process Technology National & Local Joint Engineering Laboratory, Wuhan 430062, People's Republic of China
Hubei Key Laboratory of Lipid Chemistry and Nutrition, Wuhan 430062, People's Republic of China

Shouwen Chen
Hubei Collaborative Innovation Center for Green Transformation of Bio-Resources, Environmental Microbial Technology Center of Hubei Province, College of Life Sciences, Hubei University, Wuhan 430062, People's Republic of China

Rosemarie W. Hammond
USDA ARS NEA BARC Molecular Plant Pathology Laboratory, Beltsville, MD 20705, USA

Natalia Y. Kovalskaya
USDA ARS NEA BARC Molecular Plant Pathology Laboratory, Beltsville, MD 20705, USA
Oak Ridge Institute for Science and Education, ORISE, Beltsville, MD 20705, USA

Qiancheng Zhang
School of Chemical Engineering, Inner Mongolia University of Technology, Hohhot 010051, Inner Mongolia, China

Qingsheng Qi and Junshu Wang
State Key Laboratory of Microbial Technology, Shandong University, Qingdao, China

Zhanying Liu and Jian Pang
School of Chemical Engineering, Inner Mongolia University of Technology, Hohhot 010051, Inner Mongolia, China
Inner Mongolia Energy Conservation and Emission Reduction Engineering Research Center in Fermentation Industry, Hohhot 010051, China

Index